Viral Pathogenesis

From Basics to Systems Biology

Viral Pathogenesis

From Basics to Systems Biology

Third Edition

Edited By

Michael G. Katze
Department of Microbiology
University of Washington
Seattle, WA, USA

Marcus J. Korth
Department of Microbiology
University of Washington
Seattle, WA, USA

G. Lynn Law
Department of Microbiology
University of Washington
Seattle, WA, USA

Neal Nathanson
Emeritus Professor
Microbiology, School of Medicine
University of Pennsylvania
Philadelphia, PA, USA

AMSTERDAM • BOSTON • HEIDELBERG • LONDON • NEW YORK • OXFORD • PARIS
SAN DIEGO • SAN FRANCISCO • SINGAPORE • SYDNEY • TOKYO

Academic Press is an imprint of Elsevier

Academic Press is an imprint of Elsevier
125 London Wall, London EC2Y 5AS, UK
525 B Street, Suite 1800, San Diego, CA 92101-4495, USA
225 Wyman Street, Waltham, MA 02451, USA
The Boulevard, Langford Lane, Kidlington, Oxford OX5 1GB, UK

First edition 2002 published by Lippincott Williams & Wilkins

ISBN: 978-0-12-800964-2

British Library Cataloguing-in-Publication Data
A catalogue record for this book is available from the British Library

Library of Congress Cataloging-in-Publication Data
A catalog record for this book is available from the Library of Congress

For information on all Academic Press publications
visit our website at http://store.elsevier.com/

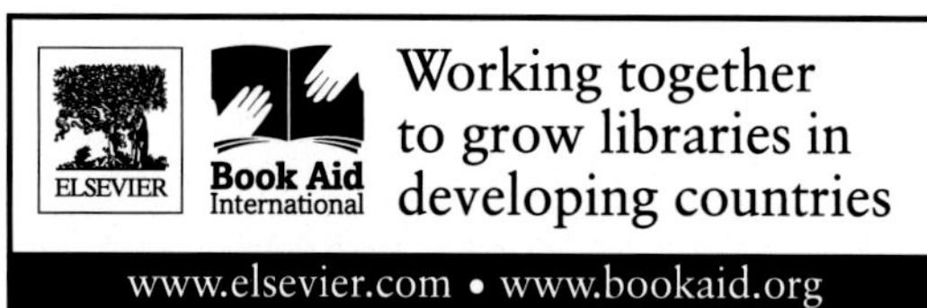

Illustrations by Wendy Beth Jackelow, MFA, CMI
Medical and Scientific Illustration
Staten Island, NY, USA

Cover Image: Provided by Wendy Beth Jackelow, MFA, CMI.
http://www.wbjackelowstudios.com/

Publisher: Sara Tenney
Acquisition Editor: Jill Leonard
Editorial Project Manager: Pat Gonzalez
Production Project Manager: Julia Haynes
Designer: Inês Cruz

Typeset by TNQ Books and Journals
www.tnq.co.in

Contents

4. Innate Immunity

Christine A. Biron

5. Adaptive Immunity

E. John Wherry and David Masopust

6. Aberrant Immunity

E. John Wherry and David Masopust

7. Patterns of Infection

Neal Nathanson and Francisco González-Scarano

Part II
Systems-Level Approaches to Viral Pathogenesis

Part III
Emergence and Control of Viral Infections

16. Emerging Viral Diseases

James W. Le Duc and Neal Nathanson

17. Viral Evolution

Adi Stern and Raul Andino

18. Viral Epidemiology

Kaitlin Rainwater-Lovett, Isabel Rodriguez-Barraquer and William J. Moss

19. Viral Vaccines

Juliet Morrison and Stanley Plotkin

Contributors

Raul Andino Department of Microbiology and Immunology, University of California, San Francisco, CA, USA

Ralph S. Baric Department of Microbiology and Immunology, University of North Carolina, Chapel Hill, NC, USA

Victoria K. Baxter Department of Comparative and Molecular Pathobiology, Johns Hopkins University School of Medicine, Baltimore, MD, USA; W. Harry Feinstone Department of Molecular Microbiology and Immunology, Johns Hopkins Bloomberg School of Public Health, Baltimore, MD, USA

Christine A. Biron Department of Molecular Microbiology and Immunology, Brown University, Providence, RI, USA

Emily K. Cartwright Emory Vaccine Center and Yerkes National Primate Research Center, Emory University, Atlanta, GA, USA

Sumit K. Chanda Sanford-Burnham Medical Research Institute, La Jolla, CA, USA

Robert W. Doms The Children's Hospital of Philadelphia and the Perelman School of Medicine, University of Pennsylvania, Philadelphia, PA, USA

Martin T. Ferris Department of Genetics, University of North Carolina, Chapel Hill, NC, USA

Denise A. Galloway Fred Hutchinson Cancer Research Center, Seattle, WA, USA

Francisco González-Scarano School of Medicine, University of Texas Health Sciences Center, San Antonio, TX, USA

Diane E. Griffin W. Harry Feinstone Department of Molecular Microbiology and Immunology, Johns Hopkins Bloomberg School of Public Health, Baltimore, MD, USA

Mark T. Heise Department of Genetics, University of North Carolina, Chapel Hill, NC, USA; Department of Microbiology and Immunology, University of North Carolina, Chapel Hill, NC, USA

Jeffery R. Johnson University of California, San Francisco, CA, USA

Marcus J. Korth Department of Microbiology, University of Washington, Seattle, WA, USA; Washington National Primate Research Center, University of Washington, Seattle, WA, USA

Nevan J. Krogan University of California, San Francisco, CA, USA

G. Lynn Law Department of Microbiology, University of Washington, Seattle, WA, USA; Washington National Primate Research Center, University of Washington, Seattle, WA, USA

James W. Le Duc Galveston National Laboratory, University of Texas Medical Branch, Galveston, TX, USA

David Masopust Department of Microbiology and Immunology, School of Medicine, University of Minnesota, Minneapolis, MN, USA

Juliet Morrison Department of Microbiology, University of Washington, Seattle, WA, USA

William J. Moss Departments of Epidemiology, International Health and Molecular Microbiology and Immunology, Johns Hopkins Bloomberg School of Public Health, Baltimore, MD, USA

Neal Nathanson Department of Microbiology, Perelman School of Medicine, University of Pennsylvania, Philadelphia, PA, USA

Alan S. Perelson Theoretical Biology and Biophysics, Los Alamos National Laboratory, Los Alamos, NM, USA

Stanley Plotkin University of Pennsylvania, Philadelphia, PA, USA

Kaitlin Rainwater-Lovett Department of Pediatrics, Johns Hopkins School of Medicine, Baltimore, MD, USA

Douglas D. Richman VA San Diego Healthcare System, San Diego, CA, USA; University of California, San Diego, La Jolla, CA, USA

Isabel Rodriguez-Barraquer Department of Epidemiology, Johns Hopkins Bloomberg School of Public Health, Baltimore, MD, USA

Monika Schneider Sanford-Burnham Medical Research Institute, La Jolla, CA, USA

Guido Silvestri Emory Vaccine Center and Yerkes National Primate Research Center, Emory University, Atlanta, GA, USA

Adi Stern Department of Molecular Microbiology and Biotechnology, Tel-Aviv University, Tel-Aviv, Israel

William C. Summers Yale University, New Haven, CT, USA

Nicholas A. Wallace Fred Hutchinson Cancer Research Center, Seattle, WA, USA

E. John Wherry Department of Microbiology, Perelman School of Medicine, University of Pennsylvania, Philadelphia, PA, USA

Priscilla L. Yang Department of Microbiology and Immunobiology, Harvard Medical School, Boston, MA, USA

Preface

The Future Is Now!

As you begin your journey through the third edition, can we suggest a few take-home lessons that you might consider while looking into this book? Let us try.

We are in the midst of a revolution in biomedical research, driven by technological advances. New and evolving methods in genomics, deep sequencing, proteomics, and other omics have driven the need for computational methods and mathematical models to make sense of the tsunami of data that are being generated. This has led a transition from reductionist approaches to systems biology, or in the context of this book, systems virology (Figure 1). What does that mean? In the last half of the twentieth century, the focus was to identify individual genes, determine how they were transcribed, and understand the function of the proteins that they encoded. Increasingly, we are looking for patterns of gene expression, involving multiple gene products, epigenetic modulation, and complex pathways, rather than the effects of individual genes. How does this impact our understanding of viral pathogenesis? In several ways as follows:

Innate immunity: Systems approaches have had particular impact on our understanding of the innate immune response.

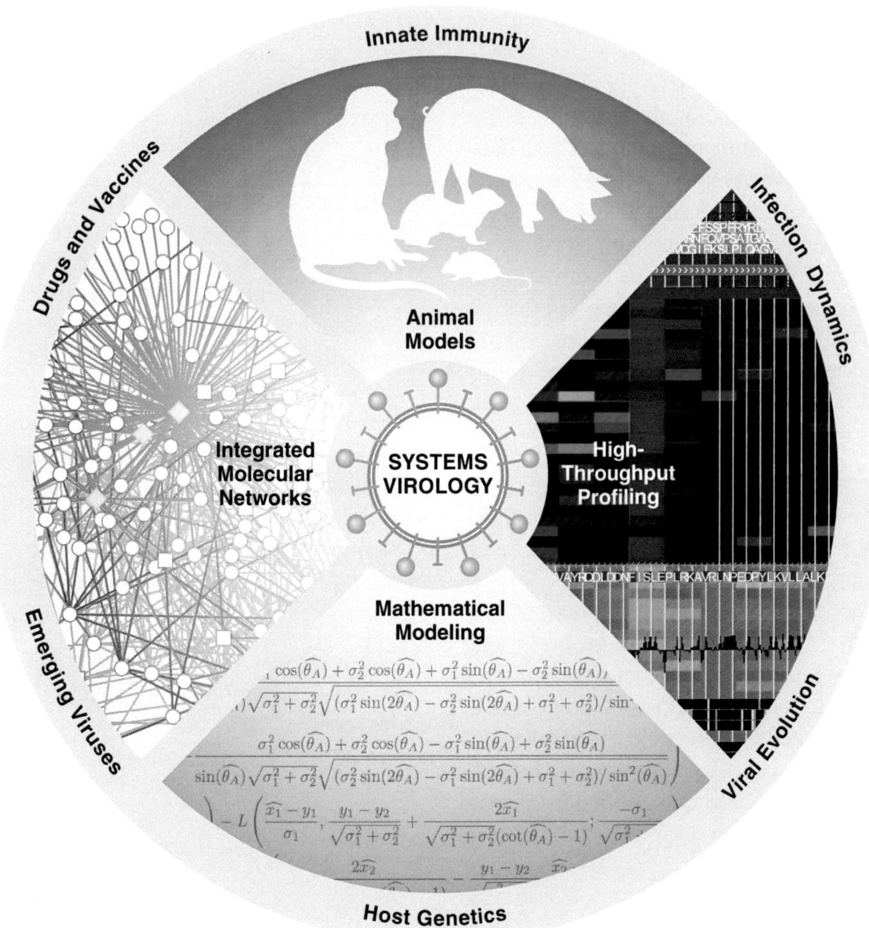

FIGURE 1 The use of systems-level approaches to study viral pathogenesis is changing our view of how viruses cause disease.

Once a viral invader has triggered type 1 interferon, a vast panoply of interferon-generated responses is unleashed, which operate simultaneously. The net effect of these responses is unpredictable and can only be understood by using global approaches to identify patterns of antiviral activities, pro- and anti-inflammatory cytokine responses, and the like. Recent studies have suggested that the long-term outcome of significant viral infections—such as HIV, hepatitis B, and hepatitis C—may be determined by the innate response in the first day after infection, even before the adaptive immune response has begun to take hold.

Infection dynamics: Systems approaches are also well suited to understanding the dynamics of the race between the virus and host, which determines the outcome of an ongoing infection. Newer methods in noninvasive data collection have made it possible to follow the evolving infection in individual living animals, rather than trying to reconstruct dynamics from animals that were sacrificed. This is critical for many viral diseases where we seek to understand why some hosts survive while others succumb, as exemplified by clinical questions raised during the Ebola pandemic in West Africa.

Viral evolution: In real life, viruses exist as swarms of genetic variants, which evolve as they replicate in individual hosts and spread among members of a population. Deep sequencing is critical to understanding the evolution of the swarm and the selection of dominant variants. These data, in turn, provide essential insights into the host response to infection. The evolution of HIV during long-term infection is an important example of this phenomenon, where recently developed methods have provided key insights into the virus–immune interaction. Sequencing approaches can also address questions such as during the 2013–2015 outbreak did Ebola virus evolve to become more readily transmissible between humans? Could this virus become established as a new human disease maintained by person-to-person transmission, as has HIV?

Manipulation of host genomes: Turning to the host, the utility of experimental animal models has exploded, driven by the ability to manipulate genomes. Knockouts, knock-ins, and earlier methods were cumbersome and slow. The CRISPR/Cas9 genome-editing system is a quantum leap that will enable a new generation of genetic experiments heretofore beyond reach. The facilitation of host genetic analysis has already spawned the new field of personalized medicine, where different treatments can be targeted to different patients. Such an approach is exemplified by current experience with hepatitis C virus, where specific virus clones and host determinants mandate different regimens for individual patients.

Emerging viruses: We are now in the era of globalization in which zoonotic viruses cross the species barrier and, with increasing frequency, become established as human infections that are maintained by person-to-person transmission independent of the original animal reservoir. Influenza has now been joined by HIV, Ebola virus, SARS, and MERS. Recent outbreaks of Ebola in West Africa (2013) and MERS in South Korea (2015) exemplify the emergence of viral diseases truly new to the human population. In response, technological advances are enabling rapid diagnosis of exotic infections and have become indispensable to understanding how these new viruses cause disease, which in turn provides a pathway to treatment and prevention.

Antiviral therapy: The development of new antiviral therapies faces the multiple hurdles of high risk, gargantuan expense, and long delay, with drug development taking over 10 years at a cost that exceeds one billion dollars. New technologies are providing a partial solution to these daunting challenges, by providing information on the virus–host interactome, expediting computational screening of large libraries of compounds, and facilitating the repurposing of approved medicines.

Vaccines: Preventive vaccines continue to provide a frontier for the control of viral diseases. Opportunities fall into several categories. First, there are diseases such as Ebola, where conventional approaches will produce effective vaccines, and the major barrier is the lack of investment in "orphan" diseases that can cause explosive epidemics. Second, there are diseases such as dengue and hepatitis C, where scientific challenges remain, but which can likely be solved by further engineering of candidate immunogens. Third, and most challenging, are viruses such as HIV and influenza that persist at an individual or population level because of the fitness of immunological escape variants. Prevention of these rogue viruses may require creative and novel approaches, such as the use of gene therapy technology to confer HIV resistance at the cellular level.

We hope that you will bring these perspectives to bear as you read through this book and wherever your research endeavors may take you. The prospects are exciting and the future is now! Happy hunting!

The editors

Part I

History and Essentials of Viral Pathogenesis

Virology, together with all of biomedical sciences, is undergoing a revolution that can be encapsulated as a transition from reductionism to systems biology. This third edition of *Viral Pathogenesis* reflects this paradigm shift and the new perspective it brings to the field. Accordingly, this edition is organized in four different parts: (I) History and essentials of viral pathogenesis; (II) Systems-level approaches to viral pathogenesis; (III) Emergence and control of viral infections; and (IV) Past and future. Part I sets forth our knowledge based on long-established methods in pathology, virology, and adaptive immunity. This section provides the background for the rest of the book, which focuses on current and future methods and discoveries in viral pathogenesis.

We begin Part I with an overview of the *human toll of viral infections*, which describes the burden of viral diseases, both the annually recurrent illnesses and the intermittent epidemics. The ongoing AIDS pandemic and the recent Ebola epidemic in West Africa have underlined the importance of viruses as current and future disease burden. The burden of viral disease is particularly important because of the potential to prevent or treat, which contrasts with many chronic diseases. This in turn provides an imperative to understand pathogenesis, often a necessary step in the pathway to prevention.

Historical roots gives a short history of virological sciences and offers a perspective on the developments that have paved the way for the evolution of our insights into pathogenesis. In most instances, new technologies have led the way to a more sophisticated analysis of viruses and disease mechanisms. Over time, focus has shifted back and forth between the virus and the host response, leading to our current view that the virus–host combination determines the outcome of infection. This historical perspective sets the stage for a look forward, which is presented at the end of the book.

Basic concepts of viral pathogenesis begins a set of chapters that lay down basic information about pathogenesis. The chapter starts with a description of virus–cell interactions and goes onto explain the sequential steps in infection of animal hosts. The information includes the role of cellular receptors, the entry and uncoating of the viral genome, the processes of transcription and translation, and the maturation and release of new infectious virions. The chapter describes for animal hosts the sites of virus entry, replication, modes of dissemination, target organs, and shedding. Various tissue-specific pathological outcomes of virus infection are briefly recounted.

Innate immunity describes the concept of pathogen-associated molecular patterns and cellular pattern recognition receptors that constitute the host's "first response" system to viral invaders. The intracellular pathways that are triggered, and the interferon and other cellular responses are outlined. The pleiotropic response to interferons is described together with the roles of cellular elements such as mononuclear cells and NK cells. Finally, this chapter explains how the innate response "tees up" the adaptive immune response, a segue to the following chapters on *acquired immunity*. The first chapter on *acquired immunity* describes humoral and cellular responses and the sequential steps in immune induction. The role of the immune response is explained, including the somewhat different mechanisms that are responsible for recovery from infection and in prevention of reinfection. The second chapter, *immune aberration*, describes how certain virus-specific immune responses can cause disease (virus-induced immunopathology), and—in contrast—how some viruses can induce transient or long-lasting immunosuppression.

The remaining chapters in Part I embellish these basic concepts with a focus on special aspects of pathogenesis. *Patterns of infection* discusses virus virulence and host susceptibility and explains why these variables must be considered in the context of the virus–host interaction. We also discuss the dynamics that differentiate acute from persistent infections, and the variables that can tip an infection either way. *Viral oncogenesis* explains how oncogenic viruses initiate the process that leads to cancer, with emphasis on the distinctly different mechanisms that are used by RNA and DNA tumor viruses. RNA viruses encode oncogenes, altered versions of normal cellular genes, while DNA viruses in one way or another interfere with the brakes on the normal cell cycle. The chapter on *HIV/AIDS* is the only chapter devoted to a single disease, because of its global burden. The complex pathogenesis of HIV is explained together with its several paradoxical aspects, as explored in the SIV experimental model in nonhuman primates. HIV also illustrates the potential for antiviral therapy and the need for a preventive vaccine, both subjects of chapters in Part III.

Animal models have become a key ingredient in the dissection of virus–host interactions, because of new and evolving methods for manipulating the genes of many experimental animals, particularly mice. This chapter describes the wide variety of laboratory animals available to the virologist, with their advantages and liabilities, and points out the unique assets of each species for specialized questions. Explained are various methods to "knockout" or "knockin" individual genes, including the ability to target gene expression to specific organs or tissues and developmental ages. Examples are provided to illustrate these issues. These genetic methods complement the analytic strengths of systems biology and markedly enhance the experimentalists' ability to explore the virus–host interactome.

Chapter 1

The Human Toll of Viral Diseases

Past Plagues and Pending Pandemics

Neal Nathanson

Department of Microbiology, Perelman School of Medicine, University of Pennsylvania, Philadelphia, PA, USA

Chapter Outline

1. INTRODUCTION

This chapter profiles some examples of past and current viral infections that have levied a high toll in human misery and mortality. Viral epidemics can be very disruptive to civil society, and—conversely—civil disasters can trigger viral epidemics. Furthermore, viruses likely play an insidious role in initiating a number of chronic diseases whose burden is ever increasing. Evolving technologies have led to much more rapid detection of viral diseases emerging anywhere in the world (Lipkin, 2013a,b). However, current efforts to reduce the toll of viral diseases are less than satisfactory, and futurists predict the emergence of new pandemics particularly of zoonotic origin. Pathogenesis—the subject of this book—informs ongoing efforts to mitigate the toll of viral disease.

Outbreaks of infectious disease have long been recognized as disruptive to civil society, and in their worst manifestations, human catastrophes. Epidemics and their toll on human society have been the subject of scholarly treatises, such as *Plagues and People* by William McNeill, novels such as *I Promesi Sposi* by Alessandro Manzoni, and artistic works such as representations of cholera by Honoré Daumier. Among major epidemics, some such as smallpox and influenza, are viral in origin. Less dramatic but perhaps more important are viral diseases of infancy and childhood. For instance, respiratory and diarrheal viral diseases among under-5-year-old children make a very significant contribution to the burden of human disease, and remain a major cause of short life expectancy in spite of the development of pediatric vaccines. As chronic diseases become more important, we are unraveling viral infections that are causal agents (such as oncogenic viruses) or are suspected to trigger immune-mediated illnesses (such as type 1 diabetes and multiple sclerosis).

New viral diseases continue to emerge, due to a wide variety of factors that are discussed in Chapter 16 (Figure 1). Viral diseases of animals that are transmitted to humans (zoonoses) are a significant cause of emerging infectious diseases, and constitute a future menace. Finally, viral diseases of animals of economic importance have an important impact on human food security. This chapter provides examples of these different ways in which viral disease takes a toll on humans and their civil societies. We conclude by a look into the future, making a few predictions about the ongoing toll of viral disease.

Viral Pathogenesis. http://dx.doi.org/10.1016/B978-0-12-800964-2.00001-X

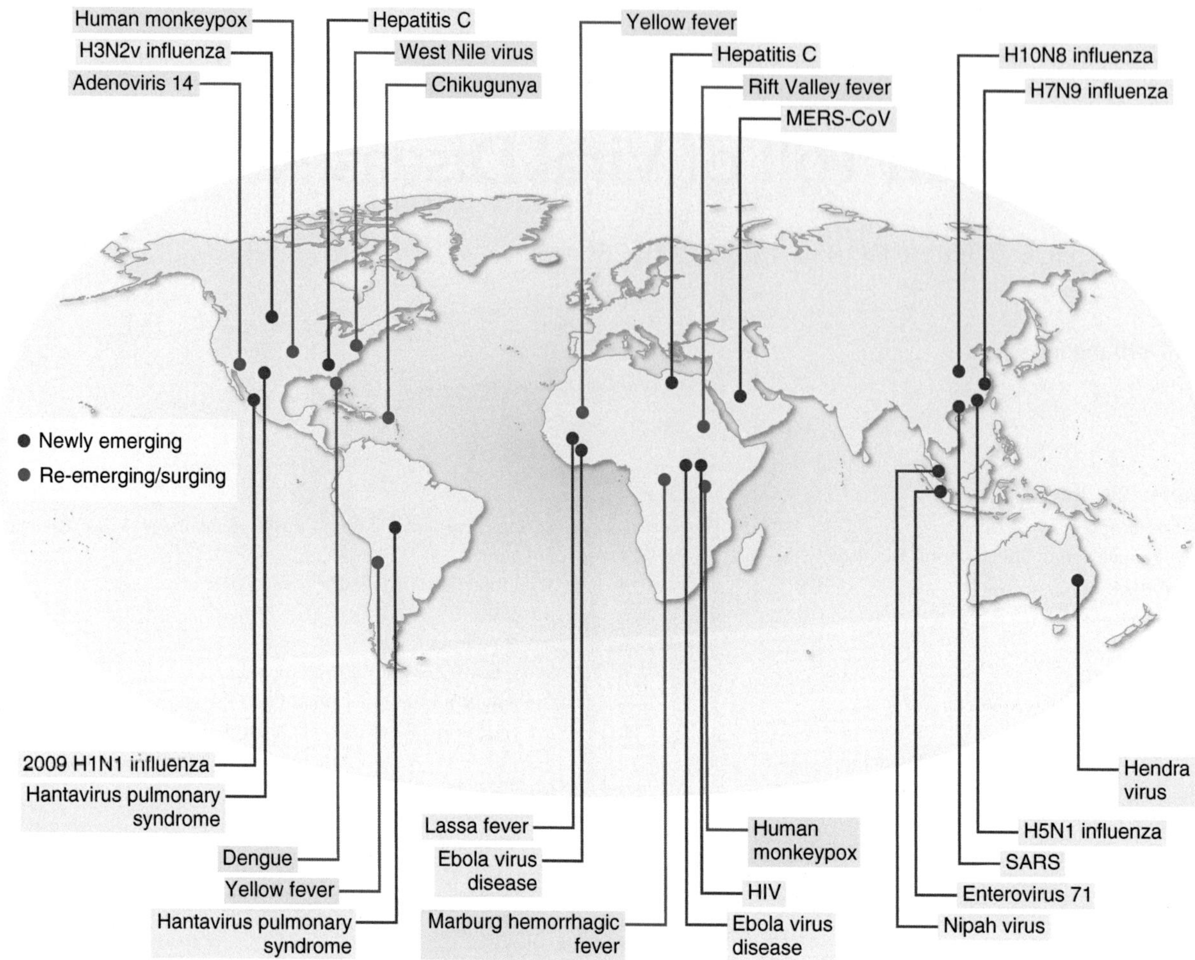

FIGURE 1 This map points out the large number of newly emerging and re-emerging viral diseases, 2014. *Courtesy of AS Fauci/National Institute of Allergy and Infectious Diseases.*

2. THE HUMAN TOLL OF SPECIFIC VIRAL DISEASES

2.1 Smallpox

Since the earliest recorded time (in ancient Egypt and subsequently), smallpox has been one of the most feared infectious diseases. In *Plagues and people*, McNeill describes a number of episodes in which smallpox altered the course of history. Introduced into a "virgin" population, smallpox is devastating. Historic reconstructions have suggested that smallpox did not exist in the western hemisphere when the Spaniards first invaded Central America. They inadvertently transmitted smallpox to the large and well-organized Mayan and Aztec populations, triggering a decimating epidemic. It was smallpox that permitted a small contingent of 600 invaders to overcome an indigenous population of millions. Even during the twentieth century, when vaccination was widely practiced, it is estimated that 300–500 million people died from smallpox or at least 2 million souls each year.

Smallpox is caused by variola virus, a poxvirus specific to humans. The virus is acquired by aerosol exposure, but spreads much less rapidly than many other viral infections since transmission usually occurs via face-to-face contact. Following infection, the incubation period is 10–20 days. Inhaled virus replicates first in the lungs, then spreads to internal organs causing a viremia and widespread dissemination. The virus replicates in skin and small dermal blood vessels, causing a rash that evolves into multiple blisters, which leave severe scars on the face and body; corneal infection may cause permanent blindness. Two different strains of variola virus circulated in humans, variola major with a mortality of 20–30% and variola minor, with a mortality of about 1%. The exact cause of death during smallpox infection is a bit murky, and has been attributed to "toxemia," that is an overwhelming infection of lungs and many other organs.

The concept of immunization against viral diseases originated with variolation, the deliberate exposure to (hopefully)

FIGURE 2 An 1802 cartoon of the early controversy surrounding Edward Jenner's vaccination theory, showing using his cowpox-derived smallpox vaccine causing cattle to emerge from patients. *Copied from Wikipedia, accessed February 2, 2014, with permission.*

small doses of virulent variola virus, a method fraught with danger. Immunization was markedly improved by the observation that vaccinia virus, a poxvirus that infected cows, produced a mild localized infection on the hands of milkmaids, who were protected against smallpox. Dr Edward Jenner, who suggested the use of vaccinia-based immunization in 1796, was first an object of derision by the medical profession (Figure 2), but is now regarded as a pioneer in preventive medicine. Vaccination is very effective, producing a high level of protection against illness and death that probably lasts a lifetime, although the degree of immunity can wane after 5–10 years.

Widespread vaccination eradicated smallpox from many countries in the second half of the twentieth century, but outbreaks continued to occur in lower income countries where many children were not immunized. Based on successful eradication of smallpox in many countries, the World Health Organization embarked on a program of global eradication in 1966. Smallpox possessed a number of attributes that made it a feasible target for eradication: it was an infection exclusive to humans, with no vector or extrahuman reservoir; all acute infections were readily detected due to the unmistakable pustular rash and there were no subclinical infections; transmission was mainly confined to direct face-to-face contact between patients and susceptible persons; the incubation period was rather long, so that outbreaks spread slowly; and it was easy to identify immune individuals who had prominent scars due to prior infection or vaccination. A key to control was the strategy of "ring vaccination" around individual outbreaks, which was more effective than population-wide immunization of children and adults. Due to heroic efforts in India and a number of other countries with limited public health programs, smallpox was eradicated by 1977, and the world declared smallpox-free in 1980 (Henderson, 2009; Foege, 2011). In spite of the outstanding record of viral vaccines as effective public health tools, even today there remain a small group of nonbelievers.

2.2 HIV/AIDS

HIV/AIDS is the subject of a later chapter in this book, so the following brief account is focused on the human toll of this devastating epidemic, the first "great plague" of the twentieth and twenty-first centuries. Although HIV was probably first transmitted from chimpanzees to humans in Africa in the 1930s, it only assumed epidemic form about 1970 in Africa and 1–2 decades later in other regions of the world (Pepin, 2011). It is estimated that AIDS has caused over 30 million deaths, and that more than 30 million other humans are currently living with HIV/AIDS, while there are more than two million new infections each year.

In some of the countries with the highest prevalence of HIV infections, it is projected that AIDS resulted in a dramatic reduction of life expectancy, an impact that may be unique among viral diseases. Life expectancy in Botswana was reduced from a non-AIDS projection of 65 years to 45 years of age, prior to a massive rollout of antiretroviral treatment. Another unusual feature of the AIDS pandemic is its concentration in young adults who are the principal engines of society. This age selection has created a vast group of orphans and decimated the incomes and integrity of families, the social unit that is a major pillar of society.

In contrast to many other viral epidemics, HIV does not "burn out" or fade away by exhausting susceptible hosts, but is perpetuated by the continual recruitment of young adults as they become sexually active. Furthermore, the virus is transmitted in many ways, by blood and blood products, by use of contaminated needles and syringes, from mothers to their newborn infants, in addition to sexual contact. Finally, HIV infections tend to concentrate in social subgroups which can be marginalized, such as sex workers, injecting drug users, and gay men. All of these characteristics have made the pandemic a daunting public health challenge.

In contrast to many other viral diseases of public health importance, HIV causes a persistent infection with an incubation period (untreated) that averages about 5 years. This makes it a potential target for antiretroviral treatment. One of the great triumphs of modern scientific medicine has been the development of a vast array of increasingly effective drug therapies, which are described in a later chapter. Currently, it is estimated that if a person aged 20 is diagnosed with a recent HIV infection and if he or she has access to optimal treatment, her/his life expectancy is about 50 years (in contrast to 5 years if untreated). The dramatic impact of antiretroviral drugs has the potential to convert a fatal infection into a chronic manageable illness. Furthermore, antiretroviral drugs have additional uses, such as prevention of mother-to-child transmission of HIV, incorporation into microbicides to protect women from exposure to HIV-infected partners, and pre-exposure prophylaxis (PrEP) for subjects with high-risk of HIV contact.

However, the great potential utility of antiretroviral drugs is dependent upon access to expensive medications, highly specialized medical care, and robust public health programs. Even in high-income countries, there is a "treatment cascade" such that no more than 30% of HIV-infected persons have optimal diagnosis, treatment, and very low HIV blood levels (Figure 3). In Africa, the locus of about 75% of the global burden of HIV/AIDS, few countries have optimal HIV control programs. A significant impediment is the price of antiretroviral drugs, although this barrier has been substantially reduced. More important, few African countries have a health system that is sufficiently robust to provide adequate diagnosis and treatment for the whole population. A reduction in the incidence of new infections is likely the best measure for the success of an HIV control program (the reproductive rate, Ro, is less than 1). HIV incidence remains very high in the more heavily impacted countries, even though some have managed to somewhat reduce the year-on-year number of new infections. (It should be noted that a reduction in new infections has not been accomplished in the United States either, where new infections have plateaued for at least 20 years.)

2.3 Influenza

Mortality associated with influenza infection is carefully monitored by the Centers for Disease Control. Over a 30-year period, 1977–2006, annual influenza-associated deaths in the United States ranged from 3000 to 49,000.

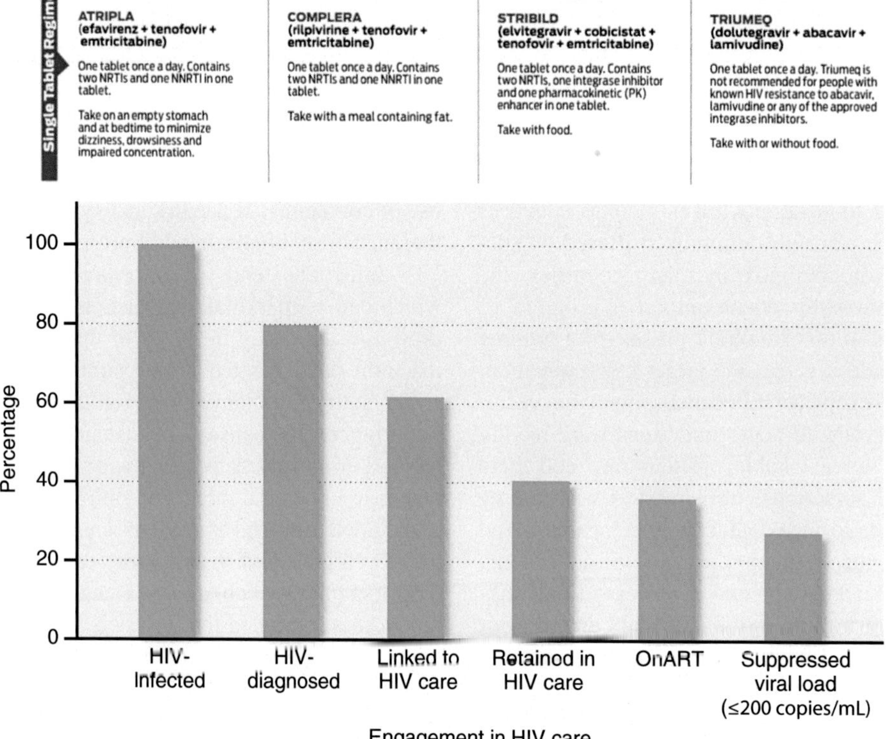

FIGURE 3 In spite of the relative availability of antiretroviral treatment in the United States, only 30% of infected individuals are on optimal viral control. This is partly due to the complexity of treatment although simplified regimens have been introduced in the past few years. Top: HIV DRUG CHART, showing some of the single-pill antiretroviral options for HIV treatment, accessed online, from POZ, Health, Life, and HIV, http://www.poz.com/, September 19, 2014. Bottom: Number and percentage of HIV-infected persons engaged in selected stages of the continuum of HIV care—United States. *Redrawn from Cohen et al. (2011), with permission.*

If it is assumed that on average, influenza caused 25,000 deaths in the United States and this death rate is applied to the global population (about 20 times the population of the United States), then there are—on average—at least 500,000 influenza-related deaths each year.

Influenza is caused by an eight-segmented negative-strand RNA virus. Influenza virions carry two external spikes, the hemagglutinin (H) and the neuraminidase (N); the hemagglutinin spike is responsible for attachment to the cellular receptor. The receptor is a sialic acid residue on cell surface proteins many of which are glycosylated. Importantly, sialic acid residues on human glycoproteins mainly have α2-6 linkages (the number two carbon on the sialic acid hexose ring is linked to the number six carbon on the galactose ring within the glycosyl moiety on the glycoprotein) while sialic acid residues on avian glycoproteins are mainly attached via α2-3 linkages. This has epidemiological implications because avian-derived type A influenza viruses preferentially attach to α2-3 linkages while type A influenza viruses that spread in humans preferentially attach to α2-6 linked sialic acid residues. Importantly, a few mutations on the viral hemagglutinin can change the sialic acid preference for the virus. There are many strains of type A influenza virus, and different strains circulate in different vertebrate species. Type A influenza virus strains are classified into subtypes by the antigenic determinants on the H and N spikes, which assort independently within the segmented viral genome. Thus, there are H1N1 subtypes, H1N2 subtypes, and so forth.

The incubation period of influenza is very short (1–4 days), and the shedding of aerosolized virus by infected patients leads to rapid spread of infection through a population. Since individuals who have recovered from infection are immune to re-infection by the same antigenic strain of virus, influenza virus exhibits a somewhat unique epidemiologic pattern. In the course of an influenza "season," usually during the winter months, a specific antigenic strain may "exhaust" the majority of uninfected members of a human population and then fade away for the lack of susceptible hosts. Under these circumstances, influenza virus is subject to "antigenic drift," whereby mutations in the H and N spike proteins generate a new virus, selected because it can escape neutralization by antibodies against the preceding epidemic strain of the virus.

Several times in a century, human influenza viruses undergo "antigenic shift" in which the prevalent H and N variant is replaced by a virus with different H and N proteins. Antigenic shifts are usually due to genetic recombination of strains of virus that are circulating in animal populations; the "new" strain of virus may have a mixture of genes derived from porcine, avian, and human viruses. Circumstantial observations suggest that genetic recombination occurs most frequently in pigs, which carry their own influenza strains and are often exposed to exogenous virus strains from both avian (wild and farmed) and human species. Less frequently, a nonrecombined animal influenza virus may jump the species barrier and become established in the human population. There are some stringent requirements for spread of a virus which represents an antigenic shift: (1) the new virus must be able to bind quite efficiently to the sialoprotein receptor on human cells, which differs slightly from the sialoprotein receptor on avian cells; (2) the new virus must be able to replicate efficiently in human cells, a property that is determined by several viral gene segments; (3) the new virus must be transmitted efficiently between humans, a property that is genetically distinct from the two preceding requirements.

Because the human population often lacks immunity to the newly emergent hemagglutinin, antigenic shifts can produce a "pandemic" with many more severe infections and deaths than are seen in a year where the prevalent strain is due to antigenic drift within the current H protein. Different variants of influenza virus vary widely in their virulence for humans, with case fatality ratios that range from 0.1% to 10%. Virulence is determined by several of the genetic segments—other than the H and N genes—of the virus acting in concert. Some highly transmissible viruses cause relatively mild disease while some highly virulent strains are poorly transmissible.

Rarely, there is a "perfect storm" in which antigenic shift is introduced by an influenza virus strain which meets all of the requirements for cellular infection, transmissibility, and virulence. It appears that this was the case only once in the last century, the "Spanish" influenza pandemic of 1918–1920 (Barry, 2004). It is estimated that the pandemic strain infected about 500 million persons, and resulted in about 50 million deaths (a case fatality ratio as high as 10%, compared to about 0.1% for a typical influenza season). This pandemic represented the largest single epidemic in recorded medical history, an unprecedented human catastrophe, which caused a transient drop in global life expectancy, an unprecedented phenomenon. Figure 4 shows a ward of influenza patients during this global pandemic.

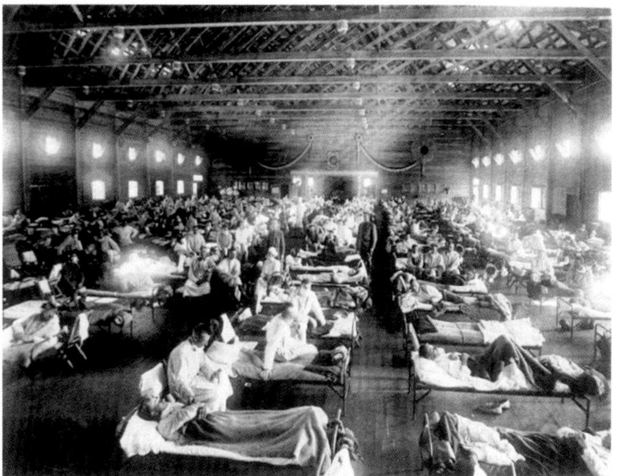

FIGURE 4 A ward of patients ill with influenza. *Camp Funston, Kansas, 1918, accessed online, September 20, 2014 https://en.wikipedia.org/wiki/Camp_Funston.*

Recent studies have reconstructed the genome of the 1918 virus from genetic fragments found in exhumed tissues of persons dying during the epidemic. The pandemic strain had an H1N1 phenotype, an antigenic pattern to which the human population was nonimmune since viruses of this type had not been prevalent in the human population for an estimated 35 years prior to the epidemic. Studies with the reconstructed virus in nonhuman primates have recapitulated the acute disease seen in humans. Diseases severity was associated with an extreme cytokine storm centered in the lungs, in contrast to the alternative theory that the virus had spread widely to many other tissues. Furthermore, studies with viral recombinants identified several viral genes (other than H and N determinants) that were associated with virulence. The unusual pathogenesis pattern may explain why the human pandemic produced a high case fatality ratio in young adults, in contrast to most influenza outbreaks where the toll is greatest in young children and elderly patients.

Of concern in 2014, there are several virulent strains of avian influenza virus that have infected a small number of humans in close contact with farmed poultry. To date these virulent animal influenza viruses have not been readily transmitted between humans, but—if they evolved to be transmissible—might have the potential to cause a pandemic, an issue discussed later in this chapter.

2.4 Measles

Measles is a ubiquitous acute transient illness of childhood. Infection is transmitted by inhalation of aerosolized virus, the virus spreads via circulating infected lymphoid cells, and many organs are involved. Although systemic, the infection is usually benign, and illness is confined to respiratory symptoms (coughing, and sneezing) and rash. Among well-nourished healthy children, measles has a low-case fatality ratio (about 0.1%). However, it is a much more serious disease among malnourished infants and children; the accompanying pneumonia may lead to a case fatality ratio as high as 10%. Prior to the introduction of measles vaccine, it is estimated that worldwide one to two million children died of measles each year (Moss and Griffin, 2012).

Measles vaccine, introduced in the 1960s, is a live attenuated virus variant that is safe, effective, and inexpensive. The widespread use of vaccine has markedly reduced global measles mortality, but the virus still causes more than 100,000 deaths each year. This reflects the failure to immunize children due to the limitations of weak health maintenance systems, particularly in low income countries. Also, unimmunized children are more likely to be malnourished and prone to serious complications of the infection.

Global eradication of measles has been discussed for many years as a public health goal. It meets many criteria for potential eradication: the vaccine is safe, effective, long lasting, and inexpensive, and humans are the only host for the virus. Endemic transmission of measles virus has been interrupted in the western hemisphere, demonstrating the plausibility of eradication in large populations. However, measles is very contagious, and transmission can be maintained in a well-immunized population where only 10% are susceptible. Therefore, very high levels of population immunity must be achieved to terminate spread of measles virus. It appears that the most realistic goal is reduction of measles mortality by maximizing immunization levels, particularly in low income countries.

2.5 The Global Toll of Viral Infections

At a global level, viral diseases take their greatest toll in developing countries, particularly in children under the age of 5 years. There are several reasons for this undue burden: A failure to immunize children with available vaccines due to weak health systems; and a markedly increased susceptibility to common childhood infections due to undernutrition which increases risk of death from infections. This relationship is evidenced in Figure 5 that plots death rates from pneumonia by countries sorted according to income.

3. EPIDEMICS AND SOCIETY

In epidemic form, viral diseases not only can decimate a population but also may have a major impact on civil society itself. Conversely, social catastrophes can produce a breeding ground for virus epidemics or impede the control of viral diseases. Several examples follow below.

3.1 Epidemic Yellow Fever in Philadelphia, 1793

Urban yellow fever is caused by a flavivirus that is transmitted by *Aedes aegypti,* a peri-domestic tropical mosquito that breeds in standing water. Both mosquito and virus are indigenous to the Caribbean and adjacent parts of Central and South America, since they cannot overwinter in temperate climates. The yellow fever epidemic of 1793 in Philadelphia, the most calamitous outbreak ever to strike an American city, was described in detail in the 1949 monograph, *Bring out your dead,* by J. M. Powell.

In 1793, Philadelphia was serving as the capital of the United States during George Washington's first term as President. Thomas Jefferson was Secretary of State, and Alexander Hamilton was Secretary of the Treasury. The new government was less than 2 years old, and its fragile structure was threatened by the ongoing French Revolution and the incipient hostility between England and France. The tension increased with the arrival of "Citizen" Genet, the new ambassador from France, who effectively fomented pro-revolutionary sentiment among the inhabitants of

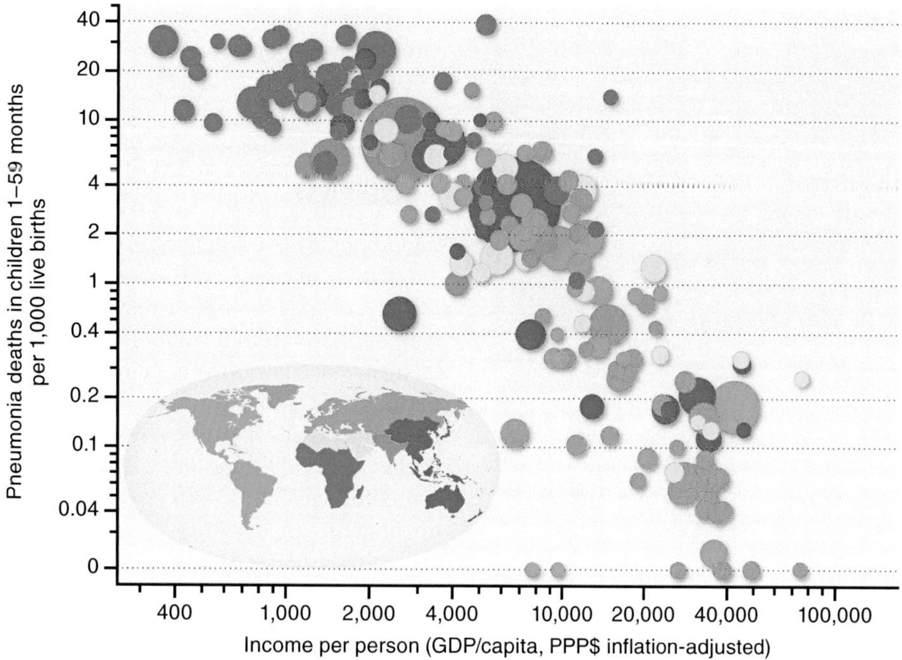

FIGURE 5 Pneumonia death rate in children under 5 years of age, plotted against GDP per capital, by country. This display uses pneumonia deaths as a proxy for viral deaths, and points out that the toll of viral diseases is highest among children in low income countries. *Redrawn from Gapminder (2014), with permission.*

Philadelphia. Washington and Hamilton wished at all costs to avoid being drawn into an entanglement in a European conflict that would provide the British with an excuse to attempt to reclaim their colonies.

In 1793, the population of Philadelphia was about 50,000 and the annual mortality was about 2%. Water was supplied by individual shallow wells, and outside most houses there was a water barrel for drinking, cooking, and other needs. Mosquitoes bred in these containers, and they were notably plentiful in the hot and dry summer of 1793. This set the stage for the outbreak.

Neither *A. aegypti* nor yellow fever virus was native to Philadelphia. They were introduced in great numbers by a major social upheaval in the Caribbean. Following the French Revolution of 1789, there was an uprising among the slaves on the sugar plantations of Dominica (present Haiti and Dominican Republic), with the slaughter of many white plantation owners. This caused a mass exodus to various ports in the United States, including Philadelphia. The yellow fever cycle was maintained on board the crowded ships, where open water barrels permitted mosquitoes to breed during the voyage, while infections were continually transmitted among the human passengers. Following the arrival of large numbers of refugees in July, 1793 in Philadelphia, cases of a very severe often fatal febrile disease were first noted near the waterfront, and then spread to the rest of the city. After a slow beginning, the epidemic gathered force and increased at a rapid rate in September, to peak in mid-October.

The dimensions of the epidemic are hard to comprehend. About 20% of the population was stricken, half of whom, about 5,000, died, the largest single epidemic rate ever experienced in an American city. Washington, Jefferson, and Hamilton were all in Philadelphia. Hamilton was stricken, causing widespread consternation in the populace, but he recovered. The nascent government ground to a halt, when much of the population fled the city. Washington wanted to remain and was only persuaded by his wife to leave, along with Jefferson. A change in the weather slowed the epidemic in mid-October, then an early mosquito-killing frost in mid-November brought it to an end. It was only by chance that this epidemic did not alter the course of American history.

3.2 Eradication of Wild Polioviruses

It is estimated that, prior to the introduction of inactivated poliovirus vaccine (IPV or Salk vaccine), each year wild poliovirus caused about one million paralytic cases worldwide. Polio was a dread disease, since it struck at random and could leave a healthy child unable to walk or otherwise severely incapacitated. The public image of President Roosevelt, who was confined to a wheelchair, drove home this fear, as well as movies showing banks of children encased in iron lungs, without which they could not breathe (Figure 6). The horror of polio was captured by Philip Roth in his novel *Nemesis*.

Following the widespread use of IPV (inactivated poliovirus vaccine) and OPV (oral poliovirus vaccine or Sabin

FIGURE 6 A ward of patients in iron lungs at Rancho Los Amigos hospital, California, during the 1953 epidemic of poliomyelitis. At first glance, this image shocks and saddens from the enormity of the problem of sick children in need of iron lungs. On closer examination, it is clear that the equipment that usually accompanied people using iron lungs, such as tracheotomy tubes and pumps and tank side tables, is not present. This scene was staged for a film. It is not historically accurate as a respirator ward, but is an example of an established photographic technique to direct the viewer's response, and advocate for vaccination for polio and other diseases. *Accessed online, September 20, 2014.*

vaccine), wild poliovirus was eradicated in the United States about 1973, and in the western hemisphere about 1990. As a result of these successes, WHO announced the goal of global eradication in 1988, at which time it was estimated that there were still about 350,000 paralytic cases annually. This initiative was so effective that paralytic polio was reduced to about 2000 cases by 2000. However, the eradication program has been stalled at that level for over a decade. Wild poliovirus was finally eliminated in India in 2011 as a result of a massive effort on the part of the government. But the virus has continued to circulate without interruption in three countries; Pakistan, Afghanistan, and Nigeria. Furthermore, for political reasons, polio vaccinators have been targeted and several have been killed by terrorist organizations. In 2013, wild virus spread from those locations to Syria as a by-product of the civil chaos in that country, and from there to Israel where it caused a "silent" outbreak recognized only by isolation of virus from sewage samples.

In Nigeria, poliovirus was long ago eliminated in the southern regions. But virus has continued to circulate in the northern regions, where the Muslim population has resisted public health immunizations sponsored by the central government which is Christian dominated. There is a current debate whether or not wild poliovirus will ever be eradicated (Nathanson and Kew, 2010). Clearly, the outcome will be determined by social and political forces rather than any scientific or public health impediments. However, writing in the Fall, 2014, there are several promising milestones: wild type 2 poliovirus disappeared in 1999; wild type 3 poliovirus has not been detected since 2012; and wild type 1 poliovirus—long endemic in three locations—was eliminated in India in 2011; it appears to be close to elimination in Nigeria in 2014; and only remains stubbornly circulating in the tribal areas of Pakistan and Afghanistan.

The foregoing examples vividly illustrate the impact of viral disease upon society and—conversely—how societal forces can influence viral epidemics and their control.

4. VIRUSES, PRIONS, AND CHRONIC DISEASE

In addition to overt outbreaks and pandemics, viruses can play a more insidious role, as instigators of chronic illnesses. In this context, the causal relationship is more subtle and may only be inferred from circumstantial data. Therefore, in many instances, the relationship is still controversial. Furthermore, research on transmissible agents that cause chronic illnesses has uncovered a novel group of infectious agents, now called "prions" (also designated "transmissible spongiform agents"). Although prions do not meet the traditional definition of viruses ("bad news wrapped in protein" according to Peter Medawar), they are included in the following discussion since they can be considered one of the simplest life forms.

It is now well established that several human viruses are major causes of specific cancers. Among these are hepatitis B virus, human papillomavirus, and Kaposi's sarcoma virus (human herpesvirus 8), which are the subject of Chapter 8. In most of these instances, the causal relationship was heavily disputed and was only established by many years of research.

Multiple sclerosis and type 1 diabetes are examples of major chronic diseases whose pathogenesis is only partially understood. In both instances, there appears to be a major immunological mechanism involved, but it is not clear what triggers this pathological response. Genetic studies of identical twins in which one twin is afflicted with multiple sclerosis, indicate that a high proportion (perhaps 50%) of the cognate twins are also stricken. These data indicate that there are significant genetic determinants of risk. However, because many cognate twins do not have the disease, it has been inferred that there is also a "trigger" event that initiates the disease process. It has been suggested that—in some cases at least—an acute viral infection acts as this postulated trigger, although direct proof has not yet been uncovered.

4.1 Prion Diseases

Scrapie is a fatal progressive degenerative neurological disease of sheep, described at least 200 years ago by sheep herders in England. The pathological hallmark is a spongiform encephalopathy. A rare sporadic fatal human syndrome, Creutzfeldt–Jacob disease, was recognized to cause similar pathological lesions. Kuru, a similar affliction, was subsequently observed as an epidemic in a small stone-age linguistic group in the eastern highlands of New Guinea. Scrapie was shown to be transmissible from sheep to sheep by intracerebral injection of brain tissue from a diseased

Sidebar 1 Lessons learned: the toll of viral infections

- Many human viruses are ubiquitous and infect a large proportion of the population, but relatively a few take a significant toll of human life or well-being. However, the relatively few viral infections that cause significant human mortality can have a devastating effect as the cause of acute pandemics or as persistent endemic diseases.
- There is no necessary relationship between the ability of a virus to perpetuate itself in the human population and its ability to cause significant morbidity and mortality. The severity of disease caused by a given virus can vary tremendously depending upon the age and health of the human population.
- Vaccine-induced immunity is the most effective way to protect humans against viral diseases, particularly for acute infections. It has been possible to develop safe and effective vaccines against many viral diseases, but a few have resisted the best efforts of modern medical science.
- Antiviral drugs are likely ineffective as weapons against acute infections but can be useful in mitigating persistent or long-incubation-period viral diseases.
- At a global level, much of the burden of viral-induced illness and death is due to the failure to deploy effective interventions rather than the lack of effective vaccines or antiviral drugs.

animal, and a parallel experiment showed that kuru could be transmitted to chimpanzees.

Early researchers assumed that the spongiform encephalopathies were caused by a new unknown group of viruses. However, years of experimentation failed to identify an RNA or a DNA associated with increasingly purified preparations of the transmissible material. Based on these observations, Stanley Prusiner proposed that the spongiform agents consisted solely of protein, in a specific molecular conformation that explained their unique properties, such as resistance to proteolytic digestion and ability to "replicate." This hypothesis, which violated the central dogma of molecular biology (information is transmitted from DNA to RNA to protein), was gradually confirmed by many experiments conducted by numerous investigators over 30 years. Examples of prions (not all of which are associated with disease states) have been described in life forms as simple as yeasts, making them a more ubiquitous phenomenon than originally recognized.

Prion diseases have taken on an increasingly important role as a cause of animal and human diseases. In the 1980s, in the United Kingdom, there was an epidemic of spongiform encephalopathy in cattle (so-called bovine spongiform encephalopathy or "mad cow" disease) that involved at least 180,000 animals. Subsequently, there was a small outbreak (more than 150 cases) of "new variant Creutzfeldt–Jacob disease" in humans, presumed to have been caused by eating meat or offal from afflicted cows. It is now postulated that several degenerative neurological diseases of humans, including Alzheimer's disease, may be caused by prions, although this is still controversial (Prusiner, 2012). If correct, prions will assume a very important role as disease agents in a globally aging population.

5. LESSONS LEARNED: THE ROOT CAUSES OF SIGNIFICANT VIRAL DISEASES OF HUMANS

From the foregoing litany of viral diseases that affect the well-being of humans, a number of generalizations can be made. Many of these concepts are developed in later chapters and are summarized in Sidebar 1. There are a multitude of viral infections of humans, but most of them are relatively innocuous and cause mild transient infections (for instance, rhinoviruses, one cause of the "common cold"). However, a few viruses cause major morbidity and mortality. In some instances, such as those viruses that are transmitted by aerosol, disease symptoms (such as coughing and sneezing) play a critical role in transmission, but in many other instances (for instance, many clinically silent enteric infections such as that caused by poliovirus) spread is unrelated to disease causation. Thus, there is no necessary relationship between the disease caused by a virus and its ability to perpetuate itself in the human population. A good example is the contrast between the two variants of smallpox virus both of which were maintained in human populations; variola major carried a 20–30% case fatality ratio while variola minor killed only 1% of its victims.

Some viruses cause acute infections while others persist. Among both groups, there are instances of serious disease. Smallpox, rabies, and polio viruses cause acute infections, while HIV, human papilloma, and Kaposi's sarcoma viruses cause persistent infections. It is noteworthy that the severity of disease caused by a given virus is determined in part by the individual human host. As an extreme example of human genetic determinants, individuals who are homozygous for the delta 32 deletion (for expression of the CCR5 protein) are resistant to infection with HIV, and there are many other human genetic determinants that influence the course of HIV infection. The age of infected individuals can also have a dramatic influence on the course of infection with a specific virus. During the epidemic of measles in a "virgin" population in the Faroe Islands in 1846, measles had a fatality ratio of 20% in infants, 10% in the elderly, but only 1% in young adults.

There are several factors that conspire to create a major virus pandemic with high mortality. These include: a population that is immunologically susceptible to the specific viral invader; a virus that can spread widely in the susceptible population; a virus variant that is highly virulent

TABLE 1 Zoonotic Viruses that Have Caused Human Illness in the Recent Past

Virus	Human Disease	Animal Reservoir	Person-to-Person Transmission
H1N1 influenza virus (type A influenza viruses)	Influenza	Wild and domestic birds; pigs	Yes Pandemic
H5N1, H7N9 influenza viruses (type A influenza viruses)	Fatal influenza	Wild and domestic birds	No
HIV (lentivirus)	AIDS	Chimpanzees	Yes Pandemic
SARS coronavirus (coronavirus)	Severe acute respiratory syndrome	Bats? other wild animals?	Yes Aborted pandemic
Middle east respiratory syndrome coronavirus (coronavirus)	Severe acute respiratory syndrome	Bats (primary reservoir)? Camels (intermediate host)?	Possible
Sin nombre virus (bunyavirus)	Severe acute respiratory disease	White-footed mice	No
Ebola virus (filovirus)	Hemorrhagic fever	Bats? pigs? nonhuman primates?	Yes Pandemic
Marburg virus (filovirus)	Hemorrhagic fever	Bats? pigs? nonhuman primates?	Limited to a few passages
Nipah virus (paramyxovirus)	Encephalitis	Fruit bats; pigs	No
Hendra virus (paramyxovirus)	Encephalitis; pneumonia	Fruit bats; horses	No
West Nile virus (flavivirus)	Encephalitis	Wild birds	No

in humans; and an infection that is not readily contained by human intervention once an ongoing epidemic is recognized. Most recent examples of severe viral diseases of humans, such as HIV/AIDS, SARS, and influenza pandemics, have been caused by zoonotic viruses. Although there are many zoonotic viruses that occasionally cross the species barrier and infect humans, relatively few of them are able to spread from person to person and infect large numbers of humans (Table 1).

6. VIRAL INFECTIONS OF ANIMALS

Many viral infections of animals can be transmitted to humans, and some of these zoonotic infections can cause serious illness in humans, such as several strains of avian influenza virus, Ebola and Marburg filoviruses, and Hendra and Nipah henipaviruses. In a few instances, such zoonotic infections can become established in human populations with devastating consequences, such as SARS (severe acute respiratory syndrome) coronavirus and HIV. These infections are dealt with in some detail in Chapters 9 and 16.

6.1 SARS

SARS appears to be a natural infection of certain species of bats, but can be transmitted to other wild animals, some of which are eaten in southeast China. SARS first appeared as an acute severe sometimes fatal respiratory disease in humans in the Guangdong province of China in 2001. Many of the early human cases occurred in persons who had contact with a variety of wild animals, such as palm civets, sold in food markets. Unusually, for a zoonotic infection, SARS then spread by aerosol from human to human. From China it was carried to Hong Kong, and thence to many distant sites, including Southeast Asia and Canada. A stringent series of quarantine steps succeeded in terminating further spread. By the end of the epidemic, in 2003, there were over 8000 recorded cases with a case fatality ratio of almost 10%.

6.2 Rinderpest

Virus diseases of animals of economic importance can also take a toll on humans, by compromising food security. Examples are rinderpest of cattle, influenza of chickens and turkeys, and foot-and-mouth disease virus in cattle, swine, and sheep.

Rinderpest (cattle plague in German) is caused by a morbillivirus that infects cattle and other even-toed ungulates but not humans. It can cause devastating epidemics that kill a large proportion of animals in a herd. In areas where cattle are a key to the economy, rinderpest epidemics have led to widespread famines. As summarized in Wikipedia "Cattle plagues recurred throughout history, often accompanying wars and military campaigns. They hit Europe especially hard in the eighteenth century, with three long pandemics which, although varying in intensity and duration from region to region, took place in the periods of 1709–1720,

FIGURE 7 Depiction of an outbreak of cattle plague (rinderpest) in the eighteenth century, The Netherlands. *Copied from Wikipedia, accessed February 10, 2014, with permission.*

1742–1760, and 1768–1786 (Figure 7). There was a major outbreak covering the whole of Britain in 1865/66. Later in history, an outbreak in the 1890s killed 80–90% of all cattle in southern Africa, as well as in the Horn of Africa." A global program to immunize cattle led to the eradication of wild rinderpest virus in 2001, a first for virus diseases of animals. As a result, the threat of epidemics has been eliminated enhancing food security for human populations.

7. SCORE CARD: ARE WE CONTROLLING MAJOR VIRAL DISEASES?

Life sciences have advanced to an extraordinary degree in the last century, and have produced a large number of preventive and therapeutic modalities that have had a major impact on human health. As example, life expectancy in the United States was about 47 years in 1900 but is now close to 80 years, a change greater than that seen in all of recorded history. Congruent with this, there are now safe and effective vaccines against more than 15 individual virus diseases of humans. This has led to a dramatic reduction in child mortality, which in turn has made an important contribution to the increase in life expectancy. Another example is antiretroviral therapy for young adults infected with HIV, which has increased life expectancy to an estimated 50 years compared to an average 5 years prior to treatment.

However, balanced against these triumphs of medical science, there still exist some outstanding current failures. Here are two examples. First, underutilization of both vaccines and antiviral drugs remains a major challenge in developing countries. It has already been noted that an estimated 150,000 unvaccinated children die unnecessarily of measles each year. HIV/AIDS is even a more serious problem. On a global level, there are still more than two million new human infections with HIV annually. And in the United States, the annual number of new HIV infections (estimated 60,000) has not dropped in the last 20 years. Second, the lack of an HIV vaccine. More than 30 after the isolation of HIV, progress toward an effective HIV vaccine is still very slow and research has yet to solve some daunting scientific challenges that impede vaccine development.

More recently, in the high income countries, an anti-vaccine movement has left some children without recommended immunizations (Offit, 2011). As an example, a 2015 measles outbreak originating in Disneyland in southern California has been triggered by children whose parents declined to have them vaccinated (Zipprich et al., 2015).

8. PANDEMICS YET TO COME: WHAT DOES THE FUTURE HOLD?

What does the future hold as to new viral diseases that could take a toll on human health? Events of the last 50 years justify some predictions, although the specifics remain hazy (Sidebar 2). In recent decades, a considerable number of new viral diseases of humans have emerged (Figure 1). Of these, the most significant are HIV/AIDS, influenza, and SARS. All of these viruses share three properties: they are zoonotic infections, transmitted from animals to humans; they produce a high level of morbidity and mortality in their human hosts; and they have adapted to spread directly from human to human. Zoonotic infections such as Sin Nombre, Ebola, Marberg, Nipah, and Hendra viruses are also highly virulent, but they have limited ability to spread from human to human; therefore, they have caused small outbreaks albeit with high mortality.

8.1 Modern Diagnostics and Global Surveillance

Since the turn of the century, there have been two important developments relevant to the rapid detection of emerging viral diseases. First, the Internet and mobile phones have facilitated the rapid exchange of health information around the world, and there are a number of open access Web sites dedicated to disease surveillance. In addition to conventional reporting of unusual disease events, it is now possible to monitor many human activities on a massive scale, such as the volume of communications on social networks like Facebook and Twitter; spurts in activity may reflect an incipient disease threat. Second, the use of evolving laboratory methods, such as PCR and next generation genome sequencing, has permitted the identification of novel viruses with lightning speed.

8.2 Influenza: Will It Cause the Next Human Pandemic?

Which viruses constitute the most likely threats in the near future? Influenza looms on the horizon as the most obvious candidate.

Sidebar 2 Viral pandemics yet to come

- Viral diseases that are "new" to humans are mainly zoonotic infections that have crossed from animal hosts to humans.
- If the recent past is prelude to the future, zoonotic viral infections can be predicted to cause new human pandemics, some of which could be devastating.
- The most dangerous emerging viruses will adapt to spread directly from human to human after they have crossed the "species barrier," and they will cause a high level of morbidity and mortality.
- Recent technical advances have vastly accelerated the ability to recognize new viral diseases of humans and identify and characterize their specific causal agents. However, the spread of emerging infections can outrace current ability to develop either vaccines or antiviral drugs.

In 2009, a novel H1N1 type A influenza virus of swine origin crossed into humans and caused a worldwide epidemic, which was estimated to have infected over 100 million people in over 200 countries, with at least 20,000 confirmed deaths. Genetic analysis of this virus showed that it contained individual genes from four different sources, avian, human, and two swine influenza virus lineages. The critical genes for the H and N proteins were derived from the 1918 pandemic strain which had persisted in pigs since that time. Although this virus was much less virulent than the 1918 influenza virus, it demonstrated the potential for recombination as a source of new influenza viruses with the ability to spread rapidly in the human population.

Wild waterfowl are infected with a large number of type A influenza viruses, which cause asymptomatic enteric infections in these birds. These viruses are regularly transmitted to domestic poultry (chickens, turkeys, geese) and pigs, where they may recombine to take on new virulence and transmission phenotypes. Most of these viruses show a preference for the α2-3-linked sialic acid receptors on avian cells and are poorly infectious for human cells which mainly express α2-6-linked sialic acid residues. However, a number of avian type A viruses have been transmitted to humans in contact with domestic poultry, including H5N1, H7N2, H7N3, H7N7, H7N9, H9N2, H10N7, and H10N8 viruses.

In particular, H5N1 and H7N9 avian viruses have been of particular concern. Type A H5N1 virus appeared as a cause of deadly outbreaks among poultry in southeast Asia about 2003, and has continued to cause high mortality in domestic birds. This virus has spread to poultry in many countries in the region, and as far west as Pakistan, Iraq, and Egypt. It has also been transmitted to humans in contact with infected poultry, and has caused over 500 reported cases with a high mortality (about 50% of reported cases). However, H5N1 virus (which preferentially binds to avian-type cellular receptors) has not spread directly from human to human.

In controversial experiments, Yoshihiro Kawaoka and Ron Fouchier independently investigated the potential for the H5N1 virus to acquire the ability to spread among humans (Herfst et al., 2012; Imai et al., 2012). An H5 gene from the avian virus (with four point mutations that changed its receptor preference from avian-like to human-like), when recombined with seven other segments from the 2009 human pandemic H1N1 virus, was able to spread by aerosol between ferrets (a surrogate for human to human spread). Alternatively, an H5N1 avian virus, with four point mutations that changed its receptor preference from avian-like to human-like, was passaged in ferrets and ultimately acquired the ability to spread between experimental animals by aerosol; however, the evolved laboratory strain was not highly pathogenic in ferrets. Both of these studies—although artificial laboratory exercises—did demonstrate the theoretical potential for a pathogenic avian virus to adapt to spread in humans.

In China, during the period March to September, 2013, the avian H7N9 virus caused more than 100 infections among humans in contact with poultry, with a case fatality ratio about 30%. Since this virus causes only mild infections in poultry (in contrast to H5N1 virus) its geographic distribution has been difficult to track. Recent studies showed that this virus preferentially attached to avian-type sialic acid receptors and was poorly transmitted by aerosol between ferrets, consistent with its failure to spread from human to human.

In summary, outbreaks in the first decade of the present century illustrate the pandemic potential of type A influenza viruses. On the one hand, the 2009 H1N1 swine virus spread widely in the human population but caused a low mortality (less than 0.1% of overt cases). On the other hand, both H5N1 and H7N9 viruses are highly virulent in humans but have not spread directly from person to person. Should a virus evolve that combined both virulence and transmissibility, it could once again create the "perfect storm" of 1918. In spite of the tremendous scientific advances since 1918 in our understanding of influenza virology and immunity, it is unlikely that an effective vaccine could be produced and administered in time to abort an ongoing global pandemic. It is this nightmare scenario that concerns influenza investigators.

8.3 Ebola Pandemic

The Ebola pandemic that started in West Africa emerged as we wrote this book. It is discussed in Chapter 16 but deserves a brief comment here. Sidebar 3 sets forth the makings of a "perfect storm" that can lead to an emerging pandemic, which—unfortunately—we have witnessed during 2014.

Sidebar 3 The making of a catastrophic viral pandemic

Step 1
Emergence of a virus new to the human population; no preexisting immunity

Step 2
The virus spreads from human to human, undergoes adaptive mutations as it spreads, and is highly virulent

Step 3
No treatment, no vaccine, no method for rapid diagnosis prior to severe illness

Step 4
The international health community underestimates the potential for a pandemic; international political leaders fail to understand the severity of problem and to respond effectively

Step 5
Inadequate facilities to treat or quarantine patients; no ability to improve patient survival; the population becomes suspicious of government advice or mandates

Step 6
Rumors spread, panic sets in locally, bystander effects—such as food shortages—arise

Step 7
Civil society begins to fray, chaos develops, and patterns of behavior increase the spread of infection

Step 8
A humanitarian disaster explodes

9. FINAL COMMENTS

Although hard to quantify, viral diseases continue to take a significant toll of human life and well-being. The relatively recent emergence of diseases like HIV/AIDS and SARS demonstrate the potential for new unpredicted viral infections to appear in epidemic and endemic form. The story of SARS, which was barely contained through the application of the ancient practice of quarantine, points out the potential dangers of new viral plagues. Likewise, the inability—to date—to successfully control the AIDS pandemic underlines the power of a viral disease to defy the formidable armamentarium of modern biomedical science. While it is possible to identify emerging viral threats with unprecedented speed and precision, some new infections spread before an adequate response can be formulated and deployed. The study of viral pathogenesis—in addition to its intrinsic interest—will continue to be of practical importance as a critical contribution toward the control of present and future viral plagues.

FURTHER READING

General

Lipkin WI. The changing face of pathogen discovery and surveillance. Nature Reviews Microbiology 2013; 11: 133–141.

Lipkin WI, Firth C. Viral surveillance and discovery. Current Opinions in Virology 2013; 3: 199–204.

Manzoni A. The betrothed (I Promessi Sposi). Richard Bentley, London, 2014.

McNeill WH. Plagues and peoples. Anchor Books, Doubleday, New York, 1976.

Morens DM. Remembering Daumier's blue period. Ecohealth 2011; 8: 527–530.

Offit PA. Deadly Choices: How the Anti-Vaccine Movement Threatens Us All. Basic Books, New York, 2011.

Wolfe N. The viral storm. Henry Holt, New York, 2011.

Smallpox

Fenner F. Poxviruses. in Clinical virology, Richman DD, Whitley RJ, Hayden FG, editors, ASM Press, Washington, DC, 2002, Chapter 17, 359–374.

Foege WH. House on fire: the fight to eradicate smallpox. University of California Press, Berkeley, CA, 2011.

Henderson DA. Smallpox: the death of a disease. Prometheus Books, Amherst, NY, 2009.

HIV/AIDS

Pepin J. The origin of AIDS. Cambridge University Press, Cambridge, UK, 2011.

Wikipedia. Epidemiology of HIV/AIDS. (accessed online February 3, 2014).

UNAIDS. Global report: UNAIDS report on the global AIDS epidemic 2013. UNAIDS, WHO, Geneva, 2014. (accessed online February 2, 2014).

UNAIDS. AIDS by the numbers. UNAIDS, Geneva, 2014 (accessed online February 2, 2014).

Cohen SM et al. Vital signs: HIV prevention through care and treatment – United States. Morbidity and Mortality Weekly Reports 2011; 60: 1618–1623.

Influenza

Barry JM. The Great Influenza: the story of the deadliest pandemic in history. Viking, Penguin Books, New York, 2004.

Glezen P, Schmier JK, Kuehn CM, et al. The burden of influenza B: a structured literature review. American Journal of Public Health 2013; 103: e43–e51.

Herfst S, et al. Airborne transmission of influenza A/H5N1 virus between ferrets. Science 2012; 336: 1534–1541.

Imai M, et al. Experimental adaptation of an influenza H5 HA confers respiratory droplet transmission to a reassortant H5 HA/H1N1 virus in ferrets. Nature 2012; 486: 420–430.

Labella AM, Mersel SE. Influenza. Medical Clinics of North America 2013; 97: 631–645.

Noymer A, Garenne M. The 1918 influenza epidemic's effects on sex differentials in mortality in the United States. Population Development Reviews 2000; 26: 565–581.

SARS

Abraham T. Twenty-first century plague: the story of SARS. Hong Kong University Press, Hong Kong, 2004.

Measles

Griffin DE, Lin WH, Pan CH. Measles virus, immune control, and persistence. FEMS Microbiology Reviews 2012; 36: 649–662.

Moss WJ, Griffin DE. Measles. Lancet 2012; 379: 153–164.

Panum PL. Observations made during the epidemic of measles on the Faroe Islands in the year 1846. Delta Omega Society, 1940.

Zipprich J, Winter K, Hacker J, et al. Measles Outbreak — California, December 2014–February 2015. MMWR 2015, 64:153–154.

Rinderpest

Wikipedia. Rinderpest. (accessed online February 3, 2014).

Poliomyelitis

Nathanson N, Kew OM. From emergence to eradication: the epidemiology of poliomyelitis deconstructed. American Journal of Epidemiology 2010; 172: 1213–1229.

Roberts L. The art of eradicating polio. Science 2013; 342: 29–35.

Yellow Fever

Powell JM. Bring out your dead: the epidemic of yellow fever in Philadelphia in 1793. University of Pennsylvania Press, Philadelphia, 1949.

Nathanson N. The emergence of infectious diseases: societal causes and consequences. ASM News (renamed Microbe) 1997; 63: 83–91.

Prions

Aguzzi A, Nuvolone M, Zhu C. The immunobiology of prion diseases. Nature reviews immunology 2013; 13: 888–902.

Prusiner SB. A unifying role for prions in neurodegenerative diseases. Science 2012; 336: 1511–1513.

Chapter 2

Historical Roots

The Family Tree of Viral Pathogenesis

William C. Summers
Yale University, New Haven, CT, USA

Chapter Outline

Starting in the nineteenth century, some diseases began to be associated with "viruses," at first simply vaguely conceived poisons, then transmissible microbes, and in the twentieth century, "filterable" agents too small to be seen by even the best microscopes. The ways these viruses (as well as other microbes) interacted with cells were at first mysterious. Cell death and host immune responses were the two primary pathogenic phenomena most often observed. Both of these important effects of virus infection, however, have become clear only with advances in basic cell biology as well as a more complete understanding of the complexities of the immune system.

1. FROM CELLS AND VIRUSES TO DISEASES

Pathogenesis, the origin of disease, is a concept that demands causal explanations. Such causal explanations, of course, depend on the knowledge and beliefs at the time of a given account of pathogenesis. Pathogenesis, depending on the era when it is considered, includes demonology, atmospheric, and astrological explanations, as well as specific chemical and microbial accounts; it includes purely descriptive accounts as well as detailed mechanistic accounts. The historical roots of *viral* pathogenesis, in particular, come to light in the past several centuries which includes concepts and ideas that, in some form or other, are still current in the twenty-first century (Table 1).

1.1 Pathogenesis in the Modern Sense

Giovanni Battista Morgagni (1682–1771) showed that the purpose of pathological anatomy is to teach us the seats of diseases, traced to the organs affected. Xavier Bichat (1771–1802) extended this approach to the tissue level, and Rudolf Virchow (1821–1902) led onward to cellular pathology. Molecular biology (not yet safely attached to any one name) is now providing a kind of chemical and biophysical pathology. It is presentist, indeed, to speak of viruses in the current sense when seeking historical roots of viral pathology prior to the latter half of the nineteenth century when disease-causing microbes were first clearly recognized. Still, it is illuminating and humbling to consider the earlier investigations and understanding of pathology of diseases, we now attribute to viruses from the time of Jenner (late-eighteenth century), Liebig (early nineteenth century), and Virchow (mid-nineteenth century).

1.2 The Concept of "Virus"

"Virus" is a term with changing meaning over the past several centuries. Originally, it was used to mean "poison" in the sense of some toxic agent (either produced from within the body by some process of degeneration, fermentation, or putrefaction, or from outside the body such as environmental sources, physical contacts, or miasmas) that could produce

Viral Pathogenesis. http://dx.doi.org/10.1016/B978-0-12-800964-2.00002-1

TABLE 1 Summary of Key Landmarks in Viral Pathogenesis

Date	Author	Milestone
1761	Morgangni	Organ pathology *(Of the seats and causes of diseases investigated through anatomy)*
1796	Jenner	Viral immunity *(An Inquiry into the Causes and Effects of the Variolae Vaccinae)*
19th C.	Many investigators	Contagionism
1801	Bichat	Tissue pathology *(Anatomie générale)*
1822	Gaspard	Fermentation/putrefaction
Mid-third of 19th C.	Many investigators	Germ theories
1837	Schleiden and Schwann	Cell theories
1840	Liebig	Fermentation and disease
1858	Virchow	Cellular pathology *(Cellular pathology, as based upon physiological and pathological histology)*
1867	Cohnheim	Inflammation/pus
1878	Pasteur	Microbes, immunization
1884	Chamberland	Porcelain filters
1892	Ivanovsky	TMV filterablilty
1898	Beijerinck Loeffler and Frosch	TMV filterablilty FMD filterablilty
1902	Reed	Transmission of yellow fever
1907/12	Harrison/Carrel	Tissue cultures
1911	Rous	Avian sarcoma virus
1928	Rivers Cowdry	Comprehensive textbook *(Filterable Viruses)* Cytology of viral inclusions
1931	Goodpasture	Chick embryo cultures
1949	Enders, Robbins, Weller	Growth of poliovirus in cell cultures
1954–57	Nagano and Kojima; Isaacs and Lindenmann	Interferon (cytokines)
1969–76	Todaro and Heubner; Martin; Stehelin, Bishop and Varmus	Viral oncogenes as cell regulatory genes
1972	Temin and Baltimore	Discovery of reverse transcriptase
1979	Crawford and Lane; Levine; Old	T-antigen/p53 interactions
1982	Prusiner	The prion hypothesis
1983–84	Montagnier and Barré-Sinoussi; Gallo	HIV as the cause of AIDS
1983–92	zur Hausen	Human papillomaviruses cause cervical cancer
1985	Mullis	Polymerase chain reaction
1989	Janeway	Pattern recognition theory of innate immunity
1998–2000	Shenk; Katze; McManus	First global gene expression profiling of host response to viral infection
2001	USA government Human Genome Project Consortium; Celera Genomics	Patriot Act: Select Agent Program Published first human genome sequence
2002	Wimmer	Synthetic poliovirus
2004	454 Life Sciences (Roche) Solexa (Illumina)	Next-generation DNA/RNA sequencing
2005	Taubenberger	Reconstruction of the 1918 pandemic influenza virus
2006–07	Frazer and Zhou; Merck, GlaxoSmithKline	Vaccine against human papilloma virus; the first licensed cancer vaccine
2014	USA government; White House; Homeland Security	Shut down gain of function experiments; a new era for virus research

disease. In reading the early literature, one must guard against reading "virus" in the modern microbial sense, because such an interpretation carries meanings not at all in keeping with the understanding of the early authors in their own time.

The concept of "virus" was constantly evolving and being used to refer to different realities over the past two centuries. This evolution was driven both by conceptual and by technical discoveries, several of which are central to the history of viral pathogenesis. In some sense, viruses are the invention of scientists. "Virus" as discussed by Edward Jenner (1749–1823) in his work on smallpox vaccination in 1778 is not the "virus" discussed by Louis Pasteur (1822–1895) in his work on rabies in 1885. Neither are these the same as "virus" described by Wendell Stanley (1904–1971) when he crystallized poliovirus in 1935.

Oversimplifying the history of the virus concept, one can identify four crucial events in this history: the realization in the latter third of the nineteenth century that many diseases attributed to "viruses" seem to be caused by minute living agents, popularly called "germs"; the discovery in the 1890s that some of these agents could be retained by fine unglazed porcelain filters and others of these agents could not (thus dividing viruses into two classes, filterable and non-filterable); the repeated failure to propagate filterable viruses outside a "host" organism and hence, the belief that this subclass of viruses were obligate intracellular parasites; and last, the visualization in the electron microscope of particulate structures with definite and specific morphologies and which could reasonably be related to the virus in question.

1.3 Virus Pathogenesis in Jenner's Time

Contagionist views of many diseases were prevalent long before the development of germ theories that provide our current conception of contagion. Some sort of material (occasionally more ethereal, however) substance was involved in the transfer of disease between individuals, from the environment, or from animals. Some scholars went so far as to assert that this material was some malevolent agent they called "virus." Their descriptions did not dwell on the agent so much as on the host responses to it. Often, the agent was a product of the diseased body itself that had the ability to provoke the disease when transferred to another individual.

Edward Jenner's work on smallpox is justifiably famous both because smallpox was a widespread and serious illness and because he investigated and substantiated a way to prevent lethal smallpox by prior infection with a more benign disease, cowpox. In his famous book on vaccination, he included an often overlooked discussion of his theoretical views on the pathogenesis of smallpox which represented some of the leading ideas of his time. He clearly explained the way the smallpox virus ("variolous virus") affected the tissues of the body and in turn produced more virus.

Although the skin, for example, adipose membrane, or mucous membranes are all capable of producing the variolous virus by the stimulus given by the particles originally deposited upon them, yet I am induced to conceive that each of the these parts is capable of producing some variation in the qualities of the matter previous to its affecting the constitution.

Further, Jenner explained the way the "variolous particles" behaved in the body when introduced by the usual routes of contagion ("casually") and when iatrogenically introduced into the skin:

What else can constitute the difference between the Small-pox when communicated casually or when brought on artificially through the medium of the skin? After all, are the variolous particles, possessing their true specific and contagious principles, ever taken up into the blood sufficiently loaded with them at some stages of the Small-pox to communicate the disease by inserting it under the cuticle, or by spreading it on the surface of an ulcer? Yet experiments have determined the impracticability of its being given this way; although it has been proved that variolous matter when much diluted with water, and applied to the skin in the usual manner will produce the disease.

Jenner was concerned as well with the origin and epidemiology of smallpox and its relation to cowpox. He was aware that cowpox in his part of England was a human affliction predominant among milk maids and not among men (Figure 1). In Ireland, however, it seemed that the distribution of the disease was different. He speculated that there was one "virus" that affected both horses and cows and was transmissible to humans but that passage through different hosts modified the virus so as to cause different recognizable pathologies:

May it not, then, be reasonably conjectured, that the source of the Small-pox is morbid matter of a peculiar kind, generated by a disease in the horse, and that accidental circumstances may have again and again arisen, still working new changes upon it, until it has acquired the contagious and malignant form…?

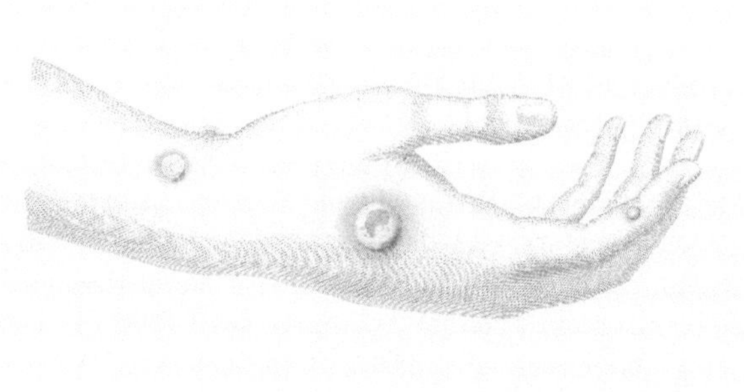

FIGURE 1 Cowpox lesions. *From Jenner (1798).*

1.4 Fermentation, Putrefaction, and Putrid Intoxication

In the late-eighteenth and early nineteenth centuries, chemical investigations of biological processes were leading to new understanding of disease. The process of fermentation, well-known from food technologies, was the usual starting point. Fermentation, in which various forms of biological materials were observed to change properties, evolve gas, produce alcohol, consume sugar, and so on, was seen as related on one hand to digestion of foodstuffs in the animal body, and on the other hand to "putrefaction," a rather vague category of happenings that occurred in many situations, including in human sicknesses. The famous chemist, Justus von Liebig (1803–1873), conceived of these processes in physico-mechanical terms and thus saw contagion and infectious disease as being transmitted by physical contact which promoted chemical transformations following the laws and principles of physics, sort of a molecular contagionist view:

> *Fermentation or putrefaction may be described as a process of transformation,—that is, a new arrangement of the elementary particles, or atoms, of a compound, yielding two or more new groups or compounds, and caused by contact with other substances, the elementary particles of which are themselves in a state of transformation or decomposition.*

No matter what one's view of the basic nature of fermentation and putrefaction, putrefying substances were seen as one way that contagion worked. Putrid materials inoculated into animals produced "putrid intoxication" suggesting that putrefaction produced toxic materials. Some of the products of putrefaction such as carbonic acid, hydrogen, and hydrogen sulfide, were tested by injection of dogs, sheep, foxes, and pigs. In some cases, such injected animals were subjected to attempts at treatment by various regimens. The work of Bernard Gaspard (1788–1871), one of the leading researchers in this field, showed that carbonic acid, hydrogen, and hydrogen sulfide were basically nontoxic, but that the main culprit in putrid intoxication was ammonia, the active, pathological, substance in the putrefying matter.

It was this work on fermentation and its cousin, putrefaction, that would provide a crucial link in the chain of understanding between microbes and pathological processes.

2. THE CELLULAR PATHOLOGY OF VIRCHOW

A key advance in understanding of pathogenesis in the early nineteenth century was the development of the cell theory of life, extensively worked out by Matthias Schleiden (1804–1881) and Theodor Schwann (1810–1882) in their famous publications of 1838 and 1839. On the basis of this new concept of living organisms, Rudolf Virchow developed a comprehensive view of pathological processes that depended on normal and abnormal cellular anatomy and physiology. In his famous text on *Cellular Pathology* he noted:

> *What Schwann, however, has done for histology, has as yet been but in a very slight degree built up and developed for pathology, and it may be said that nothing has penetrated less deeply into the minds of all than the cell-theory in its intimate connection with pathology. ...The chief point in this application of histology to pathology is to obtain a recognition of the fact, that the cell is really the ultimate morphological element in which there is any manifestation of life, and that we must not transfer the seat of real action to any point beyond the cell.*

Virchow focused our attention on the cell as the unit of pathogenesis as well as the unit of life. This mode of explanation was developed by the nineteenth-century pathologists, including, for example, his protégé, Julius Cohnheim (1839–1884), who related cells and their behavior to the process of inflammation, a cardinal sign of infection:

> *Keeping pace with this exodus, emigration, or as it is also called, extravasation of corpuscular elements there occurs an increased transudation of fluid, in consequence of which the meshes of the mesentery, or the tissues of the tongue, are infiltrated and swell. But this is not all. The extravasated colourless corpuscles distribute themselves, in proportion as their numbers increase, over a larger area, forsaking the neighbourhood of the vessels from which they were derived.*

Further, Cohnheim incorporated principles of the new germ theories into his thinking about inflammation related to infectious diseases. His linkage of toxins and infections was clearly in the tradition of nineteenth-century pathology, but his distinction between contagious and miasmatic infections showed a new refinement in understanding of viral pathogenesis:

> *By the term* ***infective diseases*** *we understand (passing over non-essential characters), such diseases as arise from the action of a poison or* ***virus****—a virus that is not always and everywhere present,* ***but is only produced by certain similarly diseased individuals, or in certain localities, or at certain seasons****. The virus is distinguished as a* ***contagium*** *or as a* ***miasma****. By the first is to be understood a poison which is developed in or upon individuals who are already diseased, or as Pettenkofer terms it,* ***endogenously****; while the name* ***miasma*** *is applied to a virus that is produced in the soil, in the atmosphere, in water,* ***in any case, externally to a diseased organism, i.e., exogenously****. Accordingly contagious diseases can only be conveyed from man to man, or from animals to man, or* ***vice versa****; whereas miasmatic diseases attack individuals who have not come into contact with persons similarly diseased.*

Virchow's cellular pathology is still the dominant paradigm going into the twenty-first century. It is the cell's response to viral infection, its death, its proliferation, or its production of bioactive substances that we use to understand and explain viral pathogenesis.

3. PASTEUR AND MICROBES

While we often, and rightly so, associate Louis Pasteur with the advent of germ theories of disease, his work was firmly situated in the mainstream context of nineteenth-century ideas of contagion, ferments, and specific causes. Again, it was the similarities between fermentations and putrefactions that suggested to Pasteur that certain contagious "diseases," first of beer-brewing, of bee-keeping, and eventually of higher animals and humans, are related to the actions of living beings too small to be seen with the naked eye. To adopt an agnostic position with respect to the biological classification of such beings, Pasteur used the generic term "microbe" (first proposed by Charles-Emmanuel Sédillot in 1878) to designate these tiny agents. This was a definition strictly based on size and invisibility, and what we now recognize as "viruses" were lumped together with other agents of disease that we now differentiate as bacteria, fungi, and protozoa.

A key aspect of Pasteur's approach was his ability to combine his knowledge of fermentation and its chemistry with the idea of putrefaction and production of disease-producing agents in what we now think of as pure cultures in nutrient solutions. In his famous work on growth of microbes in various media, he was able to establish that microbes grew and multiplied, and that by careful serial dilutions, one could establish homogeneous cultures of agents which could transmit given diseases and thus appeared to be the cause of their contagions.

Certain microbes, however, such as rabies, resisted all attempts at cultivation in nutrient media. These failures were usually attributed to technical problems, not to fundamental biological properties of such microbes. Sometimes, as in the case of hog cholera, microbes were consistently isolated from diseased animals and appeared to be capable of transmission of sickness to other animals, yet were later shown to be contaminants and not the true agent of disease.

What interested Pasteur even more, however, was the use of such microbes, which had been "attenuated" in their virulence, as agents of immunity. In this work, he followed Jenner quite closely. Just as Jenner speculated that smallpox was a variant of a disease of horses with altered virulence, and that cowpox had become weak by serial infections of nonhuman animals, Pasteur reasoned that adaptation of a microbe by "passage" through other hosts could render it "attenuated" in its virulence and hence, like the cowpox, could provoke immunity by "vaccination." Pasteur's intuition proved correct for several contagious diseases and he achieved fame and fortune for this insight.

Pasteur was first of all a practical scientist, interested in solving problems and devising interventions. Only on occasion did he engage in theoretical or mechanistic investigations, but he rather casually seemed to accept ptomaine or putrid toxin theories to explain the pathogenesis of microbial diseases. His initial explanation for the process of vaccination was based on the hypothesis that the body contained only small amounts of some essential substance or nutrient needed by the pathogen, and that the attenuated vaccine strain depleted this store of essential material thus preventing growth of the pathogen upon later challenge. This "limiting nutrition hypothesis" did, however, give way to his later theory based on combinations of the pathogen with chemical substances produced in the body, more akin to current theories based on antigen–antibody reactions.

3.1 Filterable Agents and Their Nature

The separation of the category "microbe" into our current distinction between bacteria and "viruses" resulted from advances in filtration technology. With the discovery of microbes in fermentation and in disease, it became important to have ways to remove these agents from solutions (wine, beer, water) for various purposes. A key discovery was published in 1884 by Charles Chamberland (1851–1908), a protégé of Pasteur, who found that unglazed porcelain would permit the passage of water but not bacteria. The so-called Pasteur–Chamberland filter was thus used as a device to provide a source of sterile water for laboratory, and later, home use (Figure 2). Soon other filter technologies were developed, all for similar purposes. Another common filter used diatomaceous earth (often called by the German name *kieselguhr*) and was marketed by the Berkefeld Company. Because these filters were tubular in shape, they have often been called "candle filters" (*bougie* in French).

In the course of studies on the agent causing tobacco mosaic disease, in 1892 the Russian microbiologist, Dmitry Ivanovsky (1864–1920) found that the infectivity of plant extracts was not retained by a Chamberland filter and suggested that this plant disease was caused by a tiny, unculturable bacterium. In 1898, when the Dutch microbiologist Martinus Beijerinck (1851–1931) independently repeated such filtration experiments in his attempts to culture the agent of tobacco mosaic disease, he did not accept the "tiny bacteria" hypothesis, but rather concluded that the "virus" of tobacco mosaic disease was soluble rather than particulate, and he called it a "contagious living fluid" (*contagium vivum fluidum*).

Filtration as a criterion for classifying infectious agents soon became widespread, and in a few decades, many pathogens were believed to be "filterable," that is,

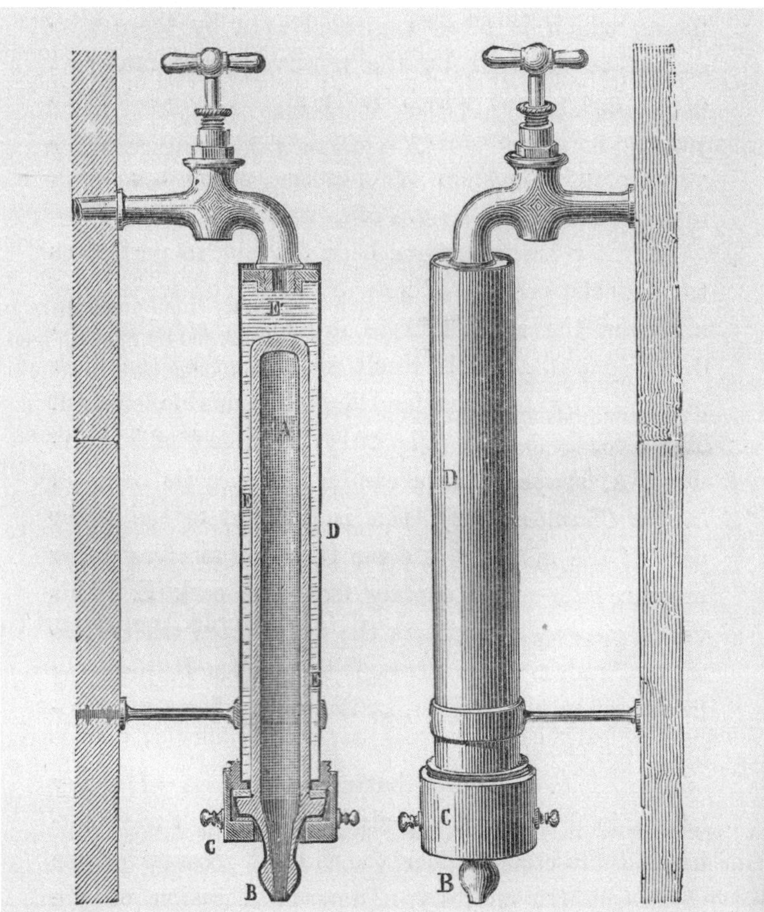

FIGURE 2 Chamberland filter for bacteria-free water supply. *From Trouessart (1889).*

filter-passing. Most textbooks by the 1920s had a chapter devoted to such filterable agents. The culmination of this trend was the famous book *Filterable Viruses*, edited by Thomas M. Rivers (1888–1962) and published in 1928. The continuing uncertainty over the biological nature of filterable viruses was summarized in this volume by Alexis Carrel (1873–1944): "A virus may be a very minute organism, or it may be a chemical substance manufactured by the cells themselves. In either case, its multiplication depends on the activity of a living tissue." This mystery tended to focus research on the nature of viruses, their biological status, and their evolutionary significance as forms between the realm of life and the realm of chemistry.

4. CYTOPATHOLOGY: RETURN TO PATHOGENESIS

Starting about 1920 and continuing for three decades, some researchers turned their attention to the cellular responses to virus infection in what might be called true viral pathogenesis. Both new histological staining methods and early applications of cell and tissue cultures to virological studies were helpful in advancing cytopathological investigations.

The main focus of cytopathology was the observation of novel structures seen in the cytoplasm and in the nucleus of infected cells, the so-called "inclusion bodies." These changes in cell architecture seemed to be related to virus infection, but their interpretation was hotly debated: Were they viruses themselves? Were they structures produced by the cells in response to viruses? Did they have any relation to virus-induced cell behaviors that were becoming increasingly well-characterized? In other words, what did inclusions have to do with pathogenesis?

Two scientists led the way during this era of virus research, Edmund V. Cowdry (1888–1975) and Ernest W. Goodpasture (1886–1960). Cowdry was a pathologist and cell biologist who believed that study of the various inclusion bodies in virus-infected cells would shed light on the pathologic processes of virus infection as well as the nature of virus growth and reproduction (Figure 3). While others had noted various types of inclusions that were associated with certain infections, for example, Negri bodies in rabies (Figure 4) and Guarnieri bodies in vaccinia and smallpox, Cowdry developed a systematic study of inclusions and proposed a classification of types of inclusions which he hoped would lead to more clarity in viral pathogenesis.

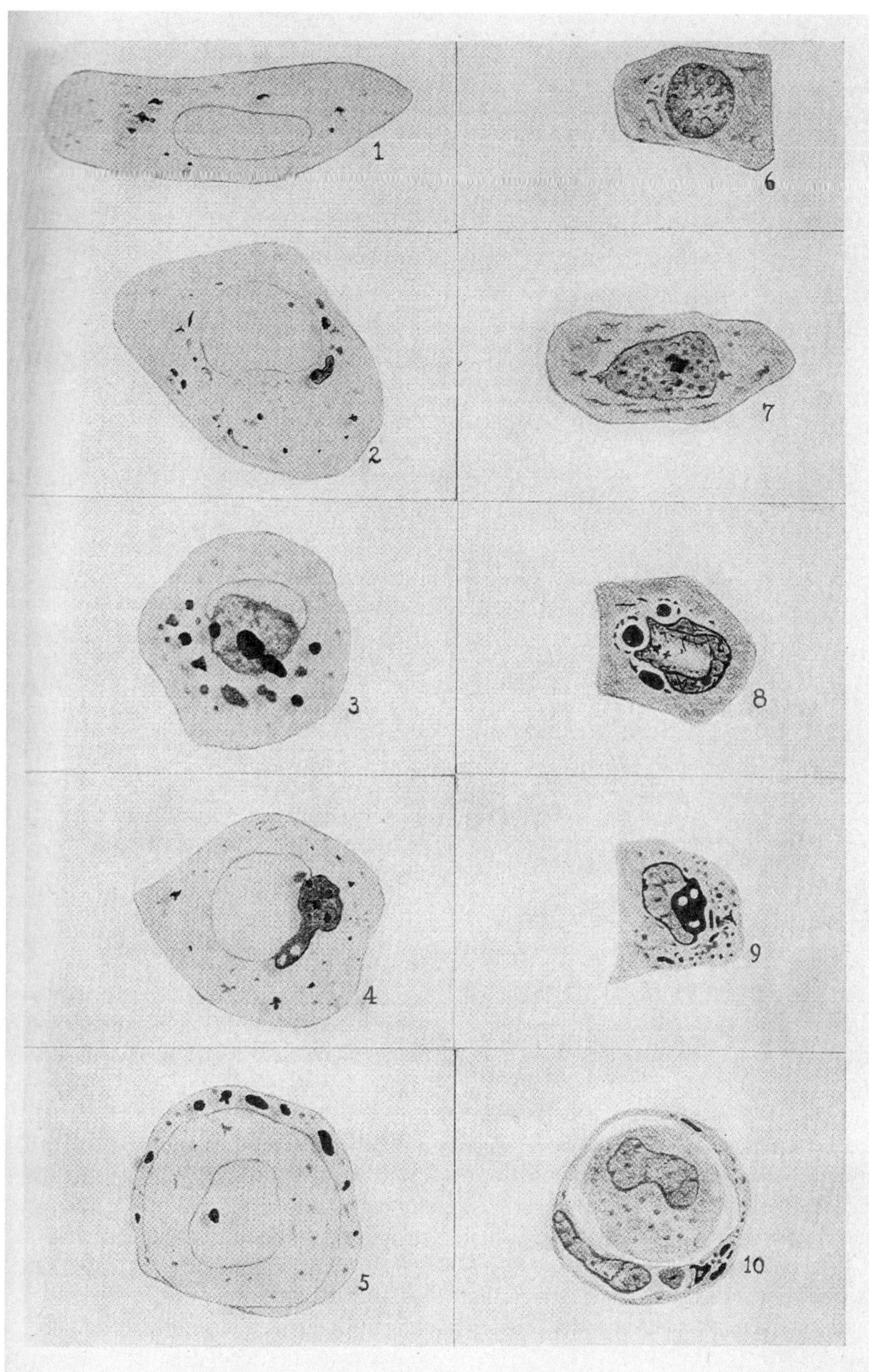

FIGURE 3 Image of vaccinia inclusion bodies in rabbit corneal cells: numbers 1–5 vital stain; 6–10 fixed and stained. 1 and 6 prior to infection with vaccinia. *From Cowdry (1928).*

He differentiated cytoplasmic from nuclear inclusions, and was skeptical of some reports of virus-associated inclusions. Some inclusions he believed were collections of the viruses themselves and other inclusions seemed to be cellular products, perhaps made in response to viral infections. Cowdry brought a chemical approach to cytopathology, asking "what is the nature of the material of which the inclusion bodies are composed" and whether this material was present in the cell prior to virus infection. He proposed the key question in viral pathogenesis when he asked: "what alterations in

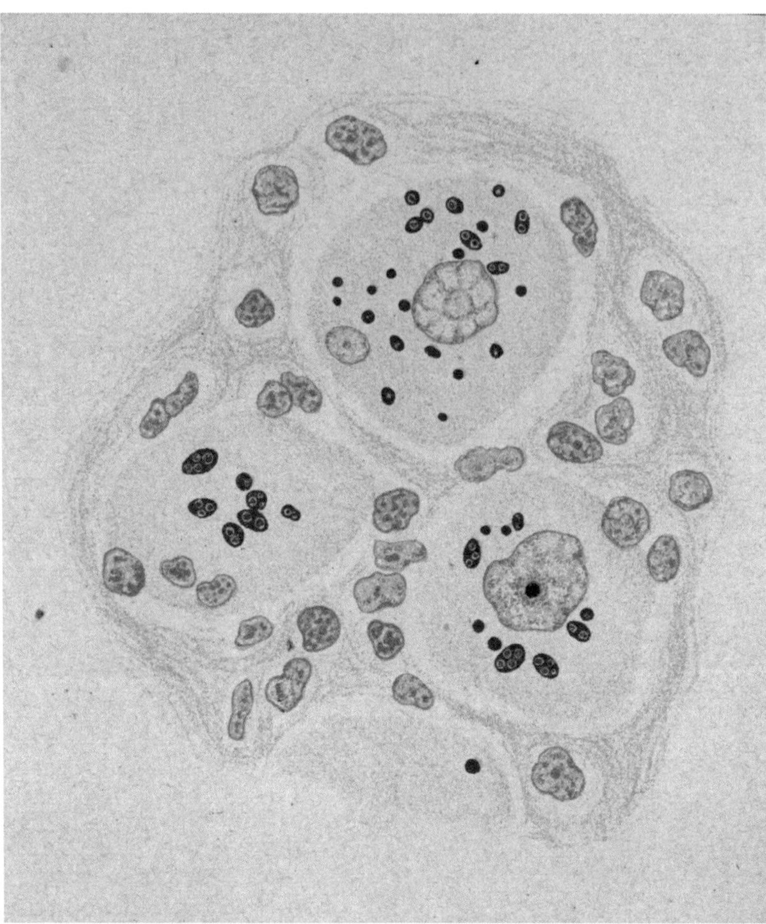

FIGURE 4 Camera Lucida image of inclusion bodies in rabies-infected dog brain cells. Now called "Negri bodies" these inclusions were first believed to be protozoal parasites. *From Negri (1903).*

cellular activity are caused by the viruses?" The unsettled nature of such inclusions was reflected in Cowdry's caution: "Many believe that the [inclusion] bodies are neither organisms *sui generis* nor combinations between organisms and cellular components but rather *reaction products* produced by the cell in response to injury caused by infective agencies which are ultramicroscopic." Others, however, were more certain about some virus-induced inclusions. In their study of herpes infections as early as 1923, Goodpasture and Teague wrote: "The presence of these intranuclear bodies means the presence of the virus, and that they represent the growth of the virus in the infected nuclei, as claimed by Lipschutz."

Cowdry is famous for his classification of inclusion bodies in virus infections or in tissues thought to be related to virus infections. In his original paper on this classification, he listed 18 cases in which his "Type A" inclusions were observed, including herpes, yellow fever, chickenpox, whooping cough, kidneys of frogs, louping ill, and "many species in the absence of disease." His "Type B" inclusions were noted in poliomyelitis, Rift Valley fever, Borna disease, and "many species unaccompanied by evidence of disease." His criteria for this classification are so vague as to be undecipherable. For example, Type A inclusions "are amorphous or particulate, but may be condensed in rounded masses." With respect to Type B, "The reaction is localized to certain [unspecified] areas of the nucleus, where acidophilic droplets make their appearance." Surprisingly, Cowdry's cytopathological descriptions of inclusions that may or may not have any relationship to virus infection became a central and persistent focus of diagnosis and investigation at least until the 1950s, and even today one occasionally finds his terminology still in use. While some cell inclusions are now recognized as manifestations of diverse cell responses to infection, this era of descriptive cytology eventually gave way to more chemical and physiological study of viral pathogenesis at the cellular level.

Two limitations on the study of viral pathogenesis were (1) the problem of specific tissue tropism that limited the study of many viruses to certain cell types from certain species, and (2) the inability to grow sufficient quantities of virus for laboratory study. Both these shortcomings were overcome in 1931 when Goodpasture, Woodruff, and Buddingh showed that the chorioallantoic membrane of the chick embryo could be productively infected by a wide variety of animal viruses. With this technical advance, it became possible to produce substantial amounts of many viruses for laboratory study and vaccinations. Like Cowdry, Goodpasture's interests centered on pathogenesis, but to some extent went beyond Cowdry's cellular focus. For example, in his work during the Great Influenza Pandemic of 1918–1919, Goodpasture explained how the as yet unidentified agent of influenza was able to cause disease:

> *The prevailing opinion is, as Wolbach has stated, that death from influenza means death from lung complication—pneumonia in some form—and that it is in the lungs that possible characteristic lesions may be found. In 1889 Leichtenstern expressed his opinion from clinical and anatomic evidences that there existed a primary pneumonia produced by the poison of influenza, and from a study of the material I have had available from the height of the epidemic last fall and subsequently I am convinced that this is true and that the etiologic factor is not any one of the numerous pathogenic microorganisms cultivated from the lungs, often in pure culture, but an unknown virus that produces the general intoxication and which may produce characteristic lesions in the lungs with or without the coincidence of other infectious agents.*

For Goodpasture, the unknown virus of influenza was pathogenic because of a "poison" that led to "general intoxication" of the patient, and thus he situated his explanation squarely in the mainstream tradition of pathology based on toxins, ferments, and similar physiological disruptions.

Although chick embryo cultures greatly simplified and advanced virological studies since the 1930s, the growth of viruses in specific cell types and from specific species in vitro had been a long-standing goal, starting with the work of Alexis Carrel in the 1920s and 1930s, but really developed fully in the work on polio by John Enders (1897–1985), Thomas Weller (1915–2008), and Frederick Robbins (1916–2003) in 1949 when they showed that tissue cultures were readily adaptable for growth of viruses in vitro. With the widespread use of in vitro virus studies, many new details of viral pathogenesis emerged. Again, it appears that scientific knowledge was dependent on new technological developments. Examples of new knowledge include studies of tissue and species tropism, a long-standing problem in virology, the distinction between persistent and latent infections, and the elucidation of the various pathways by which viruses can enter cells.

In particular, the recognition and identification of specific viral receptors on cell surfaces greatly improved understanding of viral pathogenesis at the cellular level, as did the elucidation of physiological changes in cell functions such as membrane permeability (the basis for the common description of "cloudy swelling" in gross pathology), the activation of the lysosomal and endosomal pathways upon infection with some viruses, and the membrane changes that promote syncytia formation by some viruses. All these "modern" results were made possible by the controlled laboratory infection of cells cultured in vitro.

Thus, even up to the mid-twentieth century virus pathogenesis was still dominated by Virchow's cellular pathology: pathogenesis was described in terms of damage to cells, and ascribed primarily to loss of cells and hence loss of those cell functions in physiological economy of the infected organism. Although it was accepted that some viruses, most notably the Rous sarcoma virus, resulted in cell proliferation rather than cell death, it was the death of cells that was the usual endpoint in pathogenic explanations. The development of the electron microscope in the late 1930s and its application to the study of virus structure and the interactions of viruses and cells, however, greatly advanced our understanding of the nature of viruses as microbes, of the classification of various viruses apart from the diseases they cause, and of the different ways that cells respond to virus infection.

It was the detailed study of bacteriophages, by mid-century recognized as bacterial viruses that provided the paradigm for much of virology in the latter half of the twentieth century. The notion of a generalized virus life cycle of adsorption, penetration, replication, and lysis and the subsequent elaboration of this model to include programmed gene expression, lysogenic integration, modification of cell functions such as nucleic acid and protein synthesis, as well as lateral gene transfers by so-called "transduction" were refinements in the study of bacteriophages that found ready extension to animal and plant viruses.

5. HOST RESPONSES AS PATHOLOGIES

In addition to study of virus effects on cell function, by the latter part of the twentieth-century virologists recognized virus pathogenesis in the broader contexts of the entirety of host responses. While early views described viral pathogenesis in terms of disease resulting from loss of certain essential cells, for example, death of neurons in poliomyelitis, or degeneration of liver cells in hepatitis, it became clear that processes such as cell proliferation, local inflammation, immune disorders, and even such systemic effects such as fever, were specific manifestations of the virus effects on cell physiology or the responses of the host to those cellular effects.

One of the first well-studied cases of such a host response was the discovery of interferon in 1954 (Nagano and Koijima) and 1957 (Isaacs and Lindemann), which showed that virus infection has effects beyond the infected cell itself. Of course, this discovery presaged the current understanding of the complex and varied roles that molecules termed cytokines play in virus infection and pathogenesis.

The role that viruses play in cell proliferation and carcinogenesis was recognized for nearly half a century before some clarity was brought to this particular problem of viral pathogenesis. Since the discovery by Peyton Rous (1879–1970) and then others of the oncogenic action of certain avian viruses, the mechanism of oncogenesis was debated: was the virus merely an initiator of some carcinogenic process similar to chemical carcinogens, or did the virus play a more central role? The difficulty in isolating virus from some tumors only confounded this problem. Only with the availability of molecular techniques to study genes was it possible to understand the nature of latent, integrated virus genomes, and later, virally carried cancer-causing genes.

An important example of the new viral pathogenesis of cancer is the discovery of the role of viral gene products in regulating normal cellular functions, a role termed by some as "luxury functions" meaning that they are viral functions not strictly relegated to the viral reproductive life cycle. While it was known that there were immunologically identifiable neo-antigens in some virus-induced cancers, the role and origin of such antigens were unclear. These "tumor antigens" or T-antigens were thought to be key to the carcinogenic process, however. In 1979, three research groups (Crawford and Lane, Old, and Levine) were able to use antibodies to the T-antigen of the small papova virus, SV-40 to precipitate and analyze the proteins in this complex, presumably those that stably interacted with the virus-encoded T-antigen. To the surprise of some (at least), the main component of this complex, in addition to the viral T-antigen, was a here-to-fore unknown cell protein of approximate molecular weight 53 kDa. This protein, dubbed p53, became the object of intense investigation which has subsequently shown its central role of in many crucial cell regulatory functions. This discovery provided a much more detailed account of viral pathogenesis by explaining viral carcinogenesis in terms of viral action on normal processes at the cell and molecular level.

Thus, in many cases, the "pathology" experienced by the host, such as inflammation or immune disorders, was not the simple consequence of loss of cells from viral death, but rather the result of host responses to cell products, secretions, or in some cases virus-induced proteins that were recognized and reacted against by the host. Cells did not need to be killed by viruses in order to create pathologies. As Oldstone put it, "viruses [can] alter the products of differentiated cells and cause disease without structurally altering or killing the infected cell."

5.1 The Twenty-First Century

The publication of the first human genome sequence in 2001 heralded a new era in biomedical research (Table 1). During the last 15 years, an array of new technical advances in genomics, proteomics, and other "omics" have been developed, supported by methods in computational biology. It is now possible to follow a vast number of host responses to viral infection and the cognate evolution of the viral invader. Epigenetics and sequencing of the DNA transcriptome have led to a much more sophisticated view of genome expression at the RNA and protein level. In addition to "classical" adaptive immunity, innate immunity and intrinsic host restriction factors have been recognized as players in the complex virus–host interactome. The field of viral pathogenesis is in the midst of a technical and conceptual revolution as profound as some of those that preceded it, moving from the reductionist approach that dominated the last half of the twentieth century, to a systems biology approach in the twenty-first century.

6. REPRISE

The history of viral pathogenesis is a story of conceptual advances and reformulations determined, to a significant degree by technical improvements that provided the means for ever more detailed study of both the viruses and the cells they infect. While Virchow's insight that the cell is the seat of disease remains intact after nearly two centuries, viral pathogenesis now requires our detailed understanding not only of the cell, but its interactions with the virus, as well as the interplay between the infected cell and the entire host organism. With the advent of genomics, proteomics, and many other recent technical advances, it has now become possible to describe the virus–host "interactome" in much greater detail. The field is undergoing a conceptual transition from a reductionist approach to a system biology perspective. Thus, viral pathogenesis has once again moved to a new historical level, and our evolving insights are the subject of many chapters in this 3rd edition of *Viral Pathogenesis*.

FURTHER READING

Reviews, chapters, books:

Huang AS. Viral pathogenesis and molecular biology. *Bacteriol Rev* 1977; 41: 811–21.

Hughes SS. *Virus: a history of the concept.* New York: Science History Publications; 1977.

Rivers TM, editor. *Filterable viruses*. Baltimore: Williams and Wilkins; 1928.

Waterson AP, Wilkinson L. *An introduction to the history of virology*. New York: Cambridge Univ. Press; 1978.

Weller TH. *Growing pathogens in tissue cultures: fifty years in academic tropical medicine, pediatrics, and virology*. Komaroff AL, Barlow E, editors; Canton, MA: Science History Publications; 2004.

Original Research Reports and Historical Sources:

Carrel A. Tissue cultures in the study of viruses. In: Rivers TM, editor. *Filterable viruses*. Baltimore: Williams and Wilkins; 1928.

Cohnheim J. *Lectures on general pathology, Vol 1, Ch. V.* McKee AB, translator. London: New Sydenham Society; 1889.

Cowdry EV. Intracellular pathology in virus diseases. In: Rivers TM, editor. *Filterable viruses*. Baltimore: Williams and Wilkins; 1928.

Cowdry EV. The problem of intranuclear inclusions in virus diseases. *Arch Path* 1934; 18: 527–42.

Enders JF, Weller TH, Robbins FC. Cultivation of the Lansing strain of poliomyelitis virus in cultures of various human embryonic tissues. *Science* 1949; 109: 85–7.

Goodpasture EW. The significance of certain pulmonary lesions in relation to the etiology of influenza. *Am J Med Sci* 1919; 158: 863–70.

Goodpasture EW, Teague O. Experimental production of herpetic lesions in organs and tissues of the rabbit. *J Med Res* 1923; 46: 121–38 + plates.

Goodpasture EW, Woodruff AM, Buddingh GJ. The cultivation of vaccine and other viruses in the chorioallantoic membrane of chick embryos. *Science* 1931; 74: 371–2.

Jenner E. *An inquiry in the causes and effects of the variolae vaccinae*. London: Sampson Low; 1798.

Liebig J. *Animal chemistry, or organic chemistry in its application to physiology and pathology*. Gregory E, editor. Cambridge: John Owen. 1842.

Negri A. Beitrag zum Studium der Aetiologie der Tollwuth. *Zeitschr Hyg Infektionskr* 1903; 43: 507–28.

Oldstone MBA. Viruses can alter cell function without causing cell pathology: Disordered function leads to imbalance of homeostasis and disease. In: Notkins AL, Oldstone MBA, editors. *Concepts in viral pathogenesis*. New York: Springer-Verlag; 1984. p. 269–76.

Summers WC. Inventing viruses. *Ann Rev Virol* 2014; 1: 25–35.

Trouessart EL. *Microbes, ferments and moulds*. London: Kegan Paul, Trench & Co. 1889.

Virchow R. *Cellular pathology*. Chance F, translator. London: John Churchill; 1860.

Chapter 3

Basic Concepts

A Step-by-Step Guide to Viral Infection

Robert W. Doms

The Children's Hospital of Philadelphia and the Perelman School of Medicine, University of Pennsylvania, Philadelphia, PA, USA

Chapter Outline

Viral pathogenesis seeks to understand how a virus interacts with its host at multiple levels. Key questions include the source (an infected human, animal, or insect vector), the transmission mechanism, and how the virus is shed and transmitted. Following transmission, pathogenesis is governed by the initial site of replication, whether the virus disseminates within the host, and its tropism for specific tissues and organs. In turn, these steps are dictated by the structure and replication strategy of the virus. In addition to utilizing selected synthetic biochemical pathways in the host cell, viruses frequently reprogram host cells by inducing intracellular signaling pathways that render the cell more permissive to infection. Host–virus interactions also control whether the infection is acute, chronic, latent, or transforming; how the virus interacts with the immune system; and the consequent pathophysiological response of the host. This chapter provides an overview of these basic concepts of viral pathogenesis, with emphasis on the interactions of viruses with their host cells and organisms.

1. VIRUS–CELL INTERACTIONS

1.1 Cellular Receptors and Viral Tropism

Peter Medawar described viruses as "bad news wrapped up in protein," a succinct summary of the structure of all viruses, in which the nucleic acid genome is enclosed by a protein shell (a capsid) that is sometimes further enwrapped by a lipid membrane from which project one or more viral glycoproteins. However, the news is bad only if it gets delivered: the viral genome must be introduced into the cytoplasm of the host cell, and for this to occur the virus must attach to the cell surface and then penetrate a cellular membrane to deliver its nucleic acid payload along with any viral proteins needed for subsequent replication. The ability of a virus to enter various types of cells is one of the major determinants of viral tropism.

Virus attachment factors. Initial viral attachment often occurs via relatively nonspecific, low affinity interactions with virus attachment factors. Virus attachment factors—in contrast to virus receptors—are cell surface molecules that support virus binding but are not required for virus infection. However, they can make virus attachment and infection more efficient. When infecting cells in tissue culture, the rate-limiting step in infection typically is attachment to the cell surface, as virions are in a large volume of media that minimizes chance encounters with cells. Anything that enhances this step can result in a much greater level of virus infection. The addition of a polycation such as diethylaminoethyl (DEAE)–dextran to virus-containing media typically boosts infection efficiency of tissue culture cells more than 10-fold, by serving as an electrostatic bridge, linking

Viral Pathogenesis. http://dx.doi.org/10.1016/B978-0-12-800964-2.00003-3

the virus to the cell surface. Glycosoaminoglycans, highly charged molecules on the cell surface, are well known factors that can result in more efficient virus attachment via electrostatic interactions. Multiple copies of viral attachment factors on the cell surface means that even low affinity interactions can result in high avidity binding of virus.

Cell surface lectins (carbohydrate-binding proteins) can also promote attachment of enveloped viruses to target cells as the viral proteins protruding from the viral surface are often heavily glycosylated. A well-studied example is the C-type (for calcium-dependent) lectin DC-SIGN, which binds to high mannose carbohydrates and is expressed on dendritic cells (DCs). The HIV-1 envelope protein (Env) is heavily glycosylated, and most of its N-linked carbohydrates have a high-mannose structure. HIV-1 binds avidly to DC-SIGN, resulting in efficient capture of virus by DCs (Geijtenbeek et al., 2000).

Thus, cell surface attachment factors restrict virus to the two-dimensional surface of the plasma membrane, making subsequent interactions with virus receptors more efficient. However, the importance of attachment factors in vivo is uncertain as the ratio of extracellular space volume to cells is very low and cells are in close spatial proximity. In contrast, virus receptors are critically important for virus infection both in vitro and in vivo.

Virus receptors. Virus receptors are host cell molecules, most often glycoproteins on the plasma membrane, that not only bind virions but are often essential for subsequent virus infection (Figure 1). Viral receptors are naturally occurring cellular molecules that serve physiological functions for the cell—functions that have nothing to do with infection. Viruses usually bind to their receptors with higher affinity than they do to attachment factors.

Attachment of the virus particle to its cellular receptor is conferred by a virion surface protein, often called the viral attachment protein (VAP). As a rule, there is a single VAP although other viral surface proteins often play an essential role in the steps that follow the initial attachment of virions to the cell surface. For enveloped viruses, the VAP is a surface glycoprotein that oligomerizes to form spikes

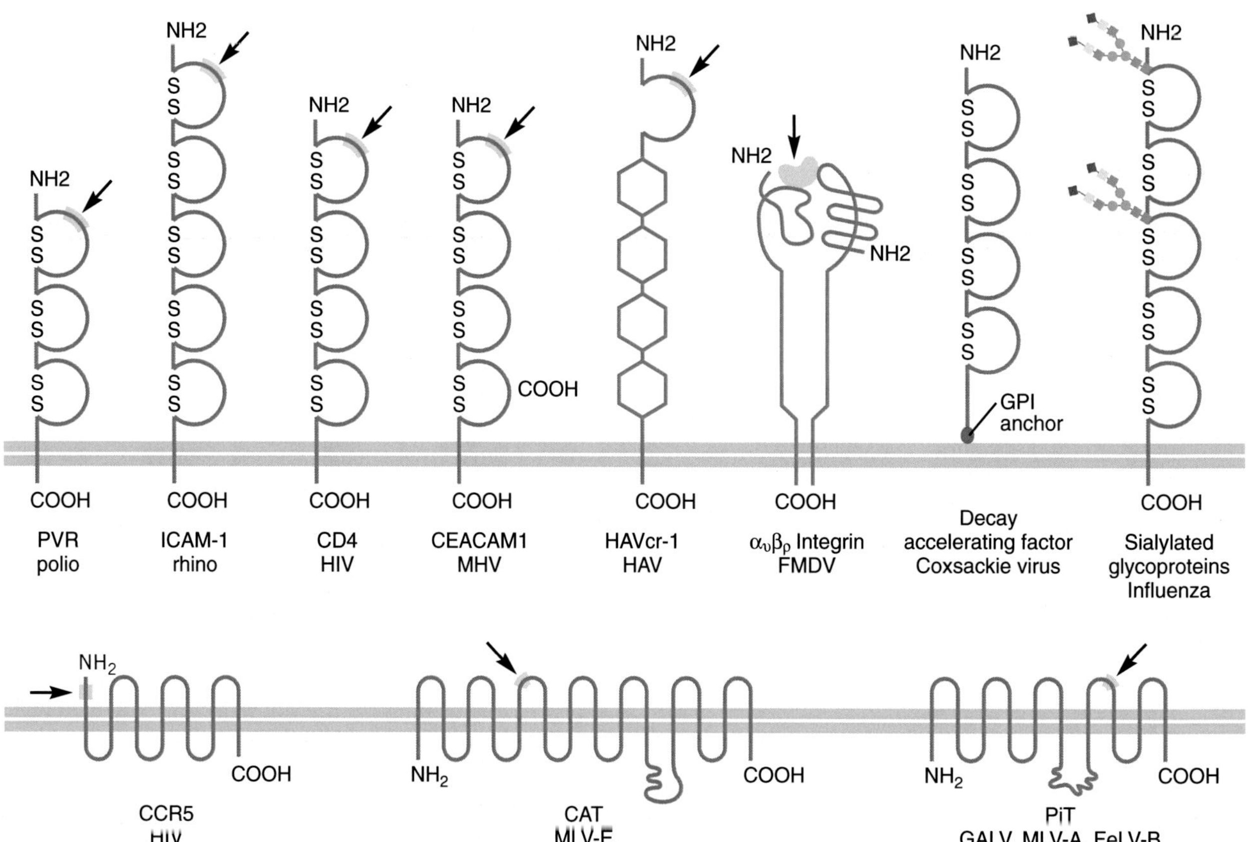

FIGURE 1 Molecular backbone cartoons of some viral receptors. Receptors diverge widely in their structure and physiological function. The amino and carboxy termini are shown, together with important disulfide bonds and the probable domains that bind virus. αvβ6: integrin chains (integrin dimers serve as receptors for many different viruses); ICAM: intercellular adhesion molecule; CCR5: chemokine receptor 5; CAT: cationic amino acid transporter; CEACAM: carcinoembryonic antigen-related cell adhesion molecule; HAVcr-1: hepatitis A virus cellular receptor 1; PiT: inorganic phosphate transporter; PVR: poliovirus receptor; Viruses: polio: poliovirus; rhino: rhinovirus, major group; FMD: foot and mouth disease virus; HAV: hepatitis A virus; HIV: human immunodeficiency virus; MHV: mouse hepatitis virus (a coronovirus); BLV: bovine leukemia virus; ALV-A: avian leukosis virus A; MLV-E: murine leukemia virus E; GALV: gibbon ape leukemia virus; MLV-A: murine leukemia virus A; FeLV-B; feline leukemia virus B.

that protrude from the viral envelope. For nonenveloped viruses, the VAP is one of the surface proteins that form the external structure of the viral capsid. While binding to virus attachment factors is often electrostatic in nature, receptor binding usually occurs via a specific domain on the VAP, most often in the form of a pocket or "canyon" that subsequently interacts with a specific domain on the cellular receptor. These domains can be defined by structural studies and mutations introduced into the cellular receptor protein or the VAP.

Antibodies neutralize virus by binding to the VAP in a way that blocks receptor binding. Since the receptor-binding domain on a VAP often lies in a recessed pocket where it is not directly accessible to antibodies, antibodies that bind to epitopes distant from the pocket can also neutralize the virus. Therefore, viral mutants may escape neutralization without affecting receptor binding. Such escape mutants allow many different virus serotypes to use the same receptor-binding domain of the VAP, as seen for rhinoviruses. Neutralizing antibody escape mutants of this kind are important for the persistence of some viruses, including HIV.

Carbohydrates can also serve as virus receptors. The best studied example is binding of the influenza virus hemagglutinin protein to sialic acid (or N acetyl neuraminic acid), a modified sugar that is found on the tips of some of the branched carbohydrate side chains of glycosylated proteins and glycosphingolipids. Different influenza hemagglutinins bind preferentially to different terminal sialic acid residues, depending on the linkage of the sialic acid to a proximal galactose or galactosamine molecule in the carbohydrate chain. Thus, human-type A influenza viruses bind most avidly to sialic acid α-2,3 galactose configurations while avian-type A influenza viruses bind best to sialic acid α-2,6 galactose. This exquisite specificity of the interaction between the VAP and its cellular receptor helps explain the species tropism of influenza viruses, since different host species express different sialic acid linkages. In turn, this distinction can determine whether an avian or porcine influenza virus can efficiently infect humans (Shinya et al., 2006).

The role of virus receptors in entry. Why are receptors essential for infection whereas attachment factors are not? While virus attachment factors merely support virus binding, virus receptors do something else: they either induce conformational changes in viral proteins that are required for membrane penetration and/or result in delivery of virions to a cellular domain or compartment that is required for entry. This distinction explains why attachment factors are not required for virus infection whereas receptors are required. A host molecule can be considered to be a virus receptor if its elimination from an animal model or one or more cell types prevents infection. The loss of a receptor may not block infection of all cell types since some viruses can utilize more than one receptor. As an example, lab-adapted strains of measles virus use CD46 as a receptor to infect cells, while wild-type strains use CD150 as a receptor to infect lymphocytes, and nectin 4 to infect epithelial cells.

The importance of receptor specificity is exemplified by recent studies of dipeptidyl peptidase 4 (DPPT4), the receptor for MERS-CoV, the cause of a newly emerging infectious disease, Middle East respiratory syndrome (see Chapter 16, Emerging viral diseases). Humans express DPPT4, but mice, hamsters, or ferrets do not and are not susceptible to infection. Molecular studies show that five amino acids in the binding domain of DPPT4 differ between human and hamster cells. Furthermore, DPPT4 on camel cells binds the MERS-CoV attachment protein. This observation provided an explanation for the apparent role of camels as an intermediate host responsible for transmission of this virus to humans (van Doorenmalen et al., 2014). Marmoset cell DPPT4 binds MERS-CoV with affinity similar to that of human cells, and marmosets are an excellent model host for this virus (Falzarano et al., 2014).

1.2 Viral Entry

Viral entry is a multistep process that follows attachment of the virion to the cell surface and results in delivery of the viral genome to the site of replication, either in the cytosol or nucleus. The key step in virus entry is penetration of a cellular membrane. For enveloped viruses, delivery of the viral genome across the lipid bilayer of the virus and a cellular membrane is accomplished by a membrane fusion reaction. For nonenveloped viruses, the viral genome is usually delivered across a cellular membrane by a pore that is formed by protein components of the viral capsid. Virus fusion and penetration proteins exist in metastable states that must be triggered in some way to undergo the needed conformational changes.

Triggers of virus entry. A number of cellular cues induce the irreversible conformational changes in viral proteins that lead to membrane fusion in the case of enveloped viruses, or membrane penetration in the case of nonenveloped viruses (Figure 2). One of the best understood is low pH. Many viruses, after binding to receptors on the cell surface, are endocytosed and delivered to endosomes where the low pH environment induces changes in viral structural proteins that mediate membrane fusion by protonating acidic residues (Helenius et al., 1980). The low pH-dependent entry of many enveloped viruses is exemplified by influenza virus. Nonenveloped viruses may also use this pathway, but instead of eliciting membrane fusion the acidic environment induces structural changes in the viral capsid that result in exposure of hydrophobic domains that insert into the cellular membrane, forming a pore through which the viral genome can pass (Figure 2).

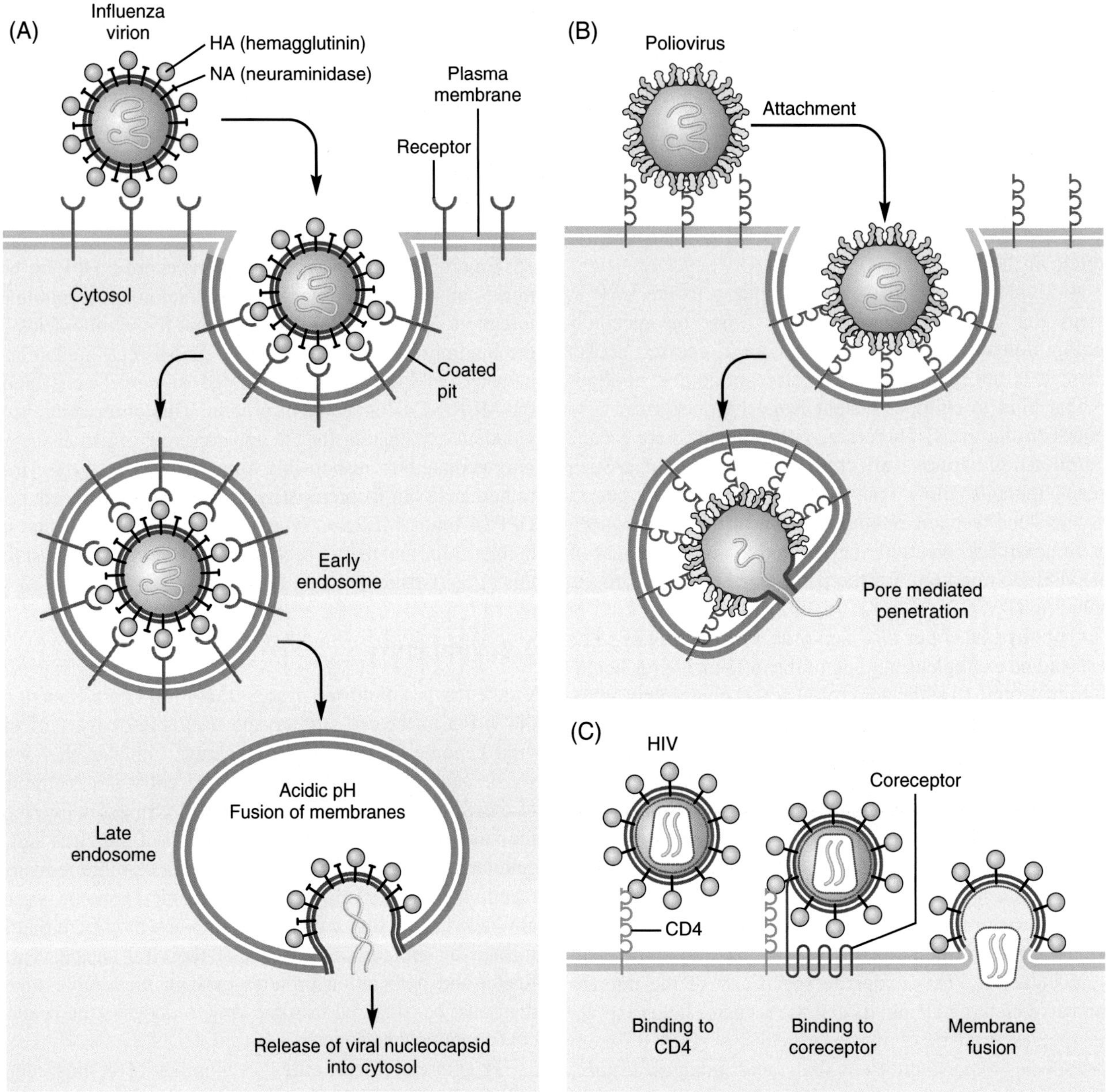

FIGURE 2 Pathways of virus entry into host cells. A. Entry of influenza virus. Key events are: attachment of the virion; internalization of the virion by endocytosis; lowering the pH (to <pH 5.5) of the endocytic vacuole leading to drastic reconfiguration of the viral attachment protein (hemagglutinin, HA1 and HA2); insertion of a hydrophobic domain of HA2 into the vacuolar membrane; fusion of the viral and vacuolar membranes; release of the viral nucleocapsid into the cytosol. This cartoon shows a nucleocapsid containing one of the eight genome segments of influenza virus. B. Entry of poliovirus. After binding to cell surface receptors, poliovirus is endocytosed and ultimately delivered to low pH endosomes. There, the low pH environment triggers conformational changes in the viral capsid that result in exposure of hydrophobic domains that insert into the endosomal membrane, forming a protein pore through which the viral genome can exit and enter the cytoplasm. C. Entry of HIV. HIV attaches to the cell surface via binding to various attachment factors such as DC-SIGN. The first required step for entry is binding of the viral envelope glycoprotein to CD4, a type 1 integral membrane protein. This binding event triggers structural alterations in the envelope glycoprotein that induce the exposure of a second receptor binding domain that engages the chemokine coreceptors CCR5 or CXCR4. Coreceptor binding triggers additional changes in envelope that enable it to elicit membrane fusion with the cell membrane.

While many viruses use low pH as a trigger to induce membrane fusion or penetration, other viruses use one or more receptors to trigger needed conformational changes for virus entry. HIV entry is an example of pH-independent entry. The HIV Env binds to CD4, a cell surface protein found on some types of T cells, macrophages, and DCs. CD4 binding induces conformational changes in Env that then enables it to bind to a second receptor, termed a coreceptor (Figure 2). The coreceptors for HIV are the chemokine receptors CCR5 and CXCR4, seven transmembrane domain receptors. Coreceptor binding induces further conformational changes in the Env protein that lead to membrane fusion (Wilen et al., 2012). Viruses can also employ other cellular cues. Binding of Rous sarcoma virus to its cellular receptor induces conformational changes that then make it responsive to low pH, while the Ebola virus glycoprotein must be cleaved by a host cell protease before it can undergo the changes needed for membrane fusion (Sakurai et al., 2015).

Other host determinants of viral tropism. While receptors and triggering mechanisms for virus entry are major determinants of viral tropism, other host factors can also influence cellular susceptibility. A number of enveloped viruses are not infectious when they bud from cells because one of the viral surface glycoproteins requires proteolytic cleavage by a host cell protease to be activated. Viruses that have Class I membrane fusion proteins (retroviruses, orthomyxoviruses, paramyxoviruses) are examples. In such instances, infectious virus is only produced by replication in cell types that express the appropriate protease, which is localized in the secretory pathway. Alternatively, some viral fusion glycoproteins may be cleaved by enzymes in extracellular fluid. In many instances, the degree of susceptibility to proteolytic cleavage is determined by a few amino acids (such as one vs several arginines) at the site of cleavage, so that mutations in 1–2 critical amino acids can alter the tissue tropism of a virus by making it more or less susceptible to this critically important proteolytic cleavage event.

A case in point is Newcastle disease virus, a paramyxovirus of birds. Virulent isolates of Newcastle disease virus encode a fusion protein that is readily cleaved by furin, a proteolytic enzyme present in the Golgi apparatus, so that the protein is activated during maturation prior to reaching the cell surface, and before budding of nascent virions. This makes it possible for virulent strains of the virus to infect many avian cell types, thereby increasing its tissue host range, and causing systemic infections that are often lethal. In contrast, avirulent strains of Newcastle virus encode a variant fusion protein that is not cleaved during maturation in the Golgi, so that nascent virus requires activation by an extracellular protease. The required protease is found only in the respiratory or enteric tracts, thereby limiting tropism to surface cells and conferring an attenuated phenotype on the virus (Panda et al., 2004). Host cell transcription factors can influence the tropism of some viruses. Papillomaviruses replicate in skin and may cause tumors, varying from benign warts to malignant cancer of the cervix. Papillomaviruses commence their replication in germinal cells that are permissive for replication of the viral genomes. However, germinal cells produce proteins that block the transcription of late structural genes of the virus. As the infected basal cells move outward and begin to differentiate, they become permissive for transcription and translation of the papillomavirus structural genes, so that complete infectious virus is only formed in cells that are about to be sloughed. Release of virus from the superficial layers of dead cells promotes the transmission of infection to new sites on the infected person and also to new uninfected hosts (Doorbar et al., 2012).

Physical factors that impact tropism. While the vast majority of human viruses replicate optimally at 37 °C, some mucosal surfaces such as the upper respiratory tract have a lower temperature, about 33 °C. Certain viruses, such as rhinoviruses that replicate in the epithelial cells of the nose and throat, have evolved to replicate optimally at 33 °C. As a result, such viruses are usually restricted in their tissue distribution by their relative inability to replicate at 37 °C, which limits their spread beyond the upper respiratory tract. The harsh environment of the gastrointestinal tract can impact virus tropism as well, due to the acid pH of the stomach, the alkaline pH of the intestine, and the destructive effects of pancreatic digestive enzymes. In general, enterotropism is limited to viruses that can survive these adverse conditions, although there are some exceptions.

1.3 Virus Structure and Replication Can Impact Pathogenesis

There are several links between viral structure and pathogenesis. Large viruses such as poxviruses or filamentous forms of viruses, such as influenza and Ebola, are simply too large to utilize clathrin-coated pits, caveolae, or other commonly used entry routes. Instead, these viruses trigger internalization by activating macropinocytosis—an example of viruses reprogramming cells to assist virus replication (Marsh and Helenius, 2006).

Another important structural feature is the surface of the virion. Enveloped viruses are not stable outside of the human body, and are typically transmitted by transfer of body fluids. In contrast, nonenveloped viruses are much more stable, and many can be transmitted by other mechanisms such as the fecal–oral route—this is how polio and many other GI viruses are transmitted. Hepatitis, from contaminated shellfish for example, is caused by hepatitis A, a nonenveloped virus that is stable outside of the human body. In contrast, hepatitis B and C viruses have envelopes, and are transmitted by sexual contact or by blood. The Caliciviruses that cause outbreaks of diarrhea on cruise ships are nonenveloped, making transmission by fomites much easier and sterilization more difficult.

Viral replication strategies are numerous, but there are several general features that provide links between replication and pathogenesis. DNA viruses that replicate in the nucleus must utilize a host cell pathway to transport their genome to the nucleus. This almost always entails interactions between the virus and the cell's cytoskeleton. RNA polymerases lack proof reading capability, and have high

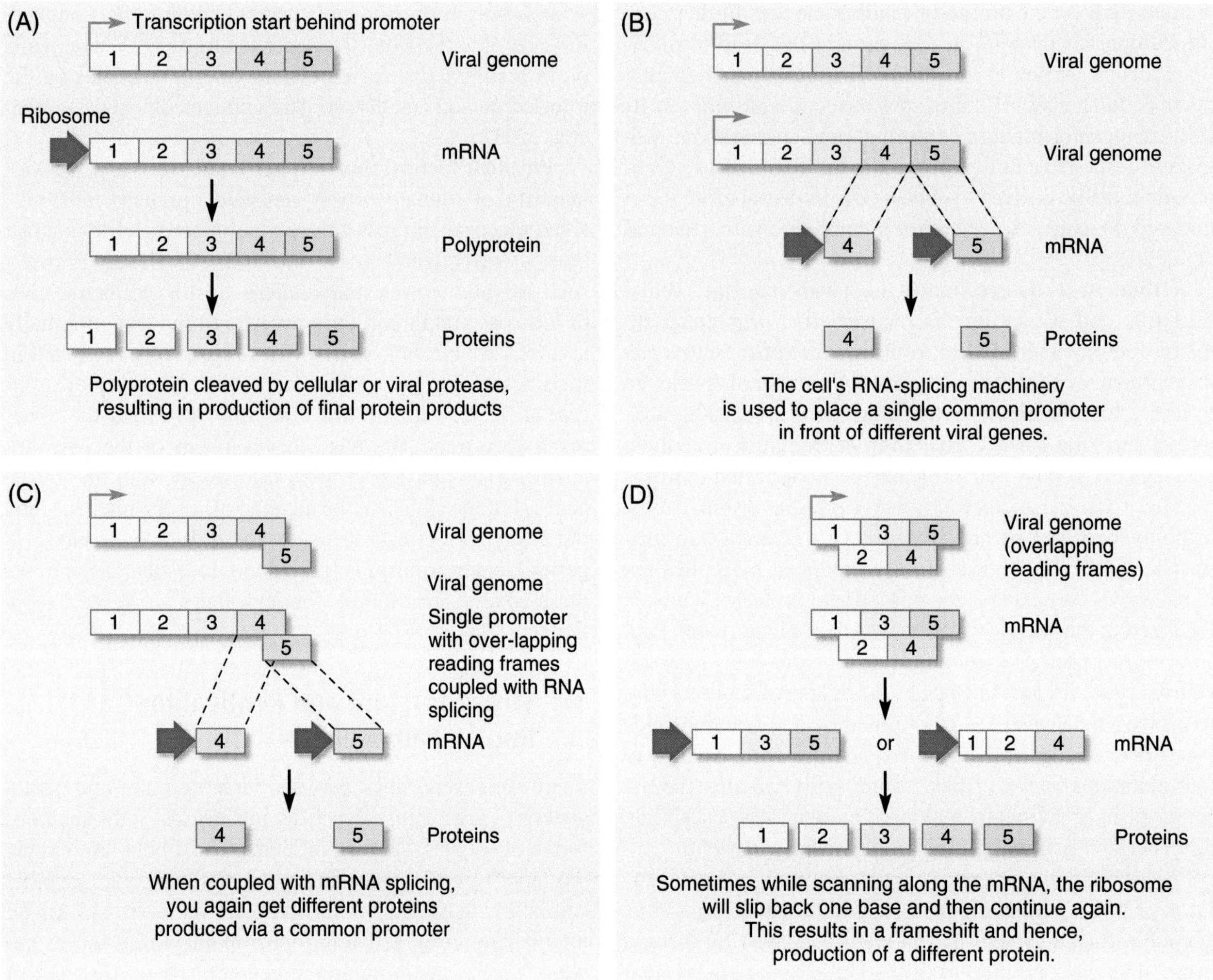

FIGURE 3 How viruses pack maximum information into small genomes. A. Many viruses produce one or more large polyprotiens that are cleaved by cellular or viral proteases co- or posttranslationally, resulting in the generation of multiple viral proteins all of which were transcribed via a single promoter. B. Some viruses utilize the cell's RNA splicing machinery, which makes it possible to place a single, common promoter in front of different viral genes. C. Some viruses use overlapping reading frames, which saves genetic space. D. Ribosomal frame-shifting is used by retroviruses. If the ribosome slips back a base and then proceeds, the reading frame is shifted and a different protein is produced.

mutation rates that enable the virus to evolve quickly in the face of new selective pressures, such as the adaptive immune response (see Chapter 17, Virus Evolution).

While only about 3% of the human genome actually codes for proteins, a very large fraction of viral genomes encode proteins. Viruses use four basic strategies to pack as much genetic information as possible into the smallest amount of genetic material (Figure 3). (1) Polyproteins: Instead of having one promoter for each viral gene, many viruses encode polyproteins. These polyproteins are translated, and cellular or viral proteases then cleave them into individual proteins. (2) Differential splicing: Some viruses use the cell's machinery to splice their genome, and a single promoter is used to transcribe different viral genes. (3) Overlapping reading frames: By having start codons in several sites, virus genes can overlap. (4) Ribosomal frame-shifting: In some viruses, a ribosome starts translating a protein and then "slips" back one base before starting again. When this happens, it is now in a new reading frame, and thus begins translating a different viral protein. These mechanisms are not mutually exclusive—HIV employs all four, for example. The net effect is to pack an impressive punch within a small genome.

2. ROUTES OF VIRAL TRANSMISSION

Most human infections result from transmission of virus from another infected human. However, some viruses are transmitted to humans from animals (bunyaviruses, arenaviruses) or via an insect vector (Dengue, West Nile viruses)—a fact that has been recognized in the case of rabies virus for hundreds of years (Figure 4). In these cases, the virus

FIGURE 4 Historical illustration showing the transmission of rabies. Arabic painting by Abdallah ibn al-Fadl, Baghdad school, 1224. *Courtesy of the Freer Gallery of Art, Washington, DC.*

has to interact with pathways present in the animal or insect vector as well in the human host. The transmission process must deliver virus to a specific site that harbors cells susceptible to infection, such as the respiratory or gastrointestinal tracts. Following transmission, the virus typically establishes a primary, local infection. Sometimes, symptoms result from replication at the primary site of infection and virus produced at this site can be spread to other hosts. Other viruses disseminate within the host, being delivered to other tissues and organs that support virus replication. Dissemination can occur by different pathways—lymphatic spread, hematogenous spread, neural spread—and reflects an impressive degree of virus–host adaptation. Upon initiation of infection, the innate and adaptive immune systems are activated and attempt to clear the virus. However, viruses often employ immune evasion strategies, and the balance between immune responses and viral evasion strategies does much to dictate the outcome (see Chapters 4, 5, and 6, Innate, Adaptive, and Aberrant immunity).

Respiratory tract. This is a common route of infection, with transmission mediated by aerosolized droplets or infected saliva and nasopharyngeal secretions. Droplet size, which is affected by temperature and humidity, plays a major role in determining the anatomic site to which the virus is delivered, with larger droplets lodging in the nose and upper airways and smaller droplets in the alveoli. Innate defenses include mucus and cilia, which trap pathogens and deliver them to the digestive tract, secretory IgA, and alveolar macrophages. Some viruses tend to infect the pharynx hence pharyngitis; others may infect epithelial cells in bronchioles, causing bronchitis. The site of initial replication can be impacted by the virus inoculum, the site in the respiratory tract to which it is delivered, and the tropism of the virus for distinct cell types within the respiratory system. Avian influenza provides a good example: In humans, there is an anatomical difference in the distribution of sialic acid linkages in the respiratory tract. The α-2,3 linkage preferred by avian influenza is found only deeper in the human respiratory pathway, while the α-2,6 linkage used by human influenza viruses is found in the upper respiratory tract, helping to explain the relative protection of humans from "bird flu."

Gastrointestinal (GI) tract. Many viruses are transmitted via the fecal–oral route where the environment is challenging—very low pH in the stomach, an alkaline pH in the small intestine, proteases, and bile detergents that will inactivate most viruses. Thus, viruses that infect the GI tract are almost always nonenveloped, and have evolved the ability to survive in the GI environment. In fact, some viruses require these conditions. For reoviruses to infect cells, their outer surface proteins have to be cleaved by host cell proteases which then enables them to bind to M cells, which are found in the epithelium that overlies Peyer's patches. The bound virus is internalized and transcytosed to the basolateral surface, enabling virus to spread beyond the epithelium (Danthi et al., 2013). There is considerable specialization, as different viruses enter and replicate in different locations throughout the GI tract, from the tonsils to the distal colon.

Urogenital tract. The urogenital tract is the preferred site of entry for a number of viruses, some of which are well adapted to infect epithelial cells and replicate locally while others use this route as a portal of entry to gain access to other tissues. Preexisting lesions can greatly increase the transmission of other viruses via breaches in the epithelium. This is a significant issue with HIV, where other sexually transmitted diseases enhance susceptibility.

Skin and mucous membranes. The intact skin is not a hospitable environment for viruses as it is covered by keratin and a layer of dying cells. As a result, infection via this route will require a break in the skin, which can be mechanical (a scratch), or the bite of an insect vector (a mosquito, tick, sandfly). Many viruses replicate in the dermis, a layer of highly vascularized tissue with fibroblasts and DCs that lies immediately below the epidermis. A smaller number of viruses replicate in the epidermis itself, notably the papillomaviruses. Direct infection of epithelial cells that line mucosal surfaces is more prevalent, though mucus and IgA serve as intrinsic barriers.

3. VIRAL DISSEMINATION AND MOVEMENT

Viruses have evolved strategies to usurp host cell pathways to move at both the microscopic scale within and between cells, and the macro scale, within the host.

3.1 Movement on the Microscope Scale

Movement of virions along the surface of a cell. The plasma membrane is structurally heterogeneous. There are tight junctions in polarized epithelial cells, lipid rafts that serve to concentrate some cell surface proteins and exclude others, and projections such as microvilli. This means that virions not only have to bind to the surface of a host cell, but must either bind or migrate to the location on the cell surface that provides molecules or endocytic pathways needed for entry. Many viruses have evolved mechanisms that result in their directed movement on the surface of the plasma membrane ("virus surfing") until they reach a location that is compatible with entry (Lehmann et al., 2005).

A notable example of this process is provided by Group B coxsackie viruses, nonenveloped virions that must bind to the coxsackie–adenovirus receptor (CAR). However, coxsackie viruses infecting the apical surface of polarized epithelial cells are faced with a conundrum: CAR is strictly localized at tight junctions and is not accessible to virions that attach to the apical plasma membrane. How then can the virus and its receptor interact? On polarized epithelial cells, coxsackie virus first binds to decay-accelerating factor (DAF), a GPI-anchored protein that is relatively evenly distributed on the apical surface of the plasma membrane. Virus binding to DAF triggers the activation of Abl kinase that in turn initiates Rac-dependent actin rearrangements, which result in rapid movement of DAF, with the associated virus, to tight junctions, where it can then bind to its primary receptor and be internalized via caveolae (Coyne and Bergelson, 2006).

Viruses can trigger their own internalization into cells. Many viruses have to be endocytosed by the host cell, and there are many pathways by which this occurs (Marsh and Helenius, 2006). Some viruses enter cells via constitutive endocytic pathways (coated pits/vesicles, caveolae), perhaps assisted by virus-induced cross-linking of cell surface receptors. However, many viruses utilize nonclathrin-dependent pathways, and transduce signals upon binding to the cell surface that causes the cell to engulf the virus by micropinocytosis, a mechanism employed by cells to internalize large particles.

Active dissemination: movement within a cell via microtubules. Once within the cytoplasm, a virus must deliver its genome to the site of virus replication. This can be a considerable distance, particularly if replication occurs in the nucleus. However, the highly viscous nature of the cytoplasm and the extensive cytoskeletal network immobilizes virions in the cytosol. To move within a cell, virions must utilize cellular pathways, often involving the microtubule network (Urnavicius et al., 2015). Interactions with microtubules can be indirect: A virus can be internalized and delivered to a vesicle that itself is transported along microtubules. Alternatively, a number of viruses interact directly with the microtubule machinery to effect retrograde (away from the nucleus) or anterograde (toward the nucleus) transport.

A particularly well-understood example of viral movement along microtubules in both directions is provided by HSV-1 which spreads from its primary site of infection at a mucosal surface by entering local nerve endings. After membrane fusion, the viral capsid is released into the cytosol and interacts with dynein, a motor protein that transports cargo toward the microtubule organizing center that is in close proximity to the nucleus. Viral capsids are thus transported, from the site of entry to a location near the nucleus, in an efficient and rapid manner over distances that can be as long as several feet in the case of retrograde movement within an axon. Following replication, newly assembled virions need to be transported in the opposite direction, which is accomplished by specific interactions to kinesin, a molecular motor responsible for anterograde transport along microtubules (Radtke et al., 2006).

Herpes zoster, like HSV-1, establishes a latent infection in sensory ganglia by moving in a retrograde fashion, from the periphery to the nerve bodies in the ganglia. Upon reactivation, which can be caused by triggers such as stress, newly synthesized particles move in an anterograde fashion along axons, delivering virus to the periphery where it forms a painful rash that involves a single dermatome (area of skin supplied by a single sensory ganglia). Thus, the lesions are typically unilateral, and can keep appearing in the same place (as cold sores do around the mouth).

Viruses can also move efficiently from cell to cell. Just as viruses can utilize cellular pathways to effect directed movement along the cell surface and through the cytoplasm, newly assembled viruses can utilize similar pathways to move to the cell surface and achieve efficient transmission between cells. Poxviruses, after assembling in the cytosol, move along microtubules to the cell surface. Some viruses preferentially bud at the site of cell–cell contact, which results in the formation of a structure termed the virologic synapse, and makes transfer of virus to the new host cell a very efficient process (Mothes et al., 2010).

Another mechanism used by viruses to effect directed transfer to an adjoining cell utilizes surface-based glycosylaminoglycans. In cells chronically infected with murine leukemia virus, newly assembled and budded virus particles are not immediately released, but are bound to glycosylaminoglycans on the cell surface. There is a concentration of viruses to the infected cell periphery, enabling them to be efficiently transmitted to neighboring cells (Mothes et al., 2010).

Finally, some viruses use actin to move between cells. Newly synthesized vaccinia virus particles have actin-binding proteins at one end of the brick-shaped virion. Actin polymerizes at the end of the virus, which propels the virion. The virus can actually be pushed out of the cell by this actin "rocket," still enshrouded by the plasma membrane. If this projection impales an adjoining cell, the membrane "shroud" then breaks off and the virus finds itself in a new host. A virus that moves in this way from cell to cell, does not enter the extracellular space, and avoids neutralization by extracellular antibodies (Cudmore et al., 1995).

3.2 Long-Range Movement within the Host

Viruses can undergo long-range dissemination within the host, utilizing the lymphatics, the blood, or the peripheral nervous system. HIV is a good example of a virus that can spread via lymphatics as cell-associated virions. During sexual transmission, HIV must cross the epithelium, perhaps through a tear or an abrasion. DCs express lectins—such as DC-SIGN and the macrophage mannose receptor—on their surface, and are among the first cell type that HIV encounters. HIV binds avidly to these lectins, and once bound to DCs, the virus is internalized and retained in an intracellular compartment. Cross-linking of DC-SIGN triggers migration of DCs to regional lymph nodes. Once in the lymph node, HIV is returned to the DC surface and is surrounded by T cells that it can infect. Thus, for all intents and purposes, HIV uses the DC as a "taxi" for transport from a mucosal surface to a more proximal lymphoid organ.

The most common mode of dissemination is via blood (viremia), and blood-borne virus can circulate either cell-free or as cell-associated virions. Poliovirus moves through the lymphatics to reach regional lymph nodes, eventually draining into the blood stream, and reaching the central nervous system (CNS). Because poliovirus circulates as cell-free virions in the plasma, humoral neutralizing antibody can prevent virus from reaching the CNS, and protect against paralytic poliomyelitis. This is the mechanism that underpins the protective efficacy of inactivated poliovirus vaccine.

Selected viruses can spread via the peripheral nervous system. The classical example is rabies virus that does not cause a viremia but only spreads from the point of infection to the CNS and then to the salivary glands. As noted above, viruses that spread via peripheral nerves utilize intracellular transport mechanisms to "hitch a ride" within the neuron.

4. VIRUS SHEDDING AND TRANSMISSION

Virus must be shed to be transmitted to a new, naive host. Viruses may be discharged into respiratory aerosols, feces, or other body fluids or secretions, and each of these modes is important for selected agents. Nonenveloped viruses are more resistant to desiccation and other environmental extremes than are viruses surrounded by lipid bilayers. For this reason, viruses transmitted by the fecal–oral route are generally nonenveloped viruses, while fragile enveloped viruses are usually transmitted by body fluids typically within a short time after release. Viruses that cause acute infections are usually shed intensively over a short time period, often 1–4 weeks, and transmission tends to be relatively efficient. Viruses such as HBV and HIV, that cause persistent infections, can be shed at lower titers for months to years but will eventually be transmitted during the course of a long-lasting infection.

Oropharynx and gastrointestinal tract. Enteroviruses may be shed in pharyngeal fluids and feces. Poliovirus replicates in the lymphoid tissue of the tonsil and in Peyer's patches (lymphoid tissue accumulations in the wall of the small intestine) whence it is discharged into the intestinal lumen. Other viruses may be excreted into feces from the epithelial cells of the intestinal tract (reoviruses and rotaviruses) or from the liver via the bile duct (hepatitis A virus).

Respiratory tract. Viruses that multiply in the nasopharynx and respiratory tract may be shed either as aerosols generated by sneezing or coughing, or in pharyngeal secretions that are spread from hand to mouth. Often, transmission is via contaminated fomites, such as handkerchiefs, clothing, or toys.

Skin. Relatively few viruses are shed from the skin. Papillomaviruses and certain poxviruses that cause warts or superficial tumors may be transmitted by mechanical contact. A few viruses, such as variola virus, the cause of smallpox, and varicella virus, the cause of chickenpox, that are present in skin lesions can be aerosolized and transmitted by the respiratory route. In fact, it is claimed that the earliest instance of deliberate "biological warfare" was the introduction into Indian tribes of blankets containing desquamated skin from smallpox cases.

Mucous membranes, oral, and genital fluids. Viruses that replicate in mucous membranes and produce lesions of the oral cavity or genital tract are often shed in pharyngeal or genital fluids. An example is herpes simplex virus (type 1 in the oral cavity and type 2 in genital fluids). A few viruses are excreted in saliva, such as Epstein Barr virus, a herpesvirus that causes infectious mononucleosis, sometimes called the "kissing disease." The most notorious example is rabies virus, which replicates in the salivary gland and is transmitted by a bite that inoculates virus-contaminated saliva. Several important human viruses, such as HBV and HIV, may be present in the semen.

Blood, urine, milk. Blood is an important potential source of virus infection in humans wherever transfusions, injected blood products, and needle exposure are common. In general, viruses transmitted in this manner are those that produce persistent viremia, such as HBV, HCV, HIV, and cytomegalovirus. Occasionally, viruses that produce acute short-term high titer viremias, such as Ebola virus or parvovirus B19, may contaminate blood products. Although a number of viruses are shed in the urine, this is usually not an important source of transmission. One exception is certain animal viruses that are transmitted to humans; several arenaviruses are transmitted via aerosols of dried urine. A few viruses are shed in milk and transmitted to newborns in that manner. The most prominent example is HIV, though visna-maedi virus of sheep and mouse mammary tumor virus can also be transmitted via milk.

Transmission. Enteric viruses are commonly transmitted by oral or fecal contamination of hands, with passage to the hands and thence the oral cavity of the next infected host. Inhalation of aerosolized virus is the major mode of transmission for respiratory viruses. Another significant route is by direct host-to-host interfacing, including oral–oral, genital–genital, oral–genital, or skin–skin contacts. Transmission may involve less natural modes such as blood transfusions, organ transplants, or reused needles. In contrast to propagated infections are transmissions from a contaminated common source, such as food, water, or biologicals. Common source transmission is quite frequent and can produce explosive outbreaks that range in size depending on the number of recipients of the tainted vehicle and the level of virus contamination. One of the most infamous was the World War II epidemic of HBV among military personnel, transmitted by a contaminated batch of 17D yellow fever vaccine.

Sexually transmitted viruses present a special situation, since the probability of spread depends upon the gender and type of sexual interaction between infected host and her/his uninfected contact. For instance, an HIV-infected male is more likely to transmit to a female partner via anal than vaginal intercourse, and that risk is reduced if the male partner is circumcised.

Transmission of arboviruses is complex, since it involves the cycle between an insect vector and a vertebrate host. There are a number of quantitative variables that determine the efficiency of vector transmission. Vertebrate host determinants include the titer and duration of viremia, while insect determinants include the competence of the vector (that is, the ability of the vector to support viral replication in several tissues and shed virus in its saliva) and the extrinsic incubation period (the interval between ingesting the virus and shedding in the saliva), as well as the distinctive feeding preferences of each insect vector. Also, there are a number of alternative patterns of viral maintenance in the vector, including overwintering of virus in hibernating mosquitoes, transovarial transmission of the virus, and venereal spread between male and female mosquitoes.

5. PATTERNS OF VIRUS INFECTION

Viruses can cause acute, chronic, or latent infections. Most viruses cause acute, self-limited infections, and most infections in turn are asymptomatic. Symptomatic infections are preceded by an asymptomatic incubation period, the length of which can be related to the inoculum (dose of the virus), the rate of replication, and many other factors. For a virus to initiate a persistent infection, a set of special circumstances must exist that permit the virus to escape the adaptive immune response. The general principles associated with each of these infection patterns are described in Chapter 7, Patterns of Infection.

6. MAJOR DISEASE MECHANISMS

Viruses can cause disease in many ways, which are discussed in subsequent chapters. Here, we will briefly review the major mechanisms involved.

Direct cytopathic effects. Many viruses kill cells, directly by lysis or by inducing apoptosis, and disease can result from loss of parenchymal cells. An example is West Nile virus which infects neurons and induces apoptosis via caspase 3, leading to encephalitis and movement disorders. Another example is Ebola virus. Individuals infected with the Zaire strain of Ebola virus typically develop a hemorrhagic fever, with loss of vascular integrity. The spike protein of Ebola virus appears to be a major culprit; it induces loss of contact with neighboring cells, which plays a role in the vascular leakage and hypotension that are characteristic of fatal Ebola hemorrhagic shock syndrome (Feldman and Geisbert, 2011).

Another example is syncytia formation. Enveloped viruses that bud from the plasma membrane deliver their glycoproteins to the cell surface. These viral proteins can bind to receptors on the surface of adjoining cells, eliciting cell–cell fusion, and syncytia formation. Respiratory syncytial virus, the major cause of viral pneumonia in young children, derives its name from the fact that it can form syncytia not only in tissue culture, but also in the lungs of infected patients.

Disease caused by antibody-mediated immunity. Some viruses, such as hepatitis B, release large amounts of antigen into the blood. Antibodies can bind to the viral antigens, leading to the formation of immune complexes that are deposited in the basement membranes of glomeruli in the kidney, leading to renal dysfunction.

Another example of antibody-mediated diseases is dengue, caused by a mosquito-borne virus that infects millions of people a year. There are four major dengue virus serotypes. Antibodies that neutralize one serotype do not neutralize the others. When a human is infected with a second serotype, antibodies produced against the first serotype bind to, but do not neutralize, the second serotype. The virus is now partially coated with non-neutralizing antibodies, enabling it to bind to cells—such as monocytes and macrophages—that have Fc receptors on their surface. This results in the efficient infection of these cells with massive release of cytokines and subsequent vascular leakage and hemorrhage (Paessler and Walker, 2013). As a consequence, there is a chance that the patient will now develop dengue hemorrhagic fever/dengue shock syndrome, which carries significant mortality.

Disease caused by virus-initiated autoimmunity. Viral antigens are sometimes similar to host antigens so that antibody or cellular responses directed against a pathogen may also cross-react with normal host molecules and cells. It is widely thought that some immunopathological diseases are the result of molecular mimicry, such as the demyelinating diseases like multiple sclerosis (MS) and Guillain–Barré syndrome (GBS). GBS is often preceded by a viral illness. It is hypothesized that a viral peptide presented by the class I pathway resembles a cellular protein (the myelin basic protein), and triggers an autoimmune response that results in demyelination.

Diseases associated with innate immunity and a cytokine "storm." Recent studies of the reconstructed 1918 strain of influenza virus suggest that the severe disease was due to an exuberant host response in the lungs that led to pulmonary edema and respiratory failure. It is postulated that an excess of proinflammatory cytokines triggered this excessive response (Tisoncik et al., 2012; Peng et al., 2014). Another example is Ebola virus disease, a hemorrhagic fever, accompanied by very high temperature and a clinical crisis. It is postulated that a cytokine storm plays a major role in acutely fatal cases.

Disease caused by virus-induced immunosuppression. Some virus infections can lead to immunosuppression, making the human host susceptible to other infectious agents (see Chapter 6, Aberrant Immunity). Measles virus suppresses the secretion of certain cytokines, leading to transient immunosuppression. Measles kills 150,000 children annually, but deaths are usually not the direct result of the virus, but rather via virus-enhanced infection with other viruses or bacteria.

Disease caused by virus-induced tumorigenesis. Oncogenic viruses are the subject of Chapter 9. Suffice it to say that RNA tumor viruses transform via different mechanisms than those used by DNA tumor viruses. However, in all instances, there is a final common pathway, with a release of the "brakes" on the cell cycle which limit the replication of normal cells. Tumorigenesis involves a multistep process that confers on virus-transformed cells; the ability to grow in the host animal.

7. REPRISE

The steps in viral infection begin with infection of individual cells followed by spread through a multicellular host organism. At the cellular level, viruses use attachment factors to associate with the cell surface and then specific cellular receptors to initiate attachment. Entry is a multistep process that involves membrane fusion of enveloped viruses or conformational changes in naked capsids, to deliver the viral genome to either the cytoplasm or nucleus. In general, RNA viruses replicate in the cytoplasm and DNA viruses in the nucleus, with a few notable exceptions, such as poxviruses and retroviruses. The virus uses viral or host enzymes to replicate its genome and for transcription and translation of viral proteins. In many instances, viruses shanghai normal cellular metabolism to support their replication and assembly. Shedding of newly synthesized virions is implemented by lysis of the infected cell to release encapsidated viruses or budding from the plasma membrane to release enveloped viruses. Host cells encode some antiviral proteins and many viruses have evolved countermeasures to evade these intrinsic defenses and to prolong the life of the virus-infected cell by blocking apoptosis or necrosis.

At the level of the host organism, there are many different routes of infection, each of which is characteristic for a specific virus. Viruses usually initiate infection via inhalation, ingestion, or penetration of the skin or mucous membranes. Virions often spread through the lymphatics to the lymph nodes where many of them replicate, to be discharged into the blood as free or cell-associated particles. Blood-borne viruses invade and replicate in various target organs, again characteristic for each virus. Spread from an infected animal or human is usually by release from either the lung, mouth and gastrointestinal tract, mucous membranes, or skin.

Viral infections range from innocuous to lethal. Some viruses are not cytocidal and do not kill infected cells; likewise many viral infections of human or animal hosts occur without apparent illness. However, viruses can cause pathological changes through a variety of mechanisms. These include direct tissue injury such as seen in smallpox, HAV hepatitis, paralytic poliomyelitis, and herpes simplex encephalitis. Virus-initiated immune-mediated diseases include HBV acute hepatitis, dengue hemorrhagic fever, and some cases of influenza pneumonia. Diseases associated with immunosuppression include measles, pneumonia, and congenital rubella. Virus-initiated cancers of humans include cervical cancer (HPV), lymphoma (HTLV-1), and hepatocellular carcinoma (HBV and HCV).

FURTHER READING

Reviews

Danthi, P., G.H. Holm, T. Stehle and T.S. Dermody. 2013. Reovirus receptors, cell entry and proapoptotic signaling. Adv. Exp. Med. Biol. 790:42–71.

Doorbar, J., W. Quint, L. Banks, I.G. Bravo, M. Stoler, T.R. Broker and M.A. Stanley. 2012. The biology and life-cycle of human papillomaviruses. Vaccine 30, Supp. 5:F55–F70.

Feldmann H, Geisbert TW. Ebola hemorrhagic fever. Lancet 2011, 377: 849–862.

Liang C, Oh B-H, Jung JU. Novel functions of viral anti-apoptotic factors. Nature Reviews Microbiology 2015, 13: 7–15.

Marsh, M. and A. Helenius. 2006. Virus entry: Open sesame. Cell 124:729–740.

Paessler, S. and D.H. Walker. 2013. Pathogenesis of the viral hemorrhagic fevers. Annu. Rev. Pathol. Mech. Dis. 8:411–440.

Radtke, K., K. Döhner and B. Sodeik. 2006. Viral interactions with the cytoskeleton: a hitchhiker's guide to the cell. Cell. Microbiol. 8:387–400.

Taubenberger, J.K. and D.M. Morens. 2006. 1918 Influenza: the Mother of All Pandemics. Emerg. Infect. Dis. 12:15–22.

Tisoncik JR, Korth MJ, Simmons CP, et al. Into the eye of the cytokine storm. Microbiology and Molecular Biology Reviews 2012, 76: 16–32.

Wilen, C.B., J.C. Tilton and R.W. Doms. 2012. HIV: Cell binding and entry. Cold Spring Harbor Perspectives in Medicine Volume: 2 Issue: 8 Article Number: a006866.

Original Contributions

Chen Y-H, Due W, Hagemeijer MC, et al. Phosphatidylserine vesicles enable efficient en block transmission of enteroviruses. Cell, 2015: 160: 619–630.

Coyne, C.B. and J.M. Bergelson. 2006. Virus-induced Abl and Fyn kinase signals permit coxsackievirus entry through epithelial tight junctions. Cell 124:119–131.

Cudmore, S., P. Cossart, G. Griffiths and M. Way. 1995. Actin-based motility of vaccinia virus. Nature 378:636–638.

Falzarano D, Feldmann H. Delineating Ebola entry. Science 2015, 347: 947–948.

Falzarano D, de Wit E, Feldman H, et al. Infection with MERS-CoV causes lethal pneumonia in the common marmoset. PLoS Pathogen 2014, 10: e1004250.

Geijtenbeek TB, Kwon DS, Torensma R, van Vliet SJ, van Duijnhoven GC, Middle J, Cornelissen IL, Nottet HS, KewalRamani VN, Littman DR, Figdor CG, van Kooyk Y. "DC-SIGN, a dendritic cell-specific HIV-1-binding protein that enhances trans-infection of T cells." Cell 100: 587–97.

Helenius, A., J. Kartenbeck, K. Simons and E. Fries. 1980. On the entry of Semliki Forest virus into BHK-21 cells. J. Cell Biol. 84:404–420.

Mothes, W., N.M. Sherer, J. Jin and P. Zhong. 2010. Virus cell-to-cell transmission. J. Virol. 84:8360–8368.

Panda, A., Z. Huang, S. Elankumaran, D.D. Rockemann and S.K. Samal. 2004. Role of fusion protein cleavage site in the virulence of Newcastle disease virus. Microb. Pathog. 36:1–10.

Peng X, Alfoldi J, Gori K, et al. The draft sequence of the ferret (Mustela putorius furo) facilitates study of human respiratory disease. Nature Biotechnology 2014, 32: 1250–1255.

Rustagi A, Gale M Jr. Innate antiviral immune signaling, viral evasion and modulation by HIV-1. J Molecular Biology 2014, 426: 1161–1177.

Shinya, K., M. Ebina, S. Yamada, M. Ono, N. Kasai and Y. Kawaoka. 2006. Avian flu: Influenza virus receptors in the human airway. Nature 440:435–436.

Sakurai Y, Kolokoltsov AA, CC, et al. Two-pore channels control Ebola virus host cell entry and are drug targets for disease treatment. Science 2015, 346: 955–960.

Urnavicius L, Zhang K, Diamant AG, et al. the structure of the dynactin complex and its interacton with dynein. Science 2015, 347.

Van Doremalen N, Mazgowics KL, Milne-Price S, et al. Host species restriction of Middle East Respiratory Syndrome coronavirus through its receptor, dipeptidyl peptidase 4. J Virology 2014, 88: 9220–9232.

Yang, Z.Y., H.J. Duckers, N.J. Sullivan, A. Sanchez, E.G. Nabel and G.J. Nabel. 2000. Identification of the Ebola virus glycoprotein as the main viral determinant of vascular cell cytotoxicity and injury. Nat. Med. 6:886–889.

Chapter 4

Innate Immunity

Recognizing and Responding to Foreign Invaders—No Training Needed

Christine A. Biron

Department of Molecular Microbiology and Immunology, Brown University, Providence, RI, USA

Chapter Outline

1. INTRODUCTION

The immune system has evolved to defend against microbial infection. It is broadly divided into "innate" and "adaptive" arms. To carry out its function, the components of the immune system deliver toxic molecules that can deter microbial replication and kill infectious organisms. Thus, the system has to be able to recognize threats as nonself. During first infections with individual microorganisms, adaptive immunity is slow to develop because it uses highly specialized receptors expressed on adaptive T and B lymphocytes. These receptors are products of genes with rearranging elements. Although the adaptive receptors provide elegant specificity for foreign antigens, they are only expressed on extremely low numbers of cells prior to exposure to particular antigenic determinants. As a result, innate immunity has a unique role in sensing and mediating protection at early times during infection. Its receptors or sensors are products of complex germ-line gene families. They recognize essential nonself determinants broadly expressed on classes of pathogens, by infected cells, or on stressed cells. Once innate sensors are engaged and activated, cascades of events are unleashed resulting in the expression of new or elevated levels of mediators produced by nonimmune and/or immune cells to deliver antiviral defenses. These innate responses also play important roles in instructing downstream adaptive immunity.

Interferons (IFNs) were the first appreciated mediators of innate immunity against viral infections. Because IFNs are produced by cells to act on cells, they are members of the large group of soluble factors called cytokines. Cytokines bind to specific receptors on cells, activating intracellular signaling pathways, which induce their biological effects. Other innate cytokines can also be induced to promote antimicrobial and inflammatory responses, including interleukin 12 (IL-12), tumor necrosis factor (TNF), IL-6, and IL-18. Major cellular contributors to innate immunity include virtually all non-T and non-B immune cells (and even subsets of T and B cells under particular conditions). In the context of viral infections, however, monocytes/macrophages, dendritic cells (DCs), and natural killer (NK) cells are particularly important. After detecting the

Viral Pathogenesis. http://dx.doi.org/10.1016/B978-0-12-800964-2.00004-5

nonself determinants characterized as pathogen-associated molecular patterns (PAMPs), the sensors of infection stimulate activation of these innate soluble and cellular responses. The sensors inducing cytokine responses include the membrane-bound toll-like receptors (TLRs) and a growing list of cytosolic receptors. Cell-surface activating receptors used by NK cells to mediate killing of virus-infected target cells also represent a class of innate sensors. The complex system of innate receptors trolls the environment within and outside of cells to detect infectious organisms, and once engaged to elicit diverse immune responses for protection.

2. INTERFERONS

A discussion of innate immunity to viral infection is best initiated with the IFNs. These soluble components of innate immunity were discovered in 1957 and immediately recognized as important for activation of host defense mechanisms against a broad range of viruses. In a classic series of experiments, Alick Isaacs and Jean Lindemann showed that if the allantoic membrane from an embryonated chicken egg was exposed to inactivated influenza virus, an activity appeared in the supernatant fluids that—when applied to an uninfected membrane—could interfere with the ability of live influenza virus to infect the second membrane; hence the name "interferon." Because the antiviral interfering activity can be elicited very rapidly in the absence of immune cells and confers resistance to many viruses, IFNs can be considered the first identified molecular executors of innate resistance. They are major mediators of broad host innate immunity against viruses.

The IFNs do not produce their effects directly within infected cells. They are secreted and spread either locally or through the circulation to other cells where they induce the production of characteristic sets of proteins that are responsible for multiple biological effects. Although the ability to induce direct antiviral defense is the major and first appreciated function of IFNs, these cytokines can also mediate a wide range of diverse immune regulatory functions.

2.1 Types of IFN

There are three different classes of IFNs, types I, II, and III (Figure 1). The first described as induced during viral infections are the type I IFNs. These cytokines are encoded by a number of different genes, including a single beta and up to 12 functional alpha genes (IFN-α/β) in the human. Virtually any nucleated cell can produce type I IFNs, if appropriately stimulated. There is a single type II IFN, IFN-gamma (γ). This IFN is an immune cell product, with the major cellular sources being NK and T cells. The most recently described are the type III IFNs. In humans, there are at least three members of this class identified either as IFN-lambdas (λs) or IL-28A, IL-28B, and IL-29. The different IFN types use different receptors, but each particular type uses a common receptor. Different IFNs induce overlapping antiviral effects, but each class has unique functions, based on the specific receptors that they use, the cell distribution of the receptors, and the preferred and alternative intracellular signaling pathways that the receptors can activate.

2.2 IFN Signaling

The type I and II IFN receptors are found on most cells, but the type III IFN receptors are expressed on a more limited range of cell types. All of these receptors are heterodimers, with extracellular ligand-binding, transmembrane, and intracellular signaling domains. When IFNs bind to their cognate receptors, a JAK-STAT intracellular signaling cascade is initiated (Figure 1). Briefly, the intracellular signaling domains of the receptors activate particular Janus tyrosine kinases (JAKs) to phosphorylate signal transducer and activator of transcription (STAT) proteins. Within minutes, the phosphorylated STAT complexes move to the nucleus where they enhance the transcription of selected genes by binding to promoter sequences. There are seven different STAT molecules, 1 through 6 with 5A and 5B. By following the pathways to their induction of antiviral functions, the common activation of STAT1 by all three IFN receptors, and the activation of STAT2 as well as STAT1 by the type I IFN receptor, were identified as preferred targets of phosphorylation. In the case of signaling through the type I and III IFN receptors, phosphorylated STAT1 and STAT2 complex with a third molecule, the interferon regulatory factor 9 (IRF9), to form the IFN-stimulated factor 3 (ISGF3) to stimulate expression of genes having IFN-stimulated response elements. Signaling through the type II IFN receptor results in activated STAT1 homodimers, also identified as the IFN-γ activation factor, to stimulate genes having IFN-γ activated sequences (GAS). Under certain conditions, the activation of STAT1 homodimers can also be a product of type 1 or III IFN signaling. The overlapping activation of STAT1 by the different classes of IFNs helps explain their common abilities to induce direct anitiviral functions.

Cytokines using JAK-STAT signaling pathways, however, can also activate alternative STAT molecules. In the case of the type 1 IFN receptor, the flexibility is remarkable with these factors conditionally activating all of the seven different STAT molecules. The choice of STATs for activation is influenced by the basal as well as dynamic modulation of relative individual STAT concentrations within particular cells. The best characterized to date are the rapid induction of elevated STAT1 after stimulation of this molecule in any cell, and the higher basal expression of STAT4 in NK cells. The flexibility in signaling helps to explain the range and the sometimes paradoxical biological effects mediated by type I IFNs. It also provides a mechanism whereby the host can use a limited number of genes to shape diverse cellular responses to cytokines.

2.3 Cell Intrinsic Antiviral Defense

IFNs are particularly important for the host defense against primary virus infection. The significant role of the type I

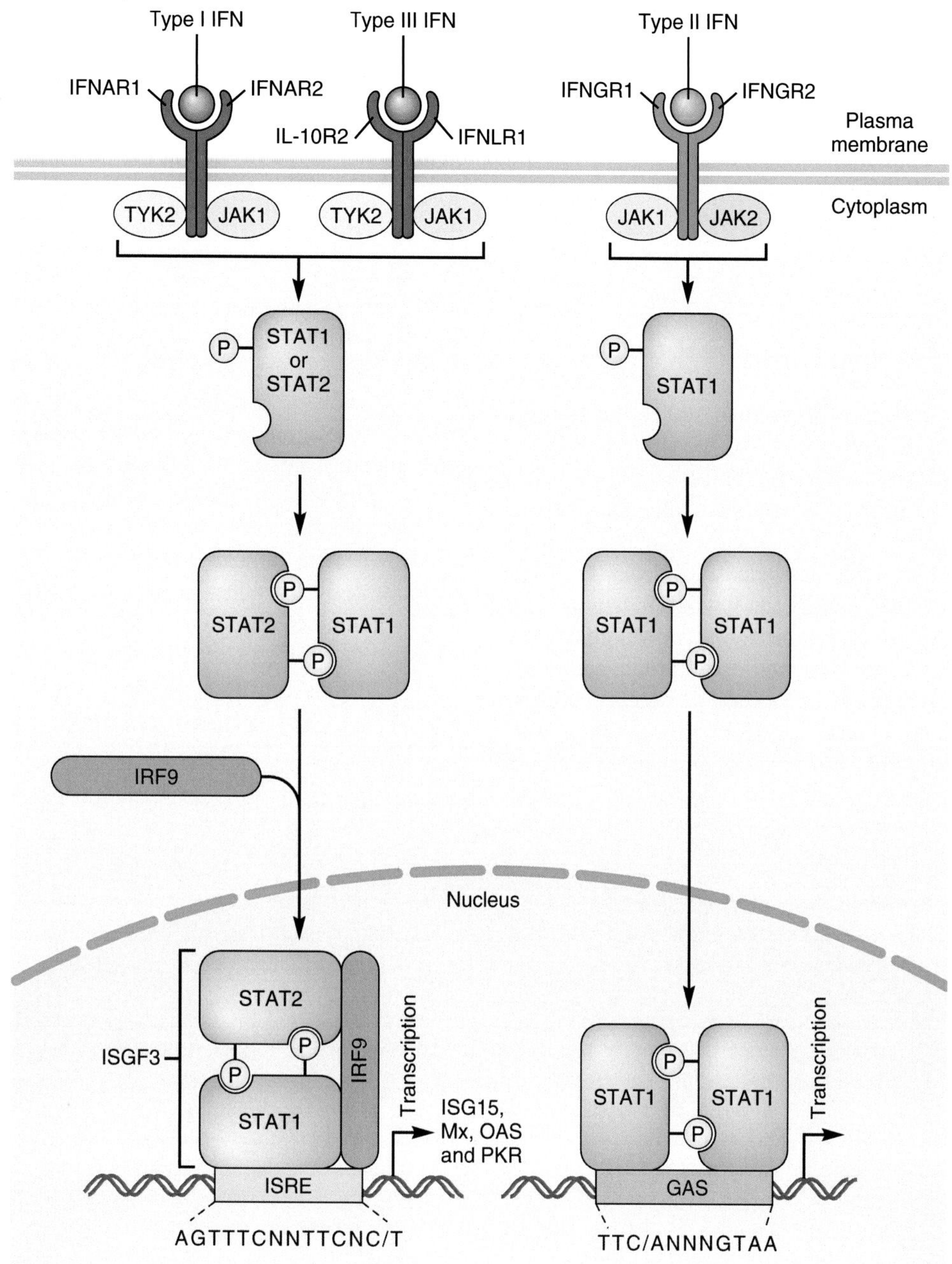

FIGURE 1 Interferons (IFNs), their receptors, and their major known signaling pathways. This diagram shows only the "classical" JAK-STAT pathway. Type I, II, and III IFNs each bind to their specific receptors, and initiate phosphorylation of STAT1 or STAT1 and STAT2; activated STAT complexes migrate to the nucleus where they bind to specific DNA sequences and transactivate downstream genes leading to the production of proteins that mediate the pleiotropic response to IFNs. IFNAR1, 2: the heterodimeric chains that make up the type I IFNA receptor; IL-10R2 and IFNLR1: the type III IFN receptor: IFNGR1, 2: type 2 IFNγ receptor; JAK: Janus-activated kinase; TYK2: tyrosine kinase; STAT: signal transducer and activator; IRF9: IFN response factor 9; ISRE: IFN-stimulated response element; GAS: IFN gamma-activated site. *(Modified from Sadler and Williams. Nature Reviews Immunology 2008, 8: 559-568, with permission.)*

IFNs can be demonstrated in animal models in which the IFN response is reduced at early times after infection by treatment with anti-IFN antibodies, or in mice whose IFN genes (or IFN receptor genes) have been "knocked out" by genetic mutation. In these instances, such animals, compared to controls, exhibit a reduced ability to contain virus infections and often show an increased incidence of illness or death. When the type I IFN response is abrogated, there is a global increase in susceptibility to most viruses, whereas a knockout of the type II IFN response has a more modest effect, which is seen with some but not other viruses. The developing understanding of the type III IFN functions suggests that these cytokines might be most important at mucosal sites. In the case of primary immune deficiencies in humans, mutations in the STAT1 gene that result in the loss of protein function are associated with increased susceptibility to different kinds of infections depending on IFNs for optimal defense. Taken together, these observations indicate that type I IFNs play a crucial role in nonspecific antiviral defense, that the type II IFN response may be one element in a multicomponent immune effector system with more importance in nonviral infections, and that the type III IFNs may act in particular locations of infection.

Viruses vary in their ability to induce IFNs and in their sensitivity to the cytokines. The ability to avoid detection for cytokine induction and to resist the antiviral effects of IFNs play important roles in viral survival. Once the type 1 IFNs are induced, they activate STAT1 and STAT2 to stimulate elevated expression of more than 100 different genes in a wide variety of cells. Many of the protein products of these genes act to block viral replication and spread. If viruses cannot escape their effects, particular IFN-induced proteins can interfere with viral entry, replication, translation, maturation/assembly, and release (Sidebar 1). Some of these molecules may preexist at low levels within cells. As the type I IFNs can also prime for their own production, elicit high STAT1 expression, and even stimulate elevated levels of innate sensors of infection, the system is clearly set up to rapidly up-regulate sensitivity for detecting infection and promoting induction of STAT1-dependent defense mechanisms.

2.4 Evading Type 1 IFN Antiviral Effects

Different viruses are inhibited to different degrees by IFNs because most viruses have evolved specific (and often multiple) mechanisms to elude the effects of these cytokines. General evasion strategies used by viruses include inhibition of the transcription and translation of cellular genes to block the expression of the antiviral effector molecules. There are also examples of specific viral IFN antagonists inhibiting the JAK-STAT and IRF3 pathways for their induction. At the earliest point of eliciting cellular antiviral defense, particular viral products act to block innate sensor recognition and function for IFN induction. The diversity of the individual and combined approaches used by different viruses, to antagonize the cellular mechanisms in place to protect against infection, is remarkable. The evolution and conservation of anti-IFN viral genes testifies to the importance of the cytokines in host defense. The immunoregulatory IFN functions make it possible for the host to use the induction of the cytokines to activate antiviral defense mediated by noninfected immune cells, and as a result, to get to pathways that cannot be easily inhibited by viral products.

2.5 IFNs and Immune Regulation

IFNs may also have indirect effects upon virus infection through their many immunoregulatory functions. Examples

Sidebar 1 IFN-mediated antiviral mechanisms

- **APOBEC3G.** The APOBEC proteins are a family of IFN-inducible cytidine deaminases that serve as a host defense mechanism against retroviruses such as HIV and hepatitis B virus. APOBEC3G associates with nascent HIV nucleocapsids and is incorporated into new virions. When these infect another host cell and the viral reverse transcription complex is formed, APOBEC3G deaminates cytidine to uridine; the dU-rich DNA transcripts are either degraded or form hypermutated defective proviruses.
- **ADAR.** IFN-inducible dsRNA-specific adenosine deaminase, ADAR, works to edit viral RNA in infected cells and impair its normal functions.
- **The 2′–5′ oligoadenylate synthetases and RNase L.** The 2–5′ oligoadenylate sythetases are a family of proteins induced by type 1 IFNs to polymerize ATP into oligomers of adenosine. These unique structures activate the RNase L enzyme to induce degradation of viral RNA.
- **Protein kinase R (PKR).** PKR accumulates in the nucleus and cytoplasm as an inactive monomer but is directly activated by viral dsRNAs. Once activated, PKR blocks protein translation in cells by inactivating eukaryotic translation initiation factor 2 alpha (EIF2α) through phosphorylation.
- **Mx GTPases.** This family of GTPase proteins (MxA and MxB in humans and Mx1 and Mx2 in mice) was first identified by studying inbred mouse strains for their sensitivity to particular viral infections. The enzymes block viral replication at different sites in the cell by trapping viral nucleocapsids, or particles at other developmental steps, leading to their eventual degradation.
- **ISG15.** The interferon stimulated gene 15 (ISG15) products, a ubiquitin homolog, is covalently linked to a wide range of newly synthesized proteins with consequences for resisting virus-mediated antagonistic effects on IFN induction and function. It can also be secreted in large amounts to act as a cytokine modulating immune responses.
- **RNA interference (RNAi).** In the mid-1990s, RNAi was discovered serendipitously when an attempt to overexpress specific plant genes (using viral vectors) instead resulted in the knockout or silencing of those genes. RNAi has been adapted for the experimental silencing of specific genes and is also emerging as a therapeutic modality. Recent studies demonstrate that RNA silencing is used by human T lymphocyte cell lines as a defense against HIV-1 and that certain cellular interfering RNAs can be induced by IFN.
- **TRIM5α.** A member of the tripartite motif (TRIM) family of proteins, TRIM5α is a restrictive element thought to act by binding HIV viral capsid protein to induce ubiquitination and proteasome-mediated degradation prior to their functional uncoding for replication.
- **Tetherin and Viperin.** These IFN-induced proteins inhibit viral replication by impairing the maturation and release of free viral particles from infected cells.

include activation of NK cell killing by stimulating elevated expression of the perforin and granzyme molecules, induction of the cytokine IL-15 to enhance NK and T cell expansion, inhibition of nonspecific T-cell proliferation, induction of T-cell resistance to NK cell mediated lysis, enhancement of long-term T-lymphocyte survival, enhancement of IFN-γ production by NK and T cells under particular conditions, and up-regulation of the major histocompatibility complex (MHC) class I molecules to render virus-infected cells more sensitive to immune surveillance by antigen-specific CD8 T lymphocytes. The pathways used by IFNs to induce these biological effects are not all understood. Some are stimulated as a result of STAT1—and a few through STAT4—activation. It remains to be seen whether or not other effects may be mediated by different intracellular signaling pathways.

3. SENSORS OF INFECTION

Although the production of IFNs in response to viral infection had been appreciated for some time, it took decades to elucidate the molecular mechanisms leading to their induction. In his seminal 1989 commentary, Charles Janeway noted the "immunologist's dirty little secret," that is that most antigens only elicit a brisk immune response when mixed with an adjuvant. Successful adjuvants contain microbial products, such as inactivated tubercle *bacillus* in Freund's complete adjuvant. Based on this, Janeway first articulated the view of "sensors" to distinguish self from nonself for the induction of innate immune responses. He hypothesized that conserved host pattern recognition receptors (PRRs) were in place to recognize repetitive PAMPs resulting from unique chemical structures on microbes. Once engaged, these sensors would be activated to stimulate innate immune responses.

It is now clear that there are many different classes of innate sensors. These receptors are the products of families of germ-line genes. They are in different cells and cellular compartments for trolling the environment to sense infection, and once stimulated, inducing appropriate innate responses. Classes of sensors include the TLRs and the C-type lectin receptors (CLRs) expressed on cell membranes. Others are cytosolic receptors with pathways inducing cytokine responses overlapping those stimulated by TLRs, and a distinct group of receptors forming complexes called inflammasomes that have a unique function in stimulating processing of preexisting proforms of a particular innate cytokine subset.

3.1 The Toll-like Receptors

The first class of innate sensors identified, the TLRs, were characterized by evolutionary comparison to sensors in invertebrates. Drosophila, like all invertebrates, does not have a system for mounting acquired antigen-specific immune responses. Instead, invertebrates depend entirely on an innate system, in which microbial challenge invokes the synthesis of a variety of antimicrobial peptides. Drosophila does encode a protein—dorsal—that is a member of the NF-κB family. Because NF-κB is a transcription factor that plays a critical role in immune induction in mammals, Jules Hoffman and his coworkers—in a jump of inspired intuition—decided to investigate whether genetic defects in the known regulatory cascade for dorsal would abrogate the ability of Drosophila to contend with microbial pathogens. It was found that flies with mutations in the transmembrane receptor Toll (at the proximal end of the dorsal cascade) were highly susceptible to fungal pathogens. The name "Toll" (german slang for "crazy") was coined for a Drosophila development mutant. Following this lead, Janeway and Ruslan Medzhitov cloned a human homolog of Toll (hence, Toll-like receptor) and showed that it was involved in the activation of NF-κB in human cells. Bruce Beutler and associates showed that mice with a genetic defect that made them unresponsive to bacterial lipopolysaccharide (LPS)—a notoriously proinflammatory microbial product—had a mutation in a TLR. This series of studies provided a mechanism—the TLRs—whereby a microbial product, broadly expressed on a class of pathogens but not on host cells, could activate immune cells, and gave credence to the PAMP/PRR hypothesis.

Characterization of the different TLRs in mammals advanced by focusing on microbial pathogens in general, with information accrued on the function of a subset of TLRs for innate immunity to viruses. Certain microbial cell wall constituents, such as LPS, lipoproteins, and peptidoglycans, are found only in bacteria and not in eukaryotic cells. Components of these unique molecules are nonself patterns that are found in classes of microorganisms. Thus, they provide the PAMPs that act as ligands for the TLR PRRs. The TLRs are expressed on a variety of immune cell types playing important roles in innate host defense against microbial pathogens. Most important are macrophages and DCs. Their expression on a specialized DC subset, the plasmacytoid DCs (pDCs), plays an important role in the induction of type I IFNs during viral infections.

Particular TLRs are found in cell surface membranes, whereas others are present in the membranes of endosomes. Cells of the innate immune system respond in several different ways when these PRRs are ligated. The TLRs usually initiate intracellular signaling pathways that lead to the elevated gene transcription and protein expression of a cascade of cytokines (such as the type I IFNs, IL-6, the pro-IL-1β precursor molecule, IL-12, and TNF). These in turn act together to orchestrate an inflammatory response against the foreign invader. In some instances, macrophages will be stimulated to phagocytose and digest foreign organisms, or to initiate chemotaxis, bringing neutrophils to the site of inflammation. An important response is the activation of DCs, followed by the binding of antigen, migration to draining lymph nodes, and up-regulation of co-stimulatory molecules, critical steps in the initiation of the acquired immune response.

The TLR family of sensors is made up of the protein products of 11 functional genes in the mouse and 10 in the human. The molecules share common structural features, including an extracellular domain containing leucine-rich repeats and a cytoplasmic domain similar to that of IL-1 receptors. Some TLRs act in concert with other cell surface molecules to bind their ligands. They signal through several intracellular pathways that predominantly involve common components such as myeloid differentiation factor 88 (MyD88) or the TIR-domain-containing adapter-inducing interferon-β (TRIF). In addition to stimulating NF-κB, activation of IRF3 and/or IRF7 by TLR signaling can result in the induction of type 1 IFNs. The molecules identified as Unc-homolog B1 (UNC93B1) homologs interact with a subset of TLRs to facilitate their trafficking within cells. Individual TLRs sense specific microbial components. TLR4 recognizes Gram-negative bacterial LPS; TLR2 recognizes peptidoglycan, a component of the cell walls of Gram-positive bacteria; and TLR5 recognizes flagellin, a major component of bacterial flagella. TLR9 recognizes unmethylated CpG (cytosine–phosphate–guanine) motifs in DNA. Because bacteria lack cytosine methylase, an enzyme found in eukaryotic cells, the unmethylated CpG motifs provide a microbial-specific ligand for this TLR.

The TLRs that are particularly relevant to a wide range of viral infections, TLR3, 7/8, and 9, are found in endosomes and endolysosomes (Figure 2). TLR3 recognizes double-stranded

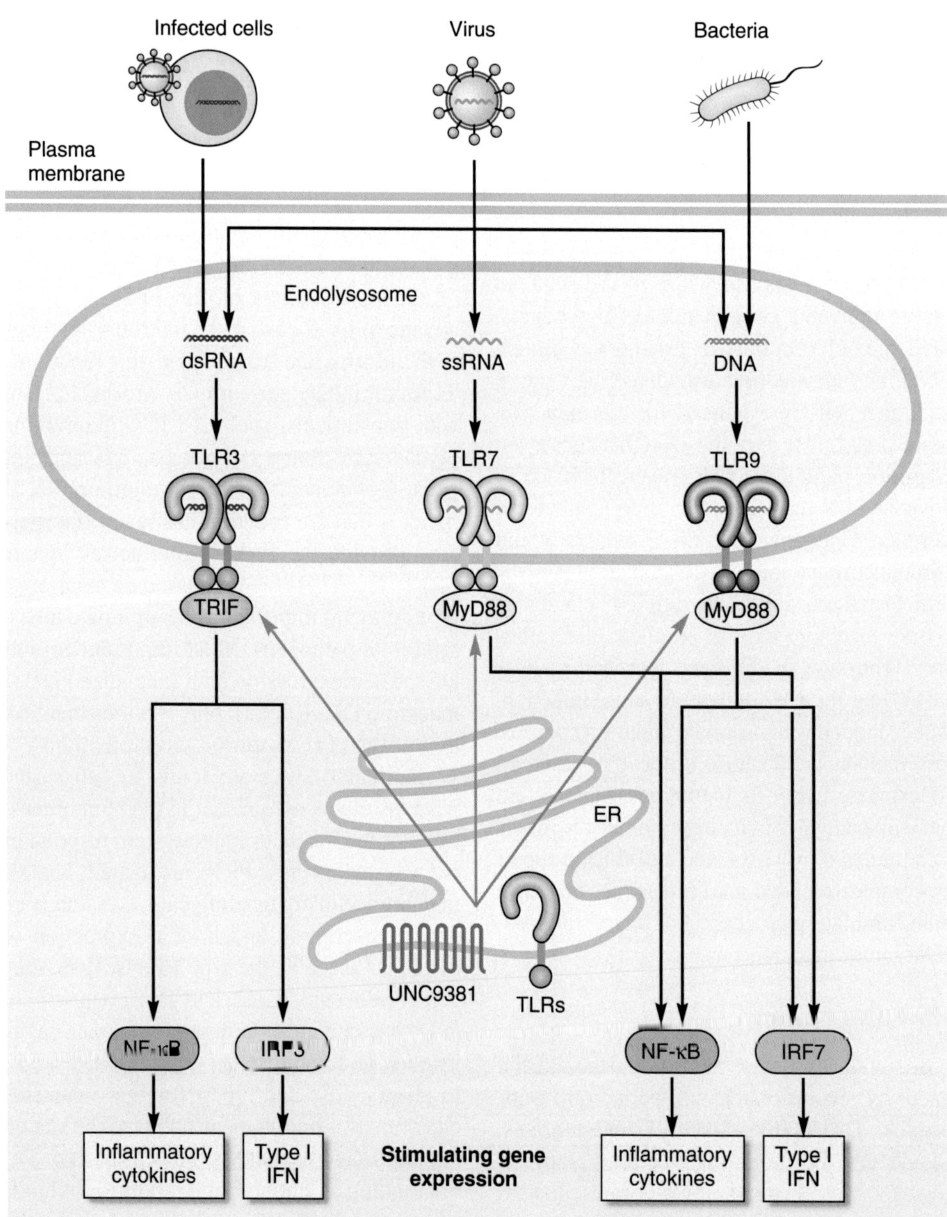

FIGURE 2 Major toll-like receptors (TLRs) sensing viral infections. The TLRs 3, 7/8, and 9 are found in membranes of endosomes and endolysomes. Exposure of these sensors respectively to dsRNA, ssRNA, or DNA resulting from viral infections stimulates their activation, through the adaptor molecules MyD88 or TRIF, of intracellular signaling pathways to result in gene induction for expression of inflammatory cytokines and type 1 IFNs. *(Adapted/Modified from Kawai T Akira S. Nature Immunology 2010; 11: 373–384, with permission.)*

(ds) RNA, an intermediate produced during the replication of many RNA viruses. TLR3 is mainly expressed on DCs and activation of DCs early in viral infection induces type I IFN, helping to explain the old observation that dsRNA is a potent IFN inducer. TLR 7/8 recognize single-stranded (ss) RNA, including viral mRNA and the genomes of ssRNA viruses. TLR9 senses DNA motifs from a variety of different sources including viral genomic material particularly rich in CG sequences.

3.2 Cytosolic Receptors

Expression of the TLRs on membranes of specialized cells provides a system for detecting microbial threat prior to cell infection and affords protection against the direct interference of IFN induction by-products of replicating viruses. These receptors cannot, however, explain the intracellular interactions leading to the induction of type 1 IFNs in infected cells. To detect nucleic acids inappropriately found in cells, there are cytosolic sensors. Members of this class of sensors recognizing viral RNA include the products of retinoic acid-inducible gene-I (RIG-I) that detect RNA with a 5' tri phosphate motif and blunt end base pairing as well as the melanoma differentiation-associated gene 5 (MDA5) products that detecting longer dsRNA. These sensors are classified as RIG-I-like receptors (RLRs) (Figure 3) and are expressed in most cells and inducible by type 1 IFNs. Once engaged, they bind to mitochondrial membranes through the interferon-beta promoter stimulator 1 (IPS1)—also known as the mitochondrial antiviral signaling protein (MAVS) adaptor molecule, also known as the interferon-beta promoter stimulator 1 (IPS1)—to stimulate pathways leading to the activation of NF-κB and IRF3, resulting in cytokine gene transcription.

There is also a growing list of known cytosolic DNA sensors. Cytosolic DNA sensors include the helicase family member DDX41, the DNA-dependent activator of IRFs (DAI), the IFN-γ-inducible protein 16 (IFI16), and the recently identified

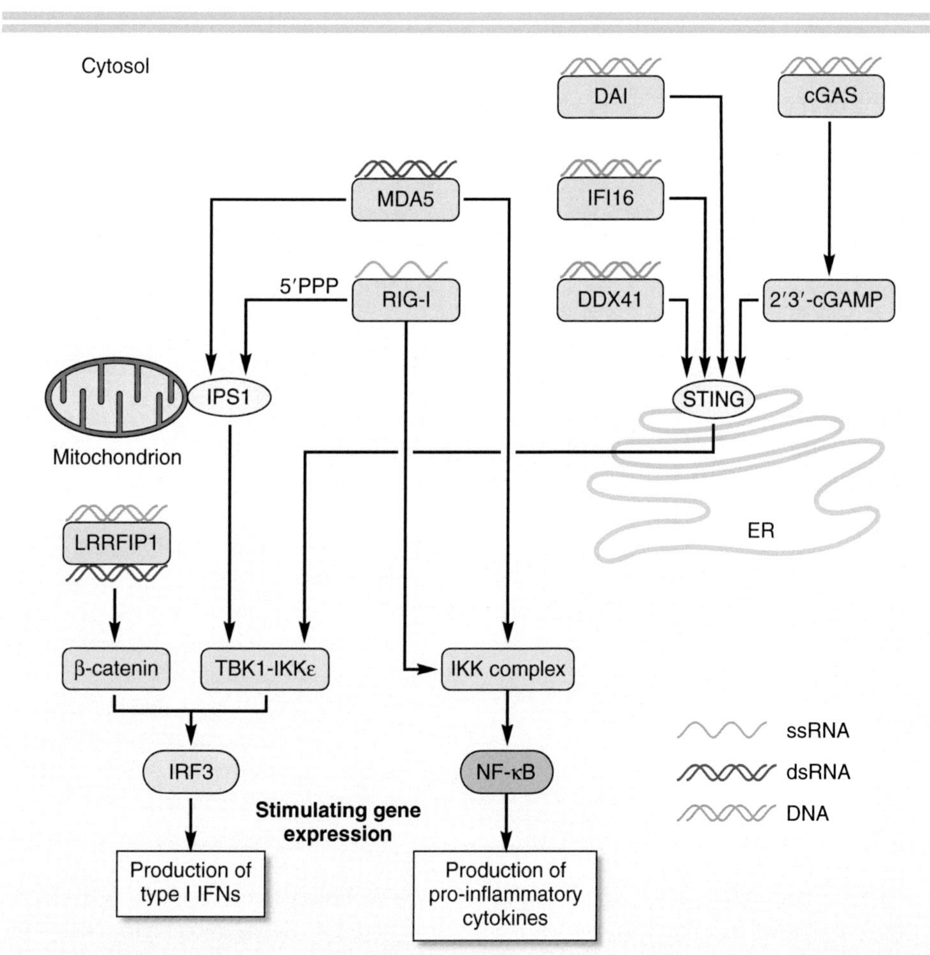

FIGURE 3 Major cytosolic receptors sensing viral infections. (A) Overview of cytosolic receptors sensing viral nucleic acids to stimulate gene expression. The RIG-1 receptor recognizing single-stranded RNA (ssRNA) and the melanoma differentiation-associated gene 5 (MDA5) receptor recognizing double-stranded RNA (dsRNA) are important in sensing viral RNA expressed in the cytosol. Once occupied, they bind to the IPS1 adaptor molecule to lead to the activation of NF-κB and IRF3 and induce the transcription of cytokine genes. There is a large list of known and candidate receptors for dsDNA inappropriately expressed in the cytoplasm, including DAI, DDX41, IFI16, and cGAS. Activated DNX41 and IFI16 along with the products of cGAS, cGAMPs, interact with STING to activate overlapping pathways stimulating cytokine genes. An independent pathway is a result of LRRFIP1 sensing.

cyclic-GMP-AMP (cGAMP) synthase (cGAS). These sensors, along with the products of cGAS function, 2′3′-cGAMP, act through the stimulatory of IFN genes (STING) protein, which is bound to endoplasmic reticular membranes, to stimulate the tank-binding kinase 1 (TBK1) for IRF3 activation and IFN induction. Another DNA receptor is the leucine-rich repeat (in Flightless 1) interacting protein 1 (LRRF1P1), but this receptor works through a different pathway, depending on β-catenin. There appear to be mechanisms for the DNA and RNA sensor pathways to be cross-activated. An example is that RNA polymerase III (Pol III) can also facilitate IFN responses to DNA by transcribing AT-rich DNA into RNA ligands for RIG-I to plug into the pathway activated by the RNA sensor.

Finally, there is a unique group of innate sensors, the NOD-like receptor (NLRs) to respond to different conditions of tissue damage or insult and stimulate the processing and release of a subset of preformed proinflammatory cytokines, that is IL-1 and IL-18. Once stimulated by appropriate ligands, these sensors become part of multiprotein complexes known as inflammasomes, that include ASC (apoptosis-associated speck-like protein containing a caspase-recruitment domain), to bind and activate the pro-caspase-1 enzyme. Functional caspase then acts to process the inactive precursors of IL-1, pro-IL-1, and IL-18, pro-IL-18. Interestingly, a related molecule, apoptosis-associated speck-like protein containing a caspase-recruitment domain (AIM2), has evolved as a DNA sensor to activate the complex and release IL-1 and IL-18 (Figure 4) during viral infection.

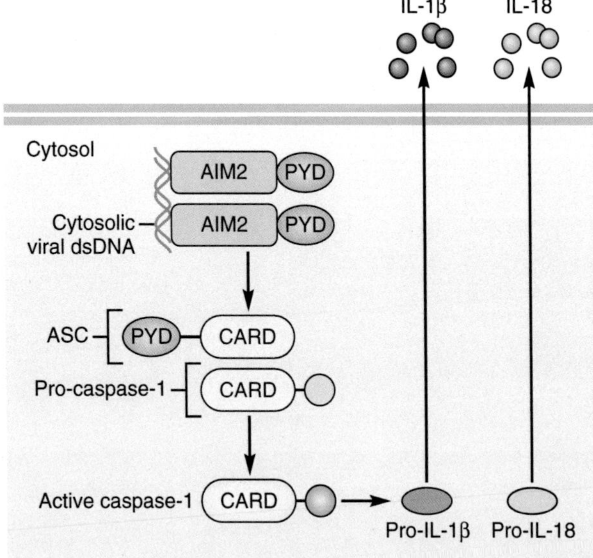

FIGURE 4 Cytosolic sensor inducing processing of inactive cytokine protein precursor molecules into active forms. The AIM2 molecule is a DNA sensor related to a different class of cytosolic receptors, that is NOD-like receptors (NLRs). Upon binding dsDNA, AIM2 interacts with the adaptor molecule ASC, to stimulate the activation of pro-caspase 1 to process and activate the inactive protein precursors of IL-1β, pro-IL-1β, and IL-18, pro-IL-18. ASC, apoptosis-associated speck-like protein containing a CARD; PYD, pyrin domain.

4. INNATE IMMUNE CELLS

Because virtually any nucleated cell can be induced to produce certain subsets of innate cytokines, they can all be considered to be part of innate immunity. The key immune cells—known to play important roles in innate defense to viral infections—are monocytes/macrophages, DCs, and NK cells.

4.1 Monocytes/Macrophages

The concept of innate host cellular defense was first introduced over 100 years ago by Elie Metchnikoff, who recognized the power of "phagocytes" to attack and destroy invading microbes. Macrophages are subsets of phagocytic cells that are derived from hematopoietic precursors in the bone marrow. They are released into the circulation as blood monocytes and enter organs to take up residence as tissue macrophages. Although phagocytic cells were first appreciated for the roles they play in defense against bacterial infections, specialized subsets of monocytes/macrophages are important filters of virus in the spleen and draining lymph nodes. While mediating this function, they produce type 1 IFNs and limit viral spread into other critical host compartments such as the central nervous system.

4.2 Dendritic Cells

In 1973, Ralph Steinman and Zanvil Cohn identified a set of accessory cells required for immune induction, which they named DCs based on their morphology. Originally, it was assumed that DCs were derived from monocytes or macrophages, although it now appears that certain DCs may be derived from a lymphoid rather than a monocytoid cell lineage. They play a central role in both innate and acquired immune responses. Several types of DCs are recognized, but information and terminology are still evolving. Because of their specialized function for producing large levels of the type 1 IFNs (one of their hallmarks), pDCs have key roles in the early defense against viral infections. They appear to develop from a lymphoid precursor, circulate as plasma-like cells, and then migrate into lymphoid tissue. In their resting or immature form, pDCs are poor antigen presenters. However, after exposure to certain stimuli, particular TLRs, they become professional antigen-presenting cells (APCs), and secrete high titers of the type 1 IFNs. If loaded with antigen peptides, they can act as APCs that present antigens to adaptive lymphocytes, but they do not routinely serve this function.

By contrast, other conventional DC subsets (cDCs) arise from circulating monocytes. They migrate to tissues, including lymphoid tissues, where they take up a dendritic morphology. In their resting state, these DCs are effective at acquiring antigens; when activated in peripheral tissues they migrate to draining lymphoid tissues and become effective professional APCs. Part of the process involves induction of the co-stimulatory molecules needed as second signals to stimulate adaptive lymphocytes. Subsets of cDCs have functions in presenting antigen on MHC class II molecules, and have pathways for taking up external antigen for presentation on MHC class I molecules.

4.3 NK Cells

NK lymphocytes were originally identified in 1975 on the basis of their ability to spontaneously kill tumor cells in mice. In contrast to the killing mediated by cytotoxic T cells, this activity was preexisting in populations from cell donors that had not been immunized with the tumor cells. Thus, they were named "natural killer" cells. Subsequent studies identified NK cells as subpopulations of lymphocytes that do not express the B-cell receptors or T-cell receptors (TCRs) for antigen and lack the CD3 molecules associated with TCRs. In humans, NK cells express CD56. Even though they do not express TCRs, NK cells do have functional properties resembling subsets of activated T lymphocytes. In appearance, they are large lymphocytes with cytoplasmic granules. These granules contain perforin—a protein that can produce pores in plasma membranes—and granzymes—that can be delivered to target cells to initiate apoptosis. They also produce a range of cytokines overlapping those produced by activated T cells.

4.3.1 NK Cell Functions

NK cells play important roles in defense against viral infections. Humans with Chédiak–Higashi syndrome, a genetic condition that causes a defect in NK cell killing, are prone to unusual sensitivities to the herpes group viruses. There is also a rare genetic condition that results in the complete absence of NK cells and profound sensitivity to the herpes group viruses. The genetic mutation resulting in this deficiency has been characterized, and individuals with mutations in this gene have similar sensitivity phenotypes. Experimentally, mice with defects in NK cells, NK cell killing, and/or particular NK cell activating receptors, either by virtue of a genetic defect or by transient antibody-mediated depletion, consistently exhibit increased susceptibility to murine cytomegalovirus. Thus, NK cells deliver their antiviral activity by eliminating cellular viral factories, particularly those supporting the replication of herpesviruses.

Type I IFNs stimulate elevated levels of NK cell cytotoxic activity in mice and humans by enhancing their expression of perforin and granzymes. Exposure to type I IFNs also elicits NK cell expression of the TNF-related apoptosis inducing ligand (TRAIL). Because receptors for TRAIL activate cell intrinsic apoptosis pathways, expression of the ligand on NK cells provides a mechanism for killing target cells expressing the death receptors that is independent of NK activating receptors. The full range of antimicrobial functions mediated by NK cells, however, is dependent on more than their cytotoxic activity. NK cells also produce IFN-γ. This cytokine can contribute to the activation of phagocytic cells for enhancing antimicrobial defense during nonviral infections, and activates the inducible nitric oxidase synthase (iNOS) enzyme to deliver antiviral effects within infected cells. In addition, IFN-γ has immunoregulatory functions for enhancing inflammation, activating APCs, driving the maturation of DCs from monocyte precursor cells, and shaping downstream T-cell responses. NK cells can also produce TNF and granulocyte macrophage colony-stimulating factor to deliver biological effects through these cytokines.

4.3.2 Responses to Innate Cytokines

Because receptor cross-linking on cell surfaces is required to initiate apoptosis pathways, NK cell-mediated killing is dependent on expression of an appropriate activating receptor on NK cells and of a reciprocal ligand on the surface of target cells, i.e., receptor-ligand pairs. The profile of innate cytokines induced during infection, however, has profound consequences for the range of NK cell responses elicited during infection (Figure 5). In addition to enhancing NK cell killing functions, the type I IFNs induce IL-15 expression, a cytokine supporting NK cell cycling and blastogenesis. When proinflammatory responses are stimulated through the engagement of TLRs, IL-12 can also be produced. When AIM2 is activated, IL-18 is released. Enhancement of the NK cell killing function induced directly by type I IFNs is delivered through STAT1-dependent signaling pathways. The NK cells, however, also have high levels of STAT4. Induction of IFN-γ is dependent on STAT4, and the type I IFNs can activate STAT4 in NK cells. This only happens for very brief periods of time, however, because the type I IFNs concurrently induce elevated levels of the STAT1 preferentially activated by their receptors. Once STAT1 levels are increased, type I IFN activation of STAT4 is limited. As a result, optimal NK cell IFN-γ production depends on IL-12 with the preferential activation of STAT4 by its receptor. The presence of IL-18 greatly enhances IFN-γ induction by either type 1 IFNs or IL-12. (Figure 5).

4.3.3 NK Cell Activating Receptors

In addition to inducing cytotoxic granule release for target cell killing, stimulation of NK cells through their activating receptors can elicit many of responses induced by cytokines (Figure 5). The NK cell activating receptors are also germ-line sensors of infection, but the gene families are distinct from those for the TLR and cytosolic sensors. They recognize ligands on the outside of the cell. NK cells use the Fc receptor, FcγRIII, to recognize Fc portions of immunoglobulin molecules on cell surfaces and mediate antibody-dependent cellular cytotoxicity (ADCC). There are,

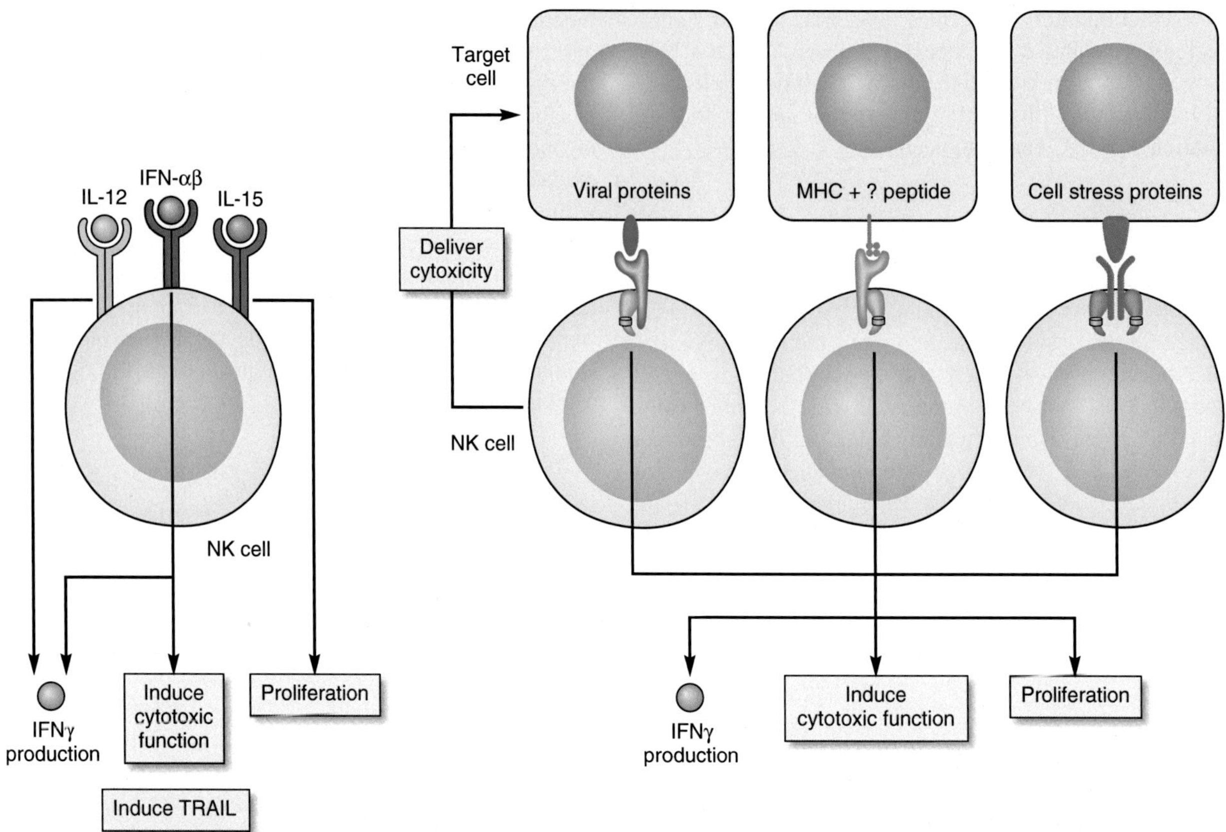

FIGURE 5 Natural killer (NK) cell responses and functions during viral infections. NK cells can be stimulated to produce cytokines, proliferate, and induced to mediate elevated killing through different pathways during viral infections. The innate cytokines IFN-αβ, IL-12, and IL-15 each have important and dominant roles in respectively activating killing, IFN-γ production, and proliferation. Because Type I IFN can only briefly contribute to IFN-γ, the combination is required for peak induction of all responses. Alternative pathways to NK cell activation are elicited through their activating receptors. These can recognize a number of different changes induced in viral-infected target cells, including viral products, changes induced by major histocompatibility complex (MHC) class 1 molecules as a result of viral infection, and host stress proteins induced during infection. These receptors are products of the NK and leukocyte receptor gene complexes. Strong stimulation through these receptors can enhance all three NK cell responses, and is required for NK cell-mediated cytotoxicity. *(Adapted from Lee SH, Miyagi T, Biron CA. Trends In Immunology 2007; 28: 252-259, with permission.)*

however, many other NK cell receptors delivering either activating or inhibitory signals. They are collectively called NK cell receptors (NKRs) and encoded by two large genetic areas identified as the natural killer and the leukocyte receptor gene complexes. Because NK cell activating receptors have short intracellular domains, their induction of signaling requires associated adaptor molecules with intracellular immunoreceptor tyrosine-based activating motifs (ITAMs) to deliver a signal to the cell (Figure 5–6). The DNAX-activating protein of 12 kDa (DAP-12) is a primary adaptor molecule used by the NKR activating receptors. The NKp46 molecule uses an adaptor molecule that is also used to signal through the TCR, CD3zepsilon. Another adaptor molecule with an alternative intracellular activating motif is DAP-10.

The NKR genes are highly polygenic, and in some cases polymorphic, with different individuals having different genes and different numbers of certain genes. The diversity provides a mechanism to recognize a wide range of ligands on target cells and as a result, recognize a wide range of conditions threatening to the host. The largest diversity is in the Ly49 genes in the mouse and the killer Ig-like receptor (KIR) genes in the human. The NKG2D and NKp46 activating receptors are conserved across species. To stimulate NK cells through these activating receptors, appropriate ligands must be expressed on the virus-infected cells. Ligands for the activating receptors can be broadly grouped into three different classes. The first are protein products of the viruses themselves. There are very few examples in this class. The m157 protein of the murine cytomegalovirus recognition by the mouse Ly49H activating receptor is the combination most clearly supported by evidence. The second class is comprised of MHC class I molecules that have been molecularly altered by infection of the target cells. This class has numerous representatives, and is an active area of investigation. The third class is comprised of host stress molecules induced on virus-infected cells. These are ligands for a particular NK cell activating receptor identified as NKG2D.

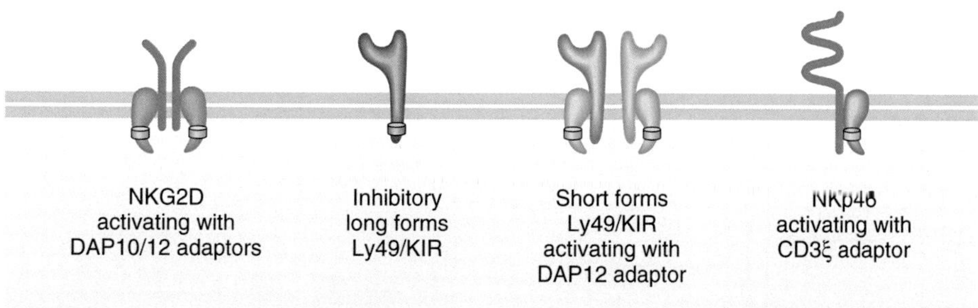

FIGURE 6 Natural killer (NK) receptors mediate inhibitory or activating signals, the balance of which determines NK cell activity. Inhibitory NK cell receptors (in red) bear an immunotyrosine inhibitory motif (ITIM, red disk) in the cytoplasmic domain. Activating receptors (other colors) associate with adaptor proteins that carry immunotyrosine activating motifs (ITAM, green disks). Representative adaptor molecules are DAP12, DAP10, and CD3ξ. *(Adapted from Vidal SM, Khakoo SI, Biron CA. Current Opinion in Virology 2011; 1: 497-512.)*

4.3.4 NK Cell Inhibitory Receptors

In contrast to the activating receptors, inhibitory receptors have longer cytoplasmic domains with their own immunoreceptor tyrosine-based inhibitory motifs to directly deliver a signal to the cell. By recruiting phosphatases, their engagement dampens phosphorylation signals delivered through the activating receptors Figure 6. Different subsets of NK cells have mixtures of activating and inhibitory receptors. Thus, the decision of whether or not NK cells "attack" is determined by the balance between the activating and inhibitory pathways at the site of target cell engagement. Inhibitory receptors recognize particular MHC class I molecules and prevent NK cells from lysing normal cells expressing these self-antigens. In contrast, cells that have reduced levels of MHC class I expression fail to engage the inhibitory molecules and are targets for lysis stimulated through the activating receptors (sometimes called the "missing self" hypothesis). Many viruses will down-regulate the expression of class I MHC by a variety of mechanisms to avoid detection by cells of the adaptive immune system. NK cells appear to provide a backup for such a condition, because infected cells that have down-regulated their class I MHC become more sensitive targets for attack by activated NK cells. However, certain viruses, particularly those with large genomes, encode genes that help to evade NK cell attack, either by expressing proteins mimicking MHC class I molecules to engage inhibitory receptors, or by blocking cell surface expression of the ligands for NK cell activating receptors.

5. DEFENSINS

Multicellular organisms produce "natural" antimicrobial peptides that provide a defense against infection, particularly by bacteria and fungi. Among the known peptides are magainin (frogs), cecropin (silkworms), and the cathelicidins and defensins (best studied in primates and other mammals). Most of these molecules are amphipathic, composed of discrete hydrophobic and cationic (positively charged) domains. Their antibacterial action is explained by the interaction of the positively charged surface of the peptide with the negatively charged heads of the outer leaflets of bacterial membranes, followed by the insertion of the hydrophobic part of the peptide into the lipid bilayer of the bacterial membrane, resulting in membrane disruption. Conversely, the outer leaflet of animal cells carries little net charge, thereby avoiding the initial binding of amphipathic peptides.

There are three subfamilies of defensins, α-defensins found in neutrophils, β-defensins found in epithelial cells, and θ-defensins (found in neutrophils of Old World monkeys). In contrast to the antibacterial action of defensins, the mechanism of their antiviral activity is less clear, perhaps because it is pleiotropic. Some members of the defensin family prevent entry of several herpesviruses, either by preventing attachment to host cells or by blocking postattachment steps in the entry pathway. CD8 T lymphocytes express α-defensins and these are active against HIV-1, although it is controversial whether they are responsible for the anti-HIV activity of CAF (CD8-associated factor).

6. COMPLEMENT AND "NATURAL" ANTIBODIES

The serum of a number of mammalian species exhibits nonspecific antiviral activity that may be mediated by either complement or by "natural" antibodies. The complement system consists of a group of proteins that circulate in the plasma. As shown in Figure 7, these proteins can participate in a biochemical cascade that results in the production of channels in a lipid bilayer, resulting in lysis of viruses or other foreign pathogens. Complement acts as part of the effector mechanism for an induced immune response. However, complement may also act as a nonspecific host defense against certain viruses. For instance, a number of animal retroviruses are lysed by human complement in the absence of antiviral antibody. p15e, a viral protein on the surface of some groups of retroviruses, binds C1q, one of

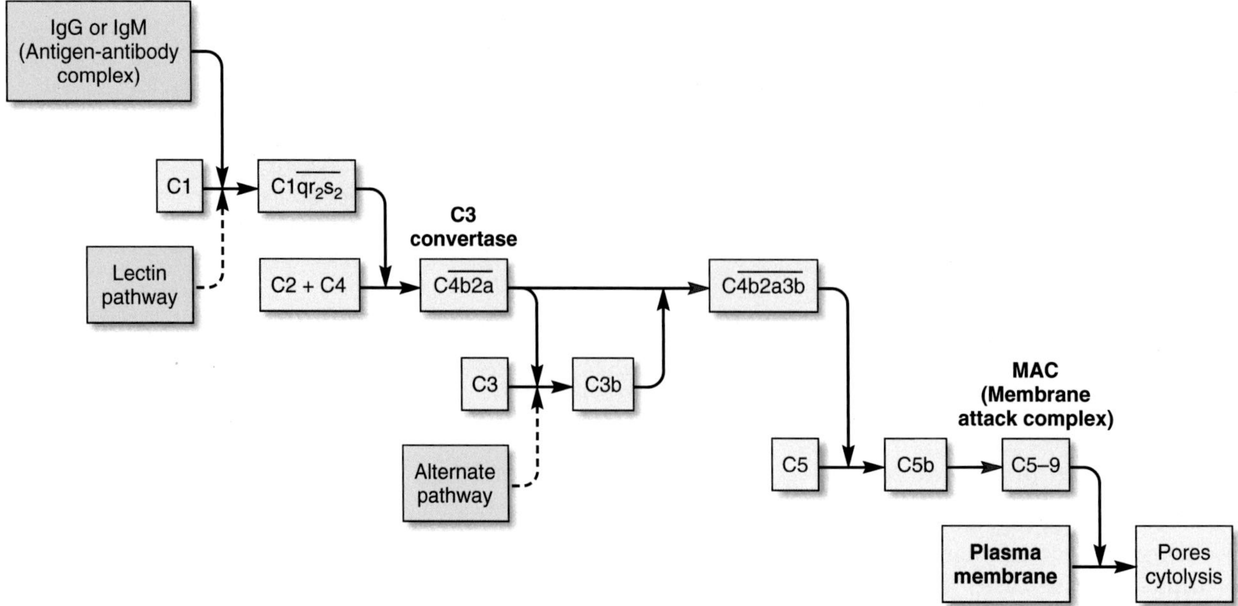

FIGURE 7 The complement cascade, an overview. The cascade can be initiated via the classical pathway, the alternative pathway, or the lectin pathway as indicated. The classical pathway is initiated by binding of antigen to antibody (either IgG or IgM) which produces a conformational change in the Fc portion of the antibody molecule enabling it to bind C1. The alternate pathway is triggered by specific carbohydrate residues on invading microorganisms that activate C3. The lectin pathway is triggered when a mannose-binding lectin (for instance, a θ-defensin) binds a cognate carbohydrate residue (on an invading microbe) followed by aggregation of enzymatic proteins that activate C2 and C4. At several steps in the cascade, a complex of proteins is formed which have catalytic activity (indicated by the overbar). The final complex of C5–9 (membrane attack complex) forms pores in the plasma membrane causing cellular lysis. *(Modified after Abbas AK, Lichtman AH, Pober JS. Cellular and molecular immunology. Saunders, 5th edition, 2003, Philadelphia.)*

the complement proteins, thereby activating the complement cascade. Complement-initiated lysis in the absence of specific antibodies has also been reported for other viruses.

"Natural" antibodies are found in the sera of many animal species and often are directed against foreign antigens to which humans or animals may be exposed. One example is antibodies against a disaccharide consisting of two galactose molecules bound in an α1-3 linkage. This digalactose is not synthesized by humans or Old World primates that lack a galactosyl transferase specific for the α1-3 bond. Because they do not see the linkage as self, primates exposed to this disaccharide—present on ingested proteins—develop antigalactose (α1-3) antibodies. Viruses grown in cells that express the digalactose incorporate this molecule into the carbohydrate side chains of their surface proteins and are neutralized by sera from Old World primates that have never been exposed to the virus itself. As a result, such "natural" antibodies have a role to play in limiting viral spread through different species.

7. RELATIONSHIP BETWEEN INNATE AND ADAPTIVE IMMUNE RESPONSES

After first exposure to a virus, the innate and adaptive arms of immunity act in a synergistic manner to defend the host against acute infection (Table 1). Innate responses can be activated within hours of sensing an invasion, while days to weeks are required for induction of adaptive responses. The cellular constituents of adaptive immunity, T and B lymphocytes, have unique receptors that are the products of rearranged genes. Adaptive immunity therefore has a much narrower focus on epitopes specific for an individual virus, but it requires a vast expansion of a small number of clonal precursor cells, which inevitably takes considerable time. The sequence of innate and adaptive responses is shown in Figure 8 for a model virus infection. For a typical acute viral infection, it requires about 1 week for the induction of acquired responses. During this interval, NK cells are activated and high levels of type 1 IFNs are secreted. These responses hold the infection in check during the induction of specific acquired immunity. The critical role of innate immunity is demonstrated in experimental models in which abrogation of specific components of the innate response potentiates viral infection.

Certain components of the innate response, however, also play a dual role because they constitute the first steps in induction of antigen-specific immunity (see Chapter 5). Typically, at the site of viral invasion, DCs recognize and bind viral PAMPs (mainly ss- and dsRNA in the case of viruses). This leads to activation of the DCs, which secrete proinflammatory cytokines such as TNF, IL-6, and IL-12; the inflammatory focus is a nexus that concentrates cells of

TABLE 1 Comparison of the Dominant Characteristics of Innate and Acquired Immunity. DC: Dendritic Cells; CD: Cluster of Differentiation; NK: Natural Killer

Property	Innate	Acquired
Specificity	BROAD PAMP-based	FINE Epitope-based
Induction time	Hours	Days to months
Requires antigen processing	No	Yes
Involves clonal expansion	No	Yes
Responder cells	Macrophages, DCs, NK cells	$CD4^+$, $CD8^+$ T cells, DCs
Effector cells	Macrophages, DCs, NK cells	$CD8^+$ T cells ($CD4^+$ T cells)
Effector mechanisms	Phagocytosis, cytolysis, chemotaxis, cytokines, chemokines,	Antibodies, cytolysis, cytokines, chemokines
Memory and effector recall	No (not generally thought)	Yes
Long-term persistence	No	Yes

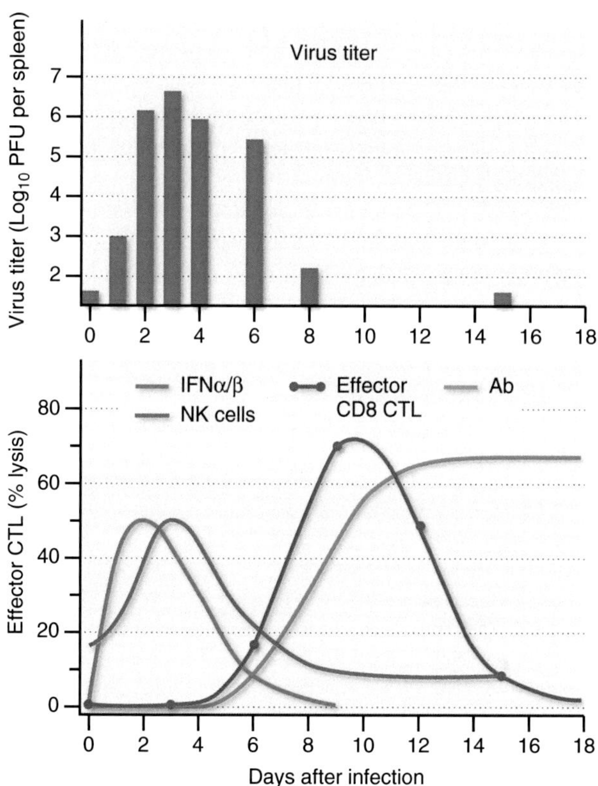

FIGURE 8 The innate responses to an acute viral infection precede the induction of antigen-specific adaptive responses. This diagram shows the sequence of IFN α/β levels, NK cell activity, virus-specific CTLs, and serum antibody (Ab) responses during acute infection with lymphocytic choriomeningitis virus. The IFN, NK, and Ab curves are drawn to an arbitrary scale. *(After Biron CA. Current Opinions in Immunology 1994, 6: 530-538; and Lau LL, Jamieson BD, Somasundarma T, Ahmed R. Nature 1994, 369: 648-652, with permission.)*

the innate system. Activated DCs then migrate from their tissue locations via afferent lymphatics to T-cell zones in draining lymph nodes. Meanwhile, DCs are processing antigen, via two separate pathways, for presentation by MHC class I and class II molecules. In addition, there is an up-regulation of co-stimulatory molecules that provide a "second signal" essential for activation of antigen-specific T lymphocytes. Also, it appears that activated DCs and type 1 IFNs play roles in overriding the potential suppression of Th cells by regulatory T (Tr) lymphocytes, and in shaping the subset of T cell responses most effective in mediating defense against the particular invading virus.

8. INNATE FUNCTIONS IN SUSTAINED OR SUBSEQUENT VIRAL INFECTION

There is still much to be learned about the interactions of the innate and adaptive arms of immunity. The ability of NK cells to use the antibody products of adaptive B lymphocytes to mediate ADCC is a long-standing example of how the innate and adaptive immune system components may interact to deliver antiviral defense. Because NK cells are present without challenge and antibodies are maintained after primary infections or vaccination, it is easy to understand how there would be opportunities for these to interact. Other circumstances supporting the overlap of innate and adaptive immunity are not clearly understood. NK cell functions in the presence of large viral burdens can include an acquired production of IL-10, a cytokine inhibiting immune responses, and the killing of APCs and T lymphocytes to regulate T cell responses to infection. Thus, in addition to their direct antiviral and positive immunoregulatory functions, NK cells have the potential to negatively regulate the magnitude of adaptive responses and protect from immune-mediated disease resulting from this arm of immunity.

The acquisition of IL-10 production by NK cells fits in a new area of research on the conditioning or training of innate immunity. Mouse NK cells conditioned during murine cytomegalovirus infection have epigenetic histone modifications that switch the IL-10 gene from a closed to open state for expression. Global genetic analysis of human effector macrophages (cells trained to enhance or negatively regulate immune responses) has also revealed epigenetic transcriptional programming shifts associated with specific functions. Thus, there is evidence that innate cells can be intrinsically changed at a genetic level to alter their function. It is not yet known how widely conditioning or training is used in different innate cell populations, how long such genetic modifications last, or the consequences of epigenetic programming in innate cells for sequential infections. Continued work probing the training and conditioning of innate cells and the interaction of innate and adaptive immunity will lead to a more complete picture of how these components work together to protect the host against the detrimental consequences of viral infections.

9. INDIVIDUAL VARIATION

The development of new approaches for probing genetic differences in the human population has made it possible to begin to understand why different individuals have different sensitivities to viral infections and virus-induced disease. Characterization of the genetic mutations (or "inborn errors") in pediatric patients with primary immunodeficiencies, resulting in extreme sensitivities to infections, have demonstrated the importance of particular functional proteins in innate cytokine receptors, STATs, and even TLRs for resistance to a narrow or wide range of infectious agents. Very recent reports are demonstrating that particular genetic variants in human can have broad effects on the composition of complex molecular circuits stimulated by TLRs and innate cytokines. Taken together, this work is revealing the consequences of genetic variation in the population that explain—in part—why humans experience infections differently.

10. REPRISE

Innate immunity is an important component of the host defense against infection. It is the only host defense system in nonvertebrate animals and synergizes adaptive immunity in vertebrates. Virtually all cells can contribute to innate immunity by producing certain innate cytokines, particularly the type 1 IFNs, and by responding to these cytokines to induce new and elevated intracellular molecular mechanisms for fighting off infections. Macrophages, DCs, and NK cells, however, are the main immune cellular constitutes responsible for innate responses. Macrophages and DCs carry PRRs that respond to PAMPs, motifs common to large classes of infectious agents but often absent in eukaryotic organisms. For viruses, ss- and dsRNA—distinct from normal cellular RNAs—are the major PAMPs. Once PRRs have been ligated, they set off intracellular biochemical cascades that lead to cellular activation. Activated cells initiate phagocytosis and secretion of many cytokines, such as type I IFN, that in turn induce inflammation and other antiviral responses. Other cytosolic receptors function to detect viruses and induce IFN production by a wider range of infected cell types. NK cells carry unique sensors in the form of activating and inhibitory receptors. The balance of the engagement of these receptors acts to protect normal cells from the detrimental effects of NK cells while activating them to kill virus-infected target cells.

Innate immunity is initiated within hours and provides a rapid array of defenses, whereas the antigen-specific adaptive immune responses are induced during the first weeks after infection. Furthermore, selected innate responses—such as the activation of DCs, initiation of antigen processing, migration to draining lymph nodes, up-regulation of co-stimulatory molecules, and the composition of the early cytokine profiles to shape downstream adaptive responses—are the essential first steps in induction of acquired immunity. Although there is a detailed understanding of their general functions during acute primary viral infections, much remains to be learned about how the innate and adaptive arms of immunity interact under conditions of sustained viral burdens and/or during subsequent viral infections, and about how genetic variations in the human population uniquely shape their responses and functions.

FURTHER READING

Books, Chapters, and Reviews

Abbas AK, Lichtman AH, Pober JS. *Cellular and Molecular Immunology, 4th edn*, Saunders, Philadephia, 2000.

Bhat N, Fitzgerald K. Regonition of cytosolic DNA by cGAS and other STING-dependent sensors. *European Journal Immunology* 2013; 44: 634–640.

Casanova JL, Holland SM, Notarangelo LD. Inborn errors of human JAKs and STATs. *Immunity* 2012; **36**: 515–528.

Carroll MC. The complement system in regulation of adaptive immunity. *Nature Immunology* 2004; **10**: 981–986.

Desmet CJ, Ishii KJ. Nucleic acid sensing at the interface between innate and adaptive immunity in vaccination. *Nature Reviews Immunology* 2012; **12**: 479–491.

Garcia-Sastre A, Biron CA. Type 1 interferons and the virus-host relationship: a lesson in détente. *Science* 2006; **312**: 879–882.

Iwasaki A, Medzhitov R. Innate responses to viral infections. In: Fields Virology, 6th Edition. DM Knipe & PM Howley, eds. Walter Kluwer/Lippincott Williams & Wilkins, New York, NY, 2013, p. 189–213.

Janeway CA Jr. Approaching the asymptote: Evolution and revolution in immunology. *Cold Spring Harbor Symposia in Quantitative Biology* 1989; **1**: 1–13.

Kallal LE, Biron CA. Changing partners at the dance: Variations in STAT concentrations for shaping cytokine function and immune responses to viral infections. *JAK-STAT* 2013; **2**: e23504.

Kanneganti T-D. Central roles of NLRs and inflammasomes in viral infection. *Nature Reviews Immunology* 2010; **10**: 688–698.

Kawai T, Akira S. The role of pattern-recognition receptors in immunity: update on Toll-like receptors. *Nature Immunology* 2010; **11**: 273–284.

Klotman ME, Chang TL. Defensins in innate antiviral immunity. *Nature Reviews Immunology* 2006; **6**: 447–456.

Lau LL, Jamieson BD, Smasundarma T, Ahmed R. Cytotoxic T cell memory without antigen. *Nature* 1994; **369**: 648–652.

Lee SH, Miyagi T, Biron CA. Keeping natural killer cells in highly regulated antiviral warfare. *Trends in Immunology* 2007; 28: 252–259.

Sadler AJ, Williams BRG. Interferon-inducible antiviral effects. *Nature Reviews Immunology* 2008; **8**: 559–568.

Turelli P, Trono D. Editing at the crossroad of innate and adaptive immunity. *Science* 2005; **307**: 1061–1065.

Vidal SM, Khakoo SI, Biron CA. Natural killer cell responses during viral infections: flexibility and conditioning of innate immunity by experience. *Current Opinion in Virology* 2011; **1**: 497–512.

Original Contributions

Bennasser Y, Le s-Y, Benkirane M, Jeang K-T. Evidence that HIV-1 encodes an siRNA and a suppressor for RNA silencing. *Immunity* 2005; **22**: 607–619.

Biron CA, Byron KS, Sullivan JL. Severe herpesvirus infections in an adolescent without natural killer cells. *New England Journal of Medicine* 1989; **320**: 1731–1735.

Brown MG, Dokun AO, Heusel JW, Smith HRC, Beckman DL, Blattenberger EA, Dubbelde CE, Stone LR, Scalzo AA, Yokoyama WM. Vital involvement of a natural killer cell activation receptor in resistance to viral infection. *Science* 2001; **292**: 934–937.

Diebold SS, Kaisho T, Hemmi H, Akira S, Reis e Sousa C. Innate antiviral responses by means of TLR7-mediated recognition of single-stranded RNA. *Science* 2004; **303**: 1529–1531.

Goldszmid RS, Caspar P, Rivollier A, White S, Dzutsev A, Hieny S, Kelsall B, Trinchieri G, Sher A. NK cell derived interferon-γ orchestrates cellular dynamics and the differentiation of monocytes into dendritic cells at the site of infection. *Immunity* 2012; **36**: 1047–1059.

Iannacone M, Moseman EA, Tonti E, Bosurgi L, Junt T, Henrickson SE, Whelan SP, Guidotti LG, von Andrian UH. Subcapsular sinus macrophages prevent CNS invasion on peripheral infection with a neurotropic virus. *Nature* 2010; **465**: 1079–1083.

Isaacs A, Lindenmann J. Virus interference. I. The interferon. II. Some properties of interferon. *Proceedings of the Royal Society London B* 1957; **147**: 258–273.

Lee MN, Ye C, Villani A-C, Raj T, Li W, Eisenhaure TM, Imboywa SH, Chipendo PI, Ran FA, Slowikowski K, Ward LD, Raddassi K, McCabe C, Lee MH, Frohlich IY, Hafler DA, Kellis M, Raychaudhuri S, Zhang F, Stranger BE, Benoist CD, De Jager PL, Regev A, Hacohen N. Common genetic variants modulate pathogen-sensing responses in human dendritic cells. *Science* 2014; **343**: 1246980.

Li Y, Lu J, Han Y, Fan X, Ding SW. RNA interférence functions as an antiviral immunity mechanism in mammals. *Science* 2013; **342** : 231–234.

Medzitov R, Preston-Hurlburt P, Janeway CA Jr. A human homologue of the Drosophila Toll protein signals activation of adaptive immunity. *Nature* 1997; **388**: 394–397.

Miyagi T, Gil MP, Wang X, Louten J, Chu WM, Biron CA. High basal STAT4 balanced by STAT1 induction to control type 1 interferon effects in natural killer cells. *Journal of Experimental Medicine 2007*; **204**: 2383–2396.

Nguyen KB, Watford WT, Salomon R, Hofmann SR, Pien GC, Morinobu A, Gadina M, O'Shea JJ, Biron CA. Critical role for STAT4 activation by type 1 interferons in the interferon-γ response to viral infection. *Science* 2002; **297**: 2063–2066.

Ochsenbein AF, Fehr T, Lutz C, Suter M, Brombacher F, Hengartner H, Zinkernagel RM. Control of early viral and bacterial distribution and disease by natural antibodies. *Science* 1999; **286**: 2156–2159.

Orzalli MH, Knipe DM. Cellular sensing of viral DNA and viral evasion mechanisms. *Annual Review Immunology* 2014; **68**:477–492.

Poltorak A, et al. Defective LPS signaling in C3H/HeJ and C57BL/10ScCr mice: mutations in the Tlr4 gene. *Science* 1998; **282**: 2085–2088.

Rathinam VA, Jiiang Z, Waggoner SN, Sharma S, Cole LE, Waggoner L, Vanaja SK, Monks BG, Ganesan S, Latz E, Hornung V, Vogel SN, Szomolanyi-Tsuda E, Fitzgerald KA. The AIM2 inflammasome is essential for host defense against cytosolic bacteria and DNA viruses. *Nature Immunology* 2010; **11**: 395–402.

Saeed S, Quintin J, Kerstens HH, Rao NA, Aghajanirefah A, Matarese F, Cheng SC, Ratter J, Berentsen K, van der Ent MA, Sharifi N, Janssen-Megens EM, Ter Huurne M, Mandoli A, van Schaik T, Ng A, Burden F, Downes K, Frontini M, Kumar V, Giamarellos-Bourboulis EJ, Ouwehand WH, van der Meer JW, Joosten LA, Wijmenga C, Martens JH, Xavier RJ, Logie C, Netea MG, Stunnenberg HG. Epigenetic programming of moncyte-to-macrophage differentiation and trained innate immunity. *Science* 2014; **345**: 1578.

Schoggins JW, Wilson SJ, Panis M, Murphy MY, Jones CT, Bieniasz P, Rice CM. A diverse range of gene products are effectors of the type 1 interferon antiviral response. *Nature* 2011; **472**: 481–485.

Steinman RM, Cohn ZA. Identification of a novel cell type in peripheral lymphoid organs of mice. I. Morphology, quantitation, tissue distribution. *Journal of Experimental Medicine* 1973; **137**: 1142–1162.

Sun L, Wu J, Su F, Chen X, Chen ZJ. Cyclic GMP-AMP synthase is a cytosolic DNA sensor that activates the type I interferon pathway. *Science* 2013; **339**: 786–791.

Tarrio, ML, Lee SH, Fragoso MF, Sun HW, Kanno Y, O'Shea JJ, Biron CA. Proliferation conditions promote intrinsic changes in NK Cells for an IL-10 Response. *Journal of Immunology* 2014; **193**:354–363.

Waggoner SN, Cornberg M, Selin LK, Welsh RM. Natural killer cells act as rheostats modulating antiviral T cells. *Nature* 2011; **481**: 94–398.

Yoneyama M, Kikuchi M, Natsukawa T, Shinobu N, Imaizumi T, Miyagishi M, Taira K, Akira S, Fujita T. The RNA helicase RIG-I has an essential function in double-stranded RNA-induced innate antiviral responses. *Nature Immunology* 2004; **5**: 730–737.

Chapter 5

Adaptive Immunity

Neutralizing, Eliminating, and Remembering for the Next Time

E. John Wherry[1], David Masopust[2]

[1]*Department of Microbiology, Perelman School of Medicine, University of Pennsylvania, Philadelphia, PA, USA;* [2]*Department of Microbiology and Immunology, School of Medicine, University of Minnesota, Minneapolis, MN, USA*

Chapter Outline

This chapter will provide a short overview of salient aspects of the adaptive immune response, with particular focus on elements relevant to viral pathogenesis. We open with a description of the cellular players, describe induction and kinetics of the antibody and cellular immune responses, and then turn to immunity as a host defense. We also weave through this chapter the utility of omics-based studies that are enhancing the understanding of how adaptive immunity influences the host response to viral infection. Specific examples of host immune responses illustrate these general principles. This short review is designed to refresh the reader's memory about the organization and function of the immune system. Students who are not familiar with basic immunology may wish to consult one of the many excellent introductory texts (see Further reading).

In the course of this chapter we will address a number of salient questions including:

- What is the difference between CD4+ and CD8+ lymphocytes?
- Why is the innate response important for induction of adaptive immunity?
- Why do we need two adaptive systems, antibody and effector lymphocytes? What are their respective roles? How do they synergize each other?
- How does the adaptive response counter the two different states of a virus, the extracellular infectious virion vs the intracellular replicating virus?
- What determines the difference between mucosal tolerance and mucosal immunity?
- What is the relationship between the microbial environment and adaptive immune responses?
- What are the differences between responding to a primary viral infection and to re-exposure a second time? what is the respective role of the two arms of the immune response in these two situations?
- How does the adaptive response clear the virus invader and "cure" the infection? How do some viruses "escape" these responses and lead to persistent infection?

1. OVERVIEW OF THE IMMUNE SYSTEM

1.1 The Players: Cells of the Immune System

Lymphocytes are responsible for both the induction and expression of adaptive immunity. There are two major classes of lymphocytes, B cells and T cells. T cells are so named because they develop in the thymus, while B cells are named after the Bursa of Fabricius, an organ found in

Viral Pathogenesis. http://dx.doi.org/10.1016/B978-0-12-800964-2.00005-7

birds (but not in mammals) where B lymphocytes undergo early maturation. In mammals, the same steps in B cell maturation occur in the bone marrow. B cells and T cells possess genetically rearranged and highly diverse antigen receptors imparting specificity to these cells.

All types of lymphocytes are constantly produced from precursor stem cells residing in the bone marrow. B lymphocytes emerge from the bone marrow and migrate directly to lymphoid tissues. Progenitor T cells migrate from the bone marrow to the thymus where they undergo maturation that prepares them to respond to an immune stimulus; they emerge from the thymus as "naïve" T cells that migrate mainly to spleen and lymph nodes.

Each B or T cell bears molecules on its surface that equip it to discharge its specialized role in the immune response (Sidebar 1). B cells carry antibody molecules that bind "epitopes" or individual immune determinants on foreign antigens. Antigens recognized by B cells may be either proteins, carbohydrates, or nucleic acids, and their epitopes consist of a small cluster of amino acids, sugars, or nucleic acids, respectively. An epitope would include about 10 amino acids, and antibodies can recognize both linear and conformational epitopes. Antibody molecules are encoded in genetic determinants that are "constructed" by rearrangement of germ line genes during maturation. Due to this exquisitely specialized developmental arrangement, antibodies can encode up to $\sim10^{10}$ different specificities.

T lymphocytes carry a T cell receptor (TCR) that is a heterodimer composed of an α and a β polypeptide. TCRs have a structure that is roughly analogous to the antibody molecule, but—in contrast to antibodies—recognizes amino acids within a peptide of 8–11 amino acids bound to a specialized groove on the surface of an MHC Class I molecule (10–30 amino acids for MHC Class II molecules). CD4+ cells carry TCRs that recognize antigen presented by MHC Class II but not MHC Class I molecules, and, conversely, CD8+ T cells recognize antigens presented by MHC Class I molecules, but not MHC Class II molecules. TCRs are formed by rearrangements of genetic determinants expressed in germ line cells, rearrangements that occur during maturation of lymphocytes. Due to this developmental process, lymphocytes can encode up to $\sim10^9$ different specificities.

- What is the difference between CD4+ and CD8+ lymphocytes?

T lymphocytes can be divided into two major categories, CD4+ and CD8+ T cells, which can be identified by their surface expresson of either CD4 or CD8 molecules that are used as markers for purposes of enumeration or cell sorting. CD4+ T cells act as "helper" cells providing signals that induce B cells or CD8+ T cells to proliferate in response to an antigen presented by professional APCs. The signals provided by CD4+ helper cells involve ancillary surface molecules that interact with cognate ligand molecules on the surface of the antigen-responding CD8+ T cells or B cells.

CD4+ cells can be divided into several subtypes defined by specific cytokines produced and functions performed, including Th_1, Th_2, Th_9, Th_{17}, T_{FH}, and T_{REG} cells. Th_1 cells secrete large amounts of IFN γ, and IL-2, drive the clonal expansion of CD8+ cells and in some cases can be directly antiviral. Th_2 cells secrete large amounts of IL-4, IL-5, and IL-13. Th_2 cells mainly play a role in parasitic infections and allergy. Th_9 cells produce IL-9 and IL-10, are thought to be closely related to Th_2 cells and also have a role in responses to parasites as well as allergic and autoimmune stimuli. Th_{17} cells make the cytokine IL-17—and in some cases IL-22—and are associated with responses to fungal or bacterial infections. T_{FH} cells express IL-21 and provide direct help to drive B-cell expansion, differentiation and antibody production through initiation and support of the germinal center reaction. T_{REG} cells express the transcription factor FoxP3 and negatively regulate responses of other T cells and B cells.

Sidebar 1 Comparison of B and T cell immune responses. CTL: cytolytic T lymphocyte

	B cell response	CD4+ T cell response	CD8+ T cell response
Type of immunity	Humoral	Cellular	Cellular
Precursor cell	B lymphocyte	CD4+ precursor	CD8+ precursor
Effector cell	Plasma cell	CD4+ helper	CD8+ CTL
Receptors recognize antigenic epitopes presented as	Linear and conformational epitopes on virions and virus-infected cells	Antigenic peptides on class II molecules	Antigenic peptides on class I molecules
Mediator molecules	Immunoglobulins (Igs)	Cytokines	Perforins, Granzymes, Cytokines
Persistence of effectors	Yes	No	No
Anamnestic (memory) response	Yes	Yes	Yes

CD8+ lymphocytes, when activated, can become cytotoxic T lymphocytes (CTL), the effector cell of the CD8+ lineage. CTL initiate a local response that includes two components, a lytic attack on target cells carrying a foreign epitope, and the production of cytokines that can attract inflammatory cells to the local site, resulting in an indirect attack on the invading parasite.

Natural Killer (NK) and Innate Lymphoid Cells (ILC) are two types of lymphocytes that are thought to lack diverse antigen receptors. NK cells are a critical cell type for early control of viral infections and tumors, whereas ILC may function in diverse microbial interactions and have a role during the tissue repair phase of some viral infections.

Other important cells involved in immune responses are monocyte/macrophages and dendritic cells (DCs). Monocytes arise from stem cell precursors in the bone marrow, and leave the bone marrow to circulate as blood monocytes, following which they may enter tissues to reside as tissue macrophages. Dendritic cells have several lineages and are found in lymphoid tissues as well as peripheral tissue sites such as skin and mucosal tissues. Dendritic cells and macrophages act as "professional" antigen presenting cells (APCs), which are specially equipped to initiate immune induction, since they express both MHC Class I and Class II molecules. Dendritic cells are physically localized to portals of pathogen entry including the skin and mucosal surfaces and express germline-encoded receptors for molecular patterns associated with pathogens.

1.2 Induction of the Immune Response

- Why is the innate response important for induction of adaptive immunity?

The innate immune response "tees up" the adaptive response. Upon viral entry, DCs become infected or take up virus particles. As a result, receptors for pathogen associated molecular patterns such as Toll Like Receptors (TLRs)—that can recognize single stranded RNA, viral DNA or other moieties—are engaged. Many classes of receptors and pathways like TLRs are now known to activate DCs and other professional APCs. These include cell surface receptors, Fc receptors that bind antibody, and intracellular sensors for viral products such as Nod-like receptors and cytosolic helicases like RIG-I. The signaling events downstream of these receptors activate DCs, and cause a cascade of critical events, including rapid reprogramming of DCs, which now leave peripheral tissues and migrate to the draining lymph node and spleen. In addition, activation of DCs by these innate signaling events causes upregulation of costimulatory proteins such as CD80 and CD86, as well as the production of inflammatory cytokines, including type 1 IFN and IL-12. Migratory DCs home to sites within spleen and lymph nodes at locations that optimize their ability to interact and efficiently initiate T- and B-cell activation. These steps, including antigen capture, trafficking to lymph nodes, and interaction with helper and effector lymphocytes, are critical first steps in induction of the adaptive immune response.

Induction of antibody (the B lymphocyte pathway) involves two complementary components. (1) It begins with the binding of foreign proteins to the immunoglobulin molecules expressed on the surface of naïve B cells. Naïve B cells are already programmed genetically to express Ig molecules with a single antigenic specificity, ie, they will recognize only a single epitope of the multitude of determinants on foreign proteins. (2) Meanwhile, as result of the innate response, professional APCs bind the same foreign protein, which is endocytosed, digested to oligopeptides, and "loaded" onto Class II MHC molecules on the surface of APCs. CD4+ lymphocytes whose TCRs recognize the specific peptide presented by the Class II molecules on the APCs—and on B cells—will be stimulated to release cytokines (IL-21). These two signals—antigen recognition via immunoglobulin and cytokine stimulus—drive the B cells to proliferate and differentiate. Mature B cells (plasma cells) will then secrete the immunoglobulin for which they have been programmed.

The T cell immune response also follows a stereotyped sequence of events, although there are many variations. The DC plays a key role in delivering three key signals leading to the full activation and differentiation of T cells.

Signal 1 is delivered in the form of antigen presentation to activate the T cell through the TCR. The DC can process protein antigens by two pathways, an exogenous and an endogenous pathway. The *endogenous* pathway may be utilized if the virus can infect the APC, and begins with the expression of a foreign protein in the cytosol of the professional APC. Viral proteins synthesized in the cytosol of the APC are delivered to the proteasome, a complex cytoplasmic organelle, which digests the protein into peptides and delivers these peptides (mainly 9–11 mers) across internal membranes into the endoplasmic reticulum, where they are loaded onto MHC Class I molecules. The *exogenous* pathway involves endocytosis of the protein, digestion into small oligopeptides in endolysosomes, and "loading" of these 10–30 mers onto the antigen-binding groove on MHC Class II molecules. MHC Class I molecules are expressed on most cells, while MHC Class II molecules are expressed mainly on professional APCs.

Signal 2 is delivered to T cells in the form of costimulation. Costimulatory receptors on T cells, such as CD27, CD28 and ICOS, bind to ligands on activated APC such as CD70, B7.1 and B7.2 or ICOSL. Signal 2 augments the TCR signal, decreases the threshold for full T-cell activation, and enhances downstream activation of cytokine genes and metabolism. CD4+ helper cells play a variable role in

CD8+ induction, and are most important for the generation of the memory subset (rather than the effector subset) of CD8+ responding lymphocytes.

Signal 3 is provided by inflammatory cytokines such as IFN-1α, IL-12, IL-1, IL-33 and others. These inflammatory cytokines can be produced by activated APC directly, or by other cells in the environment, following innate recognition of pathogens. When T cells are unable to sense signal 3 inflammatory cytokines, initial activation occurs, but the T cells are unable to survive and sustain proliferative expansion. It is thought that signal 3 is important to link the level of ongoing inflammation (a proxy for pathogen replication) to the magnitude of the effector T cell response.

If antigen-responsive T cells receive appropriate signals 1, 2, and 3 provided by DCs and/or other cells in the environment, they undergo clonal expansion, and differentiate into effector cells.

A virus infection will simultaneously induce antibody and a cellular immune response. The balance between the two responses will be influenced by the relative proportion of T_H1 and T_{FH} cells participating in the immune response. The effect of T_H1 and T_{FH} cells is mediated, at least in part, by the cytokines that they secrete IL-2 driving CD8+ T cells, while IL-21 drives B cells.

1.3 Genetic Techniques to Dissect the Immune Response

Beginning about 1990, a series of methodological advances were introduced that have made it possible to manipulate the mouse genome so as to insert or delete individual genes (see Chapter 10, Animal models). In turn, this has led to a revolution in immunobiology, since it has become possible—in a whole animal—to probe the specific role of each of the vast array of genes that control the development and function of the immune system. Recent advances in these approaches using genome manipulation with CRISPR/Cas9 technology (see Chapter 12, The virus-host interactome) have made it increasingly easy to generate genetic tools to study antiviral immunity. These techniques are now being used to analyse the immune response to viral infections in much more detail than was possible in the past. Methods that are frequently used to manipulate the activity of individual genes include:

- Transgenes. The DNA sequence of a known gene is engineered to introduce a desired change in its sequence that is randomly integrated into the recipient DNA by nonhomologous recombination, so that it does not replace the corresponding normal gene.
- "Knockout." The DNA sequence of a known gene is engineered to introduce an alteration that renders it nonfunctional. The "knockout" construct undergoes homologous recombination so that it substitutes for and "knocks out" the target gene.
- "Knockin." The same methods used to produce "knockout" mice are used to introduce an engineered gene in place of its normal counterpart.
- Inducible "knockout." The introduced "knockin" target gene construct includes an additional sequence (such as loxP in the loxP-Cre recombination system) and a gene (Cre, a site-specific recombination enzyme) that can be induced to "knockout" the target gene.
- Small interfering RNA (siRNA). A double-stranded siRNA is synthesized with a sequence identical to a sequence within the target mRNA. When introduced in vivo, the siRNA can lead to the destruction of the target mRNA and reduce expression of the corresponding protein (see Chapter 13, Host genetics).

1.4 Omics Approaches to the Adaptive Immune Response

Established human vaccines induce quite different immune responses, which is not surprising considering the variable formulations used. For instance, live attenuated viruses evoke a different response from inactivated virus vaccines. What are the intermediate steps involved in these different host responses? Pulendran and colleagues have pioneered an approach using transcriptional profiles of human blood (Li et al., 2014). By collating a vast number of blood transcriptomes, they identified several groups of responses (complement; inflammatory; pro-inflammatory cytokines; interferon α) and then compared these transcriptome profiles in subjects who received different vaccines. There were major differences in the blood transcriptome responses to an inactivated viral vaccine (influenza) and a live virus vaccine (17D yellow fever). The inactivated virus induces complement and inflammatory transcriptomes, while the live virus induces interferon transcriptomes.

Why is this significant? Traditional assays of adaptive immune responses have focused on the endpoint, that is responses to specific viral antigens. However, this approach tends to overlook "the immunologist's dirty little secret," that the response to an immunogen is very dependent upon the context of antigen presentation. The omics data identify early differences in the innate response to different immune stimuli that—in turn—have a major impact on the subsequent adaptive response. This approach has profiled the complex innate response to an immunogenic stimulus as an important determinant of the subsequent adaptive response.

The omics analytic framework is contributing to other key aspects of adaptive immunity, a few of which are briefly noted below.

- an understanding of how host genetic determinants alter the adaptive immune response, using a broad omics approach to explore differences in animals that vary in their susceptibility to a challenge virus (Rasmussen et al., 2014; Ferris et al., 2013);

- an understanding of the transcriptional pathways that distinguish functional memory CD8+ T cells from exhausted CD8+ T cells, which are critical for clearance of a virus invader and for secondary protective immunity (Doerring et al., 2012; Crawford et al., 2014);
- an understanding of the impact of aging on adaptive immunity, likely mediated in part via differences in innate responses that are best studied using omics methodology.

Omics approaches are explored in depth in Chapter 11, Systems virology and Chapter 12, the Virus-host interactome.

- Why do we need two adaptive systems, antibody and effector lymphocytes? What are their respective roles? How do they synergize each other?

2. ANTIBODY

2.1 Measures of Antibody

There are many methods to measure antibody to viruses, and both the kinetics of the response and its biological significance depend upon the assay used. The canonical assay is the neutralization test, in which antibody is tested for its ability to reduce viral infectivity. This test depends on the availability of a convenient method to measure viral infectivity, often a plaque assay. One common technique involves the use of a single viral inoculum, such as 100 PFU (plaque forming units); serial dilutions of a test antibody are tested to determine the highest dilution that will reduce the plaque count by 50%. However, neutralization tests cannot be used for some important viruses, such as hepatitis B and C viruses, that cannot readily be grown in cell culture.

There are many alternative assays that measure the ability of the antibody to bind to viral antigens, including hemagglutination inhibition, immunofluorescence, Western blot, and ELISA (enzyme linked immunosorbent assay). Of these, the most commonly used is the ELISA assay, which can readily be adapted to quantitation, automation, and rapid throughput. An antigen, either whole virus, a viral protein, or a viral peptide, is bound to a substrate, and then incubated with serial dilutions of test antibody; adherence of the test antibody is determined with a conjugated antiserum directed against immunoglobulin of the species under test.

In contrast to the ELISA, the Western blot is a qualitative test that provides information about the specificity of the test antibody. Proteins from a viral lysate are separated, often using polyacrylamide gel electrophoresis, and cross-linked to a cellulose strip. An unknown serum is tested for its ability to bind to any of the proteins on the strip, and the reaction is "developed" with a labeled antisera against immunoglobulin, as for the ELISA.

Antibody production can also be measured at a cellular level. In the ELISPOT assay, cells—including plasma cells—are prepared from blood or lymphoid tissues, and overlaid on a surface to which a target antigen has previously been bound. Antigen-specific antibody released by specific plasma cells binds to the cognate antigen, and the "focus" is developed using a variation of the method used in ELISA assays. This assay permits the counting of antibody-secreting cells (ASCs) and can be employed for studies of the dynamics of the antibody response.

- How does the adaptive response counter the two different states of a virus, the extracellular infectious virion versus the intracellular replicating virus?

2.2 Effector Functions of Antibody

Anti-viral antibodies can subserve host defenses in several different ways. First, they can bind to virus and neutralize it. There are multiple mechanisms of neutralization, but the most common is to coat the virion and prevent its attachment to receptors on permissive host cells. A minority of neutralizing antibodies bind to the virus but do not prevent attachment to the cell; instead they interfere with later steps in viral entry. Many antibodies that bind to viral proteins fail to neutralize, usually because the specific epitope is not expressed on the surface of infectious virions.

In addition to reducing infectivity of virions, antiviral antibodies can act in several other ways. Many antibodies will bind C1q, the first of the complement proteins, and initiate lysis of the virion. Complement activation also promotes inflammation, thus enhancing local immune activity. Also, many antibodies will bind to the Fc receptors on Kupffer cells, macrophages and other leukocytes. Complexes consisting of virions bound by antibodies will be more efficiently phagocytosed and degraded than free virions, and this process is sometimes called opsonization. Finally, antibodies can "arm" NK cells that carry receptors for the Fc domain of immunoglobulin molecules. The process of ADCC (antibody-dependent cell-mediated cytolysis) then results in the destruction of infected cells.

In general, the ability of antibodies to protect in vivo correlates with their neutralizing activity. However, this correlation is not absolute. The mechanism whereby non-neutralizing antibodies protect is not well understood but, in some instances, may be mediated by the clearance of virus–antibody complexes by the reticuloendothelial system.

2.3 Kinetics of the Antibody Response

Antibodies can be measured in different bodily fluids, such as blood (plasma or serum), and mucosal fluids (nasal, throat, bronchial washes, feces, semen, genital washes). There are several classes of antibodies. Serum contains IgG (the principal class), IgM, and IgA, most of which is derived from plasma cells in bone marrow, lymph nodes, and spleen. Mucosal fluids contain IgA (locally produced by plasma cells in mucosa-associated lymphoid tissues or MALT) and IgG (partly a transudate of serum IgG and partly produced in MALT).

An example of the humoral (serum) antibody response is shown in Figure 1, which illustrates that the dynamics of the responses are quite different, according to antibody class. In general, serum IgM appears rapidly (1–2 weeks) and disappears in about 3 months, while IgG appears more slowly (2–4 weeks), peaks at 3–6 months, and then gradually wanes although it may persist at submaximal levels for a lifetime.

The ELISPOT assay has been used to trace the location and kinetics of plasma cells (mature antibody-producing B cells) As shown in Figure 2, during the primary immune response most antibody-producing plasma cells reside in the spleen but, following acute infection, they are mainly found in the bone marrow. During re-infection, there is another transient increase in plasma cells in the spleen as well as expansion of the population in the bone marrow. The immune host continues to produce considerable titers of virus-specific antibody following primary infection. Persistent antibody production is likely due to several mechanisms, including longlived plasma cells in the bone marrow, and (possibly) memory B cells that continue to differentiate into newly mature plasma cells, stimulated by antigen or by cytokines in the local environment. Plasma cells can also be found at mucosal surfaces where they may be important to sustain protective antibody levels at barrier surfaces.

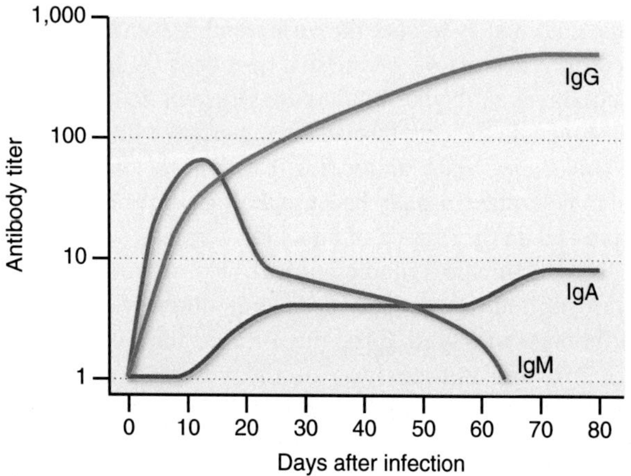

FIGURE 1 The time course of different classes of antibody response. This figure is a generalized representation of the response of human subjects to a virus infection, based on studies with live virus vaccines.

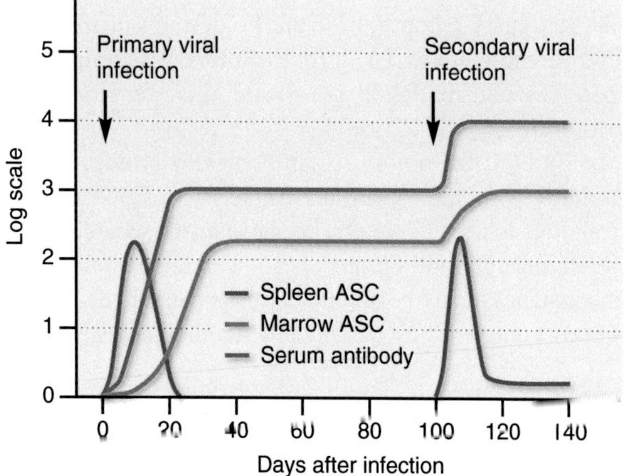

FIGURE 2 The time course of response to primary and secondary virus infection, based on studies in animal models. The figure shows the titer of circulating antibody, and relative numbers of antibody-secreting cells in the bone marrow and spleen. The homeostatic replenishment of antibody-secreting cells in the bone marrow is the main source of antibody persistence following acute infection or vaccination.

3. CELLULAR IMMUNITY

3.1 Measures of Cellular Immunity

Limiting dilution assay. The classical assay for cellular immune responses is the limiting dilution assay or LDA, introduced in the 1960s (Sidebar 2). Cells are obtained from blood, spleen, or lymphoid tissues of an animal immune to the virus under study (virus "X") and are cultured for 1–2 weeks in the presence of whole inactivated virus X, its proteins, or peptides. Under these conditions, virus X-specific CD8+ cells undergo clonal expansion, presumed proportional to their numbers in the original cultured sample. The expanded virus-specific CD8+ lymphocyte population is then tested for its ability to destroy target cells bearing the viral antigens in the context of syngeneic MHC Class I molecules.

Target cells are infected with a vector, such as a recombinant vaccinia virus, that expresses the virus X antigens; the targets will now express the antigens of virus X bound to the subject's MHC Class I molecules. The target cells are labeled with ^{51}Cr, a marker that binds to intracellular proteins. Cultured CD8+ cells are incubated with target cells, and the release of ^{51}Cr is measured as an indicator that the targets have been lysed by the CTL. Although the LDA represents a biologically relevant measure of cellular immunity, it is very cumbersome, tedious, capricious, and expensive.

Tetramer assay. The tetramer assay detects antigen-specific CD8+ cells by a reagent that consists of a single peptide molecule fixed in the antigen-binding groove of an MHC Class I molecule that is chemically coupled into a tetramer that also bears a sensitive fluorescent marker. Tetramers are necessary because the affinity of a single peptide/MHC complex is too low to allow binding of T cells. Such tetramer-labeled cells can then be enumerated using flow cytometry. It is assumed that numbers of tetramer-staining cells are proportional to the numbers of effector cells measured in the cytolytic assay.

Sidebar 2 Measures of CD8+ T lymphocyte immune responses. Methods are described in Appendix IV, Laboratory techniques commonly used in immunology, in Abbas AK, Lichtman AH, Pillai S, Cellular and molecular immunology, 8th edition, Elsevier, Philadelphia, 2015

Assay	Fresh or cultured cells	Target cell or marker	Readout
LDA (limiting dilution assay)	Cultured with antigen	Cell presenting epitope bound to syngeneic class I MHC	^{51}Cr release from target cells
CTL (cytolytic T cells)	Fresh or frozen; shortterm culture with antigen	Cell presenting epitope bound to syngeneic class I MHC	^{51}Cr release from target cells
Tetramer	Fresh or frozen	Fluorescent complex with epitope bound to syngeneic class I MHC	FACS (fluorescence- activated cell sorter)
Intracellular cytokine staining (ICS)	Fresh or frozen	Fluorescent complex bound to intracellular cytokine (such as IFN-γ)	FACS (fluorescence- activated cell sorter)
ELISPOT	Fresh or frozen	Secreted cytokine (such as IFN-γ)	Monocellular focus of released cytokine in culture dish

Intracellular Cytokine Staining (ICS). CD8+ T cells are cultured with an oligopeptide which stimulates the production of interleukins and cytokines, such as IFNγ or TNFα (tumor necrosis factor). The cells are treated with brefeldin A (to prevent secretion of cytokines) and are fixed with a chemical such as glutaraldehyde that leaves their structure intact. The cells are then permeabilized so their intracellular cytokine can be stained by a specific antibody. Lymphocytes that responded to the oligopeptide can then be enumerated using flow cytometry to measure the particular intracellular cytokine. The ICS assay has become a widely used approach, particularly where sophisticated research equipment is available. Advances in flow cytometry allow these assays to be used in conjunction with staining for many (up to 18 or more) other surface or intracellular proteins, to gain sophisticated information on the function and differentiation of individual T cells and B cells.

ELISPOT assay. The ELISPOT assay is similar to that described above for plasma cells. Lymphoid cells are plated in medium, and stimulated with oligopeptides. Those CD8+ T lymphocytes recognizing a specific epitope secrete cytokines. The surface of the culture plate has been prepared by coating with antibody that will recognize a particular cytokine, such as IFNγ, and the number of foci of secreted and bound cytokine are then enumerated to count the number of responding T lymphocytes. The ELISPOT assay is often used for high throughput of large numbers of samples, or where laboratory facilities are limited.

- How does the adaptive response counter the two different states of a virus, the extracellular infectious virion versus the intracellular replicating virus?

3.2 Effector Functions of CD8+ Lymphocytes

CD8+ cells exert their effects mainly by two mechanisms, cytolytic attack on target cells or secretion of interleukins and cytokines. When CD8+ cells—by interaction between the TCR on the CD8+ cells and the peptide on the MHC Class I molecule of the target cell—are stimulated to "attack" target cells, they release perforin, a molecule that produces channels in the plasma membrane of the target cell leading to lysis. In addition, CD8+ cells secrete granzymes (serine esterases), which pass through the channels in the target cell and trigger apoptosis. Effector CD8+ cells are not destroyed in this process and survive to kill additional "prey."

CD8+ cells also release a number of cytokines such as IFNγ, TNFα, and IL-2 as well as many chemokines, which can promote recruitment of other cells to sites of infection. These cytokines and chemokines induce a chemotactic effect that draws monocytes and other cells to the site of infection, leading to the elimination of virus and the removal of dead cells.

Hepatitis B virus infection provides an example of the effector function of CD8+ T cells, which has been studied in an HBV transgenic mouse model, where 100% of hepatocytes carry the viral genome. In this model, virus-specific CTL clear viral genes, and their mRNA transcripts and cognate proteins. Clearance cannot involve CTL killing since the mice undergo a mild disease from which they recover, and must be based on the ability of cytokines to purge intracellular viral RNA and proteins in a noncytolytic manner, a remarkable observation that remains to be explained. A significant number of persistent HBV infections of humans are "spontaneously" cleared without intervention, and it can be speculated that a similar mechanism explains this enigmatic phenomenon.

3.3 Kinetics of the Cellular Immune Response: Effector and Memory T Cells

The following description applies to acute infections in which the host clears the virus (persistent infections are discussed in Chapter 7, Patterns of infection). Virus-specific CD8+ effector CTLs appear during the first week of infection, rapidly

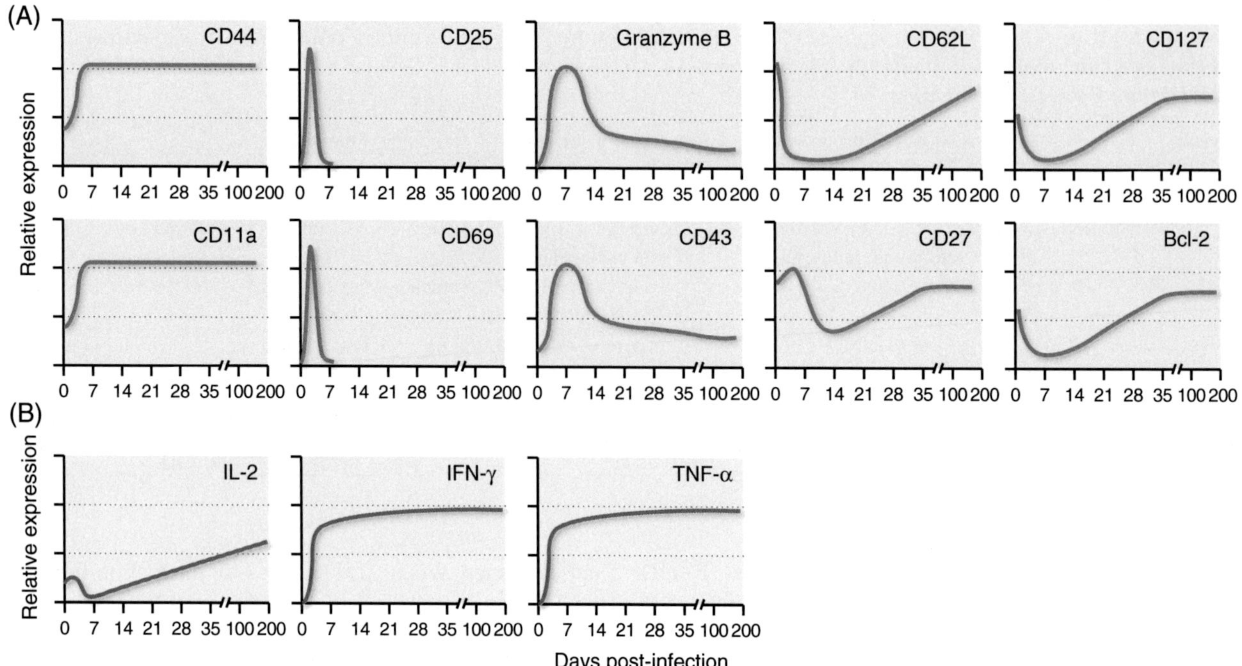

FIGURE 3 The conversion of CD8+ cells from effector to memory phenotype during and after acute viral infection, based on studies in an animal model (acute infection of mice with lymphocytic choriomeningitis virus). The upper panel (A) shows surface markers on CD8+ lymphocytes and the lower panel (B) shows expression of cytokines after 5h of in vitro antigen-specific stimulation of the same cells. Bcl-2: anti-apoptotic molecule; CD11a: adhesion molecule; CD25 part of IL-2 receptor; CD27: TNF receptor superfamily; CD43: adhesion molecule; CD44: adhesion molecule; CD62L: lymph node homing receptor; CD69: early activation marker; CD127: IL-7 receptor α chain. After Masopust D, Kaech SM, Wherry EJ, Ahmed R. The role of programming in memory T-cell development. Current opinion in immunology 2004, 16: 217–225, with permission.

increase to a peak typically 7–14days after infection, and then wane by 3–6weeks. During this clonal expansion, T cells that recognize viral antigens can expand 10–10,000-fold in 1–2weeks, representing up to 20% or more of total circulating CD8+ cells. The CD8+ peak often corresponds to the period when virus is being cleared by the host.

These effector T lymphocytes have a short half life, probably no more than a few days and then undergo apoptosis, which accounts for their rapid disappearance after the peak of viral infection. Following the acute period of infection, effector CD8+ cells drop by 10- to a 100-fold below their peak but do not completely disappear. A subset of effector CD8+ lymphocytes, expressing the IL-7 receptor **α** (IL-7R**α**), are the predestined progenitors of the CD8+ memory pool. This residual population of antigen-committed precursor CD8+ cells are usually considered to be "memory" T cells because of their ability to proliferate rapidly if the host is reinfected. CD8+ memory T cells probably persist for the lifetime of the animal, and are maintained by a low level of homeostasis controlled by the cytokines IL-7 and IL-15.

Memory and effector T lymphocytes can be distinguished from each other by a variety of surface markers, as well as by their content of intracellular cytokines (Figure 3). Memory CD8+ cells are found predominantly in lymphoid tissues and are sometimes called central memory cells (T_{CM}). A subset of CD8+ T cells are found predominantly in peripheral tissues and blood, mount a robust IFN-**γ** and TNF-**α** response, and are sometimes called effector memory (T_{EM}) cells. It appears that in response to virus reinfection, T_{EM} cells respond to peripheral infection more rapidly, while T_{CM} cells will proliferate and then differentiate, so that they produce a larger population of effector cells but with a greater lag time. A third population of memory T cells found in peripheral tissues, termed Resident Memory (T_{RM}) has recently been defined. These T_{RM} seed peripheral tissues during the effector phase of the response, but then become long-term residents of these non-lymphoid tissues and do not recirculate efficiently. The Trm pool is likely quantitatively larger than previously estimated based on newer approaches to quantify lymphocytes in tissues using quantitative immunofluorescence microscopy rather than inefficient extraction techniques.

The role of CD4+ helper cells in the induction of CD8+ memory cells is complex. Virus-specific CD4+ cells are not required for the induction of the early CD8+ effector response in many settings. However, if CD4+ cells are absent during immune induction, the functional qualities of CD8+ memory cells are impaired.

4. MUCOSAL IMMUNE RESPONSES

Mucosal tissues, which include the respiratory, gastrointestinal, and genitourinary systems, line the inner surfaces of the body. The mucosa are specialized to maintain intimate contact with the outside world. For this reason, mucosal tissues typically comprise thin permeable barriers and are vulnerable to infection. Not surprisingly, most viruses have evolved to enter the host via these routes. Following clearance of mucosal infections, memory lymphocytes and plasma cells of the adaptive immune system patrol these tissues to guard the frontlines, with the goal of preventing infections upon re-exposure right at the first site of viral contact. In fact, some estimate that mucosal tissues contain more memory lymphocytes than the entire rest of the body. Trm in mucosal tissues may be a major protective mechanism for some viruses, acting rapidly to initiate new immune responses in situ.

4.1 Mucosal Antibodies

Mucosal tissues are abundantly populated by plasma cells, which occupy both the connective tissue underlying epithelial surfaces and dedicated mucosal lymphoid tissues such as the tonsil in the oral pharynx and Peyer's patches in the intestinal wall. A high proportion of mucosal plasma cells secrete IgA, which has a number of specialized features that make it well suited to protect mucosal surfaces. Firstly, IgA does not activate the highly inflammatory complement system. Minimizing unnecessary mucosal inflammation is important to maintain the integrity of the mucosal barrier as well as critical organ functions, such as respiration and nutrient absorption. Secondly, IgA is assembled as a dimer; two IgA antibody molecules are joined together by a protein "joiner" known as a J chain. Dimers may have higher functional avidity and neutralization potential compared to antibodies that are secreted as monomers. Dimerisation also allows IgA transport into the mucosal lumen. Epithelial monolayers that line the intestinal lumen express specialized poly-Ig receptors on their basolateral surfaces. These receptors capture IgA dimers, and deliver them across the epithelium and into the mucosal lumen. A part of the poly-Ig receptor, called the secretory piece, remains attached to the secreted IgA. This secretory piece binds to the mucus glycoproteins that line mucosal tissues, allowing IgA antibodies to persist at the mucosal surface (rather than just being washed away), thus guarding these common viral entry sites. IgM, secreted as pentamers, is also produced within mucosal tissues. In addition, serum IgG leaks from capillaries into mucosal tissues, in a process referred to as passive transudation. It is estimated that humans secrete an astonishing 5 g of Ig at mucosal surfaces every day (which corresponds to $\sim 10^{19}$ antibody molecules!).

All three types of antibody contribute to protection of mucosal surfaces. IgA may also be important for preventing over-colonization and invasion by commensal bacteria that make their home in the intestinal lumen. These "good" bacteria help their hosts digest food and compete with pathogens for physical and ecological space, but must be restricted to the exterior of mucosal surfaces. Mucosal IgA antibody levels for pathogens may be short-lived, lasting only months to a few years. In contrast, serum antibody may persist for decades due to the persistence of long-lived antibody-producing plasma cells that are maintained in the specialized environment of the bone marrow.

4.2 Mucosal T Cells

Mucosal tissues also contain abundant populations of memory CD4+ and CD8+ T cells, which patrol the frontline sites of initial viral exposure. Naïve T cells mainly patrol lymph nodes because these structures collect antigens from body extremities and bring them to a central location. This mechanism optimizes detection since the number of antigen-specific T cells is extremely low for any pathogen not yet encountered. By limiting their surveillance to lymph nodes, rare naïve T cells have a chance to detect cognate antigen in a timely way.

After an infection is cleared the first time, the host retains an expanded population of memory T cells. These populations are comprised of different subsets that survey different anatomic compartments. Central memory T cells (T_{CM}) patrol lymph nodes, like naïve T cells. However, T_{CM} will not be able to recognize the invading pathogen until its antigens reach the lymph node. Thus, this anamnestic response fails to capitalize on the earliest opportunity for intervention, right at the frontline site of initial pathogen re-exposure.

While naïve and T_{CM} recirculate through lymph nodes (visiting perhaps a dozen lymph nodes each week), most mucosal memory T cells are resident. During initial T cell activation and proliferation following infection or vaccination, a fraction of activated T cells migrate to mucosal sites and never leave (Figure 4). After clearance of an infection, these cells persist, differentiate into a specialized type of resident memory T cells (T_{RM}), to distinguish them from lymph node-patrolling T_{CM}. Both T_{RM} and T_{CM} make valuable contributions to protective immunity. T_{RM} may suppress the infection quickly at the source, leading to rapid pathogen control. But if this is insufficient for complete elimination, T_{CM} provide a very powerful reserve force capable of producing a second wave of very abundant effector T cells.

- What determines the difference between mucosal tolerance and mucosal immunity?

4.3 Mucosal Tolerance

One of the most remarkable aspects of gut immunity is "tolerance," the absence of an immune response to the bacteria, proteins, and other antigens that abound in the intestinal lumen. Tolerance appears to be mediated in part by T_{REG} cells and IL-10. Humans (and mice) with mutations in this immunoregulatory pathway develop severe bowel

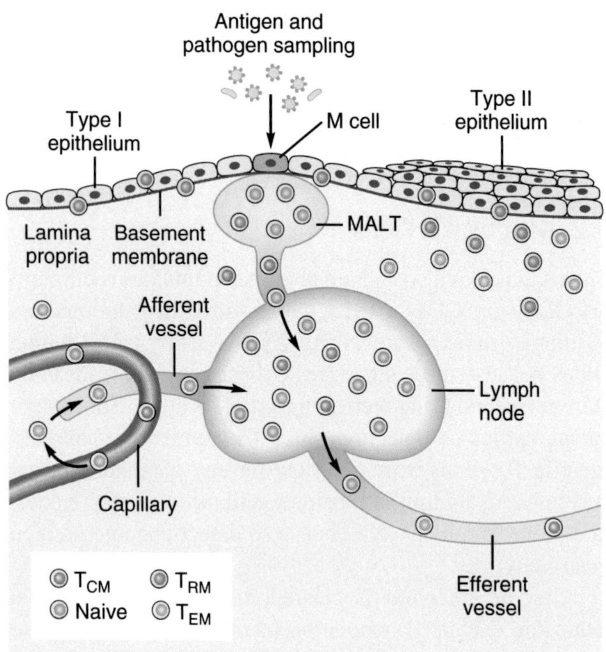

FIGURE 4 Contrasting response of lymphoid vs mucosal T cell subsets during immune induction. Following viral infection, one subset becomes central or effector memory T cells (T_{CM} or T_{EM}), and patrol lymph nodes (T_{CM}), where they can respond to cognate antigens or perhiphеral blood and visceral organs (T_{EM}). These cells upregulate CD62L and other markers as shown in Figure 3. Another set of T cells seed mucosal sites and other peripheral tissues, where local tissue signals drive differentiation into Resident Memory T cells (T_{RM}). These T_{RM} reside in tissues such as the intestinal mucosa, lungs and skin long-term, can rapidly reactivate effector functions in situ and provide local protective immunity upon reinfection. T_{RM} upregulate adhesion and retention molecules such as CD103 and CD69 that help ensure permanent tissue residence and lack of systemic migration. *Redrawn from Masopust and Schenkel (2013).*

inflammation, presumably an immune response to bacterial antigens in the gut. However, live viral vaccines that replicate in the gastrointestinal tract, do elicit a robust immune response. What is the explanation for this apparent paradox? The most likely hypothesis is that immunogenic live viruses are bound to or ingested by dendritic cells and invoke an innate response that is the pre-requisite for an immunogenic (vs a tolerogenic) response. This view is supported by the experimental observation that oral administration of a nonreplicating antigen will only induce a mucosal immune response if it is co-administered with a potent inducer of innate immunity, such as cholera toxin.

- What is the relationship between the microbial environment and adaptive immune responses?

4.4 The Microbiome and Adaptive Immunity

The postnatal development of the immune system is conditioned by exposure to the microbiome, the host of microbial species that colonize the skin and mucosal linings (Chung et al., 2012). The role of microbial environment in adaptive immunity has long been recognized since germ-free mice fail to develop a "normal" immune system, either anatomically or functionally. Apparently the "stress" offered by the microbial environment provides a needed stimulus for immune maturation, although the detailed mechanisms have yet to be elucidated.

In contrast to this benign role for the microbiome, there is a delicate balance between the mammalian host and the microbial inhabitants of the gut. When there is a breakdown in the intestinal barrier that keeps microbes at bay, and a loss of immunological tolerance, a state of chronic intestinal inflammation can lead to disease immune dysfunction. This appears to be the case in HIV/AIDS where immune dysregulation plays a key role in development of immunodeficiency. Thus the microbiome is a significant player in the pathogenesis of at least one important viral disease.

5. IMMUNITY AS A HOST DEFENSE

- What are the differences between responding to a primary viral infection and to re-exposure a second time? what is the respective role of the two arms of the immune response in these two situations?

5.1 Recovery from Initial Infection

Recovery from the initial infection should be distinguished from response to reinfection, since the relative role of various host defenses is quite different. The cellular immune response can produce a large number of effector cells in a relatively short time while the antibody response develops more slowly. Also, effector CTLs have several mechanisms for destroying or purging virus infected cells, while antibody acts most effectively on free infectious virus that has not yet initiated cellular infection. For these reasons, the cellular immune response is probably the most important component of host defense against a primary infection, although antibodies often play a synergistic role in clearing primary viral infections.

West Nile virus (WNV) provides a good example. WNV is an arbovirus that was introduced into New York state in 1999, and has since spread across the United States. It causes an acute febrile illness in humans some of whom develop encephalitis. The role of the immune response in recovery from WNV infection can be dissected in a mouse model that mimics the infection in humans. After subcutaneous injection, this virulent flavivirus produces a transient viremia, following which virus crosses the blood–brain barrier, enters the central nervous system, infects neurons, and produces acute encephalitis. In this experimental system, some mice recover and clear the virus, while others succumb, just as in humans. When CD8+ cells are deleted (by gene knockout) the mortality rises; however, viremia is rapidly terminated although the virus is not cleared from tissues. When B cells are deleted (another gene knockout) the mortality also increases, and the mice fail to clear the viremia effectively. In this example, it appears that both limbs of the immune response are necessary for an optimal host response to initial infection.

5.2 Protection Against Reinfection

Antibodies usually play a more important role than CD8+ T cells in immune protection against virus challenge, shown in studies in which immune animals are depleted of either B cells or CD8+ T cells. In the immune host, antibodies circulate in the plasma and will also be present in mucosal fluids, the antibody class and concentration depending on the prior immunizing experience. These pre-existing antibodies are particularly effective, if viremia plays an essential role in the dissemination of the infection to target tissues. Not only does prior infection provide antibody that exists at the time of reinfection but the re-exposure induces an anamnestic or recall response, so that the antibody titer rises more rapidly than after first exposure to the virus and to higher peak titers. Furthermore, restimulation leads to the "maturation" of the antibody response, due to a selection of B cells that synthesize antibodies with increased affinity for the stimulating antigen and are more effective antiviral weapons.

Poliomyelitis, an important human disease, now on the verge of eradication, illustrates these points. Polio is an enterovirus, that infects monocytes in the tonsil and Peyer's patches, lymphoid organs that associated with the gastrointestinal tract. The virus spreads to lymph nodes, is released into the blood where it causes a cell-free viremia, and occasionally invades the central nervous system where it destroys lower motor neurons, causing an acute paralysis (President Franklin Roosevelt was the most famous victim).

There are two effective vaccines, inactivated poliovirus vaccine, IPV, and oral poliovirus vaccine, OPV. Neither one provides "sterilizing immunity" but both induce neutralizing antibody and protect against paralysis. How do these vaccines mediate protection? There is an excellent monkey model of polio, that can be used to test the mechanism. If animals are treated with passive antibody (produced in humans or other primates) and then challenged with virulent poliovirus, good protection is induced, indicating that antibody alone can account for protection. Furthermore, humans with congenital defects of cellular immunity but relatively normal antibody production (such as the diGeorge syndrome) respond well to polio vaccines and are protected against paralytic disease.

Mucosal immune induction (by infection or local vaccination) may produce greater protection than systemic immunization against a mucosal virus challenge. This is illustrated by polio vaccines. Although both IPV and OPV elicit neutralizing antibodies, OPV-immunized infants are much more resistant to an oral OPV re-challenge than are their IPV-immunized counterparts. This is likely due to APCs that are resident in the mucosae. DCs from the intestinal mucosa tend to migrate to intestine-draining lymph nodes where they prime effector B and T cells that home back to the intestinal mucosa.

- How does the adaptive response clear the virus invader and "cure" the infection? How do some viruses "escape" these responses and lead to persistent infection?

5.3 Escape from Immune Surveillance

The contest between virus and host is a never ending one, but the virus has a potential advantage, since it can evolve more rapidly. Not surprisingly, viruses have developed a number of mechanisms to evade immune surveillance and persist in the infected host. These include virus escape mutants; latency in which the viral genome persists as an episome or is integrated into the host genome; tolerance, in which viruses are treated as self-antigens; immune exhaustion; and other still more byzantine strategies. Viral persistence is discussed in Chapter 7, Patterns of infection.

5.4 Passive and Active Immunization

One of the greatest contributions of medical science to human welfare is the application of immunological principles for the prevention of viral and other infectious diseases. Immune responses can be exploited in two ways, passive and active.

Passive immunity. Perhaps the most important example of passive immunity is the transfer of maternal antibodies across the placenta and via breast feeding to newborn infants, which provides a shield against potentially fatal infectious diseases for the first 6–12 months of life. For instance, in a famous nineteenth-century outbreak in the Faroe Islands, measles mortality was as high as 25% in unprotected infants (whose mothers lacked measles antibodies), compared to an expected 1% in infants with maternally conferred antibodies. Passive antibody is used for selected viral diseases, such as rabies, where it complements post-exposure vaccination for persons bitten by a rabid animal. The most current use of passive antibody is the treatment of humans in the early stages of Ebola disease, either with monoclonal antibodies (Qui et al., 2014) or plasma obtained from patients recently recovered from Ebola infection (Hayden, 2014). Studies in nonhuman primates have demonstrated an unexpected high level of efficacy, late in the course of potentially lethal experimental infection.

Active immunity. Beginning with Jenner's introduction of vaccination, immunogens have been developed that will induce protection against a large number of viral diseases. Chapters 19, Viral vaccines, is devoted to this important aspect of applied immunology, but a few brief comments are in order. (1) Experimental studies with passive antibodies have shown that antibody alone can protect against most viral diseases. Therefore, the induction of long-lasting neutralizing antibody has been the focus of most successful viral vaccines. (2) It appears that most established human vaccines do not induce "sterilizing" immunity. When challenged with a wild virus, immunized individuals undergo a subclinical infection that does not cross the threshold to clinical disease. Where there are no subclinical infections, such as HIV, it has been much more difficult to formulate an effective immunogen. (3) Most effective vaccines induce antibodies that bind close to the receptor binding site on the viral surface, and putative antibody escape mutants either cannot infect cells or are not "fit" to compete with existing wild viruses. Thus, vaccines introduced more than 50 years ago (against measles, polio, and many others) still protect against wild virus infection. There are a few exceptions, notably influenza and HIV, where antibody escape mutants are "fit." Under those conditions, it is very difficult—perhaps impossible—to formulate a universal vaccine.

6. REPRISE

Viral infection induces an adaptive antigen-specific immune response that has two main effector components, antibodies and effector T lymphocytes. Immune induction is initiated by the innate response that activates and brings together the cellular actors. This leads to antigen-specific induction that involves cooperative interaction of professional APCs (DCs macrophages, and B lymphocytes), helper cells (CD4+ T lymphocytes), and effector cells (plasma cells, CD8+ T lymphocytes).

B cells mature into plasma cells that produce antibodies, which recognize both linear and conformational epitopes on native proteins. Following a primary virus infection, antibodies are continually synthesized over a long period of time. On second exposure to a virus, there is a relatively rapid anamnestic response with an increase in antibody production. Effector CD8+ T cells express receptors that recognize antigenic linear oligopeptides, which are presented bound to MHC proteins on the surface of many cell types. In response to infection, CD8+ T cells are rapidly generated, and quickly wane and disappear once the invading virus is controlled. On reinfection, CD8+ memory cells undergo an anamnestic response and again quickly increase and decrease.

During primary viral infection, the cellular immune response plays an important role in controlling and clearing the virus, while antibody is less important. Following clearance of a primary infection, persistent antibody provides protection against reinfection. There are many variations on these general patterns, depending upon both the characteristics of the virus and of the infected host. In specific infections, the outcome is determined by the race between the viral invader and the host's innate and adaptive immune responses.

ACKNOWLEDGMENT

We thank Rafi Ahmed for previous contributions to this chapter.

FURTHER READING

Immune Response: General Reviews, Chapters, and Books

Abbas AK, Lichtman AH, Pillai S. Cellular and molecular immunology, Elsevier, Philadelphia, 8th edition, 2015.

Doering TA, Crawford A, Angelosanto JM, et al. Network Analysis Reveals Centrally Connected Genes and Pathways Involved in CD8+ T Cell Exhaustion Versus Memory. Immunity 2012, 37: 1–15.

Gitlin AD, Nussenzweig MC. Fifty years of B lymphocyes. Nature 2015, 517: 139–141.

Jameson SC, Masopust D. Diversity in T cell memory: an embarrassment of riches. Immunity 2009, 31(6): 859–871.

Leslie M. Cleanup crew. Science 2015, 347: 1058–1061.

Masopust D, Schenkel JM. The integration of T cell migration, differentiation and function. Nature reviews immunology 2013, 13: 309–321.

Matzinger P, Kamala T. Tissue-based class control: the other side of tolerance. Nature Reviews Immunology 2011, 11: 221.

Medzhitov R, Schneider DS, Soares MP. Disease tolerance as defense strategy. Science 2012, 335: 936.

Prendergast AJ, Kienerman P, Goulder PJR. The impact of differential antiviral immunity in children and adults. Nature Reviews Immunology 2012, 12: 636.

Pulendran B. Systems vaccinology: probing humanity's diverse immune systems with vaccines. Proc Natl Acad Sci USA 2014, 111(34): 12300–12306.

Seder RA, Ahmed R. Similarities and differences in CD4+ and CD8+ effector and memory T cell generation. Nature immunology 2003, 4: 835–842.

Swain SL, McKinstry KK, Strutt TM. Expanding roles for CD4+ T cells in immunity to viruses. Nature Reviews Immunology 2012, 12: 136.

Wherry EJ. T cell exhaustion. 2011. Nat Immunol 131(6):492–499.

Original Contributions

Burdeinick-Kerr R, Griffin DE. Gamma interferon-dependent noncytolytic clearance of Sindbis virus infection from neurons in vitro. J Virology 2005, 79: 5374–5385.

Chung H, Pamp SJ, Hill JA, et al. Gut immune maturation depends on colonization with a host-specific microbiota. Cell 2012, 149: 1578–1593.

Crawford A, Agelosanto JM, Ko C, et al. Molecular and transcriptional basis of CD4+ T cell dysfunction during chronic infection. Immunity 2014, 40: 289–302.

Diamond MS, Shrestha B, Marri A, Mahan D, Engle M. B cells and antibody play critical roles in the immediate defense of disseminated infection by West Nile encephalitis virus. J virology 2003, 77: 2578–2586.

Doerring TA, Crawford A, Angelosanto JM, et al. Network analysis reveals conecties genes and pathways involved in CD8-T cell exhaustion versus memory. Immunity 2012, 37: 1130–1140.

Ferris MR, Aylor DL, Bottomly D, et al. Modeling host gnetic regulation of influenza pathogenesis in the collaborative cross. Plos pathogenesis 2013, 9: e1003196.

Hayden EC. The ebola questions. Nature 2014, 514: 554-556.

Herati RS, Reuter MA, Dolfi DV, et al. Circulating CXCR5+PD-1+ response predicts influenza vaccine antibody responses in young adults but not elderly adults. J Immunology 2014, 193: 3528–3537.

Li S, Rouphael N, Duraisingham S, et al. Molecular signatures of antibody responses derived from a systems biology study of five human vaccines. Nature Immunology 2014, 15: 195–204.

Mudd PA, Martins MA, Erissen AJ, et al. Vaccine-induced CD8+ T cells control AIDS virus replication. Nature 2012, 491: 129.

Qui X, Wong G, Audet J, et al. Reversion of advanced Ebola virus disease in nonhuman primates with ZMapp. Nature 2014, 514: 47–53.

Rasmussen AL, Okumura A, Ferris MT, et al. Host genetic diversity enables Ebola hemorrhagic fever pathogenesis and resistance. Science 2014, 346: 987–991.

Salk J. Considerations in the preparation and use of poliomyelitis virus vaccine. J. American Medical Association 1955, 158: 1239–1248.

Sridhar S, Begom S, Bermingham A, et al. Cellular correlates of protection against symptomatic pandemic influenza. Nature Medicine 2013, 19: 1305.

Wilson JA, Hevey M, Bakken R, Guest S, Bray M, Schmaljohn AL, et al. Epitopes involved in antibody-mediated protection from Ebola virus. Science 2000, 287: 1664–1667.

Immune Response: Mucosal

Reviews

Cerutti A. IgA changes the rules of memory. Science 2010, 328: 1646–1647.

Iwasaki A. Antiviral immune responses in the genital tract: clues for vaccines. Nature Reviews Immunology 2010, 10: 699.

Mestecky J, Raska M, Novak J, et al. Antibody-mediated protection and the mucosal immune system of the genital tract: relevance to vaccine design. J Reproductive Immunology 2010, 85: 81–85.

Mueller SN, Gebhardt T, Carbone FR, Heath WR. Memory T cell subsets, migration patterns, and tissue residence. Annu Rev Immunol. 2013, 31:137–161

Rescigno M, Chieppa M. Gut-level decisions in peace and war. Nature medicine 2005, 11: 254–255.

Von Herrath M, Homann D. Tolerance tag team. Nature medicine 2004, 10: 585–587.

Original Contributions

Azizi A, Kumar A, Diaz-Mitoma F, et al. Enhancing oral vaccine potency by targeting intestinal M cells. PLoS Pathogens 2012, 6(11): e1001147.

Mackay LK, Rahimpour A, Ma JZ, Collins N, Stock AT, Hafon ML, Vega-Ramos J, Lauzurica P, Mueller SN, Stefanovic T, Tscharke DC, Heath WR, Inouye M, Carbone FR, Gebhardt T. The developmental pathway for CD103(+)CD8+ tissue-resident memory T cells of skin. Nat Immunol 2013 14(12):1294–1301.

Schenkel JM, Fraser KA, Vezys V, Masopust D. Sensing and alarm function of resident memory CD8+ T cells. Nat Immunol 2013, 14(5): 509–513.

Shin H, Iwasaki A. A vaccine strategy that protects against genital herpes by establishing local memory T cells. Nature 2012, 491(7424): 463–467.

Zhou Y, Kawasai H, Hsu S-C, et al. Oral tolerance for food-induced systemic anaphylaxis mediated by the C-type lectin SIGNR1. Nature Medicine 2010, 10: 1128.

Chapter 6

Aberrant Immunity

The Consequences of Overreacting or Underperforming

E. John Wherry[1], David Masopust[2]

[1]Department of Microbiology, Perelman School of Medicine, University of Pennsylvania, Philadelphia, PA, USA; [2]Department of Microbiology and Immunology, School of Medicine, University of Minnesota, Minneapolis, MN, USA

Chapter Outline

1. INTRODUCTION

Aberrant immunity refers to the ability of selected viruses to perturb the adaptive immune response in either of two distinct ways. On the one hand, the virus-specific immune response can cause pathological lesions, while on the other some viruses can suppress or sabotage adaptive immunity. How is this possible and how does it square with the protective action of adaptive immunity? In this chapter, we will present the evidence for these counterintuitive outcomes, explore their mechanisms, describe examples of human viral diseases caused by aberrant immune responses, and address the following questions.

- What is the evidence that a virus disease is caused not by the virus, but by the virus-specific adaptive immune response?
- What experiments in animal models demonstrate an immunopathological cause for a viral disease? What are the underlying mechanisms?
- What is the circumstantial evidence of an immunopathological cause for selected viral infections of humans?
- How can a virus suppress adaptive immunity? Is suppression limited to the instigating virus or is it more generic for all adaptive immune responses? What are the mechanisms of suppression?
- What human virus infections are accompanied by immunosuppression?

2. VIRUS-INITIATED IMMUNOPATHOLOGY

Most virus diseases are caused by virus–cell interactions that either lead to cytolysis or cause cellular dysfunction, as described in Chapter 3, Basic concepts in viral pathogenesis. However, sometimes viral disease is mediated by the virus-specific immune response rather than by the infection itself. This is particularly true of viruses that are relatively noncytopathic so that infected cells are not immediately destroyed. In such instances, the very same cellular or humoral immune responses to viral antigens that serve as host defenses, can also mediate a pathological response. Although, at first glance this appears paradoxical, it reflects the dynamic balance between virus and host. If the immune response clears the infection by destroying a small number of virus-infected cells the host survives with minimal symptoms and no permanent damage. If a large number of cells are infected before immune induction, the same immune-mediated destruction can cause severe or fatal pathological consequences. Every component of the immune response is capable of causing disease, although the mechanisms and manifestations will differ. Cell-mediated cytolysis, cytokine-induced inflammation, and antibody–antigen complexes are each responsible for some instances of virus-induced immunopathology (Table 1).

Viral Pathogenesis. http://dx.doi.org/10.1016/B978-0-12-800964-2.00006-9

TABLE 1 A Selected List of Immune-Mediated Viral Diseases of Humans or Animals

Host Species	Virus Family	Virus and *Disease*	Immune Modality that Mediates Disease	Immune Modality that Protects from Reinfection
Human	Hepadnaviruses	Hepatitis B *Hepatitis; Glomerulonephritis*	CD8+ cells; Antibody	Antibody; CD8+ cells?
Human	Flaviviruses	Dengue *Hemorrhagic fever/shock syndrome*	Antibody?	Neutralizing antibody
Human	Paramyxoviruses	RSV *Bronchiolitis*	CD4+ T cells	Neutralizing antibody
Mouse	Arenaviruses	LCMV *Choriomeningitis; Hepatitis*	CD8+ T cells	Neutralizing antibody CD8+ cells
Mouse	Picornaviruses	TMEV *Demyelination*	CD4+ T cells	Neutralizing Antibody
Mink	Parvoviruses	ADV *Aleutian disease (Glomerulonephritis)*	Immune Complexes	?

ADV, Aleutian disease virus; LCMV, Lymphocytic choriomeningitis virus; RSV, Respiratory syncytial virus; TMEV, Theiler's murine encephalomyelitis virus.

TABLE 2 Virus Diseases that are Immune-Mediated Meet Many of these Criteria

Phenomenon	Can Be Seen in Animal Model Infections	Can Be Seen in Human Virus Infection
Productive cellular infection may be noncytocidal	YES	YES
Under some circumstances hosts are apparently healthy while undergoing a high-titer infection	YES	YES
Infection of newborn hosts may be associated with asymptomatic high-titer persistent infection	YES	YES
Under circumstances where virus causes disease, there is an immune response concurrent with disease development	YES	YES
Pathological lesions contain virus-specific T cells, complement, or immune complexes, compatible with an immune mechanism	YES	YES
Disease associated with acute virus infection can be prevented by immunosuppression, without reducing virus replication	YES	
In hosts undergoing an asymptomatic persistent infection, disease can be produced by adoptive immunization with virus-specific T cells or antibodies	YES	

- What is the evidence that a virus disease is caused not by the virus, but by the virus-specific adaptive immune response?

There are a number of criteria that together provide evidence of a virus-induced immunopathology (Table 2). For animal models of virus-induced immunopathology, experimental interventions provide rigorous proof. For human infections, the interpretation depends more on circumstantial evidence.

- What experiments in animal models demonstrate an immunopathological cause for disease? What are the underlying mechanisms?

If the disease can be reproduced in an experimental animal, it is possible to dissect the mechanism by use of manipulations such as immunosuppression, genetic knockouts of components of the immune system, adoptive immunization, transgenic animals expressing viral genes, and the like. Experiments with lymphocytic choriomeningitis virus

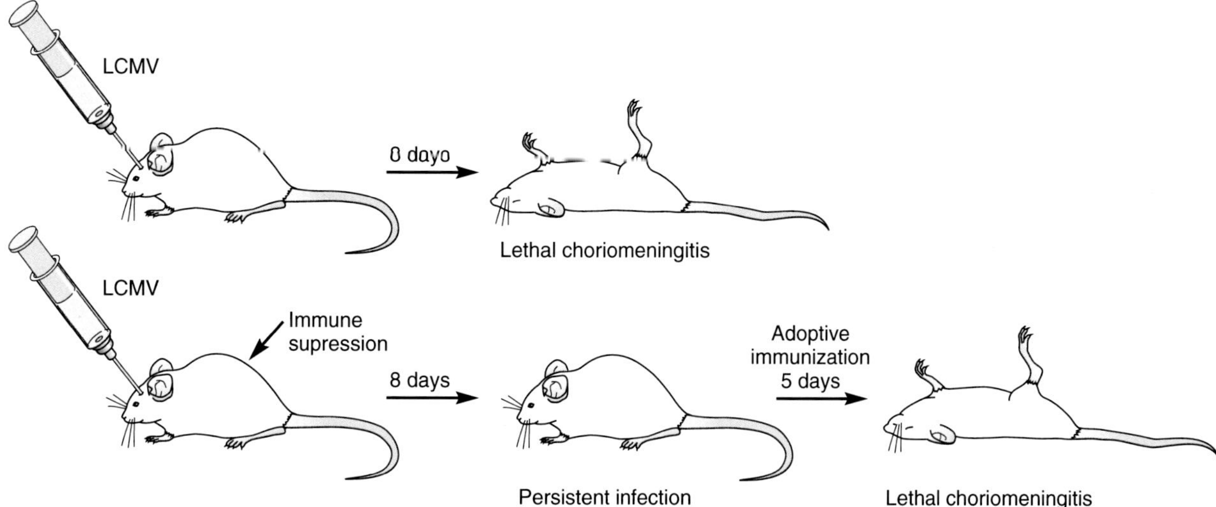

FIGURE 1 Experimental demonstration that a viral disease is mediated by the immune response, using immunosuppression and immune reconstitution. Upper panel. Mice injected by the intracerebral route with lymphocytic choriomeningitis virus (LCMV) die with convulsions about 8 days later. Lower panel. If infected mice are given a single injection of cytoxan, an immunosuppressive drug, they survive with a persistent LCMV infection. If the surviving mice are adoptively immunized with LCMV-immune spleen cells, they die 5 days later with the convulsive syndrome. *After Gilden DH, Cole GA, Monjan AA, Nathanson N. Immunopathogenesis of acute central nervous system disease produced by lymphocytic choriomeningitis virus. Journal of Experimental Medicine, 1972, 135: 860–889, with permission.*

(LCMV), a somewhat obscure arenavirus of wild mice, first led to the idea that viruses could cause immune-mediated disease, and still serves as a useful animal model of immunopathology.

When mice are infected intracranially with LCMV, they develop acute convulsive disease (Figure 1). However, if they are also treated with an immunosuppressive drug, cyclophosphamide, they remain well. Although the suppressed animals appear healthy, the virus replicates at a high level and produces a lifelong persistent infection of the brain and lymphoid tissues. These observations provide two reasons to infer an immunopathological process: first, immunosuppression protects against disease (while suppression usually enhances viral disease), and, second, high-titer virus infection does not cause illness or pathological lesions. When LCMV-immune (but not naïve) T-lymphocytes are transferred into these immunosuppressed mice, the acute illness—acute convulsive diathesis—is reproduced, strong evidence that the disease is mediated by a virus-specific immune response (Figure 1).

2.1 Cell-Mediated Immunopathology

CD8+ cytolytic T lymphocytes (CTLs) are the most important cellular mediator of virus-induced immunopathology. They attack virus-infected cells in a series of steps, including polymerization of perforin monomers to form a channel in the plasma membrane of the target cell and release of granzyme B (a serine protease) into the cytosol of the target cell, which triggers an intracellular biochemical cascade that initiates apoptosis. In addition to the direct cytopathic attack on target cells, activated CD8+ CTLs release a variety of proinflammatory cytokines (such as INF **γ**, TNF **α**, and several interleukins) that play a role in some manifestations of virus-induced immunopathology. CD8+ T cells are the mediators of LCMV disease described above, as demonstrated in adoptive transfer experiments with fractionated spleen cells and studies depleting immune subsets.

- What is the circumstantial evidence of an immunopathological cause for selected viral infections of humans?

Hepatitis B is an example of a human viral disease that appears to be mediated by the cellular immune response (Guidotti and Chisari, 2006; Trepo et al., 2014; Rehermann, 2013). The first clues about the pathogenesis of HBV-induced hepatitis come from a comparison of viral antigen and antibody markers in patients with diverse outcomes. When newborn infants are infected during the birth process, they develop a persistent asymptomatic infection. These infants have a high titer of virus in the blood, as reflected by the amount of hepatitis B surface antigen (HBsAg). They do not raise an immune response to HBsAg (no CTL or antibody) and fail to clear the infection. Asymptomatic high-titer persistent infection implies that HBV does not cause illness directly.

When adults are infected with HBV, they often develop an acute liver infection and clear the virus in few weeks as reflected in the levels of HBsAg in their blood; during the phase of virus clearance, acute hepatitis occurs (Figure 2). From the time course of liver disease, immune response, and viral clearance, it may be inferred that the same process that

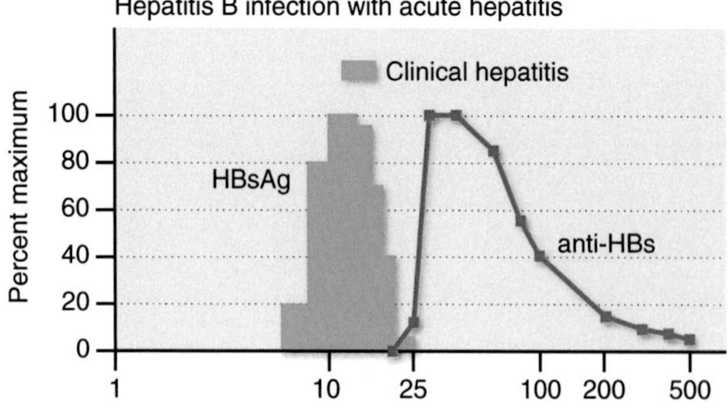

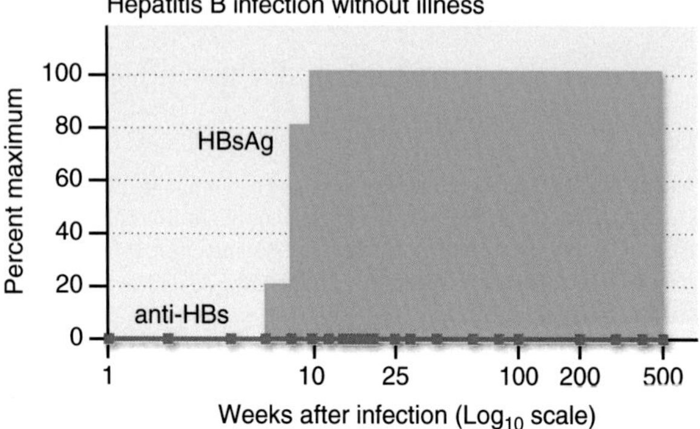

FIGURE 2 Circumstantial evidence that hepatitis B disease is mediated by the virus-specific immune response to infection. Among adults, hepatitis B infection takes several different courses. Upper panel: 20–35% of infections result in acute clinical hepatitis, during which they develop a brisk immune response, clear the virus, and recover completely. Lower panel: 2–10% of individuals undergo inapparent infection with no hepatitis but a persistent high-titer infection that persists for many years. *Redrawn from Robinson WR. Hepatitis B virus and hepatitis D virus, in Mandell GL, Bennett JE, Dolin R. Principles and practice of infectious diseases. Churchill Livingstone, New York, 1999, 1652–1683.*

clears the virus also produces disease, an example of the dual role of the immune response. In contrast, a small number of adults develop persistent HBV infections in the absence of either immune response or hepatitis, similar to the course of infection in infants (Figure 2). Again, the constellation of acute hepatitis accompanied by a brisk immune response versus asymptomatic virus persistence in the absence of an immune response, provide circumstantial evidence that HBV hepatitis is an immunopathological process. Studies in experimental mouse models of HBV infection indicate that both hepatic necrosis and viral clearance are mainly mediated by the cellular immune response (Guidotti and Chisari, 2006).

2.2 Antibody-Mediated Immunopathology

Dengue hemorrhagic fever/dengue shock syndrome (DHF/DSS) is considered to be an antibody-mediated viral disease, although this hypothesis remains speculative since the pathogenesis of DHF/DSS is not well understood. Dengue viruses can be grouped into four antigenically distinct types based on serological typing. Dengue is transmitted by *Aedes aegypti* mosquitoes and is widespread in tropical areas where the vector is indigenous. From time to time, depending upon prevalence of the vector, massive outbreaks of dengue occur in southeast Asia or in the Caribbean. During epidemics, most infections cause an acute but self-limited febrile illness with complete recovery. However, a small proportion of individuals develop shock and/or a hemorrhagic diathesis, with mortality up to 10%. Epidemiological studies indicate that DHF/DSS occurs almost exclusively in children who are undergoing a second infection with a dengue virus of a serotype different from their primary dengue infection. These observations suggest that DHF/DSS might be mediated by an immunological mechanism.

Cell culture studies show that dengue viruses replicate in blood monocytes, and infection of these cells can be enhanced in the presence of small amounts of antibodies against any of the dengue virus serotypes (Halstead, 2007; Acosta et al., 2014). Enhancement is mediated by binding of the Fc portion of the antibody molecule to Fc receptors on the surface of monocytes, which facilitates entry of virus into these host cells. It is postulated that an anamnestic response of effector T lymphocytes may lead to exaggerated cytokine production and a proinflammatory cascade leading to vascular permeability and shock syndrome (Pang et al., 2007). However, detailed understanding of the immunopathophysiology of DSS remains a challenge for future research (St John et al., 2013).

Immune complex disorders are another manifestation of antibody-mediated immunopathology. If antiviral antibodies are produced during persistent viremic infections, antigen–antibody complexes may be formed in the circulation. These immune complexes can cause immunosuppression by preventing optimal function of phagocytes in the reticuloendothelial system. In addition, immune complexes, due to their high molecular weight, may fail to transit the glomerular filter and accumulate under the basement membrane that is situated external to the glomerular capillary wall. Immune complexes can fix complement which, in turn, will generate proinflammatory cytokines that draw a variety of leukocytes into the periglomerular space. The resulting chronic inflammatory process causes scarring of the glomerulus, chronic glomerulonephritis, gradual reduction of renal function, and eventual kidney failure. Chronic glomerulonephritis is seen quite often in patients with long-term HBV and HCV infections, and likely is mediated—at least in part—by antiviral antibodies (Deray et al., 2015). A similar syndrome is seen in mice with persistent infections with LCMV.

2.3 Innate-Mediated Immunopathology

Innate immunity is the first response to a virus invader, and—like the adaptive response—it also must be modulated to serve the host well. If the innate response is too exuberant, it may exacerbate the disease consequences. The H1N1 influenza strain that caused the global pandemic of 1918 provides a significant example. Studies with the reconstructed 1918 virus have addressed the question, why did this virus cause such a high mortality, particularly in healthy adults, in contrast to other more recent strains of H1N1 influenza viruses? A comparison of the pathogenesis of both influenza strains in the macaque model have provided important clues (Kobasa et al., 2007). Two observations stand out: first, the 1918 virus spread more rapidly in the lungs. Second, and most important, it elicited a severe inflammatory response, but a lesser activation of interferon α genes, than did the recent H1N1 isolate. This excessive inflammatory response caused so much pulmonary edema and vascular congestion that it compromised pulmonary function and markedly increased mortality.

2.4 Virus-Initiated Autoimmunity

It has long been suspected that, on occasion, a virus infection might induce an autoimmune response. The following sequence of events has been hypothesized: (1) a virus infection elicits an immune response to a number of B- and T-cell epitopes on the virus-encoded proteins; (2) a few of these epitopes are also shared with one or more host proteins; (3) the virus-induced epitope-specific antibodies or T lymphocytes are capable of reacting with the cognate epitopes on the host proteins; and (4) these anti-host immune responses may elicit an autoimmune disease process.

Is there evidence of epitopes that are shared between virus proteins and host proteins? When antiviral monoclonal antibodies are isolated from mice shortly after infection with a variety of RNA or DNA viruses, a small number of these virus-specific antibodies can react with one or more tissues in uninfected mice (Srinivasappa et al., 1986). In most of these instances, the shared epitope remains to be defined. Although it seems plausible that virus infections may trigger some cases of autoimmune disease in humans, it has been difficult to develop definitive evidence for specific diseases.

3. VIRUS-INDUCED IMMUNOSUPPRESSION

In a few instances, virus infections suppress (rather than stimulate) the adaptive immune response. Immunosuppression may be "global," affecting responses to many antigens or it may be specific for the infecting virus. The mechanisms of immunosuppression are diverse (Table 3) and

TABLE 3 Examples of Human and Animal viruses that Cause Immunosuppression

Virus Group	Virus and *Disease*	Cells Infected (Lymphoreticular)	Mechanisms of Immunosuppression	Manifestations of Immunosuppression
		Human viruses		
Morbilliviruses	Measles virus *Measles*	Monocytes Thymic epithelial cells	Dysfunction of APC Perturbation of cytokine homeostasis and intracellular signaling	Reduced DTH Enhanced infections
Rubiviruses	Rubella virus *Congenital rubella*	Lymphoid cells	Fetal infection leading to tolerance	Persistent rubella infection Absence of virus-specific cellular immunity
Lentiviruses	HIV *AIDS*	CD4+ T lymphocytes Monocytes	Killing of CD4+ T cells Immune activation leading to exhaustion of CD8+ T cells	Opportunistic infections Enhanced neoplasia
		Animal viruses		
Arenaviridae	LCMV *Choriomeningitis*	Dendritic cells Monocytes	Defective immune induction	Persistent LCMV infection
Morbilliviruses	CDV *Canine distemper*	Monocytes Lymphocytes	Perturbation of cytokine homeostasis and intracellular signaling	Encephalitis Bacterial superinfections
Morbilliviruses	RV *Rinderpest*	Monocytes Lymphocytes	Perturbation of cytokine homeostasis and intracellular signaling	Lethal gastroenteritis
Lentiviruses	SIV *AIDS*	CD4+ T lymphocytes Monocytes	Killing of CD4+ T cells Immune activation leading to immune exhaustion	Opportunistic infections
Retroviruses	MuLV (Defective variant) *MAIDS*	B Lymphocytes	B, T cell dysfunction	Opportunistic infections

AIDS, Acquired immunodeficiency virus; CDV, Canine distemper virus; DTH, Delayed-type hypersensitivity; FPV, Feline panleukopenia virus; HIV, Human immunodeficiency virus; LCMV, Lymphocytic choriomeningitis virus; MAIDS, Murine AIDS; MuLV, Murine leukemia virus; RV, Rinderpest virus.

include: (1) replication of the virus in one of the cell types involved in immune induction, such as antigen-presenting cells (particularly macrophages or dendritic cells) or CD4+ (helper) T lymphocytes, leading to the induction of cellular apoptosis or aberrant production of cytokines; (2) tolerance, often associated with fetal or newborn infection, produced by clonal deletion of T lymphocytes that respond to viral antigens; and (3) exhaustion, overwhelming immune stimulation leading to loss of effector or other functions of lymphocytes and other immune cells.

Immunosuppression associated with a virus infection was first described about 100 years ago by von Pirquet who noted that patients lost their tuberculin sensitivity (skin test reaction to the antigens of *Mycobacterium tuberculosis*) during and after measles (Figure 3).

- How can a virus suppress adaptive immunity? Is suppression limited to the instigating virus or is it more generic for all adaptive immune responses? What are the mechanisms of suppression?

3.1 Mechanisms of Immunosuppression

Viral infection of monocytes, dendritic cells, and lymphocytes. A number of viruses can replicate in the lymphoreticular system, and the ability of a virus to initiate immunosuppression depends upon interactions with the cellular components of this system. Therefore, variables such as virus strain, virus dose, route of injection, age of the host, and extraneous immunosuppression will determine whether a virus induces an immunogenic or an immunosuppressive response. An immunosuppressive strain of LCMV, clone 13, has been compared with a nonimmunosuppressive strain (Armstrong strain). Clone 13 preferentially infects macrophages, dendritic cells and stromal cells, disrupting antigen presentation and vitiating immune induction. As a consequence, clone 13 produces a high-titer persistent viremia whereas the Armstrong strain causes an acute viremia that is cleared within 10 days.

It has recently been shown that virus infection of dendritic cells also plays an important role in the immunosuppressive action of measles virus. Measles virus infection inhibits the development, expansion, and migration of dendritic cells, interrupting antigen presentation, with the effects described later in this chapter.

HIV and related viruses (such as SIV, simian immunodeficiency virus) utilize the CD4 molecule as their primary receptor. As a result, HIV replicates in CD4+ T lymphocytes and monocytes/macrophages, cell types that express CD4 on their plasma membranes. T lymphocytes, if they are proliferating, undergo a cytopathic infection. The depletion of CD4+ lymphocytes, that act both as T-helper cells and, under some circumstances, as T-effector cells, contribute to functional immunodeficiency that leads to opportunistic infections.

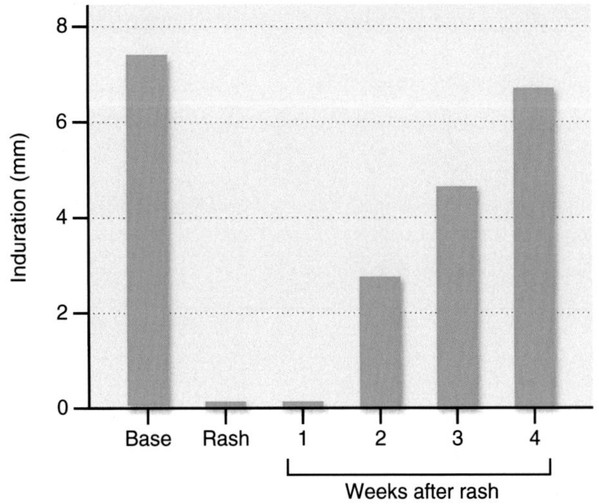

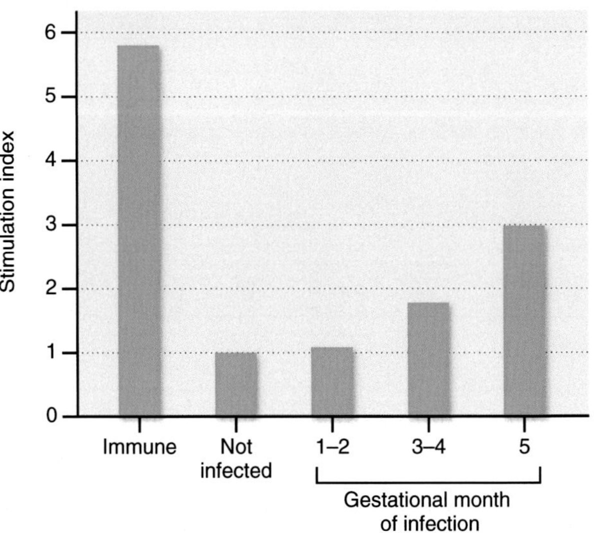

FIGURE 3 Some viral infections of human are immunosuppressive. Left-hand panel. The effect of acute measles infection upon delayed-type hypersensitivity (tuberculin skin test). Immunosuppression occurs during measles rash, and gradually returns to normal over the next month. Base: tests at 1 and 2 weeks prior to measles. *(After Tamashiro VG, Perez HH, Griffin DE. Prospective study of the magnitude and duration of changes in tuberculin reactivity during uncomplicated and complicated measles. Pediatric Infectious Diseases Journal 1987, 6: 451–454, with permission)*. Right-hand panel. Defective cell-mediated immune responses in children with the congenital rubella syndrome. Cellular responses to rubella antigen are compared in three groups: children infected postnatally (immune column), uninfected children, and children with congenital rubella syndrome who were infected at different months of gestation but tested years later. These cases of congenital rubella syndrome exhibit reduced cellular immunity but only to rubella antigens. *(After Buimovici-Klein E, Lang PB, Ziring PR, Cooper LZ. Impaired cell-mediated immune response in patients with congenital rubella: correlation with gestational age at time of infection. Pediatrics 1979, 64: 620–626)*.

Tolerance: virus as self. Normal animals are immunologically "tolerant" to self antigens. There are several mechanisms for the induction of tolerance to foreign antigens, including the thymic deletion of "forbidden" clones of T lymphocytes, or anergy caused by the peripheral activation of T lymphocytes without proper costimulation or inflammatory signals. The potential for responding to a viral antigen as "self" is demonstrated in transgenic mice that are engineered to express a viral antigen. Such animals do not produce virus antigen-specific antibodies or T lymphocytes. Tolerogenic virus infections are often characterized by several features: (1) they are most easily induced in fetal or newborn animals; (2) they are more readily induced if virus replication produces high levels of antigen; (3) tolerance is virus-specific and responses to other viruses are not impaired; and (4) tolerance often leads to virus persistence.

LCMV provides a well-studied model of tolerogenic infections in newborn mice. Such mice fail to develop LCMV-specific CTLs, even though they produce LCMV antibodies. LCMV infects the thymus of neonatal mice, and T cells that carry T-cell receptors for LCMV antigens are deleted in the thymus due to normal thymic negative selection events. A similar phenomenon likely takes place in congenital rubella, discussed below.

Exhaustion: overwhelming the immune response. Viruses that cause very high levels of infection can overwhelm or "exhaust" the developing T-cell response. Such T-cell exhaustion can be induced in adult mice infected with a high dose of selected variants of LCMV. There is an initial brisk CTL response, but these CTL fail to clear the infection. These exhausted CTL are eliminated or can persist long-term in a poorly functional state. Exhaustion of memory T lymphocytes is an important feature of many chronic human infections with high viral replication, including HBV and HCV. This issue is discussed further in Chapter 07, Patterns of infection.

- What human virus infections are accompanied by immunosuppression?

3.2 Immunosuppressive Viral Infections of Humans

A number of viruses—from many different virus families—can cause immunosuppression of varying degrees of severity and specificity in humans or animals (Table 3).

Measles. Measles is the prototype of an acute virus infection that produces global immunosuppression. The tuberculin skin test response before, during, and after measles virus infection exhibits a transient period of immunosuppression that lasts about 1 month (Figure 3). During the period of measles-induced immune suppression, there is an increased susceptibility to other infections, a transient potential for exacerbation of chronic infections such as tuberculosis, and remission of autoimmune diseases such as juvenile rheumatoid arthritis and the nephrotic syndrome.

The mechanisms of measles-induced immunosuppression are complex and probably involve at least three different cell types and pathways (De Vries et al., 2012). (1) Dendritic cells and monocytes are infected, and their antigen-presenting activity is severely compromised, with reduced activation, proliferation, and ability to traffic. (2) Circulating T lymphocytes (CD4+ and CD8+) are decreased by about 50%, due to both apoptosis and failure to proliferate. Relatively few monocytes and lymphocytes are infected, which suggests that indirect mechanisms play a role in lymphopenia. (3) There are aberrant cytokine responses; Il-4 and Il-10 production is increased (skewing the immune response from Th1 to Th2 and reducing macrophage activation) and Il-12 and TNF α production is decreased.

Rubella. Rubella (German measles) is an example of a virus that causes virus-specific immunosuppression in utero or in infancy. In children or adults, rubella is an acute infection, which produces a brisk immune response that clears the virus within a few weeks. If primary infection occurs during pregnancy, the virus can cross the placenta and infect the fetus. Fetal infections, which often produce congenital rubella syndrome (developmental malformations), usually persist throughout pregnancy and are not cleared until about 6–12 months of age. Infants or children with congenital rubella syndrome exhibit markedly diminished cellular immune responses to rubella virus antigens even when tested many years after in utero infection (Figure 3). However, such infants do raise brisk antibody responses to rubella virus. It appears that gestational infection leads to a suppression of the rubella-specific cellular—but not the humoral—immune response. Rubella is an example of a virus that induces partial tolerance, since it is "seen" as self if it is presented early enough in prenatal development.

Human immunodeficiency virus (HIV) and AIDS. The pathogenesis of AIDS is the subject of Chapter 9, HIV and AIDS, and the following sketch focuses on the mechanisms of immunosuppression. Cellular immune responses to opportunistic infectious agents are reduced in patients with AIDS compared with normal subjects while antibody titers are generally maintained until late in the illness. AIDS-associated opportunistic infections often represent a failure of CD8+ lymphocytes, the effector arm of the cellular immune response. Herein lies a paradox: CD4+ cells—not CD8+ cells—are the direct target of the virus. Likely, the failure of CD8+ cells is partly due to the loss of the helper functions provided by CD4+ cells, and partly to chronic immune activation and exhaustion of CD8+ lymphocytes themselves, which is discussed in more detail in Chapter 9. When patients with opportunistic infections receive highly active antiretroviral treatment (HAART), a remarkably rapid clearing of the opportunistic infection is often seen, implying a return of the effector component of cellular immunity.

4. VIRAL PROTEINS THAT PERTURB THE IMMUNE RESPONSE

A number of viruses encode proteins that can perturb the adaptive immune response. This is particularly true for larger viruses, such as herpesviruses and poxviruses, which have "captured" cellular genes that have evolved to act as "decoys" that interfere with antiviral host defenses. These viral proteins operate through a wide variety of mechanisms, illustrated in the examples below.

After an antiviral antibody has bound to its cognate virion, the Fc portion of the antibody molecule undergoes a conformational change that results in the binding of the C1q complement protein to the Fc domain, initiating the complement cascade. The complement cascade, in turn, can lead to virolysis or cytolysis of virus-infected cells, thereby acting as a host defense. Herpes simplex virus (HSV) encodes two envelope glycoproteins, gE and gI, that can bind to the Fc segment of the immunoglobulin molecule. This bipolar bridging (to the Fab and Fc ends of the immunoglobulin molecule) prevents activation of the complement cascade, and thereby sabotages an arm of the adaptive immune response.

CD8+ CTLs, through their T-cell receptors, recognize viral antigens presented as peptides on Class I molecules, leading to lysis of virus-infected cells. Poxviruses, adenoviruses, herpesviruses, and HIV encode proteins that down-regulate the expression of Class I molecules, which renders virus-infected cells less susceptible to antiviral CTLs. Some adenoviruses are oncogenic in animals and others are not. The oncogenic adenoviruses down-regulate MHC Class I expression, enabling them to persist and transform infected cells. Once again, the virus fights back against the adaptive immune response.

5. REPRISE

Although the innate and adaptive immune responses usually play a protective role in viral infection, a few viral diseases are immune mediated. Optimally, the *innate* immune response emphasizes interferon and other antiviral genes that call effector cells to the sites of virus replication. However, in some instances, such as infection with the 1918 strain of influenza virus, there is a dysregulation of the response, in which excessive inflammation causes fatal pulmonary edema.

If the *adaptive* immune response is sufficiently brisk, it will clear the infection with minimal cellular damage, but if the infection is widespread when antiviral effector T lymphocytes appear, then the attack on infected cells can produce severe disease. Immunopathology is most often seen with viruses that are relatively noncytopathic so that infected cells are not immediately destroyed but become targets for CTLs. Evidence for an immunopathological mechanism include: high-titer persistent infection absent any illness or virus-specific immune response; an association of an antiviral immune response with disease; and experimental immune manipulation in animal models that provides rigorous proof that disease is immune mediated. Hepatitis B is the most prominent human diseases caused by a virus-specific immune response. Infants undergo asymptomatic high-titer persistent HBV infection of the liver absent an antiviral immune response, while adults have an immune response concomitant with acute hepatitis and eventual clearance of the virus.

A few virus infections lead to immunosuppression, which can be induced through several mechanisms, including: (1) destruction of dendritic cells (or other antigen-presenting cells) or subsets of T lymphocytes; (2) infection of fetal or newborn animals leading to tolerance; and (3) exhaustion driven by overwhelming stimulation. Virus-induced immunosuppression may be limited to the infecting agent or more global, affecting the response to many antigens. HIV, measles, and congenital rubella are examples of immunosuppressive human viral infections. Measles virus interferes with several steps in induction of the adaptive immune response and causes transient global suppression. Fetal infection with rubella leads to virus-specific immune tolerance that persists into childhood. HIV attacks helper CD4+ T lymphocytes, perturbing immune induction, and causes chronic inflammation of the gastrointestinal tract resulting in exhaustion of CD8+ effector T cells, leading to the acquired immunodeficiency syndrome.

FURTHER READING

Immunopathology

Reviews, Chapters, and Books

Acosta EG, Kumar A, Bartenschlager R. Revisiting dengue virus-host cell interaction: new insights into molecular and cellular virology. Advances in virus research 2014, 88: 1–109.

Deray G, Buti M, Gane E, et al. Hepatitis B virus infection and the kidney: renal abnormalities in HBV patients, antiviral drugs handling, and specific follow-up. Advances in hepatology, 2015: doi.org/10.1155/2015/596829.

Guidotti LG, Chisari FV. Immunobiology and pathogenesis of viral hepatitis. Annual Review of Pathology 2006, 1: 23–61.

Halstead SB. Dengue. Lancet 2007, 370: 1644–1652.

Russier M, Pannetier D, Baize S. Immune responses and Lassa virus infection. Viruses 2012, 4, 2766–2785.

Rehermann B. Pathogenesis of chronic viral hepatitis: differential roles of T cells and NK cells. Nature medicine 2013, 19: 859–865.

St John AL, Abraham SN, Gubler DJ. Barriers to preclinical investigations of anti-dengue immunity and dengue pathogenesis. Nature reviews microbiology 2013, 11: 420–430.

Tisoncik JR, Korth MJ, Simmons CP, et al. Into the eye of the cytokine storm. Microbiol Molec Biol Rev 2012, 76(1): 16–32.

Trepo C, Chan HL, Lok A. Hepatitis B virus infection. Lancet 2014, 384: 2053–2063.

Original Contributions

Battegay M, Kyburz D, Hengartner H, Zinkernagel RM. Enhancement of disease by neutralizing antiviral antibodies in the absence of primed antiviral cytotoxic T cells. European Journal of Immunology 1993, 23: 3236–3241.

Chen I. Saving California's calves. Science 2015, 348:626–627

Fujinami RS, Oldstone MBA. Amino acid homology between the encephalitogenic site of myelin basic protein and virus: mechanism of autoimmunity. Science 1985, 230: 1043–1045.

Gilden DH, Cole GA, Monjan AA, Nathanson N. Immunopathogenesis of acute central nervous system disease produced by lymphocytic choriomeningitis virus. Journal of Experimental Medicine, 1972, 135: 860–889.

Guidotti LG, Rochford R, Chung J, Shapiro M, Purcell R, Chisari FV. Viral clearance without destruction of infected cells during acute HBV infection. Science 1999, 284: 825–830.

Kobasa D, Jones SM, Shinya K, et al. Aberrant innate immune response in lethal infection of macaques with the 1918 influenza virus. Nature 2007, 445: 319–323.

Oldstone MBA. Molecular mimicry and autoimmune disease. Cell 1987, 50: 819–820.

Penaloza-MacMaster P, Barer DL, Wherry EJ, et al. Vaccine-elicited CD4 T cells induce immunopathology after chronic LCMV infection. Science 2015, 347: 278–282.

Srinivasappa J, Saegusa J, Prabakhar BS, et al. Molecular mimicry: frequency of reactivity of monoclonal antiviral antibodies with normal tissue. J Virology 1986, 57: 397–401.

Immune Suppression

Reviews, Chapters, and Books

Chowdury A, Silvestri G. Host-pathogen interaction in HIV infection. Current opinion in Immunology 2013, 25: 463–369.

De Vries RD, Mesman AW, Geijtenbeek TB. The pathogenesis of measles. Current opinion in virology 2012, 2: 248–255.

Doering TA, Crawford A, Angelosanto JM, et al. Network Analysis Reveals Centrally Connected genes and pathways involved in CD8+ T cell exhaustion versus memory. Immunity 2012, 37: 1–15.

Moss WJ, Griffin DE. Measles. Lancet 2012, 379: 153–154.

Okoye A, Picker LJ. CD4+ T cell depletion in HIV infection: mechanisms of immunological failure. Immunological Reviews 2013, 254: 54–64.

Original Contributions

Ahmed R, Salmi A, Butler LD, Chiller JM, Oldstone MBA. Selection of genetic variants of lymphocytic choriomeningitis virus in spleens of persistently infected mice. Journal of Experimental Medicine 1984, 60: 521–540.

Brooks DG, Teyton L, Oldstone MBA, McGavern DV. Intrinsic functional dysregulation of CD4 T cells occurs rapidly following persistent viral infection. J virology 2005, 79: 10514–10527.

Buimovici-Klein E, Lang PB, Ziring PR, Cooper LZ. Impaired cell-mediated immune response in patients with congenital rubella: correlation with gestational age at time of infection. Pediatrics 1979, 64: 620–626.

Dubin G, Socolof E, Frank I, and Friedman HM. Herpes simplex virus type 1 Fc receptor protects infected cells from antibody-dependent cellular cytotoxicity. Journal of Virology 1991, 65: 7046–7050.

Hahm B, Triffilo MJ, Zuniga EI, Oldstone MBA. Viruses evade the immune system through type 1 interferon-mediated STAT2-dependent, but STAT1-independent, signaling. Immunity 2005 22: 247–257.

Isaacs SN, Kotwal GJ, Moss B. Vaccinia virus complement control protein prevents antibody-dependent complement-enhanced neutralization of infectivity and contributes to virulence. Proceedings of the National Academy of Sciences 1992, 89: 628–632.

Mina MJ, Metcalf CJ, de Swart RL, et al. Long-term measles-induced immunomodulation increases overall childhood infectious disease mortality. Science 2015, 348: 694–698.

Moskophidis D, Lechner F, Pircher H, Zinkernagel RM. Virus persistence in acute infected immunocompetent mice by exhaustion of antiviral cytotoxic effector T cells. Nature 1993, 362: 758–761.

Rasmussen AL, Okumura A, Ferris MT, Green R, Feldmann F, et al. Host genetic diversity enables Ebola hemorrhagic fever pathogenesis and resistance. Science 2014, 346(6212): 987–991. Doi 10.1126/science.1259595.

Tamashiro VG, Perez HH, Griffin DE. Prospective study of the magnitude and duration of changes in tuberculin reactivity during uncomplicated and complicated measles. Pediatric Infectious Diseases Journal 1987, 6: 451–454.

Tishon A, Borrow P, Evans C, Oldstone MBA. Virus-induced immunosuppression. Virology 1993, 195: 397–405.

Wherry, EJ. T cell exhaustion. 2011. Nat. Immunol. 131(6): 492–499.

Chapter 7

Patterns of Infection

Unwanted Guests—Quick Visits and Extended Stays

Neal Nathanson[1], Francisco González-Scarano[2]

[1]*Department of Microbiology, Perelman School of Medicine, University of Pennsylvania, Philadelphia, PA, USA;* [2]*School of Medicine, University of Texas Health Sciences Center, San Antonio, TX, USA*

Chapter Outline

The preceding chapters describe essential aspects of viral pathogenesis, including virus–cell interactions; viral spread within a host; and intrinsic, innate, and adaptive immune responses. This chapter extends the theme and addresses diverse patterns of viral infections that are determined by both the virus and the host. Thus, virulence or susceptibility depends upon the specific virus–host combination. This is particularly true in the case of persistent infections, which involve a delicate balance between virus and host. We will focus first on virus virulence and host susceptibility, and then turn to the complex variables that govern persistent infections. Chapters 4–6, on innate, adaptive, and aberrant immunity, and Chapters 11–15, on systems biology approaches, also provide important insights into the patterns of infection.

1. VIRULENCE DEFINED

Viral virulence (or pathogenicity) is the ability of a virus to cause disease in an infected host. Since variants of a single virus can exhibit different levels of severity, viral virulence is an important consideration in studying patterns of disease. Furthermore, understanding the subtleties of viral virulence has important practical implications, since avirulent or attenuated variants of a virus are often used as live vaccines; examples are those for smallpox, poliomyelitis, measles, and yellow fever.

The virulence phenotype depends upon many variables, including the viral strain; the route of infection; the viral inoculum; and the species, age, and genetic susceptibility of the host. To profile the differences between the virulent and attenuated viruses, it may be necessary to choose an experimental combination of host and route that is somewhere in the spectrum between the most benign and the most severe ends of the disease scale.

1.1 How is Virulence Measured?

The comparative pathogenesis of virulent and avirulent virus strains can be used to elucidate the biological mechanisms underlying their phenotypes. Differences in virulence phenotypes can be qualitative or quantitative. *Qualitative variation* may be manifested in various ways. For instance, viral clones may exhibit differences in their tropism, such that one clone replicates well in the brain while another clone replicates

Viral Pathogenesis. http://dx.doi.org/10.1016/B978-0-12-800964-2.00007-0

TABLE 1 Virus Virulence Can Be Measured in Many Ways, Depending Upon the Pathogenesis of Individual Virus Diseases

Measure of Virulence	Requirement 1	Requirement 2	Experimental Animals or Humans?
Quantitative Measures			
Number of infectious units that cause disease or death (infectious units per LD50)	Titration of virus in cell culture (PFU; TCID; other)	Death or disease endpoint titration in experimental animals	Animals
Ratio of titers to compare different strains of a virus	Titration of virus 1 (PFU per endpoint)	Titration of virus 2 (PFU per endpoint)	Animals
Number of cases of disease per 1000 infections	Count of disease cases in human or animal population	Count of infections in the same population	Humans or animals
Qualitative Measures			
Distribution and severity of pathological lesions	Lesions in different organs and tissues	Semiquantitative measure of pathological lesions	Humans or animals

well in the liver or in the gastrointestinal tract. Alternatively, different clones may spread by different routes, one clone producing viremia whereas another spreads by the neural route. In some instances, viral clones will vary in the innate or acquired immune responses that they induce, or in their susceptibility to antibody or to cellular immune defenses. *Quantitatively*, for any given expression of pathogenicity, it is possible to compare the number of infectious units required to produce a specified outcome such as mortality or other disease parameter for different viral strains (e.g., ID50 or LD50).

Quantitative measurements. Virulence can be measured by various methods that are summarized in Table 1. Death being a binary outcome, it is an ideal measure for experimental models; where the differences in virulence are not as pronounced, a more subtle analysis is useful. For example, virulent and avirulent strains can both be uniformly fatal yet vary in the average days to death of the experimental host. Other common measures of virulence are paralysis (e.g., poliovirus), changes in liver enzymes (any hepatitis virus), decrease in the proportion or number of CD4+ lymphocytes (immunodeficiency viruses), or sometimes more distant surrogate measures such as behavioral abnormalities for some neurotropic viruses.

Most comparisons of virulence use a ratio of infectious units to disease outcome. For instance, for a virus that forms plaques in vitro, calculation of the PFU/LD50 makes it possible to distinguish different degrees of virulence. More modern assays of RNA or DNA copies are useful, but they too have potential pitfalls, as they seldom measure complete copies of the viral genome, and could be affected by incomplete genomes that are incapable of replication. To assess the relative virulence of two strains of a given virus that cause similar disease, the infectious unit per 50% disease endpoint for each can be compared (second row in Table 1).

In a clinical setting, virulence can sometimes be measured, but a different approach is required. In some outbreaks, it is possible to estimate the number of infections that have occurred in a population as well as the number of clinical cases. In this instance, the ratio of cases per 1000 infections provides a measure of virulence.

Qualitative measurements. Qualitative comparisons describe differences in the localization of pathological changes or in their nature. For instance, Figure 1 shows a comparison of the central nervous system (CNS) lesions caused by five neurotropic flaviviruses. Some viruses produced severe lesions in the brain, others targeted the spinal cord, and some produced similar lesions throughout the neuraxis, reflecting the multidimensional nature of virulence.

2. HOW CAN VIRULENCE BE ALTERED? CLASSICAL METHODS

Viruses can be manipulated to alter their virulence, either deliberately or inadvertently. Classically, this was accomplished by growing a virus in either animals or cell culture (Table 2). With the advent of molecular genetics, it has become possible to introduce mutations that change virulence. These studies are most informative when the phenotypes differ as much as possible.

2.1 Selection of Viral Clones

One of the simplest methods to obtain virus strains with differing virulence is to select genetic clones of a virus and compare them using one of the measures described above. Preferably, this is done by picking plaques from a culture plate. This method was used by Albert Sabin to identify the attenuated strains of polioviruses that were developed into

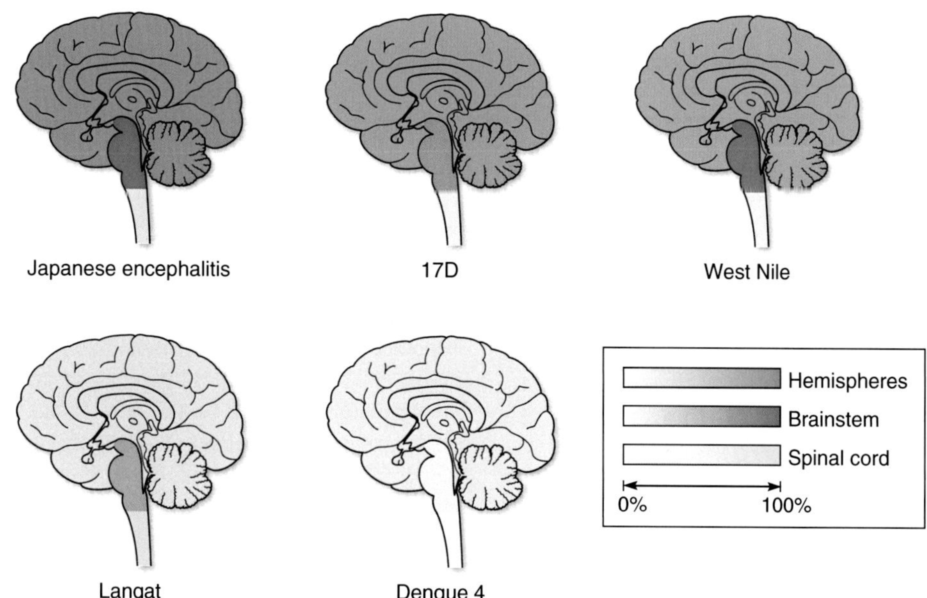

FIGURE 1 Virulence is multidimensional since it may involve different tissue- or organ-specific patterns of infection. In this example, lesion severity is a qualitative as well as a quantitative phenomenon. Nonhuman primates were injected intracerebrally with five different neurotropic viruses and the severity of lesions in different central nervous system regions were graded using a standard scale. *After Nathanson N, Gittlesohn AM, Thind IS, Price WH. Histological studies of the monkey neurovirulence of group B arboviruses. III. Relative virulence of selected viruses.* American Journal of Epidemiology *1967, 85: 503–517, with permission.*

TABLE 2 Experimental Alterations of Viral Virulence

Method to Alter virulence	Enhancement or Attenuation of Virulence
Selection of individual viral clones	Either
Passage in animals	Enhancement (usually)
Passage in cell culture at "standard" conditions	Attenuation (usually)
Passage in cell culture under restrictive conditions such as altered temperature	Attenuation (usually)
Passage in selected cell cultures such as neural cells	Enhancement (usually)
Alter viral genome by introducing point mutations that change viral genes or stop codons that inactivate viral genes	Either

oral poliovirus vaccine (OPV). If a virus cannot be plaqued, then endpoint dilution can be used to (hopefully) select individual virus clones.

2.2 Passage in Animals

In the pioneering days of virology, viruses were often maintained by serial passage in animal hosts. It was found that pathogenicity would change during the course of multiple passages, and this adventitious finding was exploited to obtain viruses of different pathogenicity. In general, during repeated animal-to-animal transmission, a virus adapts in order to replicate optimally under the conditions of passage. Yellow fever, a flavivirus, produces fatal hepatitis in monkeys; if passaged intracerebrally in mice, it will become highly neurovirulent for mice, but will lose its ability to cause hepatitis in monkeys. However in nature, where viruses have adapted to specific hosts, virulence may vary from high (rabies in raccoons or foxes, smallpox in humans) to low (SARS coronavirus in bats, rhinoviruses in humans). Similarly, complex experimental passage of a strain of simian immunodeficiency virus (SIV) led to a neurovirulent strain that is less capable of causing immunodeficiency, but can be used to study neurovirulence when co-inoculated with an immunodeficiency causing virus.

2.3 Passage in Cell Culture

With the advent of cell culture, viruses were usually maintained by in vitro passage. It was soon observed that serial transmission alters the biological phenotype, often reducing virulence for animals or humans. This observation was exploited in the deliberate search for attenuated variants that could be used as prophylactic vaccines. Experience has demonstrated several underlying principles.

- Apparently identical passage lines can yield virus stocks differing in their virulence.
- An RNA virus stock represents a "swarm" of highly related virus genotypes that may have different phenotypes.
- Passage usually selects for virus clones already present in the population that replicate preferentially and thus alter the phenotype of the virus swarm (see Chapter 17, Viral Evolution).

Historically, the failure to recognize the influence of passage upon the biological phenotype has led to some important errors in virological research. For example, intracerebral passage of poliovirus in monkeys leads to selection of variants that are highly neurotropic but have lost much of their infectivity and pathogenicity when administered by the oral (natural) route. Pathogenesis studies with these neuro-adapted poliovirus resulted in the erroneous conclusion that poliovirus was not an enterovirus but was naturally transmitted by the intranasal route, and this misapprehension led to trials of nasal astringent sprays as a method to protect children against paralysis.

More recently, passage of HIV-1 in T cell lines selected for laboratory variants that differed from wild-type virus in their ability to plaque in MT-2 cells (a T cell line), use of the CXCR4 co-receptor (not the CCR5 receptor), and inability to infect macrophages. When used for serological assays, the adapted viruses were readily neutralized by sera from patients naturally infected with HIV. These findings resulted in the mistaken conclusion that HIV-1 could be readily neutralized and the consequent prediction that it would be relatively easy to develop a prophylactic vaccine. Once it was recognized that viral isolates only maintain their natural phenotype if passaged in primary blood mononuclear cells (PBMC), it became clear that many wild type HIV isolates are very resistant to neutralization, presenting a daunting challenge for vaccine development.

2.4 Passage in Cell Culture under Restrictive Conditions

There are a number of methods that have been used to enhance the selection of attenuated virus variants from an uncloned virus stock.

Temperature-sensitive mutants. Wild type viruses will replicate well at 37 °C and often at temperatures up to 40 °C (most cell cultures do not thrive above 40 °C). Temperature sensitive (ts) variants, on the other hand, replicate well at 37 °C but poorly if at all at 40 °C. It is relatively easy to select for temperature sensitive variants and they will often exhibit an attenuated phenotype when tested in animals and a restricted infectivity range in host tissues.

Another method for the selection of attenuated mutants is passage at a low temperature, such as 25 °C, about the lowest temperature at which most mammalian cell cultures can be maintained. A cold-adapted influenza virus is temperature sensitive and exhibits restricted pneumotropism in ferrets, an animal that develops severe pneumonia after intranasal infection with wild-type human influenza viruses (Massaab et al., 1985). This attenuated strain is used as a live influenza vaccine (see Chapter 19, Viral Vaccines).

3. HOW CAN VIRULENCE BE ALTERED? GENETICS TO THE FORE

3.1 Small Changes in the Genome Can Have Large Effects on Virulence

Determinants of viral virulence may be encoded in any part of the viral genome and changes in only a few nucleotides can have dramatic effects on virulence. Over the last few decades, a large body of information has been assembled regarding these determinants, summarized in Sidebar 1. A few salient examples are described below.

Sidebar 1 Genetic determinants of viral virulence and attenuation

- The use of mutant viral clones has made it possible to identify the role of specific genes and proteins as determinants of virulence.
- There are no "master" genes or proteins that determine virulence, and attenuation may be associated with changes in any of the viral proteins as well as in untranslated genomic sequences.
- Virulence phenotypes can be altered dramatically by a change that leads to an alteration in a single amino acid, or by a single nucleotide change in a noncoding region. Variants with mutations in several critical sites may be more attenuated than those with a single point mutation. The frequency of reversion to virulence is inversely proportional to the number of discrete attenuating mutations.
- Reversion to virulence of an attenuated variant can involve back mutation at the genetic site of attenuation, but can also be produced by compensatory mutations at a different site in the same protein or even in another viral protein.
- Attenuating mutations are often host range alterations that affect replication in some cells or tissues but not others.
- Although many attenuated viral variants have been identified, only in a relatively few instances has the mechanism been identified at a biochemical or structural level.

OPV is comprised of attenuated clones of each of the three poliovirus serotypes that—in comparison with virulent wild-type polioviruses—have markedly reduced neurovirulence after direct intrathalamic or intraspinal injection in macaques. OPV viruses that have reverted to virulence after feeding to humans—such strains are routinely isolated from the stool of vaccinated children—have a small number of nucleotide differences in comparison with the attenuated parent vaccine strains. Recombinants between parent and revertant OPV viruses can therefore be used to identify the influence of individual critical nucleotides in neuropathogenesis (Minor, 1992). In type-3 OPV, there are only four important nucleotide determinants, which are located both in the nontranslated region of the genome and in the genes encoding the structural proteins.

In the case of influenza virus, changes in virulence may also be due to specific combinations of viral genes. There has long been a mystery how the 1918 influenza epidemic caused such a high mortality with an estimated 50 million deaths. An analysis using recombinant viruses—constructed through gene reassortment between the reconstructed 1918 virus and a less-pathogenic contemporary strain—indicated that virulence is multigenic (Table 3). Pathogenesis analysis showed that the 1918 virus triggers an outpouring of cells and cytokines in the lung, reflected in the profiles of gene expression. This is an instance of a host response that is a deleterious determinant of virus virulence.

3.2 Manipulating the Viral Genome to Alter Virulence

With the introduction of molecular genetics into virology, it has become relatively easy to alter the viral genome in a controlled manner. Genetic changes that can be deliberately introduced include point mutations that alter function of individual proteins, or inactivation of nonessential viral genes by introduction of stop codons or deletions of substantial gene segments. Relevant methods are described in Chapters 11–13.

For instance, HIV has four accessory genes—Nef, Vif, Vpu, Vpr—that are not essential for replication in most cell cultures. SIV strains lacking one or more of these genes are attenuated when used to infect rhesus macaques, compared to their parental counterparts. In fact, it was hoped that an HIV strain lacking the Nef gene might be used as an attenuated viral vaccine in humans. A group of patients infected with a naturally Nef-deleted strain of HIV remained healthy, absent antiretroviral treatment for many years but eventually developed AIDS, ending this approach to a vaccine; the Nef-deleted HIV strain was attenuated but still too dangerous to use in humans.

TABLE 3 Multigenic Determination of Virulence. The 1918 Influenza Virus (H1N1 Antigenic Type) Was Reconstructed and Used to Make Reassortants with a Contemporary Human H1N1 Isolate (Tx/91). Parent and Reassortant Viruses Were Tested for Virulence after Intranasal Inoculation of Adult Mice. Reassortants Encoding Many of the Genes of the Virulent 1918 Virus Showed Considerable Virulence but Were Less Lethal than the Parent 1918 Virus, Indicating That Virulence was Associated with the Full Spectrum of Viral Genes Acting in Concert

Viral Gene	Genetic Composition of Parent and Reassortant Viruses			
HA	1918	Tx/91	1918	Tx/91
NA	1918	1918	1918	Tx/91
M	1918	1918	1918	Tx/91
NP	1918	1918	1918	Tx/91
NS	1918	1918	1918	Tx/91
P1	1918	1918	Tx/91	Tx/91
P2	1918	1918	Tx/91	Tx/91
P3	1918	1918	Tx/91	Tx/91
Titer in lung (Log10 EID50 per ml)	7.5	5.2	6.0	3.0
Mortality in mice (10^6 PFU intranasal)	100%	0%	100%	0%
Survival (days)	3 days		6.2 days	

This is from Fornek, Korth, and Katze, Advances in Viral Research, 70:81–100, 2007.
After Tumpey TM, Basler CF, Aguilar PV, Zeng H, Solorzano A, Swayne DE, Cox NJ, Katz JM, Taubenberger JK, Palese P, Garcia-Sastre A. Characterization of the reconstructed 1918 Spanish influenza pandemic virus. *Science* 2005, 310: 77–80, with permission.

Another pertinent example is the H5N1 strain of influenza virus that had produced a high mortality among a few poultry farmers in China. To investigate the potential transmissibility (and therefore the virulence for the human population) of H5N1 virus, two research groups independently conducted controversial "gain-of-function" experiments. As a surrogate for human subjects, both investigative teams used ferrets. Although ferrets could be infected intranasally with wild-type H5N1 virus, they did not transmit infection to other ferrets housed in close proximity. However, when the researchers introduced a few mutations and performed animal-to-animal passages, they obtained virus mutants that were transmissible by aerosols. Recent genetic analyses have suggested that there are at least three transmissibility determinants in this H5N1 virus: basic amino acids adjacent to the cleavage site in the viral hemagglutinin; amino acid 627 in PB2, one of the viral polymerase molecules; and a short sequence in the NS1 protein that may influence interferon or other host responses.

4. ENTER THE HOST: HOW VIRULENT AND ATTENUATED VIRUSES DIFFER IN THEIR PATHOGENESIS

The sequential steps in viral pathogenesis are described in Chapter 3, Basic Concepts. The ability of a virus to move through each of these steps can impact virulence. There are examples of viruses that differ in their infectivity at the portal of entry, in their ability to disseminate, or in their replicative capacity in target organs or tissues. In many instances, the pathogenicity of a virus strain is determined by its ability to evade host defenses such as the intrinsic, innate, or adaptive immune response. Once again, it is the virus–host interaction that determines virus virulence.

4.1 Sequential Steps in Infection

Portal of entry. Respiratory viruses such as influenza, that cause severe disease, replicate well in the lower respiratory tract where the temperature is close to 37 °C. However, the rhinoviruses, which do not grow well at 37 °C, cannot infect the lung proper, and are confined to the upper respiratory tract since they are adapted to replicate at 33 °C and only cause mild illness (the common cold). As mentioned above, a cold-adapted influenza virus is sufficiently attenuated to be used as a vaccine. In this instance, attenuation is associated with the ability to replicate well at 33 °C but not at 37 °C.

Viremia. Most systemic viruses spread via the bloodstream. If viral strains differ in the duration and titer of the induced viremia, this may alter their ability to reach critical target organs and thereby influence virulence. Wild-type isolates of poliovirus vary in the degree of viremia that they produce and this correlates with their paralytogenicity after extraneural infection.

Neural spread. Some viruses spread along neural pathways rather than by viremia, and neurally spreading viruses can also be experimentally attenuated. Rabies virus is a good example of an "obligatory" neurotrope. Attenuated vaccine strains of rabies virus, obtained by passage of wild-type virus in nonneural cells, show a marked reduction in their virulence when tested by intracerebral injection in mice. The attenuated phenotype is maintained if the virus is passaged in BHK-21 cells, which come from kidney tissue, but reverts to greater virulence when the virus is passaged in cultured neural cells (usually astrocytic) or in the brains of suckling mice.

Neuroinvasiveness. West Nile virus isolates are grouped into two major lineages, based upon genetic sequence. When tested in mice, all isolates have high neurovirulence, but there are major variations in neuroinvasiveness (Table 4). West Nile virus was introduced into the United States in 1999 (see Chapter 16, Emerging Viral Diseases) and spread from New York state across the country to the West coast, with outbreaks of encephalitis in humans. Recent isolates from humans and birds in the United States are among the most neuroinvasive strains of West Nile virus, consistent with the severity of this still emerging viral disease.

TABLE 4 Different Wild-type Isolates Show Distinct Qualitative and Quantitative Patterns of Virulence. Although All Isolates of West Nile Virus are Equally Neurovirulent, They Differ in Neuroinvasiveness. Virus Strains Were Titrated in 4-Week-old NIH Swiss Mice by Intraperitoneal (ip) or Intracerebral (ic) Routes of Inoculation

Lineage	Virus Strain	Neuroinvasiveness (PFU per ip LD50)	Neurovirulence (PFU per ic LD50)
1	USA99b	0.5	0.1
	EGY50	50	0.7
	AUS91	>10,000	3.2
2	SA58	3.2	0.3
	CYP68	>10,000	0.5

After Beasley DWC, Li L, Suderman MT, Barrett ADT. Mouse neuroinvasive phenotype of West Nile virus strains varies depending upon virus genotype. *Virology* 2002, 296: 17–23, with permission.

4.2 Tissue Tropism

Variants of a single virus can differ in their relative pathogenicity for different tissues or organs, which confers a multidimensional character upon virulence. As mentioned above, in the search for attenuated poliovirus strains, Sabin used passage in cell culture to obtain the virus strains used to formulate OPV. When fed to susceptible subjects, these strains replicated as well as wild-type isolates, but (compared to wild-type isolates) had only minimal neurovirulence in monkeys.

HIV-1 is another example. All HIV-1 strains replicate well in primary cultures of peripheral blood mononuclear cells (consisting mainly of T lymphocytes). Some wild-type strains also replicate in primary cultures of monocyte-derived macrophages but not in transformed lines of T lymphocytes. What is the explanation for these differences in tropism? Macrophage-tropic strains of HIV-1 use the CD4 primary receptor very efficiently; an ability required to infect macrophages as they express CD4 at much lower levels than lymphocytes. Most primary isolates of HIV-1 use only the CCR5 co-receptor and therefore cannot infect T cell lines that express only the CXCR4 co-receptor. However, primary cultures of T lymphocytes express high levels of CD4, and both co-receptors, CCR5 and CXCR4, and can be used by all strains of HIV-1.

4.3 Host Intrinsic Response

Animal hosts have evolved a large set of constitutive non-immune defenses against viral invaders, many of which have been discovered only recently. In turn, viruses have evolved counter measures to overcome these host defenses, and these counter measures are one determinant of virulence. The best studied example is HIV-1. A comparison of HIV-1 with the enzootic SIV viruses of African nonhuman primates has identified a number of host cellular "antiviral" genes and proteins, such as APOBEC3C, tetherin, and trim5α, which constitute host intrinsic defenses. In some instances, there are cognate viral proteins that counteract these intrinsic host defenses (see Chapter 9, HIV and AIDS and Chapter 16, Emerging Viral Diseases). For example, SIVcpz of chimpanzees was acquired from African monkeys, but underwent key genetic changes to circumvent the intrinsic antiviral proteins encoded in the chimpanzee genome. In turn, these mutations permitted SIVcpz to cross the species barrier into humans, who share similar antiviral proteins with chimpanzees.

4.4 Host Innate Response

The pattern and dynamics of the innate response to a specific virus strain can play an important role in its virulence. A variety of studies have been done using "omics" approaches to understand the innate response to virus infections.

When virulent and avirulent strains of H1N1 influenza virus were compared in mice, the virulent virus initiated a faster, more robust, and sustained inflammatory response in the lungs (Figure 2), resulting in severe lesions (Korth et al., 2013). It is speculated that a similar phenomenon accounted

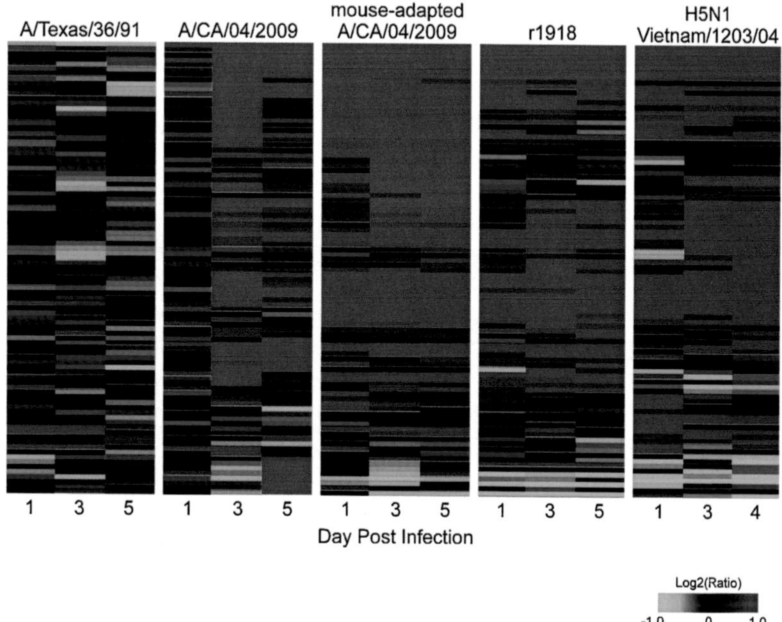

FIGURE 2 Differential induction of inflammatory gene expression by influenza viruses causing mild or severe respiratory disease. Shown are changes in inflammatory gene expression profiles over time in the lungs of mice infected with A/Texas/36/91 (a nonpathogenic seasonal isolate), A/CA/04/2009 (a mildly pathogenic 2009H1N1 pandemicisolate), or highly pathogenic mouse-adapted 2009H1N1, r1918, or avian H5N1 viruses. Expression values are represented as the average of the log2ratio of infected to respective mock-infected samples for three biological replicates per condition. Red and green indicate that gene expression is increased or decreased relative to mock, respectively. *From Korth MJ, Tchitchek N, Benecke AG, Katze MG. Systems approaches to influenza virus host interactions and the pathogenesis of highly virulent and pandemic viruses.* Seminars in Immunology. *2013; 25(3):228–39, with permission.*

for the virulence of the 1918 influenza virus in humans. Another study found that the dynamics of the inflammatory response to H5N1 and H1N1 viruses in mice plays a key role in the outcome of infection. A rapid response that quickly resolves is associated with survival and recovery. If the infection is not resolved quickly, the response may cause so much inflammation that it enhances pulmonary dysfunction and endangers the host.

Fatal Ebola infections are associated with an excess outpouring of proinflammatory cytokines (a cytokine "storm") and a disseminated intravascular coagulopathy (McElroy et al., 2014). The terminal events appear to be a shock syndrome accompanied by multiple organ failure. To construct a mouse model of Ebola disease, it was necessary to use a mouse-adapted strain of the virus (Rasmussen et al., 2014). A screen of inbred lines of mice showed that only a few selected lines were susceptible to full-blown lethal hemorrhagic shock syndrome, while other lines could be infected but recovered. Lethal outcomes were associated with inflammatory signaling and vascular permeability, markers of a cytokine storm. In this mouse model, differences in host susceptibility appear to be associated with different alleles of the endothelial tyrosine kinases Tie1 and Tek (Tie2). How these genes influence the likelihood of a cytokine storm remains to be clarified.

4.5 Host Adaptive Immune Response

Altered virulence and pathogenicity of variant viruses may be mediated through the host adaptive immune response (see Chapter 5 Adaptive Immunity). For instance, clone 13 of lymphocytic choriomeningitis virus (LCMV) differs from the Armstrong strain by virtue of its ability to replicate more rapidly in dendritic cells and macrophages (professional antigen-presenting cells). The rapid destruction of dendritic cells by clone 13 interferes with antigen presentation, thereby suppressing the immune response which, in turn, permits the virus to escape clearance. As a result, clone 13 initiates a persistent infection without acute illness. In contrast, the Armstrong strain, which does not interfere with induction of adaptive immunity, causes a benign immunizing infection with rapid virus clearance. These differences have been mapped to just two amino acids in the surface protein of LCMV. This example illustrates the delicate balance between acute and persistent infections caused by different variants of a single virus.

4.6 Viroceptors and Virokines

DNA viruses with a large genome, particularly the herpesviruses and the poxviruses, encode a number of proteins that counter host defenses. Virokines are viral proteins that mimic host cytokines stimulating cell proliferation and increasing the number of virus targets. Viroceptors are viral proteins that mimic receptors for host defensive cytokines, "decoying" them away from their intended cellular receptors. For example, vaccinia virus encodes a complement control protein that blocks the complement cascade and a tumor necrosis factor viroceptor that binds this host defense molecule. Herpes simplex virus (HSV) encodes two glycoprogeins, gE and gI, that act as an Fc receptor; the receptor binds and inactivates antiviral antibodies.

5. HOW DO VIRUSES PERSIST?

The prototypical viral infections are acute, and induce host-defensive responses that clear the virus and leave the host with long-lasting virus-specific immunity. However, many viruses are capable of persisting, often for the lifetime of the host. In order to persist, a delicate balance must be achieved so that, on the one hand, the host is not killed by the destructive effects of the virus while, on the other hand, the virus is able to evade the multitude of immune defenses that act to eliminate it. How this happens is the theme of this section.

The mechanisms of persistence range along a spectrum (Sidebar 2). At one extreme are viruses that continue to replicate at high titers over long periods of time, while at the other extreme are viruses that become latent, emerging at rare intervals to replicate for short periods of time. Between these ends of the spectrum are examples of smoldering infections that share characteristics of both replication and latency. Viruses employ a variety of strategies to escape immune surveillance, and these tend to be specific for different styles of persistence. Thus, immune tolerance often characterizes high titer persistent infections, whereas active immune responses are seen in many latent infections. Some selected examples of each style of persistence are listed in Table 5.

5.1 Immune Clearance of Acute Viral Infection

As a prelude to consideration of persistence, it is useful to briefly recapitulate the mechanisms by which the immune response controls and eliminates an acute virus infection. During the innate response, effector T lymphocytes can destroy virus-infected cells, produce antiviral cytokines, and recruit mononuclear cells to sites of viral replication and destruction (see Chapter 4, Innate Immunity). During the adaptive response, antibody neutralizes and opsonizes free infectious virions (see Chapter 5, Adaptive Immunity). In some instances, both antibody and virus-specific effector lymphocytes can purge virus-infected cells without destroying them. It is these mechanisms that a virus must evade in order to persist.

5.2 High Titer Infections

For a persistent virus to replicate at high titer, it must avoid catastrophic pathogenic effects, either because it is

Sidebar 2 Mechanisms of persistence

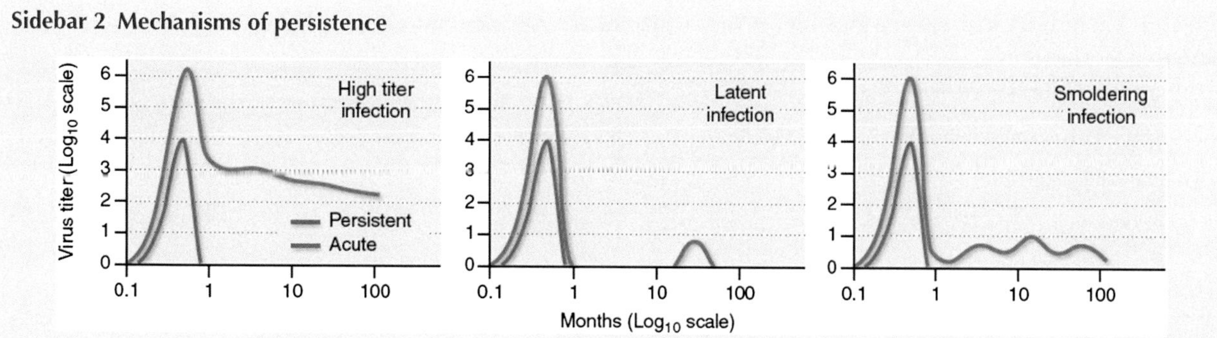

- *High titer replication* requires that virus either be noncytocidal or that there is rapid replacement of target cells by cellular proliferation. Immune surveillance is unable to eliminate the virus, due to tolerance, immune complex formation, viral mutation, or other mechanisms.
- *Latency* usually requires that the viral genome persist in a nonreplicating mode, either integrated into the genome of the host cell or as an episome, although intermittent active replication may occur. Immune surveillance may be competent to eliminate replicating virus but not latent viral genomes.
- *Smoldering infections* involve continuous productive infection and cell-to-cell transmission at a low level. Potentially effective immune surveillance is circumvented by mechanisms such as antigenic variation, infectious immune complexes, or intercellular bridges.

Modified after Johnson R. Neurotropic Virus Diseases. *Raven Press, 1985, with permission.*

not acutely cytocidal or because it attacks target cells that can be replenished regularly and effectively. Many viruses can replicate productively without causing cell death, and a number of them can cause persistent infections. In such instances, the initial dynamics resemble those of an acute infection, following which the virus titer decreases somewhat but then reaches a set point that may be maintained indefinitely or gradually decline. Examples of this pattern are hepatitis B virus (HBV), HIV, and LCMV of mice.

Immune tolerance and persistent infection with nonlytic viruses. A virus that persists at a high level has to escape an effective immune response that would control the infection. Therefore, high-titer virus persistence is often accompanied by immune "tolerance," an apparent absence of a virus-specific adaptive immunity. The mechanisms by which tolerance can be induced include deletion of "forbidden" clones of naïve T lymphocytes in the thymus, or exhaustion of peripheral virus-specific T lymphocytes in the presence of excess antigen (see Chapter 6, Aberrant Immunity). Tolerance may be limited to specific components of the effector limb of the immune response. For instance, hepatitis B persistence is characterized by absence of antibody against HbsAg but not against HBcAg. LCMV persistence is characterized by absence of cellular immune responses while virus-specific antibody is produced.

Virus-specific exhaustion of effector T lymphocytes has best been analyzed in the LCMV model of persistent infection. There is a marked difference in the properties of LCMV-specific CD8+ virus-specific memory T cells that are generated during acute versus persistent infections. Acute infections induce memory T cells that exhibit the cardinal properties of self-renewal in the absence of antigen (and high levels of receptors for IL-7 and IL-15), whereas the memory T cells from animals with persistent infections express low levels of these two interleukin receptors, and gradually disappear in the absence of antigen. In this model, the memory cells associated with persistent infection fail to differentiate into effector CD8+ T cells capable of eradicating persistent infection.

Evidence in the LCMV model for the role of immune tolerance in maintaining viral persistence is provided by the experimental termination of persistence by intravenous injection of virus-specific CD8+ T cells, obtained from animals undergoing acute infection. Similar results have been obtained with HBV, where cytolytic T lymphocytes (CTLs) specific for HBs epitopes clear virus from hepatocytes.

Lytic viruses. It is unusual for high titer persistence to be produced by a cytolytic virus, but the primate lentiviruses represent an important exception. The main target cells for these lentiviruses are CD4 lymphocytes that often undergo lytic infection. It has been calculated that the continuous destruction of CD4 cells results in a reduction of the average half-life of these cells from 75 to 25 days. However, early in the course of infection the bone marrow is able to respond to the abnormal rate of destruction by increasing the production of naïve CD4 cells at a rate sufficient to maintain a reasonable concentration of circulating CD4 cells. This permits a lytic virus to persist at a high titer for an extended period of time in the relative absence of clinical illness; eventually, the bone marrow is unable to compensate and CD4 levels drop, leading to functional immunodeficiency.

TABLE 5 A Selected List of Human and Animal Viruses that Cause Persistent Infections through Different Mechanisms

Virus family Virus example Disease	Host(s)	Site of persistence	Cytocidal in permissive cells	Immune response
High Titer Replication				
Hepadnaviruses Hepatitis B virus Cirrhosis	Human (newborn)	Hepatocytes	No	Split tolerance
Flaviviruses Hepatitis C virus Cirrhosis	Human	Hepatocytes	No	Variable
Arenaviruses Lymphocytic choriomeningitis virus Glomerulonephritis	Mouse (newborn)	Macrophages, other cells	No	Split tolerance
Latent Infection				
Herpesviruses Herpes simplex virus Cold sores, encephalitis	Human	Sensory neurons	Yes	Yes
Cytomegalovirus Pneumonitis, retinitis, hepatitis	Human	Lymphocytes	Yes	Yes
Epstein–Barr virus Mononucleosis	Human	B cells	Yes	Yes
Varicella-zoster virus Herpes zoster	Human	Sensory neurons	Yes	Yes
Smoldering Infection				
Paramyxoviruses Measles Subacute sclerosing panencephalitis	Human	Neurons, glia	Yes	Super normal
Lentiviruses HIV AIDS	Human	CD4 lymphocytes	Yes	Variable
Polyomavirus JC virus Progressive multifocal leucoencephalopathy	Human	Oligodendrocytes	Yes	Yes
Lentiviruses SIV AIDS	Nonhuman primates	CD4 lymphocytes	Yes	Variable
Lentiviruses Visna-maedi virus Pneumonitis, encephalitis	Sheep	Monocytes	Yes	Yes
Equine infectious anemia virus Anemia	Horses	Monocytes	Yes	Yes
Oncogenic Infection				
Polyomaviruses Human papillomavirus Carcinoma cervix, other sites	Human	Epidermal cells	Yes	Yes
Hepadnaviruses Hepatitis B virus Hepatocellular carcinoma	Human (newborn)	Hepatocytes	No	Split tolerance

TABLE 5 A Selected List of Human and Animal Viruses that Cause Persistent Infections through Different Mechanisms—cont'd

Virus family Virus example Disease	Host(s)	Site of persistence	Cytocidal in permissive cells	Immune response
Flaviviruses Hepatitis C virus Hepatocellular carcinoma	Human	Hepatocyte	No	Variable
Herpesviruses Epstein–Barr virus Burkitt's lymphoma	Human	B cells	Yes	Yes
Herpesvirus Human herpesvirus 8 Kaposi's sarcoma	Human	Lymphocytes	Yes	Yes
Lentivirus Human lymphotropic virus-1 Adult T cell leukemia	Human	CD4+ lymphocytes	No	Yes

In contrast to most high titer persistent infections, lentiviruses induce immune responses rather than tolerance. The immune response to lentiviruses is quite effective, as judged by its ability to rapidly contain the acute phase of infection, resulting in a reduction from peak viremia at about 6 weeks to a set point about 1000-fold lower at about 3–6 months. Once this set point is reached, a dynamic equilibrium is established between virus production and clearance. The plasma half-life of individual SIV virions is <30 min in the absence of immunity and about 10 min in infected animals with an established immune response. It has been calculated that to maintain virus titers of 10^2–10^4 infectious virions per ml of plasma requires the production of 10^{10}–10^{12} new infectious virions daily. In this instance, high titer persistence is maintained by an extraordinary rate of virus production that exceeds the rate at which a potent cellular immune response can clear virus-infected cells. See Chapter 15, Mathematical Approaches, for a discussion of how differential equations are used to model viral production and clearance.

5.3 Latent Infections

Latent infections are produced by a considerable number of human herpesviruses, including HSVs, varicella-zoster virus (VZV), Epstein–Barr virus (EBV), and cytomegalovirus (CMV). There is a characteristic sequence of events following primary infection. Initially, the virus replicates in permissive cells at the portal of entry. The virus is lytic and destroys these permissive target cells. Once immune induction has occurred, the virus is cleared and appears to be eliminated.

However, the viral genome persists in a latent form. Latency occurs in one or more cell types—such as neurons for HSV and VZV—that are distinct from the permissive cell types that support productive lytic infection. Neurons are restrictive or permissive, depending upon their physiological state. Under conditions of restriction, the virus undergoes the early steps of entry and uncoating, but further steps in replication are blocked. In some instances, the double-stranded DNA genome integrates into the host genome, while in other examples the genome persists as a nonintegrated episome, in the nucleus or cytoplasm.

If latency occurs in cell types that—like neurons—do not divide, then there is no need to replicate the latent genome. If the viral genome is integrated into the host genome, as with retroviruses, then it will be automatically replicated during the cell cycle. Episomal DNA can also be replicated by the enzymes involved in copying cellular genomes. However, there are no parallel mechanisms for RNA, so RNA viruses cannot assume a latent state unless they undergo reverse transcription to DNA intermediates.

Latency maintains the viral genome for the lifetime of the infected host. Activation of latent infections may occur at irregular intervals, or it may never occur in some infected individuals. Activation of latent genomes can be initiated by a number of stimuli, characteristic for each virus. For instance, HSV can be activated by fever, sunburn, and trigeminal nerve injury. Most of these stimuli appear to act upon the primary sensory neurons in which latent HSV genomes are maintained. Waning of the immune response can enhance the risk of activation of some herpesviruses, such as VZV.

Following reactivation of HSV, the viral genome may spread by axoplasmic transport in both centripetal and centrifugal directions. Centrifugal spread conducts the virus to the skin where it may replicate and spread, causing herpes labialis ("fever blister" or "cold sore"). After spreading for a few days, host defenses prevent further spread, and the skin lesion heals. Centripetal spread from the trigeminal ganglion conducts the HSV genome to the CNS, where, in relatively few instances, it can cause a devastating encephalitis.

Typically, viruses that cause latent infections induce innate and acquired immune responses, brisk and potent immune response that clears the initial infection. When the latent infection is activated, immune surveillance limits its spread, but virus produced during activation may be spread to another host. For instance, activation of latent VZV produces characteristic skin lesions in older adults; seronegative children exposed to virus aerosolized from these lesions can develop chicken pox, the primary form of VZV infection.

5.4 Smoldering Infections

"Smoldering" infections fall between the extremes of high titer persistence and latency. Infectious virus is produced, but at minimal levels that may require special methods for detection and isolation. Virus continues to spread from infected to uninfected cells but often at an indolent tempo. If the virus is pathogenic, it may produce a gradually progressive chronic disease. There is a detectable immune response to the virus. The ability of a virus to spread in the presence of a potentially effective immune response is a paradoxical phenomenon, and involves a variety of strategies, several of which are described below.

Immunologically privileged sites. There are a few organs and tissues that appear to favor virus persistence, particularly the brain and kidney. The brain has classically been considered an immunologically "privileged" site because immunological effector mechanisms may spare "foreign" cells in the brain (in contrast to foreign cells in other sites). There are at least two factors that account for virus persistence in the brain. First, the blood–brain barrier limits the trafficking of lymphocytes through the brain and, second, neurons express little if any MHC Class I molecules, rendering them relatively poor targets for virus-specific CTLs.

The kidney is the other major tissue that frequently harbors persistent viruses, such as JC and BK polyomaviruses, and cytomegalovirus. There is no clear explanation why virus in the kidney should be able to evade immunological surveillance, although it has been speculated that lymphocytes may not readily cross the subendothelial basement membrane to access infected glomerular epithelial cells.

Infectious immune complexes. In some instances where a virus persists in the presence of an active immune response, infectivity in the blood circulates in the form of immune complexes that are composed of infectious virions coated by virus-specific antibodies. Immune complexes can be demonstrated by the addition of anti-IgG antisera that will "neutralize" the infectivity. The molecular mechanism through which an antibody-coated virion can retain its infectivity has never been well elucidated. One possibility is that the complex is bound to Fc receptors on macrophages and internalized in vacuoles in which the complex dissociates, followed by infection of the macrophage.

Antigenic variation. During the course of persistent infection, there may be a selection for viral variants that are able to escape neutralization. Such resistant virus variants usually represent point mutations, often in the viral attachment protein. This phenomenon has been observed with several persistent lentiviruses such visna/maedi virus of sheep, equine infectious anemia virus, and very prominently HIV (discussed in Chapter 9, HIV and AIDS).

Intercellular bridges. In some instances, the process of entry of viruses into cells can be short circuited, so that a transient intercellular bridge is formed. The bridge permits the viral genome to pass from cell to cell without having to survive in the extracellular environment, thus providing a means of avoiding neutralizing antibody. This phenomenon is probably operative in subacute sclerosing panencephalitis (SSPE), a progressive fatal neurological disease (discussed in more detail below). In SSPE, a defective variant of measles or rubella virus spreads gradually from neuron to neuron in spite of extraordinarily high titers of neutralizing antibody in the extracellular fluid of the brain parenchyma.

5.5 Clinical Examples of Smoldering Infections

In its typical course, measles spreads primarily to children by the respiratory route and causes a systemic febrile infection with a rash that usually resolves in 1–2 weeks with no serious consequences. However as indicated previously, SSPE is a rare complication of measles (~1 case per 100,000 primary infections) that occurs unpredictably in apparently normal children following uneventful recovery from acute measles. Several years after measles, these children develop a progressive encephalitis that is invariably fatal in 6–12 months. Measles antigens can be detected in either biopsy or postmortem brain, and electron microscopy reveals measles nucleocapsids in neurons and glial cells in the brain.

SSPE variant viruses exhibit underexpression or defects in one or more viral proteins; these mutations inhibit budding and the production of free-infectious virions. In this example, escape from immune surveillance is associated with the selection of variant viruses that lose the ability to mature and bud properly while maintaining the gene functions for replication of the viral nucleocapsid. The incomplete virus core can still be transmitted from cell to cell, and is slowly cytocidal, leading to the clinical progressive and ultimately fatal encephalitis.

HBV infection follows different courses, depending upon age at the time of infection. Infection of adults is often a self-limited acute process, with clearance of the virus usually accompanied by hepatitis. Neonatal infection (transmission from a woman who is a chronic virus carrier to her newborn infant) usually leads to persistence (Figure 3). The infected infant has high levels of circulating HBsAg and

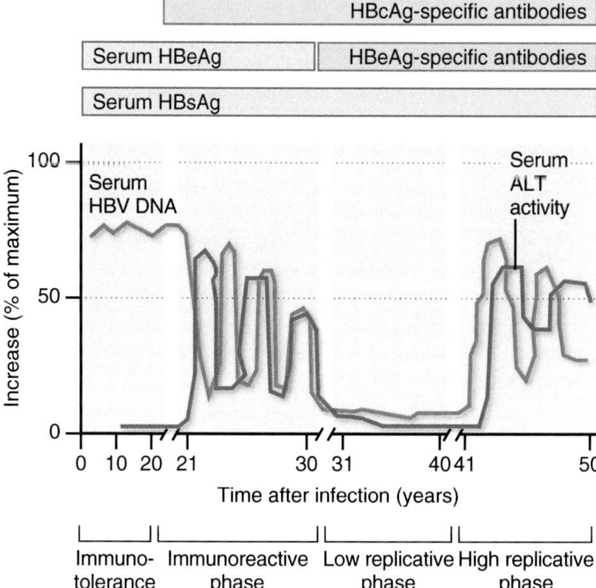

FIGURE 3 Complexities of a smoldering virus infection. The course of persistent hepatitis B virus following infection of a newborn infant. Most such HBV infections begin as an immunotolerant process with high virus titers and a minimal cellular immune response, which may last for decades to a lifetime. However, for reasons that are not known, the infection may shift into one of three other modes, a low replicative phase, an immunoactive phase, or a high replicative phase. In the low replicative phase, patients may control or clear the infection with the development of anti-HBe antibody and recovery of cellular immunity. In the immunoactive or high replicative phases, the immune response attempts to clear the virus but is only partially successful, with resultant necroinflammatory hepatitis that may progress to cirrhosis (permanent liver damage). ALT, alanine transaminase; a liver enzyme whose level in the serum reflects liver function. *After Rehermann B, Nascimbeni M. Immunology of hepatitis B virus and hepatitis C virus infection.* Nature Reviews Immunology *2005, 5: 215–229, with permission.*

serum HBV DNA. Anti-HBs (viral surface antigen) or anti-HBe (precore antigen) antibodies cannot be detected in the serum of infected infants but anti-HBc (viral core antigen) antibodies are present, a state sometimes called "split tolerance." Persistence is associated with high dose cellular tolerance against HBsAg, with only minimal levels of HBs-specific CD8+ T cells due to the deletion or exhaustion of antigen-specific T lymphocytes. Since CD8+ T cells play a major role in control or clearance of HBV, this is a major factor in persistence.

Persistent HBV high titer replication may last for decades to a lifetime. However, through an unknown mechanism, the infection may shift into one of three other modes, a low replicative phase, an immunoactive phase or a high replicative phase. In the low replicative phase, patients may control or clear the infection with the development of anti-HBe antibody and recovery of cellular immunity. In the immunoactive or high replicative phases, the immune response attempts to clear the virus but is only partially successful, with resultant necroinflammatory hepatitis that may progress to cirrhosis (permanent liver damage) and/or hepatocellular carcinoma.

6. REPRISE

Virus virulence and host susceptibility are interdependent properties that are determined by the virus host combination and that together result in the manifestations of infection. A virus that is virulent in one setting may be innocuous in another, and a host may be susceptible or resistant depending upon age, route of infection, or properties of the virus.

Virus virulence can be measured in a variety of ways, based on mortality, illness, or pathological lesions, each of which can be quantified. The virulence phenotype also may be qualitative, involving differences in the tropism of different viral variants. Wild-type isolates of a virus may vary in virulence, and virulence variants can also be selected by experimental manipulation. Attenuated variants often exhibit host range or temperature-sensitive phenotypes, and may provide candidates for live virus vaccines. The attenuated phenotype can be manifested at any step during the course of infection, from invasion, to spread, involvement of target organs, or shedding. Virulence or attenuation can be mapped to specific viral genes and individual nucleotides, and may be associated with one or many viral genes, including noncoding sequences. For viruses with large and complex genomes, virulence may be conferred by accessory genes that act as virokines or viroceptors.

A number of viruses from a wide variety of virus families, can persist for months, years, or even lifelong. In persistent infections there is a delicate balance, so that the host is not killed by the destructive effects of the virus and the virus manages to evade immune surveillance. Most persistent infections fall into one of three distinct patterns. (1) Persistence of the virus—usually a noncytocidal agent—at high titer in various tissues, often associated with immune tolerance due to deletion or exhaustion of antigen-specific T lymphocytes. (2) Latent infections, in which the virus persists as an untranslated genome that eludes recognition by the host immune response; Latent infections may reactivate periodically and reactivations are usually terminated by host immune surveillance. (3) Smoldering infections fall between the other two patterns, with the virus continuing to replicate but at a low level in the face of a brisk immune response. Evasion of immune surveillance is achieved by a number of mechanisms, such as immunologically privileged tissue sites, intercellular bridges, and antigenic variation. Persistent infections may be asymptomatic but often are associated with a wide variety of chronic diseases. As we will see in the following chapter, some persistent viral infections can also lead to cancer.

FURTHER READING

Further Reading: Virus Virulence and Host Susceptibility

Reviews, Chapters, and Books

Belser JA, Tumpey TM. H5N1 pathogenesis studies in mammalian models. Virus research 2013, 178: 168–185.

Cox JE, Sullivan CS. Balance and stealth: the role of noncoding RNAs in the regulation of virus gene expression. Annual Review of Virology 2014, 1: 89–109.

Doering TA, Crawford A, Angelosanto JM, et al. Network Analysis Reveals Centrally Connected Genes and Pathways Involved in CD8+ T Cell Exhaustion versus Memory. Immunity 2012, 37: 1–15.

Fornek JL, Korth MJ, Katze MG. Use of functional genomics to understand influenza-host interactions. Advances in Virus Research 2007, 70: 81–100.

Korth MJ, Tchitchek N, Benecke AG, Katze MG. Systems approaches to influenza-virus host interactions and the pathogenesis of highly virulent and pandemic viruses. Seminars in Immunology. 2013;25(3):228–39. http://www.sciencedirect.com/science/article/pii/S1044532312001030.

Minor PD. The molecular biology of poliovaccines. Journal of General Virology 1992, 73: 3065–3077.

Original Contributions

Cilloniz C, Pantin-Jackwood MJ, Ni C, et al. Molecular signatures associated with Mx1-mediated resistance to highly pathogenic influenza virus infection: mechanisms of survival. J Virology 2012, 86: 2437–2446.

Clark HF. Rabies viruses increase in virulence when propagated in neuroblastoma cell culture. Science 1978, 199: 1072–1075.

Engelmann F, Josset L, Thomas Girke T, et al. Pathophysiologic and Transcriptomic Analyses of Viscerotropic Yellow Fever in a Rhesus Macaque Model. PLOS neglected diseases 2014 8 (e3295): 1–16.

Herfst S, Schrauwen FJ, Linster M, et al. Airborne transmission of influenza A/H5N1 virus between ferrets. Science 2012, 336: 1534–1541.

Imai M, Watanabe T, Hatta M, et al. Experimental adaptation of an influenza H5 HA confers respiratory droplet transmission to a reassortant H5 HA/H1N1 virus in ferrets. Nature 2012, 486: 420–428.

Josset L, Belser JA, Pantin-Jackwood MJ, et al. Implication of inflammatory macrophages, nuclear receptors, and interferon regulatory factors in increased virulence of pandemic 2009 H1N1 influenza A virus after host adaptation. J Virology 2012, 86: 7192–7206.

Linster M, van Boheeman S, de Graaf M, et al. Identification, characterization, and natural selection of mutations driving airborne transmission of A/H5N1 virus. Cell 2014, 157: 329–339.

Maassab HF, De Border DC. Development and characterization of cold-adapted viruses for use as live virus vaccines. Vaccine 1985, 3: 355–369.

McElroy AK, Erickson BR, Flieststra TD, et al. Ebola hemorrhagic fever: novel biomarker correlates of clinical outcome. J Infectious Diseases 2014 210: 558–566.

Muramoto Y, Shoemaker JE, Le MQ, et al. Disease severity is associated with differential gene expression at the early and late phases of infection in nonhuman primates infected with different H5N1 highly pathogenic influenza viruses. J Virology 2014, 88: 8981–8997.

Nathanson N, Gittelsohn AM, Thind IS, Price WH. Histological studies of the monkey neurovirulence of group B arboviruses. III. Relative virulence of selected viruses. American Journal of Epidemiology 1967, 85: 503–517.

Rasmussen AL, Okumura A, Ferris MT, et al. Host genetic diversity enables Ebola hemorrhagic fever pathogenesis and resistance. Science 2014, 30 October 2014/page 1/10.1126/science.1259595.

Sabin AB, Hennessen WA, Winsser J. Studies on variants of poliomyelitis virus. Journal of Experimental Medicine 1954, 99:551–576.

Tchitchek N, Eisfeld AJ, Tisoncik-Go J, et al. Specific mutations in H5N1 mainly impact the magnitude and veolicty of the host response in mice. BMC Systems Biology 2013, 7: 69.

Tumpey TM, Basler CF, Aguilar PV, Zeng H, Solorzano A, Swayne DE, Cox NJ, Katz JM, Taubenberger JK, Palese P, Garcia-Sastre A. Characterization of the reconstructed 1918 Spanish influenza pandemic virus. Science 2005, 310: 77–80.

Watanabe T, Tisoncik-Go J, Tchitchek N, Watanabe S, Benecke AG, Katze MG, et al. 1918 Influenza Virus Hemagglutinin (HA) and the Viral RNA Polymerase Complex Enhance Viral Pathogenicity, but Only HA Induces Aberrant Host Responses in Mice. Journal of Virology. May 1, 2013;87(9):5239. http://jvi.asm.org/content/87/9/5239.abstract.

Further Reading: Persistent Infections

Reviews, Chapters, and Books

Ahmed R, Chen ISY, editors. Persistent viral infections. John Wiley & Sons, New York, 1999.

Oldstone MBA. Viral persistence: parameters, mechanisms and future predictions. Virology 2006, 334: 111–118.

Weiland SF, Chisari FV. Stealth and cunning: hepatitis B and hepatitis C viruses. J Virology 2005, 79: 9369–9380.

Original Contributions

Cattaneo R, Rebman G, Baczko K, ter Meulen V, Billeter MA. Altered ratios of measles virus transcripts in diseased human brains. Virology, 1987, 160: 523–526.

Chen M, Sallberg M, Hughes J, Jones J, Guidotti LG, Chisari FV, Billaud J-N, Milich DR. Immune tolerance split between hepatitis B virus precore and core proteins. J Virology 2005, 79: 3016–3027.

Landais I, Nelson JA. Functional genomics approaches to understand cytomegalovirus replication, latency and pathogenesis. Current Opinion in Virology. 2013;3(4):408–15. http://www.sciencedirect.com/science/article/pii/S1879625713000965.

Lederer S, Favre D, Walters K-A, Proll S, Kanwar B, Kasakow Z, et al. Transcriptional Profiling in Pathogenic and Non-Pathogenic SIV Infections Reveals Significant Distinctions in Kinetics and Tissue Compartmentalization. PLoS Pathog. 2009;5(2):e1000296. http://dx.doi.org/10.1371%2Fjournal.ppat.1000296.

Derek DW, Gardner CL, Sun C, et al. RNAviruses can hijack vertebrate microRNAs to suppress innate immunity. Nature 2014, 506: 245–249.

Poole E, Wills M, Sinclair J. Human Cytomegalovirus Latency: Targeting Differences in the Latently Infected Cell with a View to Clearing Latent Infection. New Journal of Science. 2014;2014:10. http://dx.doi.org/10.1155/2014/313761.

Thimme R, Wieland S, Steiger C, Ghrayeb J, Reimann KA, Purcell RH, Chisari FV. CD8+ T cells mediate viral clearance and disease pathogenesis during acute hepatitis B virus infection. J virology 2003, 77: 68–76.

Wherry, E.J. T cell exhaustion. 2011. Nat. Immunol. 131(6):492–499.

Chapter 8

Viral Oncogenesis

Infections that Can Lead to Cancer

Nicholas A. Wallace, Denise A. Galloway
Fred Hutchinson Cancer Research Center, Seattle, WA, USA

Chapter Outline

1. INTRODUCTION

Infectious agents are associated with an estimated 20% of human cancers. Moreover, increased tumor incidence in immunocompromised individuals suggests that nononcogenic pathogens are cofactors in additional types of cancers. Of these transmittable carcinogens, the vast majority are viruses. We begin our discussion of infections that can lead to cancer with a brief introduction to the mechanisms of virus-induced transformation, oncogenesis, and a few of the key cellular proteins and pathways related to these processes. We then discuss representative examples of the different kinds of oncogenic viruses.

1.1 Cellular Immortalization, Transformation, and Oncogenesis

Oncogenesis is a multistep process through which otherwise normal cells are transformed to become cancer cells. Typically, a normal cell will grow for a finite number of population doublings before undergoing senescence, a permanent exit from cell cycle progression. In culture, primary cells reach this limit after undergoing mitosis about 50 times. Because they are no longer capable of proliferating, senescent cells eventually die off. Immortalization, or the ability to bypass this restriction, is central to oncogenesis. When the number of cell doublings is no longer limited, cells can continue to grow unencumbered for an infinite number of divisions.

Despite the decreased restraint on replication, immortalized cells must undergo further changes before they are considered truly transformed cells. Specifically, they must also lose contact inhibition, so that close contact with another cell no longer limits cellular proliferation. Once contact inhibition has been overcome, cells must also achieve anchorage-independent growth, demonstrated by the ability of a cell to form a colony while suspended in soft agar. Finally, a cell is considered truly transformed if it is capable of inducing a tumor when transplanted into an animal.

Although alteration of many cellular pathways is required to achieve the phenotypes associated with transformation, reduced cell cycle control, reduced or lost apoptotic response, and loss of genome stability are each hallmarks of transformed cells. Many oncogenic viruses achieve this deregulation by targeting the p53 and pRB tumor suppressor proteins, as discussed later in this chapter.

Viral Pathogenesis. http://dx.doi.org/10.1016/B978-0-12-800964-2.00008-2

1.2 Discovery of Oncogenes

The discovery of oncogenes, genes that when overexpressed or mutated can drive tumorigenesis, can be directly traced to the discovery of Rous sarcoma virus (RSV), an oncogenic virus that causes tumors in chickens. While working at Rockefeller Institute in New York in the early 1900s, Peyton Rous discovered that a cell-free supernate prepared from a homogenate of certain chicken tumors was capable of inducing similar tumors in other chickens. This discovery would ultimately win him the Nobel Prize in Physiology or Medicine. In combination with the work of many other laboratories, it eventually led to the concepts of oncogenes as well as infectious agents capable of inducing cancer. (See Chapter 21 "Breakthrough: Nobel Prize discoveries in viral pathogenesis" for a more detailed description of this discovery.)

1.3 Types of Oncogenic Viruses

The diversity of viruses is staggering. Oncogenic viruses are generally classified into RNA and DNA tumor viruses, in part because they use different mechanisms to cause tumors. These two larger groups are subdivided into smaller sets, with RNA oncogenic viruses divided on the basis of their mechanism of transformation, and DNA oncogenic viruses clustered on the basis of the relative size of their genomes.

2. ONCOGENIC RNA VIRUSES

Most of the known oncogenic RNA viruses are retroviruses, although there are exceptions, such as hepatitis C virus (HCV). Retroviruses are enveloped viruses with a single-stranded RNA genome of positive polarity that replicate via a DNA provirus that is generated by reverse transcription. Most oncogenic retroviruses are animal viruses, with the exception of human T-lymphotropic virus (HTLV-1). Oncogenic retroviruses transform cells in one of three ways (Table 1 and Figure 1): (1) acute retroviral transformation, which occurs rapidly (over a period of weeks) after infection, requires a coinfection with a helper retrovirus, and produces polyclonal tumors; (2) nonacute retroviral transformation, which occurs over a period of years, does not require a helper retrovirus, and produces clonal tumors; and (3) trans-acting retroviral transformation, which also occurs over a long period of time and does not require a helper virus, but results in oligoclonal tumors. The mechanistic details of each of these types of retroviral transformations are discussed in the following sections.

2.1 Acute Retroviral Transformation

Acute retroviral transformation occurs rapidly and requires the coinfection of a nonautonomous retrovirus and a "helper virus." In lieu of viral genes, acute transforming retroviruses express mutated forms of cellular oncogenes (v-onc), acquired from the host genome (Figure 1(A)). As a result, helper viruses are required to provide the viral replication machinery. *V-oncs* carry mutations from their cognate normal cellular counterparts, which result in a permanently active oncogene that allows these retroviruses to induce tumors with high efficiency. As a corollary of efficient induction, the transformation process also leads to polyclonal tumors.

RSV is the first and archetypal example of an acute transforming retrovirus (Martin, 2004). In 1976, work from the Bishop and Varmus laboratory at University of California, San Francisco showed that a cDNA probe from *v-Src* could hybridize with cellular DNA from multiple uninfected avian species. This seminal discovery led to the hypothesis that *v-Src* had a cellular homolog, which was extended by the demonstration that the viral oncogene, *v-Src*, could phosphorylate cellular proteins. Cloning and sequencing of the cellular *Src*, *c-Src*, revealed that the primary differences between the two versions of Src was two activating mutations found in *v-Src*. As with many oncogenes, constitutive activation of *Src* promotes cellular proliferation, cell survival, and angiogenesis.

TABLE 1 Features of the Different Types of Transformation by Retroviruses

Type of Transformation	Time to Tumor Formation	Clonality of Tumors	Genetic Mechanism of Transformation	Replication Competence of Oncogenic Virus
Acute	Weeks	Polyclonal	Viral oncogenes (v-onc)	Nonautonomous; helper virus required
Nonacute	Years	Clonal	Insertion of viral genome upregulates cellular oncogenes (c-onc)	Autonomous
Transacting	Years	Oligoclonal	Viral accessory genes upregulate cellular genes	Autonomous

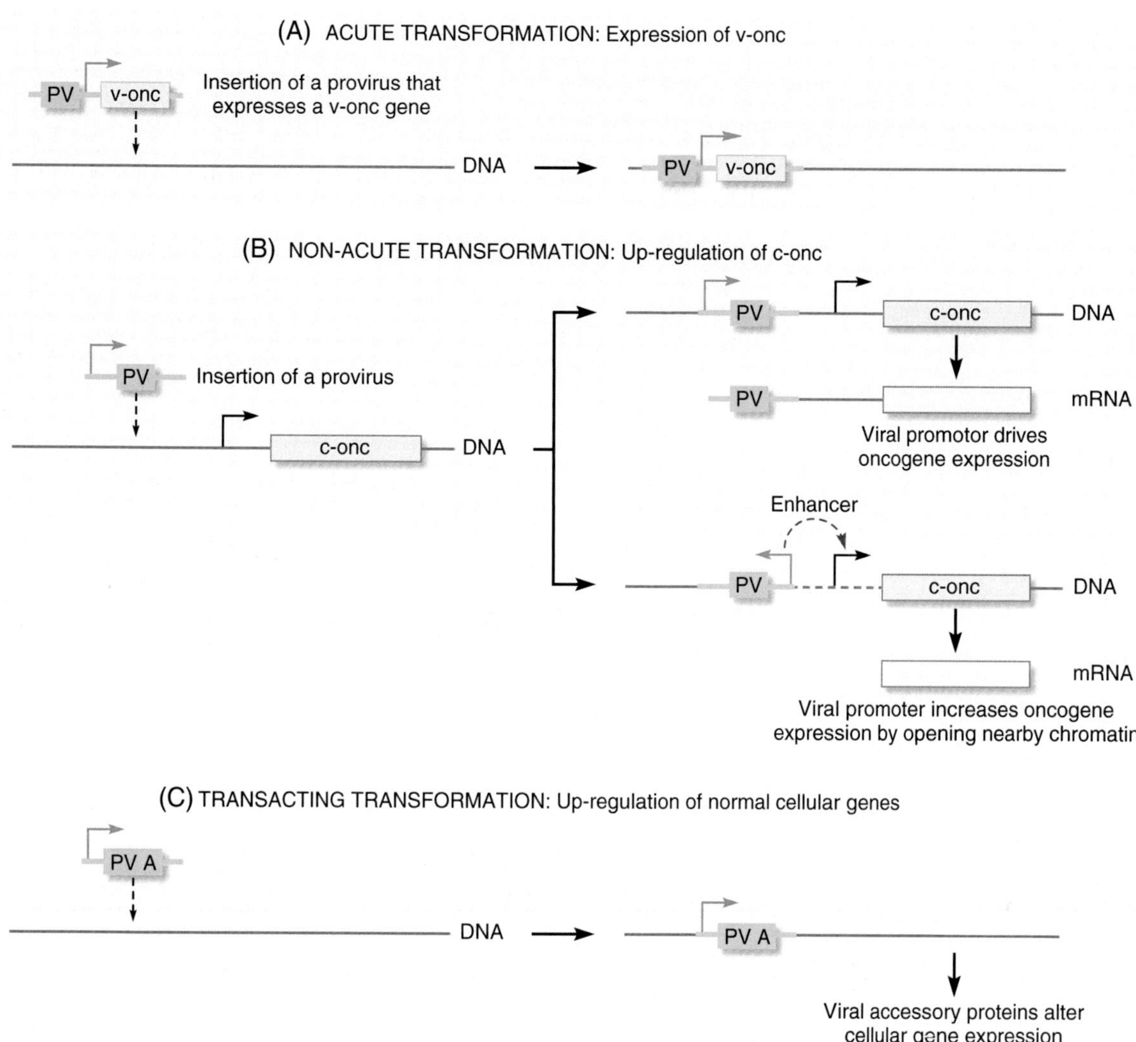

FIGURE 1 Mechanisms of retroviral transformation. (A) Acute retroviral transformation (expression of v-onc) occurs when a retrovirus inserts a copy of a v-onc gene into the host cell genome. This activated oncogene is then capable of driving tumorigenesis. (B) Nonacute retroviral transformation (up-regulation of c-onc) occurs when the insertion of a provirus (PV) occurs near a cellular oncogene (box labeled c-onc). If the PV inserts with its promoter (blue arrow) oriented toward the cellular oncogene, the viral promoter can drive up-regulated transcription of the oncogene. If the PV inserts its promoter away from the oncogene, it can increase transcription from the oncogene enhancer (black arrow) by "opening" up nearby chromatin. (C) Trans-acting retroviral transformation occurs when a retrovirus expresses an accessory protein that alters the expression of cellular genes. Their increased expression results in tumorigenesis.

The discovery that this viral gene was derived from a cellular gene led to the idea that overexpression of cellular genes could lead to cancer. Although the principle of cellular oncogenes had its origins in research on chicken viruses, the mechanism is also operative for viruses of mammals. Thus, its discovery represented a substantial advance in the understanding and treatment of human disease (Table 2).

2.2 Nonacute Retroviral Transformation

Nonacute retroviral transformation is a slow process, reflective of the indirect mechanism of action, where random insertional mutagenesis results in increased expression of an oncogene (Figure 1(B)). Although there are no known nonacute transforming retroviruses in humans, representatives of this group infect mice, chickens, and cats. Murine leukemia virus (MuLV) is a representative of these nonacute transforming retroviruses. MuLV reverse transcribes and integrates a proviral copy of its genome into host cells, resulting in an infection that can last the life of the infected cell. This lifelong infection affords the virus opportunity to replicate and integrate numerous copies of its genome into the cellular genome.

Transformation induced by MuLV infections occurs when one or multiple copies of the MuLV proviral DNA

TABLE 2 Representative RNA Tumor Viruses and Their Cognate Oncogenes

Virus	Animal Host	Oncogene and Its Function
Abelson murine leukemia virus	Mouse	Abl (tyrosine kinase)
AKT8 murine leukemia virus	Mouse	Akt (serine–threonine kinase)
Avian erythroblastosis virus	Chicken	ErbA (transcription factor)
Friend murine leukemia virus	Mouse	Fms (macrophage colony stimulating factor)
Hardy-Zuckerman-4 feline sarcoma virus	Cat	Kit (stem cell receptor)
Harvey murine sarcoma virus	Mouse	h-Ras (Ras signaling)
Moloney murine leukemia virus	Mouse	Ets (transcription factor)
Murine sarcoma virus 3611	Mouse	Raf (serine–threonine kinase)
Reticuloendotheliosis virus	Chicken	Ref (NF-κB pathway)
Rous-associated virus 1 Avian erythroblastosis virus	Chicken	ErbB (epidermal growth factor)
Rous sarcoma virus	Chicken	Src (tyrosine kinase)
Simian sarcoma virus	Primate	Sis (platelet-derived growth factor)
HTLV-I	Human	Tax (promotes expression of cell cycle; blocks apoptosis)

inserts proximal to regulatory sequences controlling expression of host oncogenes. This can result in an up-regulation of expression of the nearby gene in two separate ways. If the provirus insertion is oriented so that it is transcribed toward the target oncogene, the viral promoter can drive transcription of the cellular gene. Alternatively, if the provirus inserts in the opposite orientation, it can still result in increased expression through up-regulation of enhancer elements. Enhancer elements increase expression of a nearby oncogene by opening the surrounding chromatin, making it more conducive to gene expression.

Due to the infrequency of oncogenes in the host genome, a multitude of nontransforming insertions occur for every transforming event, and amplification of oncogene expression by proviral insertional mutagenesis is highly unlikely. This inefficiency results in the slow process that produces the clonal tumors. However, the random integration of the provirus in tumor cells has been exploited to discover a large number of cellular oncogenes.

2.3 Trans-acting Oncogenic Retroviruses

Like nonacute retroviral transformation, this mode of transformation is mediated by increased expression of cellular genes, but not necessarily cellular oncogenes. Trans-acting transformation by oncogenic retroviruses results from viral accessory proteins, which activate cellular genes that promote proliferation and protect against apoptosis (Figure 1(C)).

HTLV-1 illustrates this mechanism of transformation. In addition to *gag*, *pol*, and *env* genes, HTLV-1 expresses four additional genes (*tax*, *p12*, *p13*, and *p30*) that assist in viral replication. With respect to transformation, HTLV *tax* is the most important gene. HTLV Tax acts as a strong enhancer of gene expression that is capable of driving the expression of cellular and viral genes. Through this stimulation of gene expression, HTLV Tax promotes cell cycle progression and also blocks apoptosis. As a result, HTLV infection induces transformation in two distinct but synergistic ways. First, by increasing cell proliferation, HTLV Tax decreases the fidelity of genomic replication. Second, by blocking cells from undergoing programmed death, HTLV Tax attenuates the use of apoptosis as a "last ditch" mechanism to prevent the propagation of cells carrying potentially oncogenic mutations. HTLV-1-induced adult T-cell lymphomas are an extremely rare but very aggressive form of lymphoma.

2.4 HCV and Liver Cancer

HCV is a blood-borne virus that was discovered in 1989 through the combined efforts of scientists led by Daniel W. Bradley at the Center for Disease Control and Michael Houghton at Chiron Corporation, who were searching for the cause of "nonA nonB" viral hepatitis. Because blood screening for the virus was not optimized until 1992, many people were accidentally infected with HCV-contaminated blood transfusions. HCV is the most common blood-borne

TABLE 3 Representative DNA Tumor Viruses of Humans and Their Mechanisms of Transformation

Virus Family	Virus and Disease	Primary Transforming Genes/Proteins Mechanisms of transformation
Adeno	Adenovirus No naturally occurring disease (Tumors in experimental animals)	E1A; E1B Inactivation of tumor suppressor proteins
Polyoma	Merkel cell polyomavirus (MCPyV) Merkel cell carcinoma	Not known
Papilloma	Human papillomavirus (HPV) Cervical carcinoma (and many other cancers)	E6; E7 Inactivation of tumor suppressor proteins
Hepadna	Hepatitis B virus (HBV) Hepatocellular carcinoma	X protein Inhibition of DNA repair Modification of Ras signaling
Herpes	Epstein–Barr virus (EBV) Burkitt's lymphoma	EBNA 1; EBNA 3C Inactivation of tumor suppressor proteins
Herpes	Kaposi's sarcoma herpesvirus (KSHV; HHV8) Kaposi's sarcoma	LANA; v-cyclin; v-Bcl2 Inactivation of tumor suppressor proteins

human pathogen in the United States, and over three million chronic infections lead to more than 15,000 deaths per year. As discussed in Chapter 20 (Antiviral therapy), the recent development of new drugs has increased the ability to terminate persistent HCV infections.

HCV-associated liver cancer develops over the course of decades, and only 5% of HCV infections result in cancer, which reflects an indirect mechanism of tumorigenesis. Unlike the oncogenic mechanisms associated with the viruses discussed above, HCV promotes hepatocellular carcinoma indirectly by enabling persistent infection and inducing inflammation in the liver. Persistence is promoted by reduced immune activation, which is mediated by a viral protease (HCV NS3/4A) that cleaves and disables two host immune proteins, mitochondrial antiviral signaling protein and TIR-domain-containing adapter-inducing interferon-β. Although the viral protease is capable of blunting the immune response enough to allow persistent infection, HCV infections still induce chronic inflammation and liver damage. This results in a continual proliferation of hepatocytes to replace those destroyed by inflammation, which in turn increases the risk of oncogenic mutations.

3. ONCOGENIC DNA VIRUSES

The discovery of oncogenic DNA viruses that infect mammals occurred in the 1930s, when Richard Shope described a cell-free tumor extract (cottontail rabbit papillomavirus) that could induce tumors in rabbits. Research on tumor viruses in experimental animals ultimately led to the discovery of the oncogenic activity of small tumor viruses in the polyoma and papilloma virus families, multiple members of which cause a range of human tumors. Research in this field has also led to the development of prophylactic vaccines. Hepatitis B virus (HBV) vaccine protects against virus-initiated hepatocellular carcinomas, and vaccines against the most common types of oncogenic human papillomaviruses (HPVs) protect against numerous anogenital tract carcinomas.

There are multiple families of DNA viruses that infect and cause cancer in humans (Table 3). Reflecting their taxonomic diversity, oncogenic DNA viruses use numerous mechanisms of transformation. In the following discussion, these viruses are grouped by their relative genome size (large and small) as well as by mechanism of transformation. Common to many oncogenic viruses is that relatively few infections result in overt cancer. This disparity may reflect the fact that there is evolutionary pressure on these viruses to be relatively nonpathogenic, thereby ensuring that persistently infected hosts can survive to transmit infection.

The transforming proteins from DNA viruses usually destabilize or otherwise impair a class of cellular proteins known as tumor suppressor proteins. Tumor suppressor proteins act as barriers to transformation, and the inhibition or loss of these proteins increases the likelihood of transformation. p53 and pRB, the most important tumor suppressor proteins, are part of separate signaling

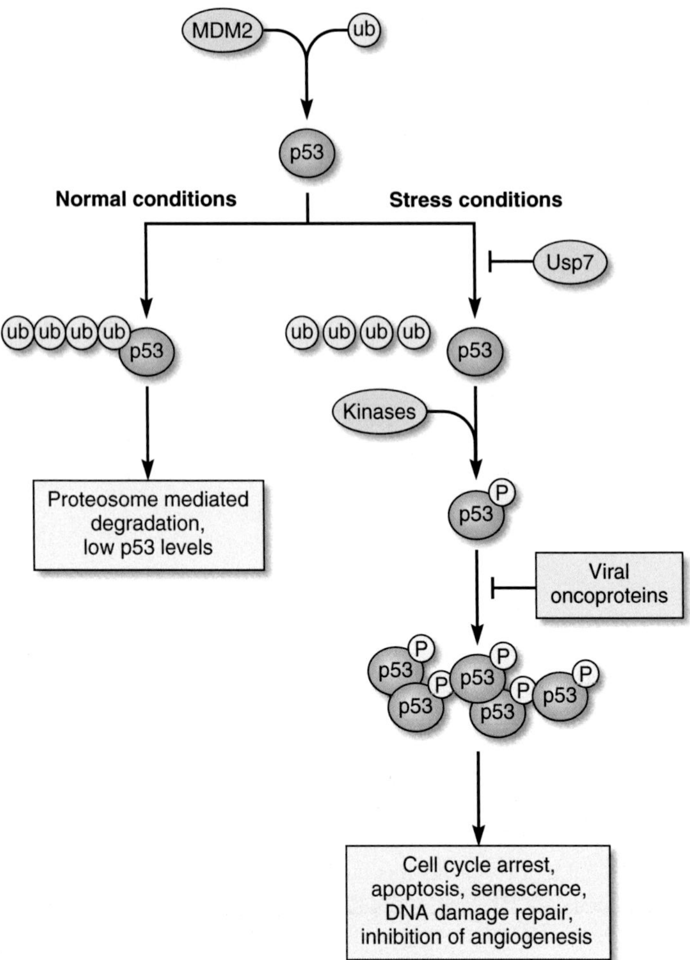

FIGURE 2 The p53 signaling pathway. The "master" protein, p53, is a tumor suppressor. Under normal conditions, the amount of p53 protein in cells is maintained at a low level by the ubiquitin ligase, MDM2. In these conditions, MDM2 polyubiquitinates p53, targeting it for proteasome-mediated degradation. In response to cellular stress, including DNA damage and errors in mitosis, ubiquitin is removed from p53 by Usp7. In addition, cellular kinases (ATM, ATR, DNApk, and others) counteract MDM2-mediated degradation by phosphorylating p53. This stabilizes p53 and allows the tumor suppressor to accumulate and drive expression of its target genes, resulting in multiple antioncogenic activities (cell cycle arrest, apoptosis, senescence, DNA damage repair, and inhibition of angiogenesis). Viral oncoproteins bind p53 and interfere with p53-mediated growth arrest, thus removing a critical brake on the cell cycle and contributing to the transformation or immortalization of cells.

pathways that regulate both apoptosis and cellular propagation (Figures 2 and 3). Many of the oncogenes of DNA tumor viruses act directly or indirectly to abrogate the action of these key tumor suppressor proteins; in turn this loosens the rigid control of the cell cycle and also enhances virus replication. The loss of tumor suppressor activity is not limited to virus-associated cancers, as p53 is the most often mutated (or otherwise abrogated) gene in tumor cells, and the loss of pRB results in retinoblastoma, the cancer for which the protein is named.

3.1 Small Oncogenic DNA Viruses

Small DNA tumor viruses, such as HPVs and polyomaviruses, transform cells in culture primarily through expression of viral oncogenes. Unlike the oncogenes of acute transforming retroviruses, these viral oncogenes are not mutated cellular genes and are not acquired through recombination between viral and cellular DNA. Furthermore, these oncogenes are required for the viral life cycle. Frequently, small oncogenic DNA viruses rely on cellular polymerases that are only expressed during the S-phase of the cell cycle. Oncogenes from these viruses tend to promote cell cycle progress into S-phase by abrogating checkpoints that would pause cell proliferation in response to deleterious stimuli. Also important to the replication of these viruses is their ability to inhibit apoptosis, thereby preventing their host cells from dying. The promotion of propagation, together with restriction of the apoptotic response, results in the accumulation of cells with destabilized genomes. These properties are responsible for transformation by small oncogenic DNA viruses.

3.1.1 Human Papillomaviruses

HPV is a major cause of disease in humans. In addition to the plantar, common, and genital warts caused by low-risk HPVs, HPVs can induce tumorigenesis in the oropharynx, anus, penis, cervix, vulva, and vagina. Although robust screening has dramatically reduced the frequency of most of these malignancies in the developed world, globally about 250,000 women die each year from HPV-associated cancer.

The HPV family is a very large group of viruses that infect epithelial surfaces. Based on sequence homology,

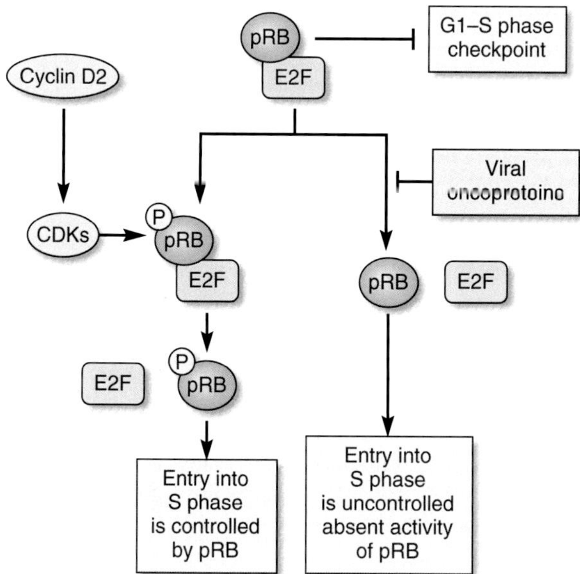

FIGURE 3 The pRB Pathway. The retinoblastoma protein, pRB, is a tumor suppressor. The pRB pathway regulates progression from the G1-phase to the S-phase of the cell cycle by binding and inhibiting E2F transcription factors. Phosphorylation of pRB can cause it to dissociate from E2F transcription factors freeing them to drive progression of cells into S-phase. The phosphorylation of pRB is mediated by cyclin-dependent kinases (CDKs). As their name suggests, activation of these kinases requires expression of cyclin D2. Several viral oncoproteins bind and sequester pRB, abrogating its braking action and promoting cellular proliferation and tumorigenesis.

there are 5 genera of HPV with close to 200 different viral types. However, relatively few types of HPV are capable of transforming cells. Only 15 HPVs are termed high risk (HR-HPV) on the basis of their ability to both infect the anogenital and oral tracts and increase the chance of oncogenesis. Two viral oncogenes, *E6* and *E7*, are important for transformation and can immortalize cells in culture. They drive the process of HPV-associated oncogenesis in vivo, and the resulting tumors are dependent on their expression for continued proliferation.

Although E6 and E7 are capable of immortalizing cells, they are not sufficient for the full transformation phenotype and tumorigenesis. HPV induces cervical cancer over a period of decades during which cells expressing the viral oncogenes gain additional mutations that result in transformation. E6 and E7 directly contribute to the genome destabilizing events that lead to transformation. For instance, both E6 and E7 impair the ability of cells to repair damaged DNA by interacting with cellular proteins responsible for that repair. Further, these viral oncogenes also increase the likelihood of erroneous mitotic events that can result in aneuploidy and polyploidy. E7 induces these cancer-associated abnormalities directly, whereas E6 allows these atypical cells to continue replicating. The continued expression of E6 and E7 over the course of two–three decades results in substantial cellular genomic instability.

An event common to many HPV-associated tumors, and likely induced by the viral-mediated reduction in cellular genomic fidelity, is the integration of the HPV genome into the host genome. Integration of the circular HPV genome into the linear host genome requires a break in both genomes. Frequently, in HPV-associated tumors, the disruption of the HPV genome is seen in the *HPV E2* gene. E2 regulates the promoter that drives both E6 and E7 expression, and inactivation of E2 leads to increased expression of E6 and E7 and a demonstrable increase in proliferation.

E6 exerts much of its influence on the cell through its association with a cellular E3 ubiquitin ligase, E6AP. This interaction allows E6 to mark cellular proteins for proteasome-mediated degradation by bringing E6AP to the targeted protein, resulting in polyubiquitination. E6 uses this mechanism to alter steady state levels of a number of host proteins, the most important of which is the tumor suppressor, p53 (Scheffner et al., 1990). As previously discussed, p53 is the center spoke of a number of signaling pathways that dictate the cellular response to a variety of stimuli including damaged DNA or faulty mitosis.

E6 also activates telomerase, the ribonucleoprotein responsible for adding DNA sequences to the ends of chromosomes at telomeres to offset the loss of bases at chromosome termini during replication. Activated telomerase is seen in a wide array of cancers as it assists in the immortalization of cells. E6 can also directly interfere with DNA repair. E6 interacts with BRCA1 and BARD1, proteins central to the high fidelity repair of double-strand DNA breaks, as well as XRCC1, an enzyme involved in the repair of single-strand DNA breaks. These interactions impair the activity of some of these proteins, and contribute to the genomic instability seen in HPV-associated tumors.

E7 contributes to both immortalization and transformation in a number of ways. Most notable is the ability of E7 to destabilize the tumor suppressor protein, pRB (Dyson et al., 1989). Typically, pRB regulates cellular proliferation by restricting transcription factors (such as E2F) that promote progression through the cell cycle, and disabled pRB has reduced activity. As a consequence, there is significantly less restraint on the propagation of cells infected with HPV. E7 also interacts with the DNA damage repair protein, BRCA1, and impairs some of its activities, thereby destabilizing the genome of its host cell.

3.1.2 Polyomaviruses

The polyomavirus family name is derived from the initial characterization of murine polyomavirus (MuPyV), the archetypical virus in this family, which was discovered by Ludwik Gross in 1953. Much like RSV, MuPyV was first characterized as a cell-free extract that could transmit tumors to another animal. The name "polyoma" comes from the fact that these cell-free extracts resulted in multiple kinds of tumors. The virus also transforms cells in culture, and the role of this family of small DNA tumor viruses in tumorigenesis has been extensively studied (Benjamin, 2001).

The 5.3 kb genome of MuPyV encodes three proteins involved in both viral replication (large, middle, and small T-antigen) and transformation. The major transforming protein expressed by MuPyV is the middle T-antigen (Mu-MT). Mu-MT is a membrane protein that binds and activates c-src and other c-src family proteins (Courtneidge and Smith, 1983). Permanent activation of c-src is a powerful transforming event that allows Mu-MT to transform cells in culture.

The contribution to tumorigenesis of MuPyV small T-antigen (Mu-sT) is linked to its ability to bind the phosphatase PP2A (Pallas et al., 1990). PP2A is a ubiquitously expressed phosphatase that modulates the activity of numerous cellular signaling pathways by dephosphorylating proteins in these pathways, most notably β-catenin, c-Myc, Akt, and p53. Through its binding to the phosphatase, mu-sT inhibits PP2A activity allowing a single viral protein to modulate multiple cellular pathways.

The MuPyV large T-antigen (Mu-LT) promotes the initiation of viral DNA synthesis and is required for viral replication. In addition to its role in replicating the MuPyV genome, mu-LT promotes cell cycle progression by disrupting the G1-S checkpoint. The viral protein stimulates proliferation in a manner similar to E7, by binding pRB and abrogating the activity of the tumor suppressor. Further, mu-LT binds and inactivates the transcriptional coactivators p300 and CBP. These two histone acetyltransferases regulate the expression of a wide number of cellular genes by relaxing chromatin structure near promoters, thereby increasing the accessibility of transcription factors to these regions.

Advances in technology have recently facilitated the discovery of several new polyomaviruses that infect humans (Sidebar 1). The most notable of these novel polyomaviruses, the Merkel cell polyomavirus (MCPyV), was discovered in the laboratory of Patrick Moore and Yuan Chang at the University of Pittsburgh in 2008. MCPyV is integrated into the genome of approximately 80% of Merkel cell carcinomas, a rare but very aggressive skin cancer.

3.1.3 Hepatitis B Virus

Baruch Blumberg received the Nobel Prize for the discovery of HBV as a cause of acute hepatitis. His work, in combination with the efforts of Irving Millman, led to a test that could screen blood supplies for the virus and a vaccine that protects against HBV infection. HBV causes both acute and chronic infections, but it is chronic infections that are associated with hepatocellular carcinoma. Globally, there are an estimated 350 million persons living with chronic HBV infection, which increases their risk of liver cancer by 100-fold. Primary cancers of the liver are very deadly, with a 5-year survival near 10%. Worldwide, an estimated 500,000 people die of HBV-associated cancer each year.

HBV is a small circular DNA virus that has an unusual genome structure because it is partially double stranded and has a gap in the full-length strand. Four genes are encoded by the virus; core (C), surface (S), polymerase (P), and X. Several factors play a role in the pathway to liver cancer. Chronic HBV infection leads to liver inflammation, scarring, and cirrhosis, and these pathological changes always precede hepatocellular carcinoma. HBx is a regulatory protein which can increase transcriptional activity of many genes: this contributes to the persistence of HBV infection and likely to transformational steps in carcinogenesis. Also, HBx blocks immune responses that could contribute to control or clearance of the persistent infection. Finally, HBx interferes with DNA repair and apoptosis, and causes epigenetic changes, all leading to genetic instability. These actions of HBx are essential for the development of hepatocellular cancer, although the exact pathways remain to be elucidated.

3.2 Large Oncogenic DNA Viruses

Several large oncogenic DNA viruses are associated with human disease. Members of this group have large genomes (>150 kb) compared with small DNA tumor viruses (<10 kb). The most notable members, Epstein–Barr virus (EBV) and Kaposi's sarcoma-associated herpesvirus (KSHV or HHV8, human herpesvirus 8), belong to the gammaherpes family of viruses. Large oncogenic DNA viruses share similarities in their mechanism of transformation with many other DNA tumor viruses. They encode accessory proteins that can immortalize cells by promoting cell cycle progression or blocking apoptosis, often by impairing the tumor suppressor proteins, p53 and pRB. Unlike the small DNA tumor viruses, their oncogenes may not be required for virus replication.

3.2.1 Epstein–Barr Virus

Working as a physician in central Africa, Denis Burkitt first described Burkitt's lymphoma, the type of cancer that would eventually bear his name. Burkitt's observation that the cancers were concentrated around warm rainy parts of the region led him to hypothesize that the tumor was caused by a parasite. In 1964, Anthony Epstein and Yvonne Barr observed by electron microscopy that cells from Burkitt's lymphomas contained herpesvirus particles. Since its initial connection to Burkitt's lymphoma, EBV has been further linked to infectious mononucleosis as well as nasopharyngeal carcinoma and Hodgkin's lymphoma.

Although about 90% of adults have evidence of current or past EBV infections, the incidence of cancer is low,

Sidebar 1 Next-generation sequencing and the detection of tumor viruses

Next-generation sequencing (NGS), a sequencing-by-synthesis approach that has replaced the first-generation Sanger sequencing method, rapidly produces enormous volumes of sequence data. The technology is having a dramatic impact on virology in research and clinical settings alike, where it is being used for viral genome sequencing, the detection of viral genome variability and quasispecies, and for the discovery of new viruses, including tumor viruses. The use of NGS, together with an approach referred to as digital transcript subtraction (a method to subtract human sequences from data sets, leaving nonhuman sequences for further analysis), led to the discovery of a polyomavirus (MCPyV) present in tumor samples from approximately 80% of Merkel cell carcinomas. This was the first polyomavirus shown to be associated with a human cancer. Similar approaches have led to the discovery of new papillomaviruses, including novel types associated with squamous cell carcinoma. There are now over 200 established HPV types, and over 200 additional putative HPV types have been identified by NGS.

NGS is also being used to map the integration sites of retroviruses and retroviral vectors used for gene therapy. As was discussed earlier in this chapter, retroviral integration into the host genome can cause malignant transformation through the activation of host proto-oncogenes or the inactivation of tumor suppressor genes. Similarly, NGS is being used to identify and map the position of endogenous retroviruses (EVRs), which are now estimated to comprise as much as 8% of the human genome. Because NGS can be used to measure transcription (as discussed in more detail in Chapter 11: Systems Virology), it is also being used to study the expression of transcripts from EVR sequences. Although differential expression of EVR sequences in normal and tumor tissues has been detected, the involvement of EVRs in human cancer etiology remains unclear. Finally, NGS-based metagenomics is being used to characterize the human virome (the entire population of viruses that colonize the human body) and in diagnostic settings for the detection of disease-causing viruses in patients with unexplained illnesses.

suggesting that transformation is a rare event in the natural course of an EBV infection. However, the number of EBV-associated tumors is significant, as 200,000 new cases are estimated to occur annually.

The linear EBV genome is over 170 kb long and encodes more than 90 genes. This abundance of genetic material fuels a complex life cycle (Figure 4) that includes both lytic and latent phases, with specific genes expressed exclusively in one or the other parts of the life cycle. Since productive infection with EBV is cytolytic, transformation is associated with an abortive or latent infection cycle (Figure 4).

EBV is able to modulate the checkpoints that restrict cellular proliferation. During latency, the virus expresses genes that reduce the activity of pRB, and as a result, bypass the G1/S cell cycle checkpoint. EBV nuclear antigen 3C (EBNA 3C) appears to target pRB both directly and indirectly. Indirectly, EBNA 3C promotes G1/S transition by enhancing the cyclin-dependent kinase activity necessary to begin replication. EBNA 3C forms a complex with a cellular E3 ubiquitin ligase (SCF^{Skp2}) that allows the viral protein to target SCF^{Skp2} to pRB, leading to the ubiquitination and degradation of the tumor suppressor protein (Knight et al., 2005). The combined result of direct and indirect inactivation of pRB is that EBNA 3C expression allows cells to bypass G1/S arrest in response to a number of stimuli that would otherwise remove the cells from the cell cycle.

EBV nuclear antigen 1 (EBNA1) also contributes to the deregulation of the cell cycle by antagonizing p53. Typically, low steady state levels of p53 are maintained by the MDM2-mediated polyubiquitination of p53, resulting in degradation of p53 (Figure 2). In response to stimuli that elicit cell cycle arrest, ubiquitin is removed from p53 by Usp7/haUsp. This stabilizes p53 and allows a rapid increase in cellular p53 protein levels, resulting in cell cycle arrest. EBNA1 decouples this normal p53 state by competitively binding UspSP7/haUsp and abrogating its ability to stabilize p53. Finally, EBNA 3C also disrupts the G2/M checkpoint by interacting with Chk2. This interaction prevents Chk2 from becoming properly activated and inducing an arrest at the G2/M boundary.

The p53 protein can also induce apoptosis in response to certain stimuli. During the latent portion of the EBV life cycle, EBNA1 prevents p53-induced apoptosis. EBNA 3C also participates in the viral attenuation of p53-induced apoptosis by associating with p53. This two-pronged approach to preventing p53-induced apoptosis underscores the importance of continued viability of the host cell for virus replication. Since apoptosis serves as a mechanism to remove damaged, potentially oncogenic cells, this can allow survival of cells that have acquired tumorigenic mutations. EBV also encodes a number of proteins that suppress virus-specific adaptive immune responses. EBNA1 and BLNF2A interfere with antigen processing, and BGLF5 represses the synthesis of HLA Class 1 molecules required for antigen presentation.

In transformed cells from Burkitt's lymphoma, there may also be an activating translocation of the Myc oncogene. In addition, the TCF-3 transcription factor undergoes activating mutations, and the ID3 protein, which normally acts as a brake on TCF-3, is mutated to lose its potency. Up-regulated *TCF-3* increases the activity of genes that are expressed in rapidly dividing B cells, thereby enhancing the proliferation of centroblasts, the putative origin of Burkitt's lymphoma cells.

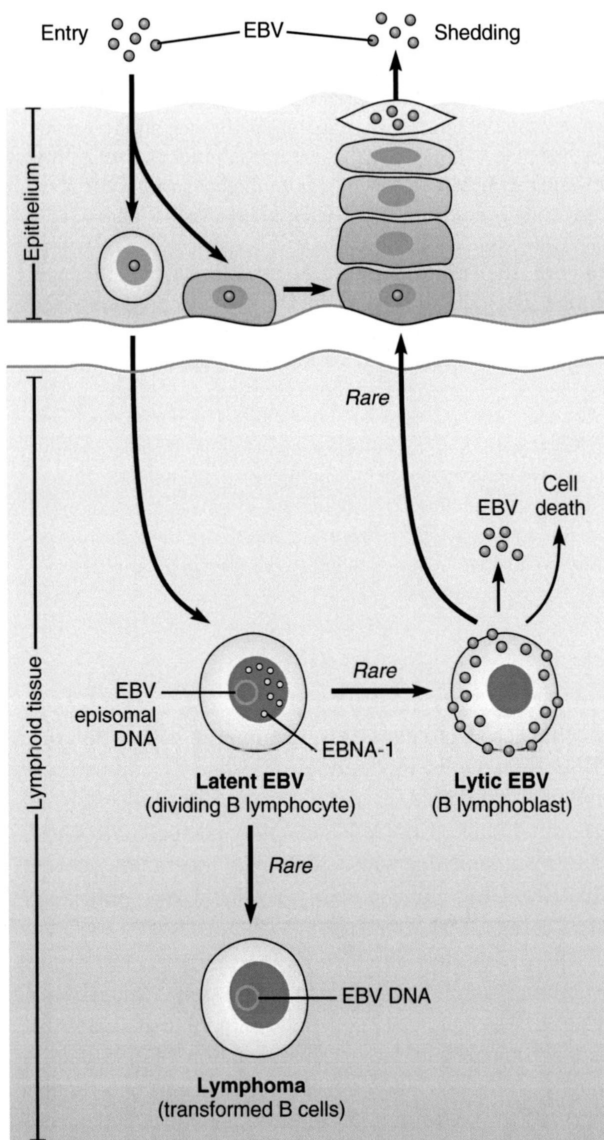

FIGURE 4 Life cycle of Epstein–Barr virus (EBV). EBV can infect epithelial cells and B cells. Epithelial cell infection results in both latent and lytic infection. If an epithelial cell is latently infected, the virus will propagate along with its host cell. A lytic infection of an epithelial cell results in the cell's death and release of EBV particles. EBV infection of a B cell usually results in a lifelong latent infection. Most commonly, the infected and activated B cell becomes a dividing memory B-cell, and during the latent portion of the EBV life cycle, EBNA1 prevents p53-induced apoptosis. Rarely, this latent infection can become lytic resulting in cell death and release of viral progeny. Other rare events can transform this memory B cell into different types of lymphoma.

3.2.2 Kaposi's Sarcoma-Associated Herpesvirus

Kaposi's sarcoma was first identified in 1872 by Moritz Kaposi, who described an unusual angiogenic tumor of the skin as a rare cancer in Jewish and Mediterranean people. During the AIDS epidemic, it was noted that HIV-infected gay men had a much greater risk than other AIDS patients of developing this unusual cancer. Based on the epidemiological associations (gay men with AIDS), it was hypothesized that KS might be caused by an infectious agent distinct from HIV. In 1994, a team led by Patrick Moore and Yuan Chang used the method of representational difference analysis to identify KS-specific DNA fragments in cells cultured from KS lesions. These DNA fragments were found to be similar to sequences of known human herpesviruses, leading to sequencing of the genome and its identification as a new member of the gammaherpesvirus family (Chang et al., 1994).

Kaposi's sarcoma is primarily a vascular tumor. This reflects the ability of KSHV to interfere with the differentiation of endothelial cells, which proliferate inappropriately. There are many broad similarities between KSHV and EBV, including the size of the viral genomes and the cellular pathways disrupted by the viral oncogenes. KSHV disrupts the G1-S checkpoint, although this occurs through novel mechanisms. KSHV encodes a protein (v-cyclin) that mimics the cellular protein cyclin-D2 (Li et al., 1997). v-cyclin is a stronger activator of cell cycle progression than its cellular analog and its expression can promote proliferation in the face of inhibitors of cyclin-dependent proliferation.

The latency-associated nuclear antigen (LANA) of KSHV antagonizes pRB-mediated cell cycle regulation. LANA binds to and inhibits pRB (Hu et al., 2002). Additionally, LANA promotes cellular proliferation by increasing the activity of the transcription factor, E2F. As a result, LANA-expressing cells continue to multiply even in the presence of inhibitors of cell cycle progression.

LANA also binds to p53 and inhibits the ability of p53 to induce apoptosis. KSHV expresses other antiapoptotic proteins, including v-Bcl2, the most-well studied KSHV protein. v-Bcl2 prevents apoptosis by blocking the formation of Bax oligomers, which would activate a checkpoint in the cell cycle. Since v-Bcl2 is under the control of a viral promoter, its expression is no longer properly regulated, and as a result, command of apoptosis is usurped. The triumvirate of viral transformation often involves inhibition of cell cycle regulation, attenuation of apoptosis, and induction of genomic instability. LANA serves in all three capacities in KSHV-induced transformation.

4. REPRISE

Oncogenic viruses are divided into two broad classes, RNA and DNA viruses. Most RNA tumor viruses are retroviruses, which replicate via a DNA provirus generated by their reverse transcriptase. Retroviruses transform cells in three distinct ways. Acute transformation involves two different viruses, a replication-deficient transforming virus that encodes a v-oncogene (a mutated form of a cellular gene) and a nontransforming replication-competent helper virus, which enables replication of the transforming oncovirus.

The resulting tumors are polyclonal. Nonacute transformation occurs when an integrated provirus acts as an insertional mutagen, up-regulating a cellular oncogene. The resulting tumors are monoclonal reflecting the inefficiency of the transforming process. Trans-acting transformation is dependent on accessory viral proteins. These trans-acting proteins are essential to viral replication and contribute to the transformation of their host cell by activating cellular proteins that drive proliferation or hinder apoptosis. The resulting tumors are oligoclonal and slow to develop.

Although a number of oncogenic retroviruses are capable of inducing tumors in animals, their relevance to human disease is mainly their role in the discovery of oncogenes. Oncogenic DNA viruses, in contrast, are an etiological factor in a significant portion of human cancers. Recent technological advancements have facilitated the identification of DNA oncogenic viruses and the types of tumors they produce, and these numbers are likely to grow.

Oncogenic DNA viruses use a great variety of transformation mechanisms. Small oncogenic DNA viruses transform via viral oncogenes that do not have cellular homologs. Large oncogenic DNA viral genomes encode oncogenes that are either similar to or distinct from cellular genes. By disabling two major tumor suppressor proteins, p53 and pRB, many oncogenic DNA viruses avoid apoptosis and ensure continued proliferation of their host cell. In addition, tumorigenic DNA viruses often induce cellular genomic instability, either actively through the inhibition of DNA damage repair, or indirectly through increased cellular proliferation.

An additional indirect mechanism of viral-associated oncogenesis is the induction of chronic inflammation. As cells proliferate to replace cells destroyed by chronic inflammation, they may acquire tumorigenic mutations. Inflammation-induced tumorigenesis is a very slow process occurring over the period of multiple decades. Both HCV and HBV induce hepatocellular carcinoma through this mechanism, and cause the majority of human hepatocellular cancers.

Prophylactic vaccines against several important oncogenic DNA viruses (such as HPV and HBV) have been developed, and there have been recent advances in the treatment of HCV infections. The recent explosion of research-driven knowledge about tumor viruses will undoubtedly lead to the prevention or cure of many virus-associated cancers.

FURTHER READING

Chapters and Reviews

Benjamin TL. Polyoma virus: old findings and new challenges. Virology 2001; 289: 167–73.

Braoudaki M, Tzortzatou-Stathopoulou F. Tumorigenesis related to retroviral infections. Journal of Infection in Developing Countries 2011; 5: 751–8.

Chen HS Lu F, Lieberman PM Epigenetic regulation of EBV and KSHV latency. Current Opinion in Virology 2013, 3: 251–259.

Feitelson MA, Bonamassa B, Arzumanyan A. The roles of hepatitis B virus-encoded X protein in virus replication and the pathogenesis of chronic liver disease. Expert Opinion on Therapeutic Targets 2014, 18: 293-306.

Griffin L, Damania B. KSHV: pathways to tumorigenesis and persistent infection. Advances in Virus Research 2014, 58: 111–159.

Horner SM. Activation and evasion of antiviral innate immunity by hepatitis C virus. Journal of Molecular Biology 2014, 426: 1198–1209.

Koike K. The oncogenic role of hepatitis C virus. Recent Results in Cancer Research 2014, 193: 97–111.

Lohmann V, Bartenschlarger R. On the history of hepatitis C virus cell culture systems. Journal of Medicinal Chemistry 2014, 57: 1627–1642.

Martin GS. The road to Src. Oncogene 2004, 23: 7910–7.

Motavaf M, Safari S, Saffari Jourshari M, Alavian SM. Hepatitis B virus-induced hepatocellular carcinoma: the role of the virus x protein. Acta Virologica 2013, 57: 389–396.

Murata T, Sato Y, Kimura H. Modes of infection and oncogenesis by the Epstein-Barr virus. Reviews in Medical Virology 2014, 24: 242–253.

Ojala PM, Schulz TF. Manipulation of endothelial cells by KHSV. Seminars in Cancer Biology 2014, 26: 69–77.

Ott JJ, Steves Ga, Groeger J, et al. Global epidemiology of hepatitis B virus infection: new estimates of age-specific HBsAg seroprevalence and endemicity. Vaccine 2012, 30: 2212–2219.

Rickinson AB. Co-infections, inflammation and oncogenesis: Future directions for EBV research. Seminars in Cancer Biology 2014; 26: 99–115.

Schmitz R, Cerbelli M, Pittaalluga S, et al. Oncogenic mechanisms in Burkitt lymphoma. Cold Spring Harbor Perspectives in Medicine 2014, 3.

Original Contributions

Chang Y, Cesarman E, Pessin MS, et al. Identification of herpesvirus-like DNA sequences in AIDS-associated Kaposi's sarcoma. Science 1994; 266: 1865–9.

Li K, Foy E, Ferreon JC, et al. Immune evasion by hepatitis C virus NS3/4A protease-mediated cleavage of the Toll-like receptor 3 adaptor protein TRIF. Proceedings of the National Academy of Sciences of the United States of America 2005a; 102: 2992–7.

Chapter 9

HIV and AIDS

Science Wrestles with 10,000 Nucleotides—Points but No Pin

Guido Silvestri, Emily K. Cartwright
Emory Vaccine Center and Yerkes National Primate Research Center, Emory University, Atlanta, GA, USA

Chapter Outline

1. INTRODUCTION

Over the last 30 years, HIV/AIDS has caused one of the largest human epidemics our planet has ever experienced. As a complex and enigmatic example of viral pathogenesis, HIV and SIV—the nonhuman primate-equivalent simian immunodeficiency virus—have been the subject of more intensive study than any other viral disease. Since AIDS was first described in 1981 and HIV discovered in 1983, tremendous progress has been made in our understanding of the evolutionary origin and basic virology of the agent, the complexities of its interaction with the host immune system, and the mechanisms responsible for its transmission, early dissemination, and pathogenesis.

This chapter begins with a description of HIV virology, including genome organization, life cycle, innate host defenses, humoral and cellular responses, and the sequence of viral and immunological events that lead to development of AIDS. It then turns to the dynamic aspects of infection, including viral and cellular turnover, chronic immune activation, and opportunistic infections. We next describe the experimental model of SIV in nonhuman primates, which has provided many insights into the pathogenesis of HIV/AIDS. The chapter finishes with a discussion of vaccine development and hopes for a cure.

The pathogenesis of HIV presents a number of challenging questions, including: (1) How does the virus persist in the presence of an active immune response? (2) How does HIV kill cells? (3) How does an infection of CD4+ lymphocytes lead to a dysfunction of CD8+ cells that are not susceptible to HIV infection? (4) Do humans have intrinsic defenses against HIV? And how does the virus circumvent these? (5) Is the disease just due to the virus or does the host response play a role? If so, how does this happen? (6) What can we learn about the pathogenesis of HIV from antiretroviral therapy (ART)? (7) If African monkeys can tolerate SIV infection and remain healthy, why does HIV cause fatal illness in humans? (8) Why do not we have an HIV vaccine? (9) Considering the potency of highly active antiviral therapy, what are the impediments to a cure? In the course

Viral Pathogenesis. http://dx.doi.org/10.1016/B978-0-12-800964-2.00009-4

of this chapter, we will attempt to address these provocative questions.

2. VIROLOGY

HIV and SIV are lentivirus members of the *Retroviridae* family of viruses. SIVs are the endogenous lentiviruses of monkey species, and most—perhaps all—species of African monkeys have their "own" strain of SIV. These SIVs are transmitted horizontally, likely by sexual intercourse, grooming, fighting, or other close contact. Although they cause lifelong persistent infections in their natural hosts, with high virus titers, they do not appear to cause disease.

There are two major groups of human lentiviruses, HIV-1 and HIV-2, both of which originate from transmission of SIVs from nonhuman primates. HIV-1 was derived from a chimpanzee virus (SIVcpz), while HIV-2 was acquired by infection with a sooty mangabey virus (SIVsmm).

This chapter is based mainly on studies of HIV-1 rather than HIV-2, because HIV-1 is responsible for most cases of AIDS, and most HIV research is dedicated to this group.

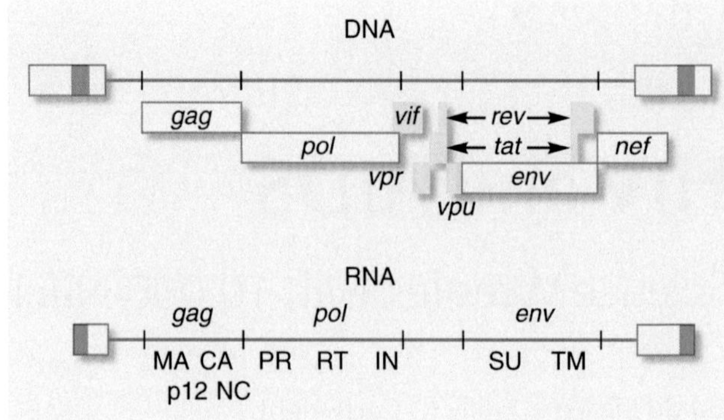

FIGURE 1 Organization of the HIV-1 genome. The RNA genome is about 9 kb long and is bounded at both ends by a noncoding repeat (R) region that encloses the three major coding genes, the *gag* (group antigen), *pol* (polymerase), and *env* (envelope) genes that are transcribed in three different reading frames. The diagram indicates the position of the major proteins encoded by each of these genes, including the MA (matrix), p12, CA (capsid), NC (nucleocapsid) proteins of gag, the PR (protease), RT (reverse transcriptase), IN (integrase) enzymes of pol, and the SU (surface) and TM (transmembrane) proteins of env. HIV-1 also encodes six nonstructural accessory proteins (tat, rev, nef, vpr, vpu, and vif) whose open reading frames are shown.

2.1 Genome Organization

The HIV genome is a single-stranded, positive sense RNA. It has three major genetic loci in common with all retroviruses: *gag* encodes the core proteins that encapsulate the RNA genome, *pol* encodes the enzymes reverse transcriptase (RT), integrase, and protease, and *env* encodes the envelope glycoprotein. However, HIV is distinguished from other retroviruses by several characteristics: (1) HIV possesses six accessory genes that encode nonstructural proteins that aid the virus in various ways, increasing both infectivity and virulence. Two of these genes (*tat* and *rev*) are required for in vitro replication. Four (*vpr, vpu, vif,* and *nef*) accessory genes are not absolutely necessary for replication in cell culture systems but at least some (such as *nef*) are required for full virulence in vivo (Figure 1). Of note, the repertoire of accessory genes is different in some SIVs that do not possess *vpr* and/or *vpu* but possess an SIV-specific gene *vpx*. (2) The surface envelope protein (gp120) binds CD4, a molecule that is found on a subset of T lymphocytes and monocytoid cells (macrophages, microglia, and dendritic cells (DCs)), making HIV a predominantly lymphotropic retrovirus. (3) The preintegration complex (PIC) of reverse-transcribed viral DNA and associated proteins can be imported across the nuclear envelope, thus allowing infection of nondividing cells. (4) HIV is a lifelong chronic infection causing an immunodeficiency that is associated with a number of opportunistic infections and cancers. (5) HIV is a strictly exogenous virus, resulting from a cross-species transmission from SIVcpz, and there are no HIV sequences within the human genome.

2.2 Clades and Viral Diversity

HIV-1 can be classified in three subgroups, M (main), N (new), and O (outlier), each originating from a different cross-species transmission event. Extensive genomic studies have identified at least nine major subtypes (clades) within the M group of HIV-1 that cause >95% of HIV-1 infections. Clades and recombinants between the clades appear to have arisen in Africa, where HIV-1 first emerged. There is no conclusive evidence that certain subtypes are associated with better or worse prognosis; some subtle differences have been noted, but each subtype is spread readily in humans. Outside of the African continent, certain clades predominate (such as clade B in North America) presumably due to a founder effect when the virus was originally introduced.

3. VIRUS–CELL INTERACTIONS

3.1 Receptors and Viral Tropism

The primary receptor for HIV and SIV is CD4, an immunoglobulin superfamily molecule expressed on two major cell types: the CD4+ subset of T lymphocytes and cells of the monocyte lineage. HIV also needs the help of a coreceptor to gain entry to the cell. There are two molecules that function as primary coreceptors for HIV: CCR5 and CXCR4. CCR5 is a β-chemokine receptor expressed on memory CD4+ T lymphocytes and macrophages and its expression is especially high in mucosal tissues. CXCR4 is the receptor for stromal cell-derived factor 1 (SDF-1) and is expressed on CD4+ T lymphocytes, bone marrow resident cells, and—at very low concentrations—on macrophages.

HIV-1 virus strains vary in their ability to bind to coreceptors and this affects their cellular host range. Viruses isolated from patients may be roughly classified into three groups, (1) those that utilize mainly CCR5 (often called R5 viruses), (2) those that utilize mainly CXCR4 (X4 viruses), and (3) those that utilize both CCR5 and CXCR4 (R5X4, or "dual-tropic" viruses). All HIV-1 strains can replicate in primary T lymphocyte cultures that express both CCR5 and CXCR4. However, monocytes are mainly susceptible to R5 viruses, and T cell culture lines are mainly susceptible to X4 viruses. Most viruses isolated from HIV infected individuals are CCR5 tropic. In the case of SIV, and in particular the well-studied SIVmac and SIVsmm strains, other coreceptors such as CXCR6, GPR15, and GPR1 can be used, both in vitro and in vivo.

3.2 Viral Variation and Evolution

Retroviruses exhibit a high rate of mutation due to the absence of a proofreading mechanism in RT. The HIV genome is 10^4 bases long and the error rate of RT is approximately 1 base mismatch per 10^4–10^5 nucleotides, or up to 1 mutation per genome replication. In an HIV-infected subject, ~10^{10} new virions are produced daily, so that each base in the genome undergoes mutation many times each day. Although HIV infection is often initiated by a selected few virions, the viral genome quickly evolves into a quasispecies or swarm of genetically related viruses. During in vivo infection, this rapid mutation rate can generate variants with a selective advantage due to their replication potential and/or ability to escape host defenses. In the presence of ART, this high mutation rate can also lead to the emergence of drug-resistant variants.

Due to the rapid rate of mutation, viruses isolated early in infection have different characteristics than those present in chronic infection. Differences include the extent of glycosylation, replication fitness, resistance to host defenses, and cytopathic properties. This change in phenotype is believed to be a result of differential requirements for establishing versus maintaining a chronic infection. For example, an early virus may need to be resistant to type I interferons and other innate defenses, at the expense of replication capacity. However later in infection, it may be more advantageous for the virus to replicate quickly or to become resistant to adaptive immune responses rather than to maintain resistance to innate defenses, which will most likely be absent at that point.

As mentioned above, the majority of isolates obtained early in HIV-1 infection are R5 macrophage-tropic viruses. The importance of R5 viruses in transmission is underlined by the finding that individuals who are homozygous for a genetic deletion of CCR5 are very resistant to infection (discussed below). The preferential transmission of R5 viruses can, at least in part, be attributed to the expression of CCR5 on CD4+ T cells that are present in the lamina propria of mucosal tissues. Another factor may be the differential ability of DCs passively to capture and transfer R5 viruses more efficiently than X4 viruses.

How do HIV infections persist in the presence of an active humoral and cellular response?

Because HIV replication inevitably produces large numbers of viral variants, it is not surprising that immunological escape mutants are constantly selected during the multiyear course of a persistent infection. Since CD8+ lymphocytes play a crucial role in the control of HIV infection, viral mutations in epitopes recognized by CD8+ lymphocytes have a selective advantage. Also, mutations accumulate in epitopes to which patients have developed neutralizing antibody responses. Escape of HIV from antibodies is possible because many neutralizing epitopes are located on the viral spike at sites distant from the CD4 binding site, so that mutants retain the ability to infect CD4+ T lymphocytes.

3.3 Virus Entry and Early Replication Events

The entry of HIV into permissive host cells is a multistep process that involves the receptor CD4 and a coreceptor (CCR5 in most cases). Initially, the viral attachment protein (gp120) binds to CD4, which triggers a conformational change that leads to coreceptor binding. Coreceptor engagement triggers a second conformational change in the transmembrane protein gp41, which brings the hydrophobic fusion peptide at the N terminus into proximity with the plasma membrane. The fusion peptide inserts into the plasma membrane, forming a prehairpin configuration. A third conformational change closes the hairpin, leading to fusion between the viral envelope and the cellular membrane. The PIC composed of RT, Vpr, matrix, and integrase associated with the proviral DNA, is formed after the viral genome has been reverse transcribed. The PIC is transported to the nucleus and if the infected CD4+ T lymphocyte is actively dividing, viral replication proceeds at a maximal rate. If the infected cell is resting, the provirus can enter the nucleus and integrate, but virion production will be relatively limited (Figure 2). Studies of the interaction of HIV proteins with cellular proteins have revealed a very large number of protein–protein interactions (Jager et al., 2012); these have yet to be integrated into a comprehensive biochemical description of HIV replication.

Various cell types of myelomonocytic lineage—tissue macrophages, brain microglia, and DCs of the skin and central lymphoid tissue—are also involved in the early events of HIV infection. Some of these cells are permissive to infection while others, though nonpermissive to infection, are hijacked by HIV to disseminate systemically and/or favor *trans*-infection of CD4+ T cells. Mature DCs are of particular importance in vivo because they bind to and

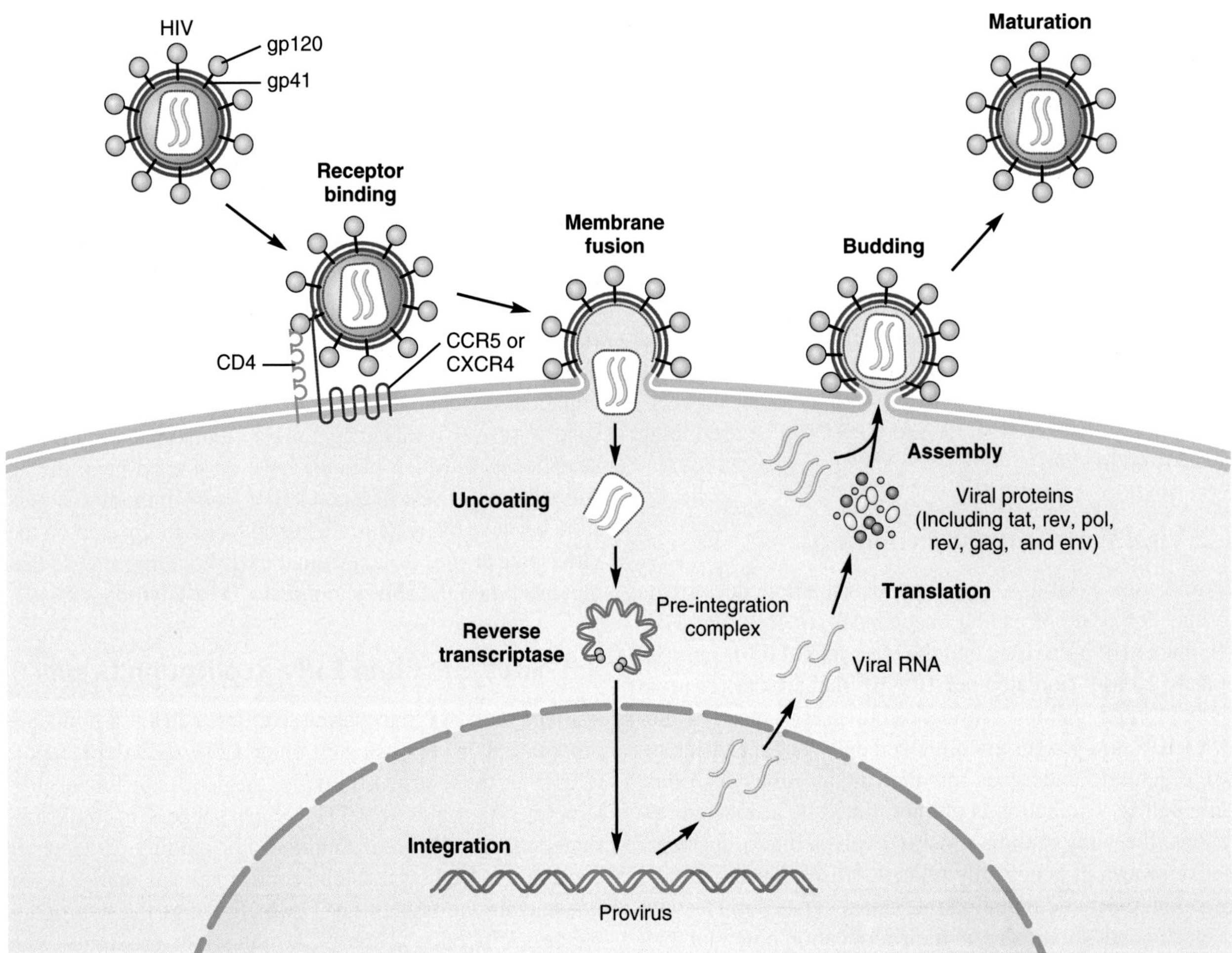

FIGURE 2 HIV life cycle. The gp120 protein binds to the CD4 receptor. Binding to CD4 triggers a conformational change that leads to binding of gp120 to the co-receptor (CCR5 or CXCR4). Binding to the co-receptor leads to a second major conformational change, this one in the TM (gp41) protein, which unfolds to expose and insert the fusion sequence at its N terminus into the plasma membrane of the cell. The two membranes fuse, leaving gp41 on the external surface. After uncoating, the pre-integration complex is transported to the nucleus and the viral genome undergoes reverse transcription. Once integrated into the genome, transcription of viral genes can begin. The first and most important proteins to be made are tat and rev. Tat aids in the efficiency of transcription from the LTR and rev allows for unspliced mRNAs (full-length viral genomes) to leave the nucleus. Together, these genes allow for transcription and translation of the structural and accessory proteins, including pol and env. Once enough gag and env has accumulated at the plasma membrane, the full-length viral genome traffics and assembles with these proteins. Gag assembles as a polyprotein in immature virions that are released from the cells. In the nascent virions, gag is cleaved by the viral protease and mature virions are formed.

sequester virions at their extracellular surface or within endosomes, though they have limited capacity for virus replication. A lectin molecule, DC-SIGN, which is expressed on the surface of DCs, has been implicated in the binding and trapping of HIV particles. Virus captured by DCs at mucosal sites can drain to lymph nodes, which are active sites of target cell activation and proliferation.

How does HIV kill cells?

Cell death in vitro. HIV or SIV can kill CD4+ lymphocytes by several mechanisms. In permissive and productively infected cells, cell death is primarily due to apoptosis, initiated in part by the Tat protein, and mediated through Fas signaling and caspase 3. Abortive infection of nonpermissive resting CD4+ T cells is thought to initiate an alternative mechanism of cell death, termed pyroptosis. IFI-16, an innate DNA-sensing molecule, recognizes partially reverse transcribed HIV DNA in the cytosol and initiates pyroptosis through caspase-1. In this form of highly proinflammatory cell death, the cellular contents of the dying cells are released into the extracellular matrix, unlike apoptosis where fragments of dead cells are contained in apoptotic bodies. However, the vast majority of cells that die during HIV infection are not productively infected at all. CD8+ T cells, B cells, and natural killer (NK) cells, none of which are susceptible to HIV infection, show increased in vivo rates of cell death during pathogenic HIV and SIV infections.

TABLE 1 Host Restriction Factors and Their Viral Counterparts, a Prime Example of the Evolutionary Arms Race between Host and Pathogen. The Factors Described Below Are by No Means an Exhaustive List, but Have Undergone the Most Research

Restriction Factor	Life Cycle Stage	Mechanism	Viral Inhibitor	Virus
TRIM5α	Early: post-entry	Interaction with viral capsid disrupts uncoating	None: restricted by mutations in capsid	HIV-1, HIV-2, SIV
APOBEC3G (A3G)	Early: reverse transcription	Cytidine deaminase: hypermutation of viral genome	Vif: targets to host proteasome	HIV-1, HIV-2, SIV
SAMHD1	Early: reverse transcription	Hydrolysis of dNTPs in cytosol	Vpx: targets SAMHD1 to host proteasome	HIV-2, SIV
BST-2/Tetherin	Late: post integration	Prevents virion release from infected cell	Vpu Env Nef All cause down regulation of BST-2 from cell surface	HIV-1 HIV-2 SIV

How does an infection of CD4+ lymphocytes lead to a dysfunction of CD8+ cells that are not susceptible to HIV infection?

Cell death in vivo. In addition to the direct virus-induced cytopathic effects seen in cell cultures, immune-mediated destruction of infected cells occurs in vivo. HIV-infected patients mount a cellular immune response in which CD8 lymphocytes through their T cell receptors recognize and lyse infected (and perhaps uninfected) cells that present viral peptides in the context of Class I HLA (see below). However in vivo, only a small minority (<1%) of CD4+ lymphocytes are infected at any time, which suggests that, in addition to destruction of infected CD4+ cells by CD8+ T lymphocytes, other mechanisms probably play a role in CD4+ lymphocyte depletion. HIV or SIV infection causes a generalized activation and increased turnover of all immune cell populations, and a high proportion of these activated cells undergo rapid apoptosis. This activation-induced cell death, that is often dependent on receptors such as CD95 and TRAIL-R, and uses caspase-8 as the main effector mechanism, involves large numbers of uninfected cells, including many cells of non-CD4 lineage (i.e., CD8+, NK, and B cells). Furthermore, HIV infection may interfere with the regeneration of CD4+ lymphocyte populations at the level of bone marrow, thymus, and lymph nodes (see below for more details).

Do humans have intrinsic defenses against HIV?

3.4 Host Restriction Factors

A number of intrinsic cellular defense factors target various stages of the viral life cycle. Both HIV and SIV encode viral accessory proteins whose main role is to target and degrade these specific host restriction factors, described in Table 1. The restriction factors that have been investigated in greatest depth are TRIM5α, APOBEC3G, SAMHD1, and BST-2/tetherin.

Postviral entry restriction: TRIM5α. TRIM5α is a host protein that acts through direct binding to the intact HIV capsid. The association of TRIM5α with viral capsid induces premature uncoating and degradation of the viral capsid, and halts reverse transcription via a proteasome-dependent mechanism (although some recent studies have suggested an alternative mechanism, using host accessory proteins). The affinity of TRIM5α from different primate species for the capsids of different primate lentiviruses determines the patterns of species specificity. For example, HIV-1 does not replicate efficiently in the cells of Old World monkeys because the TRIM5α of monkeys (such as rhesus macaques) very efficiently recognizes HIV-1 capsid and exerts a potent antiviral effect.

Defense against foreign DNA: APOBEC3G and Vif. Another important restriction factor for HIV replication is APOBEC3G (A3G). A3G is a member of the APOBEC (apolyprotein B mRNA editing enzyme, catalytic polypeptide-like) family of proteins. A3G induces hypermutation in the first minus strand of HIV proviral DNA. The catalytic subunit of A3G will deaminate cytidine (C) residues, converting them to uridine (U). Because the process of reverse transcription will use the minus strand of DNA as template for the positive strand, this C to U mutation will force a guanosine (G) to adenosine (A) mutation in the positive strand. In this way, A3G halts the viral life cycle at the early stages of reverse transcription. It is essential that A3G be packaged into budding virions and associated with the PIC in order to exert its antiviral effect in the next cell. The viral protein

Vif (viral infectivity factor) binds to A3G and recruits an E3 ubiquitin ligase complex, which targets A3G for degradation by the proteasome. A3G from some Old World monkeys is able to avoid degradation by human Vif, while human A3G confers resistance to some SIV strains.

Degradation of amino acids: *SAMHD1 and Vpx.* HIV-1 has limited replicative capacity in cells of the myeloid lineage, including monocyte-derived macrophages and DCs, as well as resting cells, which has been attributed to the presence of SAMHD1. The antiviral activity of SAMHD1 lies in its capacity to hydrolyze dNTPs (deoxyribonucleotide triphosphates) present in the cytosol of virally infected cells. This activity limits the amount of dNTPs available to the virus during reverse transcription, thus halting the virus at this stage in the life cycle. HIV-2 and SIV replicate well in these partially restrictive cell types, most likely because they encode the accessory protein Vpx. In the presence of Vpx, SAMHD1 is sequestered and targeted for proteasomal degradation. *In vitro* addition of Vpx to cultures of HIV-1 markedly improves the replication capacity in macrophages and DCs.

Preventing virion release: *BST-2 (Tetherin).* BST-2, or tetherin, is a host restriction factor that acts at the virion release, or budding, step. BST-2 is a transmembrane protein with a short cytoplasmic tail that is constitutively expressed in several cell types, and can be induced by pro-inflammatory type-1 interferon signaling on CD4+ T cells. It exerts its antiviral effect by "tethering" virions to the plasma membrane, preventing dissemination to other cells. This restriction factor is not unique to HIV-1, as BST-2 inhibits the dissemination of many enveloped viruses. Each primate lentivirus, HIV-1, HIV-2, and SIV, has at least one mechanism for BST-2 evasion. HIV-1 utilizes Vpu, HIV-2 uses Env, and several SIVs uses Nef. Each of these proteins in their respective infection settings will bind and down regulate expression of BST-2 from the cell surface.

These intrinsic defense factors may serve a second role, since some of them also act as pattern recognition receptors (PRRs), which assist in inducing the innate immune response to lentivirus infection.

4. SEQUENCE OF EVENTS IN HIV INFECTION

4.1 Transmission, Portal of Entry, and Sequential Spread of Infection

HIV is transmitted by three major routes: via sexual contact (accounting for >90% of infections worldwide), vertically from mother to child, or by injection of blood and blood products, most commonly through sharing contaminated needles and syringes. Mucosal secretions and blood contain both cell-free and cell-associated virus. The relative contribution of cell-free versus cell-associated virus is difficult to parse during natural infection since the inoculum contains both of these elements. Cell-free virions are infectious as shown by their ability to initiate experimental infections in primates. Studies of breast milk indicate that cell-associated HIV-1 is associated with early postpartum transmission, while later postpartum transmission is more strongly associated with cell-free virus. The details of HIV transmission by sexual contact or other mucosal routes are not entirely clear, specifically how the virus breaches the epithelial or mucosal barrier in order to reach susceptible CD4+ T lymphocytes or macrophages.

4.2 Early Target Cells

Our knowledge about the early events in infection comes for the most part from studies conducted using SIV in nonhuman primate models (Figure 3). Infection can be initiated by nontraumatic application of SIV to the vaginal mucosa or the tonsillar surface, and there are several ways in which virus could cross intact epithelial barriers. HIV (or SIV) may bind to the processes of DCs that extend into the mucosal lumen. In the tonsil and rectum, transmission may occur when the virus transits through epithelial or M (microfold) cells to reach underlying mononuclear cells. Epidemiological studies show that the presence of other sexually transmitted diseases, such as genital herpes, increases the risk of HIV infection. This may be due both to ulcerations and abrasions, that is openings in the epithelial barrier, and to increased number of target cells at the site of infection.

Following intravaginal infection of nonhuman primates, a small number of foci of infection form in the submucosa. In humans, mucosal transmission of one or a few genetically related virions usually establish the systemic infection in the recipient, despite the large number of viruses present in the inoculum from the donor. This is termed the "founder effect" and these transmitted founder viruses tend to differ from the dominant variant in the donors, with increased replicative capacity and a higher resistance to type-I interferon.

The initial infected cells are mucosal CD4+ T cells that may or may not show high expression of HIV/SIV co-receptor CCR5 (Figure 3). At early time points, up to ~90% of infected cells are "resting" CD4+ T lymphocytes, with small contributions from activated CD4+ T cells and infected macrophages or DCs. Important for dissemination is the ability of DCs in the mucosa to bind and concentrate virus on their surface via DC-SIGN (or other receptors), and then carry it to draining lymph nodes. Within a few days after infection, the local lymphoid tissue is heavily infected, first the draining lymph nodes (<1 week), and then the distant nodes, spleen, and circulating mononuclear cells (1–3 weeks).

In addition to the regional lymph nodes, the greatest and most sustained loss of CD4+ T cells in the body during HIV

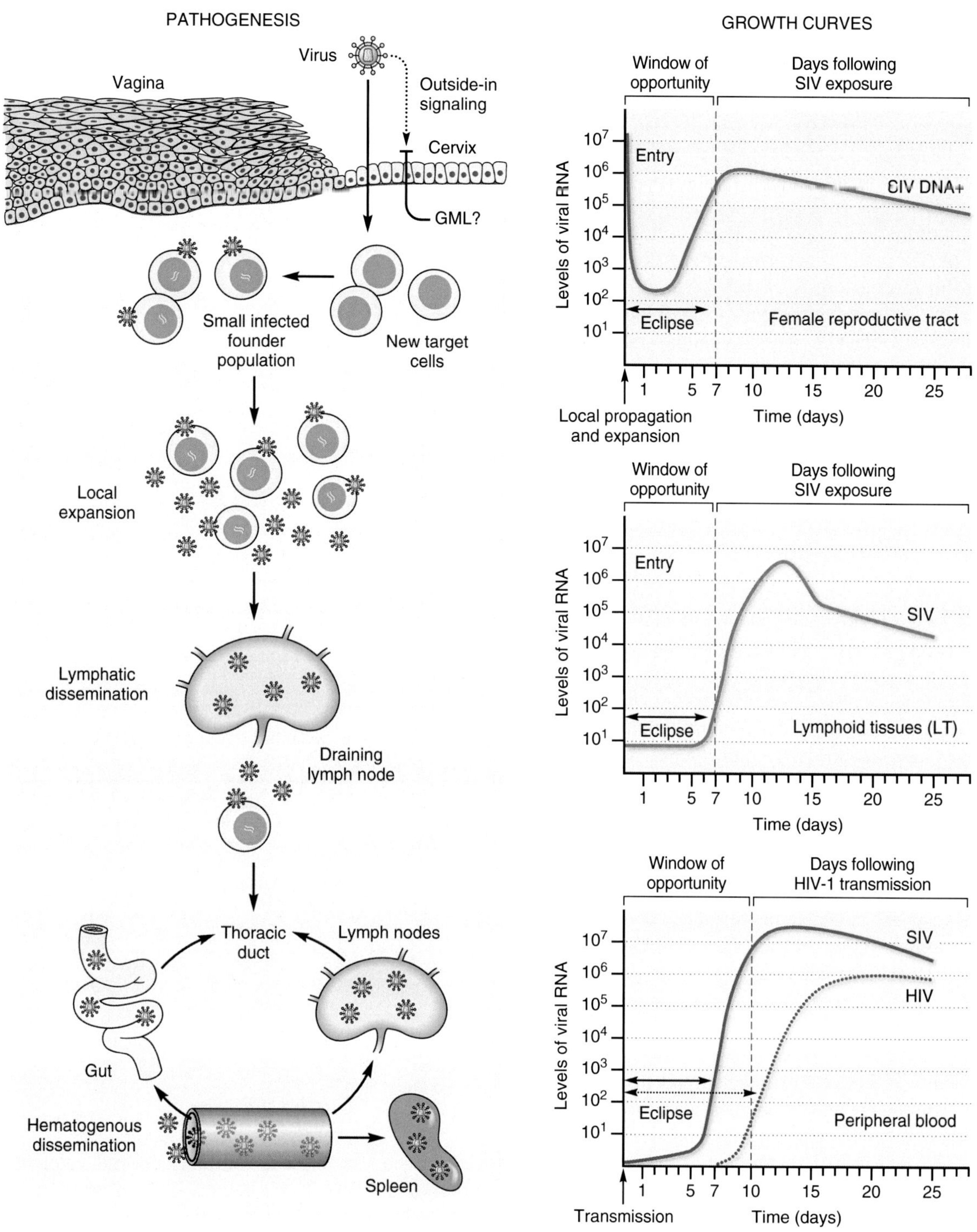

FIGURE 3 Cartoon of the early stages in vaginal transmission of SIV, to show the role of the major target cells, CD4+ T lymphocytes (resting and activated), dendritic cells, and macrophages. *After Haase AT (2011), with permission.*

and SIV infection involves predominantly CD4+CCR5+ T cells in the gastrointestinal-associated lymphoid tissue (GALT). A subset of CD4+ T cells termed Th17 cells, based on their production of IL-17, have been implicated as a preferential target for HIV/SIV-mediated depletion in the GALT. As Th17 cells are important for maintaining mucosal barrier integrity, their loss early in pathogenic HIV/SIV infection may be responsible in part for the microbial translocation and systemic immune activation observed during the chronic stages of these infections.

4.3 Viremia, CD4+ T Cell Depletion, and Disease Outcome

Due to the ability of the virus to rapidly and constantly infect new target cells—whose in vivo life span has been estimated in the range of 1.0–1.5 days—HIV/SIV infections are typically associated with chronic high levels of virus in blood and selected tissues. Viremia is usually measured as viral RNA copies per ml of plasma and is used to monitor the kinetics of HIV disease course and progression to AIDS. Another useful marker to predict the course of disease is the number of CD4+ T lymphocytes in the blood, which is usually inversely related to plasma viremia, and is a harbinger of the functional loss of immune responses during clinical AIDS. Other important indicators of the course of infection are the levels of CD4+ T cells in lymph nodes and the GALT, as well as markers of the host antiviral immune responses and the degree of chronic immune activation.

4.4 Clinical Disease Course

In most HIV-infected patients, in absence of ART, there is an acute phase of infection (about 2 months duration) with high levels of plasma viremia. During the acute phase of infection, a mononucleosis-like syndrome occurs in 50–75% of patients, accompanied by a peak in viremia and an acute drop in the CD4+ cell count in the blood.

The acute phase is followed by a subclinical, asymptomatic phase (clinical latency) with lower levels of viremia but progressive loss of CD4+ T cells. During the transition to the subclinical phase, there is induction of a cellular and humoral immune response, which is associated with a substantial drop in plasma viremia. However, in most cases the immune response is unable to completely suppress the virus, and viremia typically stabilizes 4–6 months after infection at a level often called the virus "set point." Set point viremia is maintained by a high level of virus replication, as discussed later in this chapter.

The clinical outcome of HIV infection is closely related to the set point. In a cohort of infected patients, 90% of the quartile with the highest set points progress to AIDS in 5 years, while <10% of the quartile with the lowest set point has developed AIDS in the same period of time. Once the set point is reached and viral load stabilizes, the subclinical phase can last anywhere between 1 and 20 years. During the subclinical phase, patients remain relatively healthy and do not present with overt signs of immunodeficiency, despite high rates of virus replication and a slow decline of CD4+ T cells. This dynamic process determines the next steps in infection and explains the diversity in the duration of clinical latency. Absent ART, the subclinical phase is almost invariably followed by an accelerated phase of clinical AIDS that lasts 1–2 years before death.

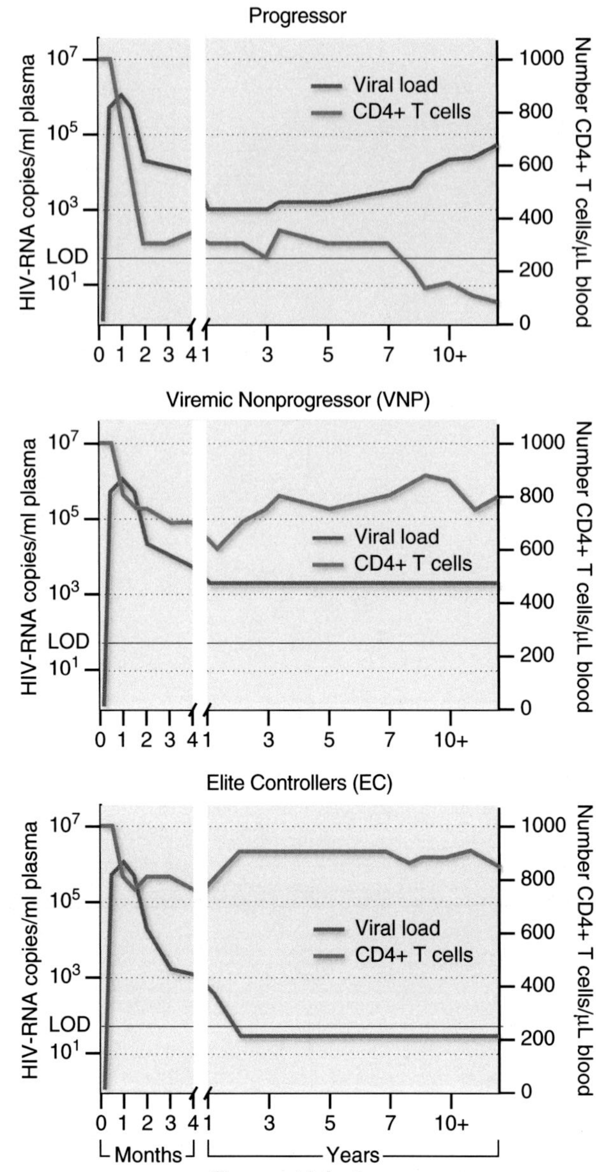

FIGURE 4 Schematic representation of HIV-controller status based on viremia and CD4+ T cell counts: progressor, viremic nonprogressor (VNP), and elite controllers (EC).

HIV-infected individuals fall into several categories based on their CD4+ T cell counts, level of viremia, and time to AIDS (Figure 4). Rapid progressors generally have high viral set points, rapid depletion of peripheral CD4+ T lymphocytes, and develop AIDS in 1–3 years from the initial infection. A rare subset (<2%) of patients called long-term nonprogressors, slow progressors, or HIV controllers (HIC), remain symptom-free for at least 10 years, but often much longer, in the absence of ART, with stable CD4 counts (average 500 cells/μL), and very low levels of plasma viremia. Another, even rarer subset

(<1%), are individuals who have undetectable viral load in the absence of ART. These patients are often called elite controllers (ECs). Another extremely rare subset of HIV-infected individuals are the so-called viremic non-progressors (VNPs), in which stable CD4+ T cell counts are maintained despite relatively high levels of viremia (Figure 4). These VNPs show features similar to those observed during nonpathogenic SIV infection of natural hosts, such as the sooty mangabeys (SMs), including the absence of chronic immune activation and low level of infected central-memory CD4+ T cells. Finally, very slow disease progression is also seen in patients infected with HIV-2, more than half of whom remain AIDS-free throughout their lives. The ability of the host to contain an HIV infection indefinitely is central to understanding the dynamics of infection and is discussed further below.

4.5 Opportunistic Infections and Neoplasms

The drop in CD4+ lymphocyte levels (normally >1000 cells/μL blood) below a critical threshold (~200 cells/μL blood), is often accompanied by a rise in virus set point, and signals the advent of AIDS-defining illnesses. Constitutional symptoms include fever, fatigue, malaise, lymphadenopathy, gastrointestinal symptoms such as diarrhea, weight loss. Opportunistic infections, such as oral candidiasis (caused by *Candida albicans*) and hairy leukoplakia of the tongue (caused by EBV infection of epithelial cells), are early evidence of immunodeficiency. Opportunistic infections are caused by a wide spectrum of parasites, including protozoa (such as *Toxoplasma gondii*), fungi (such as *C. albicans* and *Pneumocystis carinii*), and viruses (such as cytomegalovirus (CMV), herpes simplex (HSV), and varicella zoster (VZV)). In healthy persons, these infections are common, but are held in check (clinical latency) by the cellular immune response, and their emergence reflects a decline in cellular immune function.

The spectrum of AIDS-associated opportunistic infections differs geographically, reflecting the relative prevalence of different agents. For example, *M. tuberculosis* is much more important as a manifestation of AIDS in developing than in industrialized countries. Relative to the general population, AIDS patients are also at increased risk for a selected number of neoplasms. Among these are polyclonal B cell lymphomas, such as Burkitt's lymphoma (caused by EBV), cervical carcinoma (associated with HPV), and Kaposi's sarcoma (associated with HHV8). It is not clear why specific neoplasms are particularly associated with AIDS, but probably reflects the compromise of immune surveillance mechanisms that are particularly important for control of these cancers.

5. IMMUNE RESPONSE TO HIV

5.1 Innate Immune Response: The Host Fights Back

During HIV infection of CD4+ T helper lymphocytes and monocytes, there is a vigorous cellular innate response (see Chapter 4, Innate Immunity). HIV ssRNA, dsRNA, dsRNA/DNA duplexes, and dsDNA are produced during infection and have the potential to act as PAMP (pathogen-associated molecular patterns) that are recognized by PRRs (Rustagi and Gale, 2014). Specifically, the molecules RIG-1 (retinoic acid-inducible gene 1), cGAS (Cyclic GMP-AMP Synthase), and tetherin have been identified as HIV PRRs. In addition, plasmacytoid DCs recognize endocytosed HIV virions through their TLR7 PRRs. Following HIV recognition, IRF3 (interferon regulatory factor 3, a master switch in the innate system) is up-regulated, leading to production of IFNβ. In turn, this triggers a host of interferon-stimulated genes that exert a variety of antiviral activities. In addition, inflammatory cytokines such as type I interferons and IL-15, recruit and activate NK cells.

In return, the HIV-1 accessory proteins, Vif and Vpr, can cause the degradation of cellular IRF3, while Vpu blocks the expression of genes induced by IRF3. In aggregate, these HIV accessory proteins reduce activation of innate responses. Paradoxically, in the case of HIV and SIV infection, it is possible that some aspects of the innate immune response mediated by DCs may favor virus transmission and early dissemination, through the recruitment of activated CD4+ T cells that are the preferential targets for HIV and SIV infection.

The innate response may play a critical role in the long-term outcome, which can be determined within the first few days of infection. Early in HIV infection there are high levels of IFN in patients' serum, in spite of the ability of HIV to antagonize some elements of the innate response. This appears to reflect the activation of plasmacytoid DCs through IRF7. The ability of HIV to induce excessive early innate immunity likely plays an important role in immune exhaustion, which is a hallmark of the pathophysiology of HIV infection (Benecke et al., 2011).

5.2 Antibody Responses

HIV proteins are highly immunogenic and, if purified and used as experimental immunogens, can induce robust cellular and humoral immune responses. However, in HIV patients, the magnitude of the immune response does not correlate with the ability of the host to control an existing infection.

Most patients develop detectable antibodies against HIV-1 within 2 months of infection, with highest reactivity against the Gag—particularly p24—and Env (gp120 and

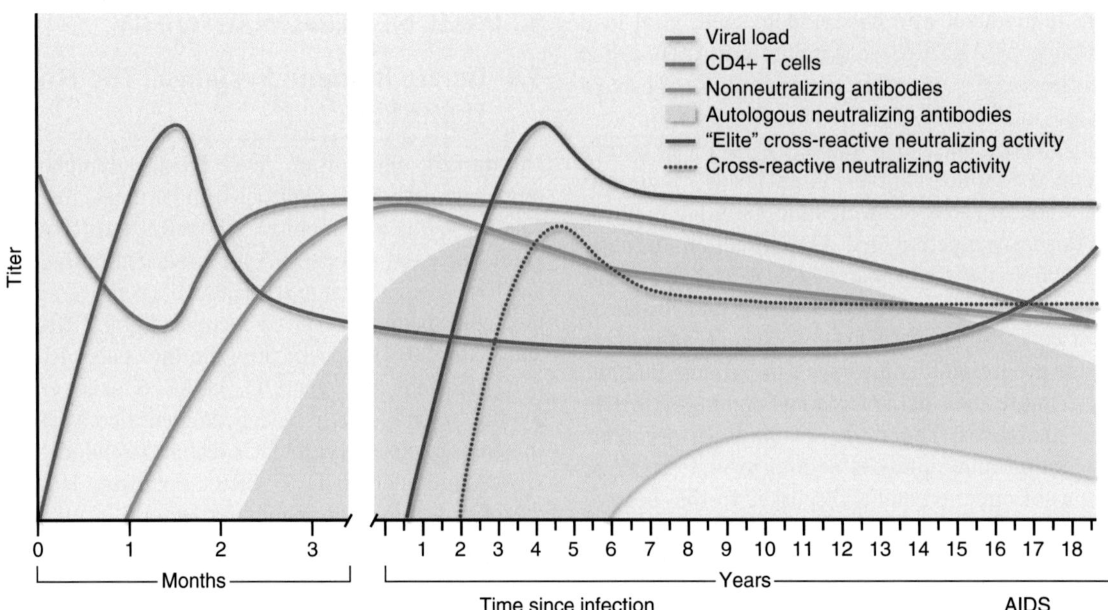

FIGURE 5 Schematic representation of HIV-1 env antibody responses during natural infection with HIV. The nonneutralizing antibody response develops around 4 weeks post-infection and remains at relatively high titers for the rest of the infection. The shaded area represents the early autologous neutralizing antibody response, which does not develop until about 2 months post-infection. These strain-specific neutralizing antibodies continuously develop during infection, because the virus can mutate away from neutralizing epitopes. The "elite" cross-reactive neutralizing antibodies represent a distinct cohort of elite neutralizers. These individuals develop broadly neutralizing antibodies very early in infection and maintain them at high titers throughout infection. This distinction is independent of controller status. The broken line is when most individuals develop cross-reacting neutralizing antibodies. As mentioned previously, this is usually 2–3 years post seroconversion and is believed to be too little and too late to control the persistent infection. *Based on data in Euler Z and Schuitemaker H. Cross reactive broadly neutralizing antibodies: timing is everything.* Frontiers in Immunology, *2012 http://dx.doi.org/10.3389/fimmu.2012.00215.*

gp41) proteins (Figure 5). These early antibodies are nonneutralizing, instead mediating some level of protection by mechanisms such as complement and antibody-dependent cell-mediated cytotoxicity. It takes 3–6 months to develop neutralizing antibodies to the transmitted founder virus, considerably longer than for many other viral infections. These autologous neutralizing antibodies are very strain specific and provide no protection against superinfection with another strain of HIV. Also, there is usually a rapid selection for virus mutants which can escape these narrowly neutralizing antibodies. During the course of infection, the accumulation of many autologous antibodies broadens the neutralizing capacity of patient's serum. In addition, some HIV-infected patients make broadly neutralizing monoclonal antibodies (bnMAbs), usually late in infection. This has important implications for design of protective vaccines, which is discussed in a later part of this chapter.

5.3 Cellular Immune Responses

CD8+ lymphocytes. Infected patients raise robust cellular immune responses to HIV. In most patients, the immunodominant CD8+ T cell response is to the most abundantly produced protein, HIV-gag, though responses to other epitopes may actually mediate better viral suppression. It is well established that CD8+ T lymphocytes play an important role in mediating control of viral load during acute and chronic HIV and SIV infection. The generation of HIV-specific CD8+ T cells coincides with the reduction of virus replication observed during the transition from the acute to the chronic phase of the infection. The best studied mechanism by which CD8+ T cells mediate control of virus replication is through peptide:HLA (p:HLA) displays on infected cells. HIV-specific CD8+ T cells recognize these p:HLA protein complexes and kill the target cells. As described above, CD8+ T cell escape mutations in the HIV and SIV genomes frequently occur early in infection, showing that cellular immunity exerts strong pressure on the virus.

Is the disease just due to the virus or does the host response play a role? If so, how does this happen?

However, these cellular responses are a double-edged sword. A potent T cell response involves activating CD4+ T cells, rendering them susceptible to productive HIV infection. Such infected CD4+ cells are a target for cytolytic CD8+ T cells, and their elimination may simultaneously lower the viral set point and deplete the CD4+ reservoir.

The role of CD8+ T lymphocytes in control HIV and SIV replication is complex. Studies using experimental depletion of CD8+ lymphocytes in SIV-infected rhesus macaques showed that the half-life of SIV-infected cells is similar in animals that were CD8 depleted and those that were mock depleted, thus suggesting that a significant portion of the in vivo antiviral role of CD8+ T cells is mediated by noncytolytic mechanisms. In this regard, an important function of CD8+ lymphocytes is the secretion of β-chemokines, which may play a role in the control of HIV. Whatever the mechanism, be it cytolytic or noncytolytic, the CD8+ T cell response of the individual is not sufficient to completely control virus replication, and eventually this CD8+ T cell-mediated control is lost due to a combination immune escape and functional exhaustion.

How does an infection of CD4+ lymphocytes lead to a dysfunction of CD8+ cells that are not susceptible to HIV infection?

During HIV infection, there is a generalized dysfunction of effector CD8+ T cells, which plays a central role in the acquired immunodeficiency syndrome. This dysfunction is associated with a state of immune activation that is likely induced in part by translocation of microbes or microbial products from the intestinal lumen. As noted already, immune activation leads to exhaustion, dysfunction, and depletion of CD8+ lymphocytes absent any direct HIV infection. Evidence for an indirect mechanism of CD8+ dysfunction is its absence in monkeys naturally infected with high titers of their endogenous lentiviruses, discussed below.

CD4+ lymphocytes. Defining the role of HIV-specific CD4+ T cell response is even more complicated than elucidating the role of HIV-specific CD8+ T cells. Activating CD4+ cells is critical to their role in inducing anti-HIV immunity, but at the same time makes them prime targets for this insidious virus that replicates best in activated CD4+ cells. Actively infected CD4+ cells are subject to direct viral killing and also potential targets for the host antiviral immune response. The complex role of CD4+ lymphocytes—as both viral targets and immune mediators—is illustrated by the observation that experimental depletion of CD4+ T lymphocytes in SIV-infected macaques results in higher virus replication and accelerated disease progression.

6. PATHOGENESIS

6.1 The SIV Model

As noted above, each species of African monkeys has its "own" strain of endogenous lentiviruses, which do not appear to cause disease. However, when transmitted to Asian macaques, which are not infected with their own lentiviruses, these African SIVs can cause simian AIDS. This was discovered when rhesus macaques, housed in proximity to some SMs (an African monkey species) developed an AIDS syndrome. The causal virus was then identified as SIVsm, the endogenous virus of the African mangabeys. This serendipitous finding has been exploited to develop a monkey model. This monkey model resembles HIV/AIDS in that both represent diseases that followed cross-species transmission of lentiviruses. There are many variants of the SIV-macaque model; the clinical course of these infections ranges from relatively nonvirulent to acutely lethal, depending upon the strain of SIVsm, the route of infection, and the age and species of monkeys. The description of pathogenesis that follows below is based on studies in both the monkey model and humans.

6.2 Virus Turnover

At the end of the acute phase of HIV and SIV infection, once viral set point is established and CD4+ T cell counts decline slowly, the infected individual enters the period of clinical latency. During clinical latency, the virus is undergoing rapid death and then replacement from infected cells, with a half-life of cell-free HIV virions of 28–110 min (Figure 6). Patients have detectable viremia of anywhere from 10^2–10^6 HIV-RNA copies per milliliter of plasma, and an estimated 10^9–10^{11} infectious virions are produced per day. The rate of replication depends on the number of infected cells present in the individual patient.

What can we learn about the pathogenesis of HIV from antiretroviral therapy?

Virus turnover can best be elucidated by using ART as an experimental tool. Upon initiation of ART, there are at least three phases of viral decay (Figure 7). If it is assumed that no new cells are being infected and residual plasma viremia is due solely to production of virus from already infected cells, it is possible to estimate the half-life of productively infected cells (see discussion in Chapter 15, Mathematical models). The first and most rapid phase of viremia decay is most likely due to the death of infected, activated CD4+ T cells with a half-life of approximately 1–2 days. The second phase is more gradual, probably resulting from the death of infected macrophages, with a half-life of about 2 weeks. The third phase of decay is postulated to come from cells that have a half-life of about 6 months, consistent with the life span of resting CD4+ memory T cells. Based on these data, it is possible to determine the sources of plasma viremia: infected CD4+ T cells are the main source of viremia, while infected macrophages make a minor contribution because of their low levels of virus production as well as lower frequency of infection (Figure 6).

6.3 T Cell Homeostasis and Immune Dysfunction

Under physiological conditions, T cell homeostasis replaces the cells that die as a result of senescence and

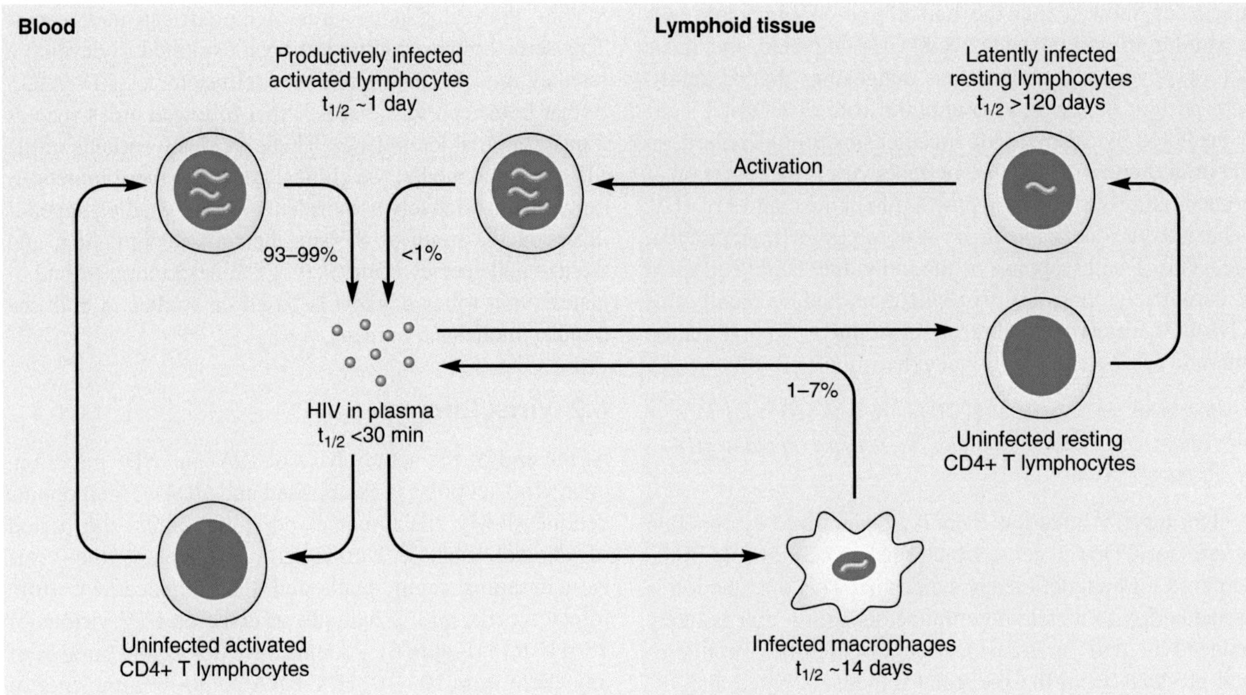

FIGURE 6 Kinetics of HIV or SIV production and clearance in vivo. Plasma virus is produced mainly by infected activated lymphocytes with a modest contribution from infected macrophages. Latent infection of resting lymphocytes is unimportant during active infection but presents an impediment to eradication by HAART (highly active antiretroviral therapy). $t_{1/2}$: half-life.

activation-induced cell death. In HIV-infected individuals, due to various pathogenic mechanisms, such as infection, cell death (apoptosis or pyroptosis), and antigen-driven proliferation, this homeostatic condition is heavily perturbed. Both CD4+ and CD8+ T cell compartments are affected. An increase in proliferating CD4+ T cells is due to several factors. First, there is significant depletion of CD4+ T cells in peripheral blood, lymphoid tissues, and mucosal sites due to both direct viral infection and activation-induced death. This cell loss may trigger the release and/or increase the availability of homeostatic cytokines such as IL-7 and IL-15 thus resulting in increased turnover of the remaining CD4+ T cells. Another important mechanism of increased T cell proliferation during HIV infection is the antigen-driven cell activation that occurs in response to HIV as well as other antigens (i.e., reactivating viruses, opportunistic pathogens, etc.).

A second contributor to this nonhomeostatic T cell proliferation is the translocation of microbial products from the intestinal lumen to the systemic circulation. Microbial translocation occurs as a result of HIV-associated mucosal immune dysfunction. It induces activation of various innate immune response mechanisms, mediated by Toll-like receptors as well as other proinflammatory molecular pathways, which ultimately increase the level of T cell activation and proliferation.

The third potential contributor of disrupted CD4+ T cell homeostasis during pathogenic HIV/SIV infections is the failure of the T cell regenerative compartment, which include the thymus, bone marrow, and secondary lymphoid tissues. During HIV infection the production of naïve T cells by the thymus—which is indirectly reflected in the frequency of circulating T lymphocytes that are TREC positive (T cell receptor excision circles, a marker of recent thymic TCR rearrangement)—is reduced, even when CD4+ T lymphocyte counts are >500 per μl. The establishment of collagen deposition and fibrosis in the lymph nodes also reduces the de novo formation of both naïve and central-memory T cells.

How does an infection of CD4+ lymphocytes lead to a dysfunction of CD8+ cells that are not susceptible to HIV infection?

Immune perturbation also explains one of the paradoxes of HIV/SIV pathogenesis: How does an infection of CD4+ T cells also cause a reduction in CD8+ T cells, which are not infected by the virus? Apparently, continual immune activation leads to shortened cell life, and exhaustion of the CD8+ T cell population. Since CD8+ cells are—in most cases—the effectors that control latent microbes, their loss plays an essential role in opportunistic infection. Collectively, these mechanisms induce a complex perturbation of T cell homeostasis that is the iconic hallmark of HIV and SIV immunodeficiency.

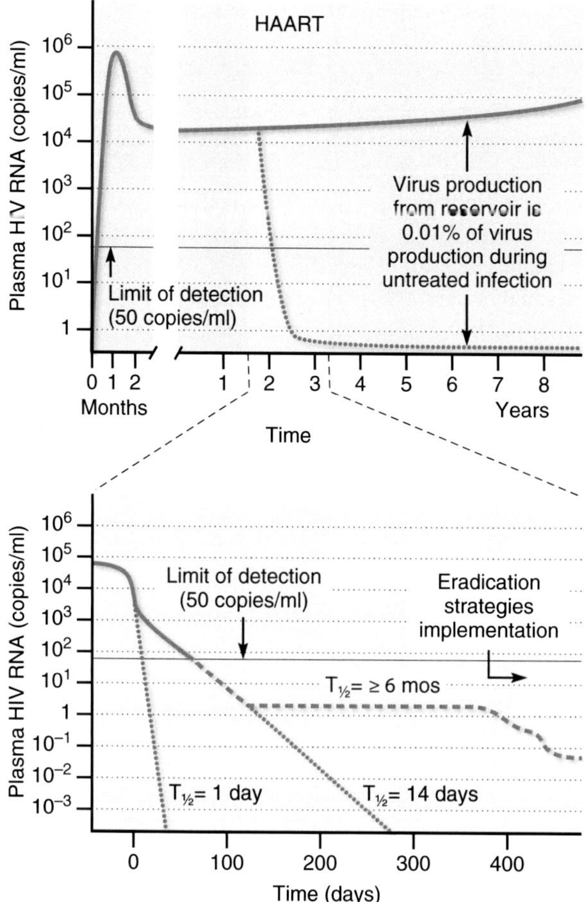

FIGURE 7 The reduction in plasma viremia following initiation of highly active antiretroviral therapy can be divided into three phases: early rapid decrease of ~100-fold due to disappearance of infected activated CD4+ lymphocytes; a slower phase due to disappearance of longer lived infected macrophages; and a plateau of persistent infection that may be below the level of detection, with a roughly estimated half-life of 6 months or greater. *Adapted from Eisele E and Siliciano RF (2012), with permission.*

6.4 Nonpathogenic SIV Infections: Lessons from the Natural Host

If African monkeys can tolerate SIV infection and remain healthy, why does HIV cause fatal illness in humans?

As already noted, most SIVs cause relatively benign, nonpathogenic, infections in their natural hosts. Indeed, SIV-infected African monkeys in captivity do not have shorter life spans than their uninfected counterparts. The two "natural host" species that have been most extensively studied are African green monkeys (AGMs), which are infected with SIVagm, and SMs, which are infected with SIVsmm. The nonpathogenic SIV infections of natural hosts share some characteristics with pathogenic SIV infections (Table 2). For example, both pathogenic infection of rhesus macaques (RM) and nonpathogenic infection of SM produce high titer viremia and significant loss of mucosal CD4+ T cells. Additionally, both pathogenic and nonpathogenic infections induce high levels of innate and adaptive immune activation during acute infection, with an inability of the adaptive immune system to effectively control virus replication.

However, some aspects of the infection are clearly distinct in the pathogenic and nonpathogenic infections. First, natural SIV hosts maintain healthy levels of peripheral CD4+ T cells, with the specific preservation of both central-memory CD4+ T cells and mucosa-based Th17 cells. Most infected CD4+ cells are the shorter lived, more differentiated subset of effector-memory CD4+ T lymphocytes. Second, in natural SIV hosts, the robust innate and adaptive immune response, that is generated during the acute phase of infection, is rapidly controlled and does not lead to chronic immune activation. Third, natural SIV hosts appear to maintain mucosal barrier integrity with low levels

TABLE 2 Features of SIV in Natural and Nonnatural Hosts

Phenotype	Natural Host (SMs, AGMs)	Nonnatural Host (Humans, RMs)
AIDS	No	Yes
Levels of peripheral CD4+ T cells	Healthy	Low
Viral load	High	High
Virus cytopathicity	Yes	Yes
Host immune control	Ineffective	Ineffective
Depletion of mucosal CD4+ T cells	Yes, stable	Yes, progressive
Mucosal immune dysfunction/microbial translocation	No	Yes
Chronic immune activation	No	Yes
Pattern of infected cells	$T_{EM}>T_{CM}$	$T_{CM}>T_{EM}$
Vertical transmission	Rare	Frequent

AGM, African Green Monkey; RM, Rhesus Macaque; SM, Sooty Mangabey; TEM, Effector Memory T Cells; TCM, Central Memory T Cells.

of microbial translocation despite a significant depletion of mucosal CD4+ T cells. The absence of significant microbial translocation may be a key factor allowing the resolution of immune activation in these animals.

Natural hosts also maintain their lymph node architecture and avoid lymphoid tissue fibrosis, thus allowing for formation of proper germinal centers and a more effective naïve and central-memory T cell regeneration in the paracortex. It is likely that all of these factors contribute to the nonpathogenic nature of SIV in the natural host.

7. GENETIC DETERMINANTS OF HOST SUSCEPTIBILITY TO HIV

There are a number of host genetic determinants that influence susceptibility to infection or the rate of progression to AIDS (Table 3). These determinants mainly fall into two categories, those that alter co-receptor availability and those in HLA loci. The HLA alleles appear to act by determining the selection of HIV peptides for immune presentation. Certain HLA alleles (i.e., HLA B27 and B57) are associated with delayed progression to AIDS and better virus control, whereas others are associated with more rapid disease progression. Similar trends are seen in rhesus macaques SIV, with Mamu-B*08 and B*17 haplotypes associated with control and Mamu-B*01 associated with progression. Co-receptor genes determine the level of expression of either co-receptors or co-receptor ligands (such as chemokines) that can reduce receptor availability for HIV attachment. The most prominent mutation is the Δ32 deletion in CCR5 that abrogates the expression of that gene. In its homozygous form, CCR5Δ32 markedly reduces the risk of infection and, in its heterozygous form, reduces the rate of progression to AIDS.

More recently, genome-wide association studies (GWAS) have been used for a broad, unbiased look at the human genome and correlates of HIV-associated disease (van Manen et al., 2012). This approach independently confirmed the finding that HLA-B*57 is associated with virus control and slower disease progression in HIV-infected individuals. To further investigate possible correlates, complete analyses of the gene expression profile have been performed, using both microarray studies and direct RNA sequencing (Bushman et al., 2013).

GWAS studies have compared various groups of HIV-infected patients, including rapid progressors, ECs, various sets of slow progressors, and uninfected subjects (van Manen et al., 2012). Most of the genetic sites identified have fallen within the MHC loci, where they may define more specific genetic determinants than identified in conventional genetic studies. A few additional loci appeared to be associated with interferon genes, innate immunity, replication within macrophages, or SNPs without a clear functional attribution.

8. VACCINES

Why do not we have an HIV vaccine?

There are several daunting scientific obstacles to formulating an effective vaccine against HIV infection. First, HIV is 100% fatal, and infection is often initiated by a single virion. Therefore, protection requires "sterilizing" immunity. This contrasts with most viral vaccines, which induce

TABLE 3 Genetic Determinants That Are Correlated with Susceptibility to HIV Infection and Progression to AIDS

Genetic Locus	Genetic Context	Biological Effect	Influence on Progression to AIDS
CCR5	Homozygous Heterozygous	Δ32 mutation in CCR5 abrogates or reduces CCR5 expression	Protects from infection Delayed progression
CCR5	Homozygous	P1 mutation in promoter for CCR5	Accelerates progression
CCR2	64I mutation Heterozygous	Unknown	Delayed progression
CCL3L1	Low copy number	Chemokine ligand for CCR5	More susceptible to infection
SDF-1	G801A SNP	Ligand for CXCR4	Delayed progression
HLA-B*35	Homozygous	CD8 T cells are more restricted	Accelerates progression
HLA-B*07	Homozygous	CD8 T cells are more restricted in their recognition of viral epitopes	Accelerates progression
HLA-B*57	Homozygous	CD8 T cells are more cross-reactive, able to see greater range of viral epitopes	Delayed progression, better viral control
HLA-B*27	Homozygous	More likely to present gag epitopes that are structurally constrained	Delayed progression, better viral control

only partial protection (vaccine-protected subjects undergo a subclinical infection on exposure to wild virus). Second, most virus vaccines induce antibodies against a single antigenic strain of wild virus (or in the case of poliovirus vaccines, three strains), and herd immunity does not select for escape mutants. However, HIV has evolved into a wide variety of escape mutants, so that the "narrow" neutralizing antibody induced by a single HIV strain will not protect against other antigenic variants of the virus.

Despite these obstacles, a tremendous effort has been made by the HIV/AIDS research community to develop and test vaccine candidates. Several large-scale clinical trials of candidate AIDS vaccines have failed to show protection—or, even worse, a trend toward increased risk of infection. So far, only one large-scale trial has shown a significant, albeit very modest, degree of protection from infection in a population of low-risk individuals (Rerks-Ngarm et al., 2008). The RV-144 trial enrolled over 16,000 subjects in Thailand and employed a prime-boost strategy in which the prime was a canarypox vector encoding HIV gag, pol, and env genes (ALVAC) and the boost was a recombinant gp120 (AIDSVAX). Follow-up studies implicated antibodies to certain regions of V1 and V2 loops of gp120 as potential correlates of the limited protection observed in this trial.

However, there are some encouraging observations. The potential protective efficacy of neutralizing antibodies has been demonstrated in a number of experimental studies of both narrow and broad neutralizing anti-HIV antibodies and antisera. Utilizing "passive" protection, monkeys are treated with antibody prior to a challenge with a potentially lethal infection with SHIV (hybrid SIV with an HIV envelope), and shown to be protected by virtue of sterilizing immunity.

Importantly, there are some neutralizing epitopes on the HIV envelope protein that are conserved across many HIV isolates (Klein et al., 2013). These epitopes have been detected by the isolation of human monoclonal antibodies with broad neutralizing activity (bnMAbs) against a wide range (>50%) of primary HIV isolates. bnMAbs are directed against a number of sites on the HIV envelope: V1/V2 loops, N-linked glycans, the CD4 binding site, and the membrane proximal external region (MPER) of gp41. However, the heavy chains of these bnMAbs tend to have long complementarity determining regions (CDRs) and extremely high rates of somatic hyper-mutation, making these antibodies vastly different than their germline counterparts. Furthermore, these relatively rare antibodies are usually obtained from patients with long-term infections, and their induction appears to require continuous antigenic stimulus. Thus the problem remains: How to induce these neutralizing antibodies with a specific immunogen that would serve as a vaccine? Indeed, no candidate HIV vaccines have elicited these broadly neutralizing antibodies in either humans or NHPs. Currently, there is an active search for such an immunogen and it remains to be seen if one can be devised.

An "end run" around this problem has been adopted by several investigators, who are using gene therapy approaches to confer the ability to produce bnMAbs upon an individual's plasma cells. Bone marrow cells are removed, transfected with vectors expressing the genes for a selected bnMAb, and infused into the donor (Schnepp and Johnson, 2014). Although promising results have been obtained in mice and nonhuman primates, it is unclear whether such an approach could be scaled to the population level.

9. HIV RESERVOIR AND CURE

Is it possible to "cure" HIV? Considering the potency of highly active antiviral therapy, what are the impediments to a cure?

There has long been an interest in developing a bona fide cure for HIV infection. However, the natural history of HIV infection has created what may be insurmountable barriers to that goal. There is dramatic evidence of HIV persistence in patients who have achieved effective control of infection (with no HIV RNA or DNA detectable in blood); once antiretroviral drugs are stopped, the virus always rebounds within 1–2 months. The HIV genome is latent in resting memory CD4+ lymphocytes, which are "invisible" to both the host immune system and antiretroviral drugs. Such latently infected CD4+ T cells can persist for long period of times and even divide without activating the virus, thus increasing the size of this "virus reservoir" pool. Occasional blips of virus production can occur during suppressive ART, due to the intermittent activation of such latently infected cells, which might lead to new rounds of infected cells. These observations suggest that HIV may continue to replicate at very low levels even during effective antiretroviral treatment. Furthermore, there is the possibility that ART does not completely shut down replication in "hidden" reservoirs such as the brain, kidney, and testis.

The persistent reservoir of latently infected cells appears to be utterly resistant to any attempt at ART intensification. Therefore, new strategies to eliminate latently infected cells will be essential to achieve a virological cure of HIV infection. Various tactics have been proposed to activate HIV expression in latently infected cells with the hope that these cells can then be eliminated either by direct HIV-induced death, by HIV-specific cytolytic T cells, or by other more arcane strategies. These approaches are now being tested in the SIV model but to date none have been successful.

Since HIV infection is an immunological disease, a true functional cure of the infection will also require the reconstitution of a fully functioning immune system. However, significant residual immunological abnormalities (i.e., low levels of immune activation, signs of immune senescence, residual fibrosis of the lymphoid tissue) are often seen in individuals in which virus replication has been suppressed by ART for long periods of time. This emphasizes the importance of devising novel, immune-based interventions, in addition to the state-of-the-art virological treatment, to treat the "residual disease" observed in a large proportion of ART-treated HIV-infected individuals. The difficulty in reversing the immunological effects of persistent HIV infection also provides a rationale for early treatment with HAART, a much debated issue in the AIDS community.

10. REPRISE

HIV is a lentivirus that targets CD4+ T lymphocytes through its receptor attachment site, and thereby mounts a direct attack on the adaptive immune system. Furthermore, as a retrovirus, HIV persists because reverse transcription results in the integration of DNA genomic copies which remain latent in resting CD4+ T cells. In addition, actively replicating virus can persist due to its ability to generate mutants that can escape anti-HIV antibodies or effector T lymphocytes.

Persistent infection is accompanied by chronic excessive immune activation, driven in part by translocation of intestinal microbes. Eventually, this results in immune exhaustion, explaining the paradoxical loss of uninfected CD8+ T lymphocytes. Immunodeficiency leads to opportunistic infections and neoplasms, which are invariably fatal absent ART. Studies of persistent SIV infection in naturally infected African monkeys show that high lentivirus titers by themselves do not account for the pathophysiology of HIV; it's not the virus alone but also the host response (chronic immune activation in part) that explains AIDS in either humans or simian models.

Detailed molecular analysis of HIV replication has enabled the development of a panoply of effective antiretroviral drugs. As a result, HIV infection has been altered from an almost invariable death sentence into a chronic infection that can be successfully managed for very long periods of time. However, HIV continues to represent an inordinate burden on humanity. In 2015, the virus is spreading at unacceptable rate (2 million new infections annually), and more than 30 million persons are living with HIV/AIDS. There still is no virological cure for infected individuals, and we lack an effective vaccine. The scientific obstacles to developing a vaccine or a cure are daunting, and the way forward remains unclear. There is an ethical imperative for continued research to solve both of these urgent global health challenges.

FURTHER READING

Reviews, Chapters, and Books

Barouch DM, Deeks SG. Immunologic strategies for HIV-1 remission and eradication. Science 2014, 345: 169–176.

Benecke A, Gale M, Jr., Katze MG. Dynamics of innate immunity are key to chronic immune activation in AIDS. Curr Opin HIV AIDS 2012; 7: 79–85.

Berger EA, Murphy PM, Farber JM. Chemokine receptors as HIV-1 coreceptors: roles in viral entry, tropism, and disease. *Annual Reviews in Immunology* 1999, 17: 657–700.

Bushman FD, Barton S, Bailey A, et al. Bringing it all together: big data and HIV research. AIDS 2013, 27: 835–838.

Charoudi A, Bosinger SE, Vanderford TH, Paiardini M, Silvestri G. Natural SIV hosts:showing AIDS the door. *Science* 2012, 335: 1188–1193.

Chowdury A, Silvestri G. Host-pathogen interaction in HIV infection. *Current Opinion in Immunology* 2013, 25: 463–369.

Coffin J, Swanstrom R. HIV pathogenesis: dynamics and genetics of viral populations and infected cells. *Cold Spring Harbor Perspectives in Medicine* 2013, doi: 10.1101/cshperspect.a012526.

Eisele E, Siliciano RF. Redefining the viral reservoirs that prevent HIV-1 eradication. Immunity 2012, 37: 377–388.

Etienne L, Hahn BH, Sharp PM, Matsen FA, Emerman M. Gene loss and adaptation to hominids underlie the ancient origin of HIV-1. *Cell Host Microbe,* 2013, 14: 85–92.

Euler Z, Schuitemaker H. Cross-reactive broadly neutralizing antibodies: timing is everything. Frontiers Immunology 2011: 3; 215: 1–11.

Fraser JS, Gross JD, Krogan NJ. From systems to structure: bridging networks and mechanisms. Molecular Cell 2013, 49: 222–231.

Klein F, Mouquet H, Dosenovic P, Scheid JF, Scharf L, Nussenzweig MC. Antibodies in HIV-1 Vaccine Development and Therapy. Science 2013, 341: 1199–1204.

Haase AT. Perils at mucosal front lines for HIV and SIV and their hosts. *Nature Reviews Immunology* 2005, 5: 783–792.

Haase AT. Early events in the sexual transmission of HIV and SIVand opportunities for intervention. Annual Review of Medicine 2011, 62: 127–139.

Schnepp BC, Johnson PR. Adeno-associated virus delivery of broadly neutralizing antibodies. Current Opinion in HIV & AIDS. 9(3):250–6, 2014 May.

Law GL, Korth MJ, Benecke AG, et al. Systems virology: host-directed approaches to viral pathogenesis and drug targeting. Nat Rev Micro 2013; 11: 455–66.

Okoye A, Picker LJ. CD4+ T cell depletion in HIV infection: mechanisms of immunological failure. *Immunological Reviews* 2013, 254: 54–64.

Pepin J. The origins of AIDS. Cambridge University Press, Cambridge, UK, 2011.

Rustagi A, Gale M Jr. Innate antiviral immune signaling, viral evasion and modulation by HIV-1. J Molecular Biology 2014, 426: 1161–1177.

Ryan CJ, Cimermancic P, Szpiech ZA, et al. High-resolution network biology: connecting sequence with function. Nature Reviews Genetics 2013, 14: 865–874.

Sharp, P.M. and Hahn, B.H. Origins of HIV and the AIDS pandemic. *Cold Spring Harb. Perspect. Biol.,* 1: a006841, 2011, 1–22.

Shea PR, Shianna KV, Carrington M, Goldstein DB. Host genetics of HIV Acquisition and Viral Control. *Annual Reviews of Medicine* 2013, 64: 203–17.

Strebel K. HIV accessory proteins versus host restriction factors. *Current Opinion in Virology* 2013, 3: 692–699.

Van den Berg LM, Geijtenbeek TB. Antiviral immune responses by human Langerhans cells and dendritic cells in HIV-1 infection. Advances in experimental medicine and biology 2013, 762: 45–70.

West, Jr AP, Scharf L, Scheid JF, Klein F, Bjorkman PJ, Nussenzweig MC. Structural Insights on the Role of Antibodies in HIV-1 Vaccine and Therapy. Cell 2014, 156: 633–648.

Original Contributions

Barrenas F, Palermo RE, Agricola B, Agy MB, Aicher L, Carter V, Flanary L, Green RR, McLain R, Li Q, et al. Deep transcriptional sequencing of mucosal challenge compartment from rhesus macaques acutely infected with simian immunodeficiency virus implicates loss of cell adhesion preceding immune activation. J Virol 2014, 88: 7962–7972. http://www.ncbi.nlm.nih.gov/pubmed/24807713.

Buchbinder SP, Mehrotra DV, Duerr A, et al. Efficacy assessment of a cell-mediated immunity HIV-1 vaccine (the Step Study): a double-blind, randomised, placebo-controlled, test-of-concept trial. Lancet 2008, 372: 1881–1893.

Deng K, Pertea M, Rongvaux A, et al. Broad CTL response is required to clear latent HIV-1 due to dominance of escape mutations. Nature 2015, 517: 381–384.

Eisele E, Siliciano RF. Redefining the Viral Reservoirs that Prevent HIV-1 Eradication. *Immunity* 2012, 37: 377–388.

Hammer SM, Sobieszczyk ME, Janes H, et al. Efficacy Trial of a DNA/rAd5 HIV-1 Preventive Vaccine. New England Journal of Medicine 2013, 369: 183–192.

Huthoff H, Malim MH. Cytidine deamination and resistance to retroviral infection: towards a structural understanding of the APOBEC proteins. *Virology* 2005, 334: 147–153.

Kluge SF, Mack K, Iyer SS, Heigele A, Learn GH, Usmani SM, Sauter D, Joas S, Hotter D, Pujol FM, Bibollet-Ruche F, Plenderleith L, Peeters M, Sharp PM, Fackler OT, Hahn BH, Kirchhoff F. Nef proteins of epidemic HIV-1 group O strains antagonize human tetherin. *Cell Host Microbe* 2014, 16: 1–12.

Letko M, Silvestri G, Hahn BH, Bibollet-Ruche F, Gokcumen O, Simon V, Ooms M. Vif proteins from diverse primate lentiviral lineages use the same binding site in APOBEC3G. *J. Virol.,* 87.11861–11871, 2013. PMCID: PMC3807359.

Rerks-Ngarm S, Pitisuttithum P, Nitayaphan S, et al. Vaccination with ALVAC and AIDSVAX to Prevent HIV-1 Infection in Thailand. New England Journal of Medicine 2008, 361: 2209–2220.

Richman DD, Wrin T, Little SJ, Petropoulos CJ. Rapid evolution of the neturalizing antibody response to HIV type 1 infection. Proceedings of the National Academy of Sciences 2003, 100: 4144–4149.

Goulet M-L, Olagnier D, Xu Z, et al. Systems Analysis of a RIG-I Agonist Inducing Broad Spectrum Inhibition of Virus Infectivity. PLoS Pathogens 2013, 9: e1003298.

van Manen D, van 't Wout AB, Schuitemaker H. 2012. Genome-wide association studies on HIV susceptibility, pathogenesis and pharmacogenomics. Retrovirology 9 (Aug 24): 70, 4690–9–70.

Chapter 10

Animal Models

No Model Is Perfect, but Many Are Useful

Victoria K. Baxter[1,2], Diane E. Griffin[2]

[1]*Department of Comparative and Molecular Pathobiology, Johns Hopkins University School of Medicine, Baltimore, MD, USA;* [2]*W. Harry Feinstone Department of Molecular Microbiology and Immunology, Johns Hopkins Bloomberg School of Public Health, Baltimore, MD, USA*

Chapter Outline

1. INTRODUCTION

Full evaluation of the interactions between a virus and its host requires a living organism, because simpler systems cannot simulate either the host response or the evasion strategies of the invading virus. Advances in *in vitro* and *in silico* techniques now allow researchers to examine the intricacies of a virus, especially on a molecular level. Both primary cell cultures and immortalized cell lines derived from animal tissues are commonly used to study a virus' effect on the biology of a cell. However, an intact living organism is required to fully evaluate the interactions between a virus and its host. The innate and acquired immune responses play key roles in the course of a viral infection and can only be studied in an animal model (or a human host). Furthermore, animal models are required for testing therapeutics and evaluating vaccines.

For viral infections of humans, ethical considerations necessitate the use of animal models to address many aspects of the virus–host interaction. Even for infections of animals, it is not always feasible to use the natural host, and it may be necessary to use a different species as a model. The model animal should not be expected to identically mimic the disease seen in the subject being modeled, but rather to act as a surrogate in understanding the infection process and clinical disease produced. This information can then be compared to what is observed in the natural host. Because a single animal model may often not be able to duplicate every feature of a virus infection exactly as it would occur in the natural host, different species may be used to study different aspects of pathogenesis. For example, a rhesus macaque may be used to recapitulate the clinical disease seen with a virus that naturally infects humans, while a mouse may be used to study the immune response to virus infection at the cellular level.

As virus research has evolved over the last century, the use of animal models has increased dramatically (Figure 1). Many different species are used in virus research, from small rodents such as mice and rats to larger species such as guinea pigs, chickens, ferrets, and nonhuman primates. Prior to the 1970s, the rat, hamster, and the guinea pig were the most common animal models used in virus

Viral Pathogenesis. http://dx.doi.org/10.1016/B978-0-12-800964-2.00010-0

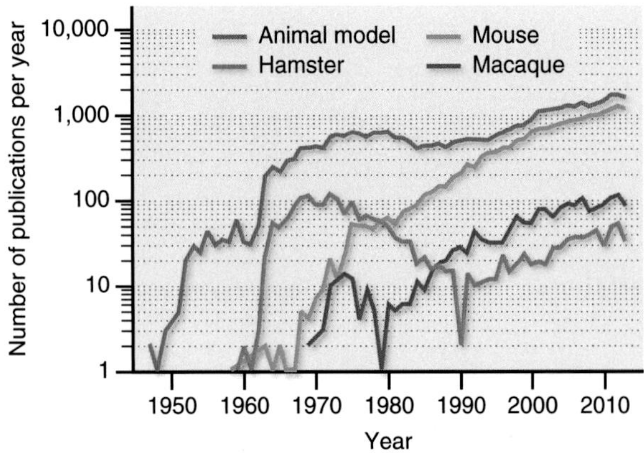

FIGURE 1 Trends in animal models used in virus research, 1940–2013. A Pubmed search was conducted using the search command "virus" + "animal model" +/− "species." Data are presented as the number of publications per year retrieved by each search.

studies. However, with advances in transgenic and knockout technology, mice have quickly become the dominant species used. In 2013, virology publications involving mice outnumbered publications for all other species combined. For many other animal species, full genome sequencing has recently increased thc information available for analysis of the host response to virus infection. Also, the use of certain selected species has increased in use in recent years due to their fit as models for particular infections. For example, the ferret has proved susceptible to respiratory viruses, such as influenza and SARS coronavirus (SARS-CoV).

The choice of an animal model for a virus study requires careful consideration. Each species possesses advantages and disadvantages that can vary with the virus being studied. Understanding the basic biology and unique characteristics of different animals can help researchers select the ideal model for their virus and experimental goal. In this chapter, we will introduce the animals commonly used in virus research and point out the considerations required when developing and using an animal model, together with some of the potential pitfalls. Recent technical developments have made it possible to manipulate the genetic background of experimental animals, which has opened exciting new vistas for animal models. Other new methods have also markedly expanded the information obtained from animal experimentation. For instance, it is now possible to visualize viruses in the same living animal repeatedly over time and look deep into normal and infected tissues with three-dimensional imaging. Finally, we will describe some of the applications of animal models for the development of new therapies and vaccines.

2. COMMONLY USED SPECIES

The following brief descriptions provide some basic information about the more frequently used (and readily available) species of experimental animals (Table 1). More detailed information is available in several of the references provided.

2.1 Mouse

There is a rich history of biomedical research using mice (*Mus musculus*) as experimental models. French biologist Lucien Cuénot studied mouse coat color genetics in 1902, and demonstrated Mendelian inheritance in mammals for the first time. Mice used in research today were bred from "fancy mice" in the early part of the twentieth century. Clarence Cook Little developed the first inbred dilute brown agouti (DBA) mouse in 1918, and today there are hundreds of available inbred strains of mice, with varying amounts of interindividual genetic variation. Over the last century, the use of mice as an animal model has exploded, making these animals the most popular species used today in virus research.

Mice have many advantages as research animals and are arguably the most cost-efficient vertebrate in current use. Multiple animals can be housed in individual cages and occupy relatively little space. They are prolific breeders with a short life cycle, as little as 9 weeks between generations for certain strains. There are a wide variety of reagents and tools available for mice, providing researchers with the ability to study almost any aspect of the immune response. In addition, mouse and human species genomes share many similarities, such as the approximate number of protein-coding genes and regions of conserved orthologous sequences. Of particular relevance to studies of pathogenesis, the murine histocompatibility complex is well described, and many aspects of innate, cellular, and humoral immune responses were originally described in mice and subsequently identified in humans. Together, these characteristics have made the mouse the preferred animal for deciphering the host immune response to many virus infections.

Despite these advantages, host-range limitations make mice resistant to many human viruses. For example, dipeptidyl peptidase 4 (DPP4), the cellular receptor used by the newly emergent Middle East respiratory syndrome coronavirus (MERS-CoV), is absent in mice. Therefore, a virus native to another species must often be adapted to mice through multiple passages. This technique selects for viruses with mutations that increase virulence, but may also alter pathogenesis. In the case of SARS-CoV, 15 passages of the Urbani strain in young BALB/c mice resulted in a virus (MA15) that is lethal in mice. Aspects of disease seen in severe human cases of SARS were reproduced by MA15 as the result of six coding mutations associated with host adaptation. On the other hand, adaptation of yellow fever virus to mice altered the disease from hepatitis to encephalitis, and the mouse-adapted virus was encephalitogenic (not hepatotropic) in nonhuman primates.

TABLE 1 Select Animal Model Species and Their Commonly Studied Virus Families

Species	Commonly Studied Virus Familes
Mouse	Poxviridae, Herpesviridae, Adenoviridae, Hepadnaviridae, Parvoviridae, Retroviridae, Arenaviridae, Bunyaviridae, Orthomyxoviridae, Paramyxoviridae, Filoviridae, Rhabdoviridae, Reoviridae, Coronaviridae, Arteriviridae, Togaviridae, Flaviviridae, Caliciviridae, Picornaviridae
Guinea Pig	Herpesviridae, Arenaviridae, Orthomyxoviridae, Paramyxoviridae, Filoviridae
Hamster	Arenaviridae, Bunyaviridae, Filoviridae, Paramyxoviridae, Flaviviridae
Woodchuck	Hepadnaviridae
Prairie Dog	Poxviridae
Cotton Rat	Adenoviridae, Orthomyxoviridae, Paramyxoviridae
Ground Squirrel	Hepadnaviridae, Rhabdoviridae, Togaviridae
Ferret	Orthomyxoviridae, Paramyxoviridae, Coronaviridae
Chicken	Poxviridae, Retroviridae, Orthomyxoviridae
Macaque	Poxviridae, Herpesviridae, Adenoviridae, Papillomaviridae, Polyomaviridae, Hepadnaviridae, Retroviridae, Arenaviridae, Bunyaviridae, Orthomyxoviridae, Paramyxoviridae, Filoviridae, Rhabdoviridae, Reoviridae, Coronaviridae, Arteriviridae, Togaviridae, Flaviviridae, Caliciviridae, Picornaviridae
Pig	Poxviridae, Herpesviridae, Parvoviridae, Circoviridae, Orthomyxoviridae, Paramyxoviridae, Filoviridae, Rhabdoviridae, Reoviridae, Coronaviridae, Arteriviridae, Flaviviridae, Caliciviridae, Picornaviridae
Ruminant	Poxviridae, Herpesviridae, Papillomaviridae, Retroviridae, Bunyaviridae, Paramyxoviridae, Reoviridae, Flaviviridae, Picornaviridae

Over the last few decades, numerous techniques to alter the mouse genome have been developed. With the sequencing and availability of more than 100 laboratory and wild-derived inbred strains of mice, researchers are now able to map genetic loci associated with disease susceptibility and identify quantitative trait loci underlying phenotypic variation. Methods to create knockout and transgenic mice have become increasingly accessible, and mice with spontaneous or engineered mutations are readily available for study. These important advances are described later in this chapter.

2.2 Other Commonly Used Species of Small Animals

2.2.1 Guinea Pig

The guinea pig (*Cavia porcellus*) is the only New World rodent commonly used in research. The Andean Incans in Peru originally domesticated the guinea pig to use as a food source and for sacrificial offerings. Dutch fanciers introduced them to Europe in the sixteenth century and bred them to create several colors and hair-coat varieties. They are more docile than smaller rodents, and their relatively low maintenance costs make them preferable to larger, more expensive nonrodent species. Although not as well characterized as the mouse, the guinea pig immune system shares many characteristics with that of humans. Despite this, few molecular and immunologic guinea pig reagents are commercially available, posing a hindrance to their use for viral pathogenesis studies.

Possibly the greatest advantage guinea pigs possess is the ability to recapitulate the gross and histologic pathology seen with many human viral diseases. They are most commonly used to study DNA viruses and negative-sense RNA viruses. Guinea pigs are highly susceptible to several arenaviruses and filoviruses and have been used in studying the pathogenesis of human hemorrhagic fevers including those caused by Ebola, Marburg, Junin, and Lassa viruses. These viruses generally replicate to high titers in immune-competent animals, and the clinical disease, hematologic profile, and pathology produced are similar to those observed in humans. Because of this, the guinea pig is a viable model for testing the efficacy of potential therapeutics. Guinea pigs are also popular models for several respiratory pathogens, including respiratory syncytial virus and influenza virus. These animals reproduce many of the characteristics of viral replication and pathology seen in humans and have been used to study aerosol infection and transmission efficiency between individuals.

2.2.2 Hamster

Hamsters were once one of the most commonly used animal models in virus research, but since 1973 their use has decreased in favor of the more genetically manipulable mouse. The Syrian or golden hamster (*Mesocricetus auratus*) is the most common hamster species used. Most of today's laboratory animals originated from a single litter whose progeny was imported into the United States in 1938, creating little individual genetic diversity. An advantage of hamsters is their extremely low rate of spontaneous disease compared to other animals, combined with their susceptibility to many viruses.

The Syrian hamster is often used for studying emerging RNA viruses, particularly hemorrhagic fever viruses. Infection of hamsters with nonadapted strains of Lassa fever virus, Rift Valley fever virus, or yellow fever virus produces pathological findings similar to those seen in humans. Hamsters infected with mouse-adapted Ebola virus are the only rodents that consistently develop the coagulopathies and vascular leakage seen in human infections. The hamster is currently the only animal model to reproduce many aspects of hantavirus pulmonary syndrome (caused by Sin Nombre virus), including the incubation period and disease pathology seen in humans.

In addition to reproducing clinical disease seen in humans, hamsters are particularly adept at amplifying many viruses, including lymphocytic choriomeningitis virus. In fact, federal regulations state that laboratory hamsters cannot be housed outdoors, lest they come in contact with wild rodents and become persistently infected; they are the only species with this stipulation. Because of their susceptibility, hamsters are commonly used for studying viral persistence and shedding.

The hamster immune system differs substantially from the human immune system, and there are only a limited number of immunologic reagents for hamsters; this has severely curtailed their use for studies of the immune response to infection. Due to the founder effect of the laboratory hamster population in the United States, very little alloantigenic variation exists.

2.2.3 Other Rodents

Other rodent species, including wild (nondomesticated) animals, are used for special situations. These species include the wood rat (*Neotoma* spp.), the deer mouse (*Peromyscus* spp.), the woodchuck (*Marmota monax*), the black-tailed prairie dog (*Cynomys ludovicianus*), the cotton rat (*Sigmodon hispidus*), and ground and rock squirrels (*Spermophilus* spp.).

2.3 Ferret

The ferret (*Mustula putorius furo*) is a member of the ancient and diverse *Mustelidae* family that includes weasels, minks, and martens. Although supposedly domesticated over 2000 years ago, the ferret was not identified as a potential model for biomedical research until the early twentieth century. They are an attractive "large animal" model for studying the pathogenesis of viruses in that they are a smaller and more cost-effective species than nonhuman primates. However, ferrets are large enough so that human neonatal equipment can be used to evaluate clinical parameters, such as temperature and blood pressure. In addition, the ferret respiratory system shares many anatomical and physiological features with humans, making the ferret a suitable model for the study of respiratory viruses.

Although ferrets were first used to study influenza in 1933, only recently has their use increased as their value has been recognized for the study of emerging respiratory pathogens such as henipaviruses, coronaviruses, and respiratory viruses. Oronasal challenge with henipaviruses produces both respiratory and neurological disease along with disseminated vasculitis, resembling the response of humans to these agents. When challenged with SARS-CoV, ferrets demonstrate severe alveolar damage and edema, though generally not to the extent seen in humans and nonhuman primates. The ferret's greatest impact on virus research lies in its contributions to the influenza field (Peng et al., 2014). It is an important model for understanding pathogenesis and evaluating potential vaccines. With the emergence of influenza strains with pandemic potential, including H1N1 and H5N1, the ferret has become an invaluable tool to predict transmission in humans (Figure 2), a major public health concern (see further discussion later in this chapter and Chapter 7, Patterns of infection).

2.4 Chicken

While rarely used as a model for studying human viruses, the chicken (*Gallus gallus*) holds an important place in history for its contributions to the field of tumor virology. In 1911, Peyton Rous first reported isolation of a "filterable agent" from a sarcoma in a Plymouth Rock hen that could be experimentally transmitted to other chickens. Further studies of this retrovirus have led to other pioneering discoveries in oncology, including *Src*, the first recognized oncogene.

Chickens have played a crucial role in unraveling the cellular basis of the adaptive immune system (Gitlin and Nussenzweig, 2015). Max Cooper in Robert Good's laboratory discovered that the progenitors of antibody-producing plasma cells were differentiated in the Bursa of Fabricius in the chicken, while the progenitors of cellular immunity were differentiated in the thymus. This discovery, which clearly distinguished B and T cells for the first time, was a key step in our current understanding of the immune system.

Chickens also have made valuable contributions to vaccine development. In the 1930s before the era of cell

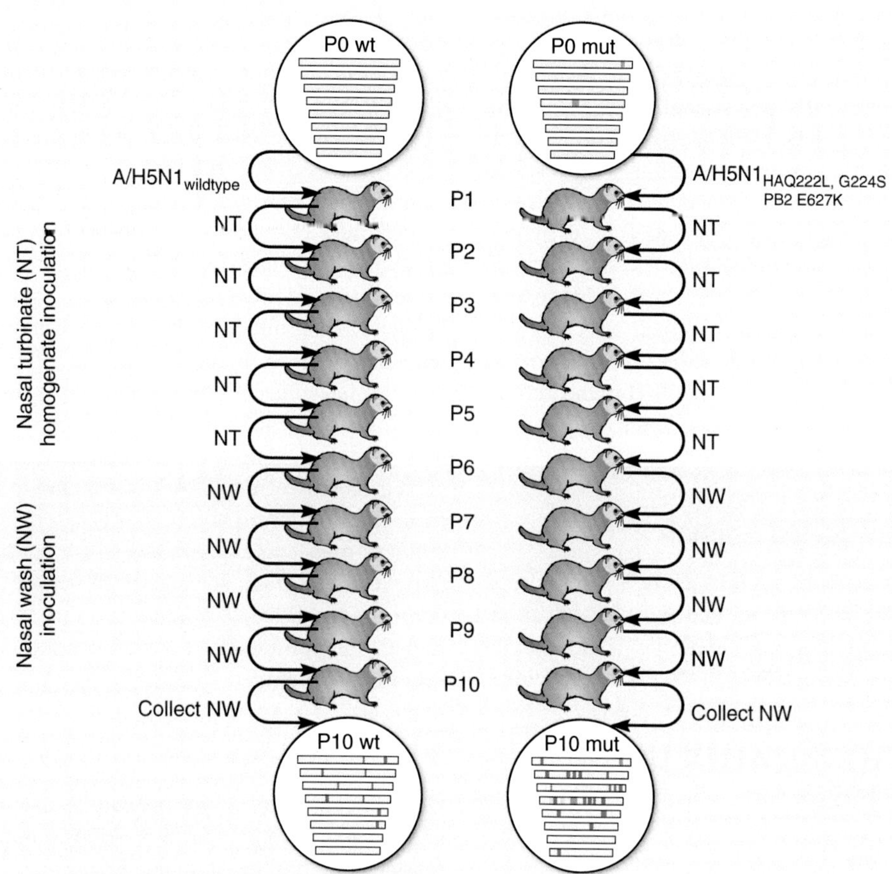

FIGURE 2 Unique animal models can be critical for pathogenesis studies of specific viruses. Ferrets are an important experimental host for pathogenesis studies of influenza, in part because they respond to influenza type A virus infection in a manner similar to humans. In addition, ferrets differ from mice and many other experimental animals because they have the same distribution of sialic acid receptors as do humans, with α-2,6-linked receptors in the upper respiratory tract and α-2,3-linked receptors in the lower respiratory tract. In the experiment illustrated above, a human isolate of influenza type A H5N1 viruses was genetically modified to introduce mutations that would change its receptor preference from α-2,3 to α-2,6 sialic acid receptors. Both wild-type and genetically modified strains were passaged in ferrets 10 times, to determine if they could be adapted to spread by aerosol from animal to animal. The modified (but not the wild-type strain) was aerosol-transmissible after passaging in ferrets. *Redrawn after Herfst et al. (2012).*

culture, Alice Woodruff and Ernest Goodpasture reported the successful propagation of fowlpox virus in the chorioallantoic membrane of chick embryos. Their technique allowed for the cultivation of uncontaminated virus for the first time, revolutionizing the field of virology. Subsequent work with embryonated chicken eggs led to vaccine development for several viral diseases including yellow fever and smallpox. Even though cell culture has largely replaced embryonated chicken eggs for virus production, the vast majority of influenza vaccines are still produced in embryonated chicken eggs.

2.5 Nonhuman Primates

Nonhuman primates share many anatomical, physiological, and immunological characteristics with people and tend to be more susceptible than other animals to infection with human viruses. Old World primates include baboons, many species of African monkeys, Asian macaques, and great apes such as chimpanzees, while New World primates include marmosets, tamarins, spider, and squirrel monkeys. Because of their closer phylogenetic relationship to humans, the Old World primates, particularly macaques and chimpanzees, are more commonly used to study virus infections.

Asian-origin macaques, specifically rhesus, cynomolgus, and pigtail macaques, are commonly used in virological studies. Macaques have made contributions to countless studies involving viruses, particularly in developing vaccines, where a higher order mammal is often required for preclinical trials. They have been used to study global diseases such as AIDS (discussed in more detail later), childhood diseases such as poliomyelitis and measles, tropical diseases such as yellow fever and dengue fever, and potential bioterrorism agents including smallpox and Ebola virus. Due to their close relationship to humans, findings from these studies often also benefit nonhuman primates that are naturally susceptible to many of these diseases.

The chimpanzee requires special mention for its historical contributions to viral pathogenesis and vaccine research. With over 99% shared genetic identity, they are our closest animal relatives. Research using these animals has resulted in vaccines for hepatitis A and B and has increased our understanding of HIV, respiratory syncytial virus, cytomegalovirus, and hepatitis C virus infections. Chimpanzees are the only great apes used in biomedical research, but federal legislation has significantly diminished their role. In 1995, the NIH enacted a moratorium on breeding chimpanzees in captivity, and in 2013 all but 50 of the NIH-owned chimpanzees were retired from research. Certain high-impact, noninvasive studies are still permitted.

2.6 Large Domesticated Animals

Large domesticated animals, such as cows, sheep, horses, camels, and pigs, are used for specialized research problems. Their description is beyond the scope of this chapter but may be found in reference books (see Further reading).

3. CONSIDERATIONS WHEN USING ANIMAL MODELS

3.1 Genetics

The genetic background of the host animal plays an important role in the phenotype produced by a virus infection. In outbred populations, individual animals commonly differ in their responses to virus infection. For example, in pigtail and rhesus macaques, some major histocompatibility complex alleles can present an immunodominant epitope of the SIV Gag protein, leading to lower viral loads than that seen in animals with other alleles.

Likewise, the immune response to pathogens and outcomes after infection can differ greatly among inbred mouse strains. For example, mice that have functional Mx1 proteins (e.g., wild *Mus spretus* mice) are less susceptible to influenza virus infection than those with nonfunctional proteins (e.g., C57BL/6 and BALB/c). Different mouse strains can also have biased immune responses that can affect susceptibility to a virus infection. For example, C57BL/6 mice tend to skew toward a Th1 response whereas other strains, such as BALB/c and DBA/2 mice, tend to have a predominant Th2 response. Therefore, it is essential to use the same strain of mouse in experiments where consistency of a phenotype is important, such as for "knockout" and "knockin" comparisons. Because many substrains exist within a strain, it is also important to use mice from the same vendor for a set of experiments.

3.2 Age and Sex

Host susceptibility to a virus can be markedly affected by age and sex. For many viruses, young animals have a higher mortality than older animals (Figure 3). For other viruses, older animals may exhibit increased susceptibility, such as with human coronaviruses (see Chapter 16, Emerging virus

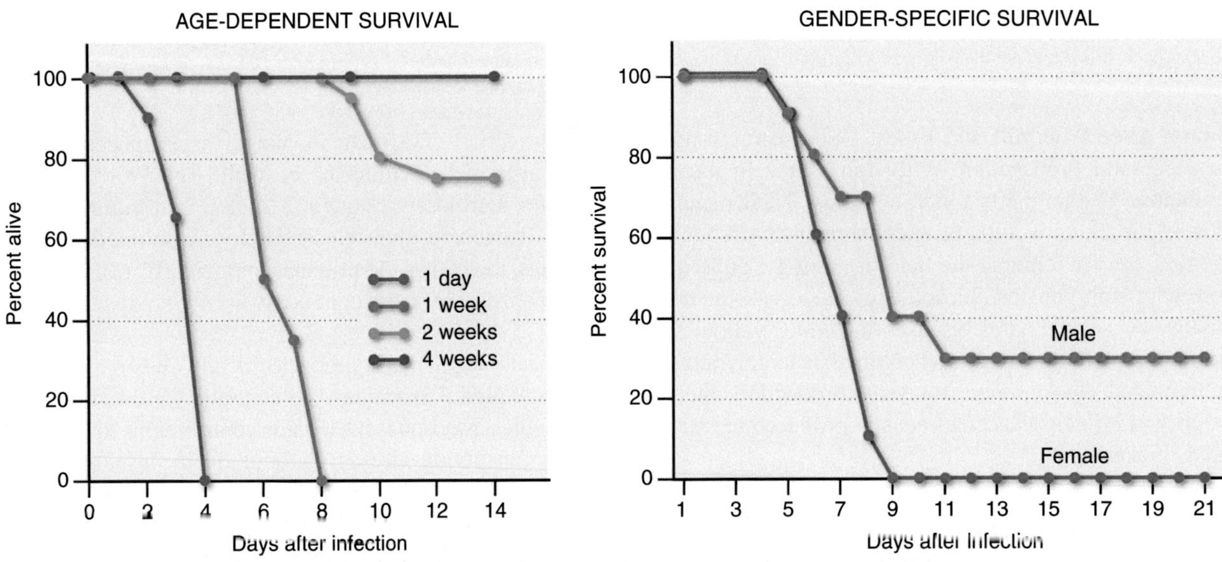

FIGURE 3 Age and sex influence susceptibility to virus infections. Left panel. Age-dependent survival in mice infected with Sindbis virus. Mice infected with Sindbis virus at a younger age have a higher mortality than mice at an older age of infection. Mice infected at 1 day of age (red line) and at 1 week of age (blue line) show 100% mortality, while mice infected at 4 weeks show no mortality. Mice infected at 2 weeks of age (purple line) have an intermediate rate of mortality Griffin (1976). Right panel. Survival curves for male and female mice infected with 10^3 TCID50 of an H1N1 strain of type A influenza virus. The LD50 was 11-fold higher for male than female mice in this experiment. *After Lorenzo et al. (2011).*

diseases). The sex of the animal often affects the outcome of virus infection. Male mice and humans are more susceptible to myocarditis caused by Coxsackie virus infection, and females have higher morbidity and mortality when infected with influenza A virus. These confounders should be identified, studied, and controlled whenever possible.

3.3 Virus Strain and Route of Infection

Different strains of the same virus can have different courses of pathogenesis, tissue tropisms, and varying disease severity (this variable is discussed in detail in Chapter 3, Basic concepts, and Chapter 7, Patterns of infection). These differences can often be exploited to identify viral determinants of virulence or attenuation. The route of infection, such as intraperitoneal versus intranasal, intracerebral or subcutaneous, can also affect the course of infection and nature of disease (Belser et al., 2013).

3.4 Coinfections

As more natural pathogens are discovered in laboratory animal species, the possibility of confounding infections affecting a virus phenotype increases. Dramatized in Richard Preston's *The Hot Zone*, a quarantined shipment of cynomolgus macaques in Reston, Virginia, created significant concern when it was discovered that the monkeys were infected with a previously unidentified strain of Ebola virus. However, the monkeys were also infected with simian hemorrhagic fever virus, an arterivirus, which exacerbated the disease severity and contributed to their deaths.

Coinfection has the potential to markedly affect the immune response to an experimental virus infection. Immunosuppressive viruses, such as simian retrovirus in macaques, can result in an altered immune response to experimental infection and disease from opportunistic secondary infections. Subclinical infection can alter the immune response of immune-competent animals and have an impact on virus studies. A number of natural infections can spread silently in colonies of laboratory mice and affect experimental outcomes. Examples include viruses such as lymphocytic choriomeningitis virus, lactate dehydrogenase virus, mouse hepatitis virus, Theiler's murine encephalomyelitis virus, murine norovirus, and mouse parvovirus, bacteria such as *Helicobacter*, *Pasteurella*, and *Staphylococcus* species, and parasites such as pinworms and fur mites. Many commercial vendors sell and animal facilities maintain "specific-pathogen free" animals, and sentinel programs monitor resident colonies for subclinical infections and infestations.

3.5 Practical Considerations

Cost can influence the selection of animal models. Smaller rodent species are less expensive than larger animals. From a research standpoint, the availability of immunologic reagents and tools needs to be taken into account. As noted, the extensive arsenal of antibodies and molecular reagents available for mice, in addition to their genetic manipulability, make them very attractive models. However, as full genome sequences become available for more species, reagents and technologies available for those animals will expand.

3.6 Ethical Considerations

Ethical considerations are important when planning the use of an animal model. The "3 R's" govern the humane use of animals in research: Replacement, Reduction, and Refinement. Whenever possible, animal experiments should be replaced with nonanimal techniques. This includes using animal cells in tissue culture (relative replacement) and in vitro techniques or computer simulations in place of animals (absolute replacement). The number of animals used for an experiment should be reduced to the minimum needed to obtain significant results. Refinement refers to improvements that minimize pain and distress for animals and allow them to participate in natural behaviors. Whenever possible, social species, which include mice, rats, dogs, and most nonhuman primates, should be housed together in socially compatible groups. When appropriate, animals should also be provided enrichment, such as nesting material for rodents, balls and other toys for dogs, or swings for nonhuman primates.

In the United States, there are Federal regulations regarding the care and use of laboratory animals. These regulations are implemented at the institutional level by Institutional Animal Care and Use Committees (IACUC). The IACUCs evaluate animal protocols prior to the initiation of a study to ensure that the use of laboratory animals conforms to government standards. Distress and pain can be powerful confounders of experiments, and optimal welfare is more likely to produce reliable, consistent results. Laboratory and husbandry staff should be properly overseen and trained in the care of laboratory animals. Compassionate care is both ethical and supportive of the highest quality of animal research.

4. EXAMPLES TO ILLUSTRATE THE SELECTION OF ANIMAL MODELS

4.1 Influenza

Disease resulting from influenza A virus infection in humans can range from a mild, self-limiting febrile illness to a fulminating lethal disease with severe acute respiratory syndrome. Vaccines produced each year for seasonal influenza reduce morbidity and mortality, but are far from optimal. With the emergence of new strains with pandemic potential, such as H1N1, H5N1, and H7N9, animal models are needed to investigate pathogenesis and immune responses and to develop new vaccines.

Waterfowl are the natural reservoir for influenza A virus, and certain avian strains can infect several mammalian species, including pigs, dogs, horses, and ferrets. Of increasing public health concern is the transmission of highly pathogenic influenza strains from birds to humans. Host restriction is highly dependent on the linkage of the sialic acid (SA) receptors found on respiratory cells. Like humans, ferrets and nonhuman primates have both α2,6- and α2,3-linked sialic acid receptors in the lower respiratory tract, but only α2,6-linked SA receptors in the upper respiratory tract. In contrast, birds and mice predominantly have α2,3-linked sialic acid receptors throughout the respiratory tract. Avian influenza isolates preferentially use α2,3-linked receptors, whereas mammalian isolates use α2,6-linked receptors. If there is a mutation in the receptor-binding site of an avian isolate, it could potentially jump species and infect mammals. If such a strain of influenza virus to which humans have no preexisting immunity crosses the species barrier, there is the potential for a pandemic, such as occurred in 1918.

When designing influenza virus studies, there are several choices of animals, including mice, ferrets, and nonhuman primates. In practice, each of these species is used for certain studies.

Mice are the most frequently used animal model for influenza studies, even though they are not a natural host of this virus. Many mouse strains, including C57BL/6 and BALB/c, have mutations or deletions in the antiviral *Mx1* gene, rendering the protein nonfunctional and the animal more susceptible to influenza virus infection. However, for productive infection, influenza viruses usually must be passaged to select for isolates that replicate efficiently in mice. Although upper respiratory signs such as nasal discharge and coughing are not seen in mice, animals do develop weight loss, lethargy, and dyspnea when infected with highly pathogenic strains of influenza virus. Histopathology tends to mirror that seen in humans.

Ferrets have become the gold standard model for influenza research because they have the same distribution of influenza virus receptors that occur in humans (see above). Ferrets develop rhinitis, fever, coughing, sneezing, appetite loss, and weight loss, similar to that seen in humans. Infection with highly pathogenic influenza virus can result in severe disease. Ferrets can transmit influenza to other ferrets and are used in transmission studies of seasonal isolates and emerging strains such as H5N1. Also, they are used to study the secondary bacterial infections that can complicate influenza.

Macaques are also used in influenza research, especially for vaccine studies. Cynomolgus macaques replicate seasonal influenza viruses in their lungs and upper respiratory tracts without developing severe clinical signs. However, when infected with virulent 1918 H1N1 or H5N1 influenza type A viruses, they tend to develop the more severe clinical signs and pathology seen in humans. Although expensive and cumbersome, macaques are selected for preclinical studies of either vaccines or antiviral drugs, as the best predictor of efficacy in humans. The investment is justified as a screen prior to initiating very expensive and time-consuming human trials.

4.2 SIV

HIV causes a disease with a very complex pathogenesis, much of which can only be studied in an animal model. Also, animal models are essential for the development of therapies, functional cures, or the ever-elusive HIV vaccine. Despite species-specific virus-host restrictions, animal models have been developed and used extensively in HIV research, through modification of the virus or studies with homologous animal lentiviruses. The following account is very brief, and a more comprehensive exposition can be found in Chapter 9, HIV/AIDS.

The SIV macaque model has become the premier model for HIV research. It was found by happenstance that strains of SIVsmm cause an AIDS syndrome when transmitted to Asian macaques, although SIVsmm is not pathogenic in sooty mangabeys, its natural host. The natural routes of HIV transmission (sexual contact, mother-to-child, and so forth) can be used to infect macaques with SIV. However, there are differences between this SIV model and HIV in humans. SIV-infected macaques progress to end-stage disease faster than HIV-infected humans, and the species of macaque and specific isolates of SIV influence the course of infection.

HIV and SIV are only about 50% identical at the nucleotide level, and neutralizing antibodies have limited cross-reactivity, indicating that SIV-infected macaques are not appropriate models for HIV-1-based vaccine testing. This problem has been partially circumvented through the development of chimeric simian-human immunodeficiency viruses (SHIVs), which can be used in preclinical vaccine and therapeutic trials.

The lack of an effective small animal model has seriously impaired HIV research. Rodents are not permissive to HIV and do not possess an analogous lentivirus of their own. Therefore, much effort has been expended to develop "humanized" mice with human immune system components. The first engineered humanized mice, such as hu-PBL-SCID and SCID-hu thy/liv mice, did not support long-term productive HIV-1 infection, despite reproducing certain aspects of viral replication and pathogenesis. Recent improvements in humanized mice using NOD/SCID/γ_c knockout (NSG), NSG-BLT, and Rag2/γ_c knockout mice have resulted in models that can support a productive HIV-1 infection and induce an HIV-1-specific immune response, allowing for potential candidate vaccine and therapeutic testing. It is unknown whether the responses seen in humanized mice will predict the responses of humans in clinical trials.

5. GENETIC INTERVENTIONS IN THE MOUSE: THE FUTURE OF ANIMAL MODELS

A great strength of the mouse as an animal model lies in the ability to manipulate its genome. As part of the Human Genome Project in 1990, the *Mus musculus* genome was sequenced as one of the first model organisms. The full genome of the C57BL/6 mouse was published in 2005, and today the genomes of over 20 other inbred strains and outbred stocks have been sequenced. This achievement has helped identify many genes for researchers to target, greatly aiding in the production of knockout and transgenic mice.

5.1 Inbred Mouse Strains

An inbred line is a mouse strain that has undergone at least 20 consecutive generations of brother/sister mating making the animals genetically identical at virtually every locus except for the sex chromosomes. This is beneficial for biomedical researchers in that genetic homozygosity decreases variability, leading to the need for fewer animals to obtain statistically significant results. Furthermore, inbred mice accept grafts from homozygous donors, a condition that has been critical for a vast number of immunobiological studies, some of Nobel Prize quality.

The Jackson Laboratory in Maine, one of the largest suppliers of mice in the United States, currently has over 200 inbred mouse strains available. Most strains also have multiple substrains due to the distribution of parent strains to different researchers and vendors. They in turn established their own colonies, eventually resulting in genetic divergence from the progenitor strain. Although this may not seem like a major problem to many researchers, the genetic discrepancies between substrains can result in phenotype differences, and so the same substrain should be used for a set of experiments.

Trait differences can be further analyzed by using established inbred lines to create recombinant inbred lines. To do this, inbred strains are crossed to make F1 and F2 generations, and those progeny are then intercrossed and inbred for 20 generations. This allows phenotypes of different traits to be mapped to a chromosome, and the larger the family of recombinant inbred strains, the greater the power and resolution of the mapping. To expand the variety (and genetic diversity) of inbred mouse strains, the Collaborative Cross Consortium (2012) is cross-breeding eight founder lines of inbred mice, as described in Chapter 13, Host genetics. To mimic the genetic diversity of a human population, outbred stocks are also available from commercial vendors.

Another ongoing approach to gene mapping is the introduction of mutations in individual genes, which can be linked to phenotypes. While previously limited to yeast and other nonvertebrate models, the accessibility of forward genetic technology in mice has dramatically increased over the last few years. Whole exome sequencing can be applied to the progeny of mice exposed to N-ethyl-N-nitrosourea (ENU) or other mutagens to identify all mutations produced and help elucidate the genetic basis of both Mendelian and complex traits (Moresco and Beutler, 2013). Further discussions of the use of genomic technology to study viral pathogenesis are presented in Chapters 12 and 13, The virus-host interactome, and Host genetics.

5.2 Transgenic Technology

While most inbred strains had been developed by the mid-twentieth century, the gene knockout and transgenic technology that has set mice apart from other models has emerged more recently. Rudolf Jaenisch and Beatrice Mintz published the first account of a genetically modified mouse in 1974, where they injected simian virus-40 viral DNA into a mouse blastocyst and showed that it was present in every cell of the resulting animal. In late 1981, four different groups reported that plasmid DNA injected into the pronuclei of fertilized mouse eggs had integrated into the host genome with stable germline transmission, thus producing transgenic mice (Figure 4). The first mouse lacking a gene through targeted mutation ("knockout") was generated in the late 1980s (Figure 5), a feat for which the 2007 Nobel Prize for Physiology or Medicine was awarded to Mario Capecchi, Martin Evans, and Oliver Smithies. Through this technology, thousands of strains of genetically modified mice have now been produced.

Virologists have used genetically modified mice to overcome some host-range limitations. To confer permissiveness to virus infection, mice that lack immune response or other host defense genes are often used. "Knocked out" genes commonly code for antiviral proteins or cytokines, such as interferons or proteins involved in immune signaling, or for genes involved in the development or maturation of immune cells, such as recombination activating gene-1 (*Rag1*). By removing antiviral proteins or populations of immune cells, researchers can identify the immune components necessary for virus control and clearance. This method presents a powerful approach to elucidating the intricacies of virus pathogenesis and the host immune response.

Another advance is the conditional knockout mouse (Figure 6). This technology, which is based on the Cre-Lox recombination and tetracycline-controlled transcriptional regulation systems, allows for targeted gene elimination or activation in either a specific cell or tissue or at a certain

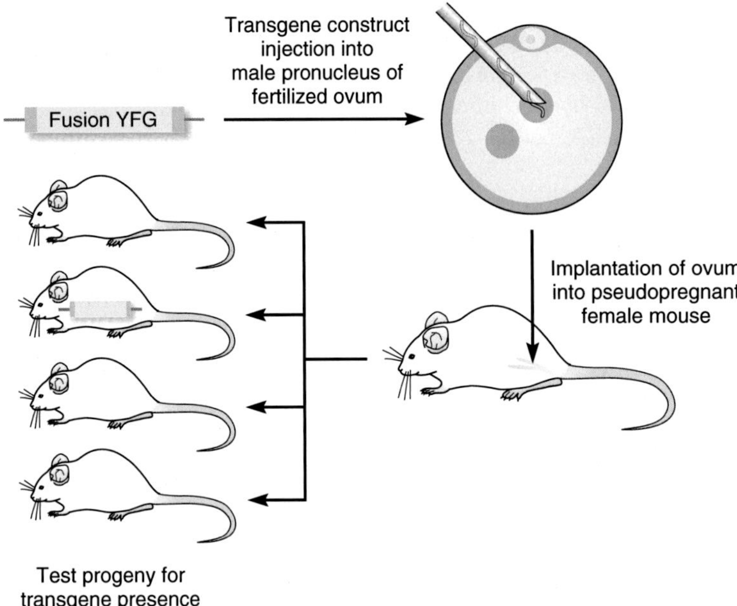

FIGURE 4 Pronuclear injection for the generation of transgenic mice. A transgenic construct is injected into the male pronucleus of a fertilized mouse ovum. The ovum is then implanted into a pseudopregnant female mouse. Once the resulting offspring are born, they can be genotyped for the presence of the desired transgene and further propagated.

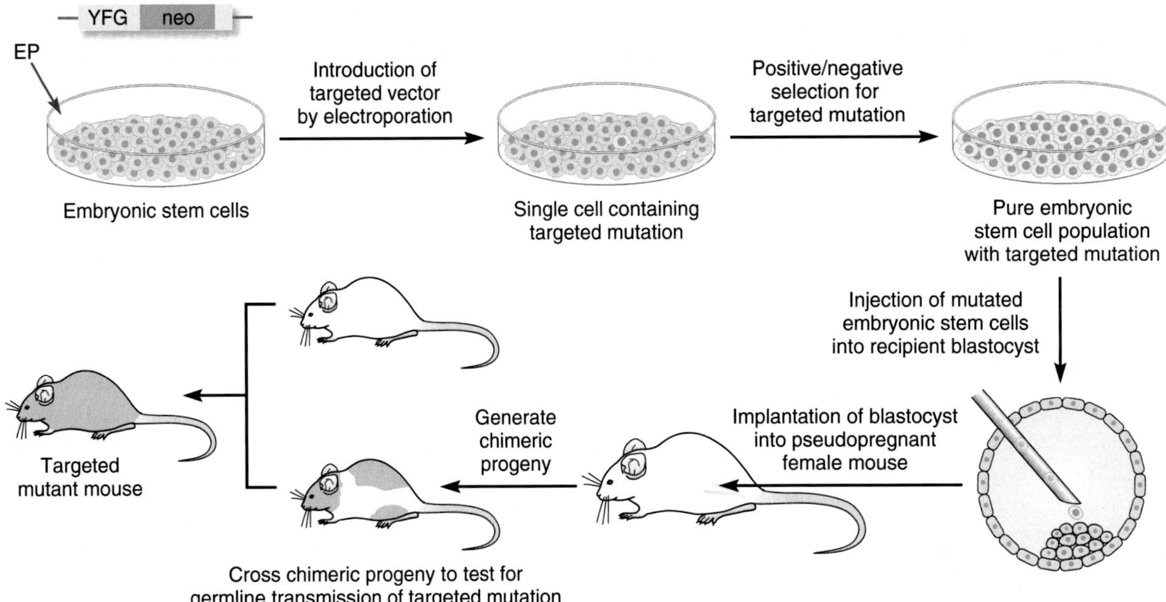

FIGURE 5 Production of "knockout" mice. A DNA vector containing a genetically disrupted gene, usually with a selectable marker, is introduced into stem cells from the blastocyst of a mouse by electroporation. The new DNA sequence is incorporated into the chromosomes of some of the stem cells in place of the original gene by homologous recombination. The stem cells containing the altered gene are then selected through a combination of positive and negative selection, and those cells are propagated until a uniform population is obtained. The mutated stem cells are injected into the blastocyst of a mouse of a different coat color than that of the mouse from which the stem cells originated. The blastocyst is implanted in a pseudopregnant mouse, and resulting chimeric progeny (as indicated by a parti-colored coat) are crossed with a mouse with the same coat color as that of the recipient blastocyst. If the germ cells of the chimeric mouse contain the targeted mutation, the resulting progeny's fur will be the same color as the coat of the mouse strain from which the stem cells originated. Genotyping is used to confirm the presence of the cassette disrupting the desired gene.

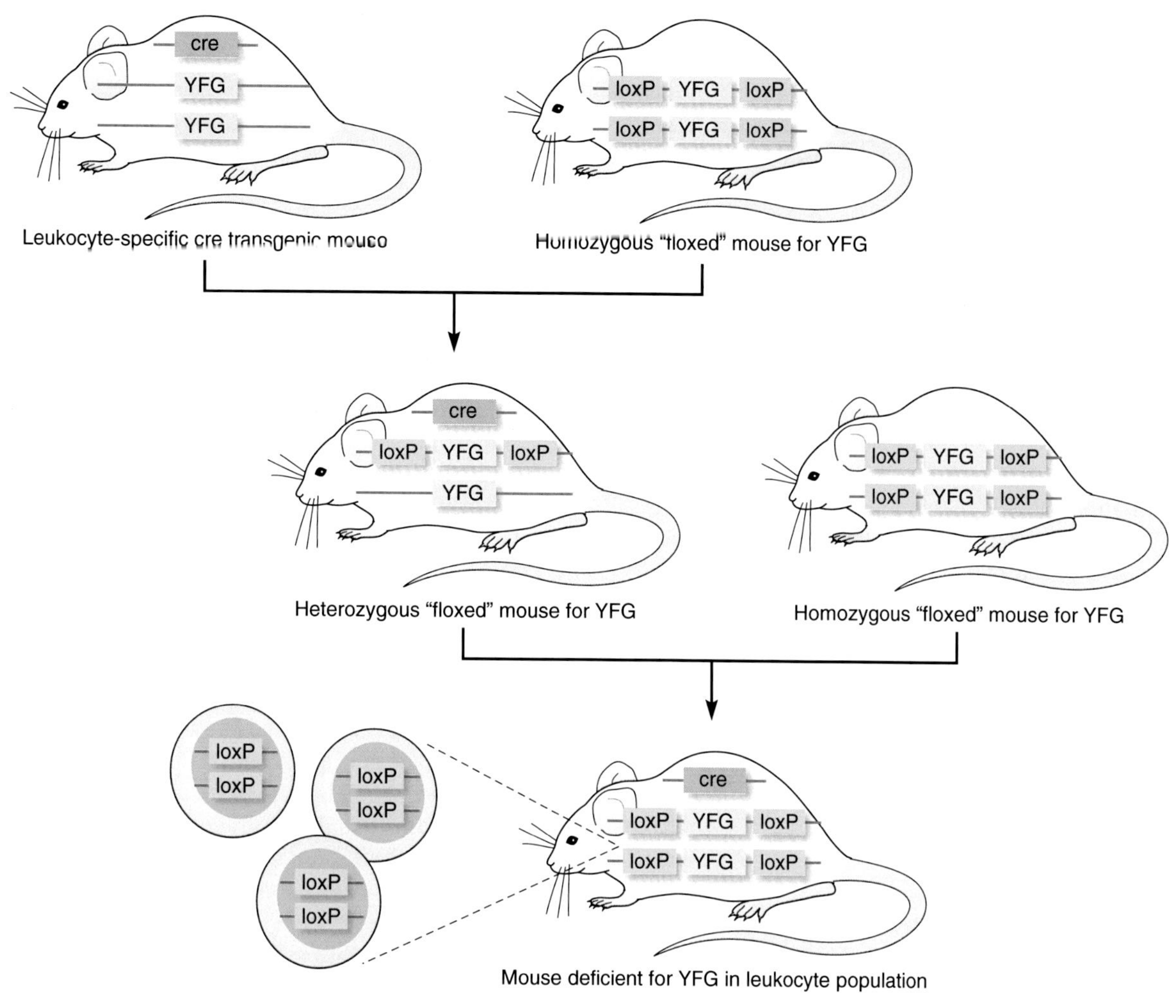

FIGURE 6 The cre/lox recombination system to develop conditional knockout mice. A mouse expressing a transgene for cre recombinase under the control of a tissue-specific promotor, such as CD4 for CD4+ T cells, is crossed to a mouse homozygous for a gene of interest flanked with loxP sites ("floxed" strain). Resulting progeny is automatically heterozygous for the floxed gene, and genotyping is performed to determine cre transgene status. A cre-positive mouse from the original cross is then crossed to another mouse homozygous for the floxed gene. Approximately, 25% of the progeny will be homozygous for the floxed gene and hemizygous/heterozygous for the cre transgene. These mice are incapable of transcribing the floxed gene in the tissue lineage containing the cre transgene and thus are the conditional knockouts.

time in development. As removal of many genes is lethal in mice during the embryonic or fetal stages of development, deleting a gene after that stage allows for the study of such knockout animals. Removal of a gene from a certain cell line can also help elucidate the role of that gene during a virus infection. However, this method can be time-consuming, labor intensive, and expensive, as interbreeding between multiple different strains is often required to obtain mice with the desired genotype.

Mice often lack the cellular receptor for a human virus (such as DPP4 used by MERS-CoV), or the mouse homologue is sufficiently different to prevent virus entry into mouse cells. In these cases, the human gene for the cellular receptor can be introduced into the mouse genome, and the subsequent human protein is then translated in the mouse, allowing for virus binding and entry into the cell. Transgenic mice can also be used to study the specific host adaptive immune response to a human virus. For instance, T cell receptors, specific for a peptide epitope from the virus of interest can be introduced into the mouse genome. When mice are challenged with the virus, it is possible to characterize the focused immune response resulting from activation of cells with that receptor.

The development and use of CRISPR/Cas9 technology (Harrison et al., 2014) to simultaneously induce targeted mutations in multiple genes has greatly facilitated the analysis of gene function at a structural level, as well as the interactions between different genes (described in Chapter 13, Host genetics). Other recently developed approaches for inducing targeted mutations include hit and run, double replacement, and PiggyBac recombinase systems, and the use of engineered nuclease pairs such as ZFNs or TALENs. These advances allow virologists to apply techniques, previously limited to virus manipulation, to animal models to better elucidate viral pathogenesis and the host immune response.

Although advances in the genetic manipulation of mice far exceeds that available for other species, the process can be very time and labor intensive. Several attempts to knock out genes or insert transgenes must generally be made before a genetically manipulated animal is successfully produced. To create mice congenic for a mutation on the desired background strain, multiple generations of backcrossing are required. Using traditional techniques, the process can take 2–3 years if no problems are encountered, but more commonly takes up to 5 years. This timeline has been significantly decreased with the advent of more advanced genetic screening technology. While ten generations of backcrossing are traditionally required to create a congenic mouse, the use of speed congenics has reduced this timeline to as little as five generations. This process uses microsatellite markers or single nucleotide polymorphisms to select for progeny that not only possess the desired genetic manipulation, but also the greatest amount of genetic information from the desired background strain. Advances in DNA sequencing technology and computational analysis have further decreased the amount of time and labor required to produce genetically manipulated mice. New developments in mouse genetics are constantly being published, and as these technologies are fully validated and optimized, the availability of genetically manipulated mice to better study viral pathogenesis will further improve.

Researchers should be cautious when interpreting data using genetically modified mice, because these artificial models are far removed from infection in the natural host. In those cases where mice prove to be insufficient for the study at hand, another animal model should be considered. Genetic modification is starting to become more commonplace in other species such as rats, pigs, ferrets, and macaques, but the process is more difficult and less well understood in these species.

6. OTHER NEW METHODOLOGIES

A host of new methods have been developed that enhance the utility of animal models, providing information that was previously unavailable. Two of these will be noted briefly: systems biology and imaging.

The host response consists of a series of antiviral and immune programs. These orchestrated molecular events, captured through transcriptomics, proteomics, and metabolomics, can provide heat maps that capture large groups of data to elucidate the host response to infection. Influenza provides an example; a comparative cross-species transcriptomic analysis found that the 2009 pandemic H1N1 influenza virus elicits differential expression of proinflammatory genes in the lung following infection of nonhuman primates, mice, and pigs (Go et al., 2012). When these inflammatory responses in the lung are overly exuberant, they may endanger the host, as shown in a comparison of highly pathogenic avian influenza virus infections (H5N1 and 1918 strains) with influenza viruses of lower virulence (Tisoncik et al., 2012; Peng et al., 2014).

Bioimaging is undergoing a renaissance through a panoply of new methods. For instance, it is possible to introduce a fluorescent genetic label into a virus that permits its visualization during infection of a living animal (Golding and Zaitseva, 2014). Repeated imaging of a single animal during the course of infection provides a dynamic view of virus replication and dissemination. Counterpart studies of T lymphocytes, dendritic cells, and macrophages capture the movements and interaction of cells that participate in the innate and adaptive immune responses (Germain et al., 2012). Furthermore, new methods for visualizing living or fixed tissues can be used to construct three-dimensional views of tissues whose architecture is preserved.

7. ANIMAL MODELS FOR EMERGING VIRUSES

Developing an animal model that accurately mimics the disease presentation of a new virus infection is an important first step in studying an emerging virus. The first animal usually evaluated is the macaque, because nonhuman primates are closely related to humans and often exhibit a similar disease outcome. However, different species sometimes more appropriately model the observed disease, such as common marmosets for the coronavirus-induced Middle East respiratory syndrome (MERS), or Syrian hamsters for hantavirus pulmonary syndrome induced by Sin Nombre virus. Because many emerging viruses must be studied at high containment levels, such as BSL-3 or BSL-4, it is often desirable to use smaller animal species that are easier to house and handle.

More often than not, emerging viruses are zoonotic, with one or more animal species acting as a reservoir. Examples include passerine birds for West Nile virus, field mice or rats for the hantaviruses, pteropid fruit bats (flying foxes) for the henipaviruses, and camels for MERS-CoV. These animals act as long-term carriers and transmitters of the virus and generally do not develop disease. However, it is informative to understand how the virus establishes persistence or latency in the reservoir species while facilitating virus shedding or transmission.

8. VACCINE DEVELOPMENT

The development of vaccines requires the use of animal models (see also Chapter 19, Viral vaccines). The United States Food and Drug Administration (FDA) requires that all new investigational drugs be tested in "relevant species" to ensure safety before beginning clinical trials in humans. Animals are used at several stages of vaccine development, including the host immune response to vaccination, efficacy against viral challenge, and selection of optimal vaccine formulations and delivery routes.

Animals have been involved in vaccine development for several hundred years. Edward Jenner observed that milkmaids, who often were naturally exposed to cowpox from their cows, were resistant to smallpox, leading to the development of vaccination in the late-eighteenth century. In 1885, Louis Pasteur reported the successful creation of an attenuated vaccine for postexposure prophylaxis against rabies by desiccating infected spinal cords from rabbits that had been inoculated with serially passaged rabies virus. The poliomyelitis vaccine was developed as a result of 40 years of research using nonhuman primates. Where species-specificity prevented use of an animal model for vaccine development, such as human papillomavirus, related animal viruses were used as surrogates. In this case, virus-like particle vaccines containing papillomavirus L1 capsid protein were tested in rabbits, cattle, and dogs infected with species-specific papillomaviruses before starting clinical trials in humans.

In choosing an animal model for the development of a vaccine, the species should reflect the course of natural virus infection in humans as closely as possible. The same organs should be affected, the pathology should be comparable, and the immune response should be similar. Additionally, disease should occur in a high proportion of animals to decrease sample size, and the route of virus infection should be the same as it is in humans. Although mice are often considered the ideal animal model for immunogenicity studies, other animal species may better fit these criteria and be a more appropriate choice for vaccine studies.

Animals are required for immunogenicity testing of candidate vaccines. This includes characterizing the specificity, level, durability, class and subclass of antibodies produced, and the cellular immune response. For most viral vaccines, the ability to induce neutralizing antibodies has proved a good predictor of efficacy. Once a candidate meets some criteria of immunogenicity, the protective efficacy of the vaccine should be evaluated by viral challenge. Challenge dose should be as low as possible to cause infection and should be administered by the same route as in natural infection. Although historically commonplace, using death as an endpoint has fallen out of favor, and if possible, morbidity and clinical signs should be used to evaluate vaccine efficacy.

In addition to evaluating immunogenicity, the safety of a vaccine must be tested in an animal model prior to clinical trials. The viral antigen itself and adjuvants, both alone and in combination with the antigen, may cause adverse effects. Unfortunately, a vaccine that meets desired criteria in an animal model may not be adequately immunogenic or safe in humans.

There are cases where human efficacy studies for a vaccine may not be feasible because of the high mortality rate or infrequency of infection. In these situations, the FDA may permit two relevant species to be used in place of humans to evaluate the effectiveness of a vaccine for licensure. This process is known as the "Animal Rule" and further points to the importance of animal models as contributors to vaccine research.

9. REPRISE

Pathogenesis studies require animal models for several reasons: (1) for human virus diseases, another host is required for experimentation; (2) in vitro or in silico systems cannot simulate the complex response of a living host, with its innate and adaptive immune response; and (3) only an in vivo system can reproduce the sometimes unpredictable response of a natural host. Pathogenesis studies in animal models are also of great importance to applied science, since they constitute an essential foundation for the development, assessment, and production of drugs and vaccines to treat and prevent important human and animal viral diseases.

Choosing an animal model is a complex decision, involving scientific and practical considerations. In many instances, it may be best to use several animal models to address different aspects of pathogenesis or to test candidate treatments or preventive interventions. The mouse has become the default animal for many virus infections because there is a vast scientific database and a large set of reagents, and because it is the least expensive animal model. However, there are many situations where other animal models are required to best address experimental questions.

The use of animals in research is required for the production of new drugs and vaccines and plays a critical role at several stages in the development process. Small animals may be used to screen large numbers of candidate drugs or immunogens for potential efficacy and also for unwanted toxicity. Larger species, particularly nonhuman primates, may provide models that better simulate human disease, where promising products can be tested to select those qualified for human trials. In special instances, the Animal Rule may be applied, where data from two animal species substitute for human trials.

Animal-based research is entering a new and exciting phase because of a variety of evolving methods to select mice with specific and variable genetic backgrounds or to manipulate the mouse genome. It is now possible to delete or inactivate specific genes ("knockout"), to insert new genes ("knockin"), or to introduce specific mutations, using an armamentarium of techniques. These methods enable the assessment of individual host genes in the response to viral infection. Recent developments in "omics" and systems biology have made it possible to record a vast number of discrete host responses during a single infection, enormously enhancing the dissection of a very complex process. Another technical advance is the new set of imaging methods to visualize both viral invader and host responses in the living animal or to capture three-dimensional images of intact functioning tissues. Together these new technologies are opening an expanded vista of virus-host interactions, which will take viral pathogenesis studies to an increasing level of sophistication.

FURTHER READING

Chapters, Books, and Reviews

Baker DG. Natural pathogens of laboratory mice, rats, and rabbits and their effects on research. Clinical Microbiology Reviews 1998; 11:231–266.

Barnard DL. Animal models for the study of influenza pathogensis and therapy. Antiviral Research 2009; 82: A110–A122.

Burns DL. Licensure of vaccines using the Animal Rule. Current Opinion in Virology 2012; 2: 353–356.

Collaborative Cross Consortium. The Genome Architecture of the Collaborative Cross Mouse Genetic Reference Population Genetics, 2012, 190: 389–401.

Fox G. Laboratory Animal Medicine, 2nd ed. Academic Press, San Diego, 2002.

Germain RN, Robey EA, Cahalan MD. A decade of imaging cellular motility and interaction dynamics in the immune system. Science. 2012; 336:1676–81.

Gerner MY, Kastenmuller W, Ifrim I, Kabat J, Germain RN. Histo-cytometry: a method for highly multiplex quantitative tissue imaging analysis applied to dendritic cell subset microanatomy in lymph nodes. Immunity 2012; 37:364–76.

Golding H, Zaitseva M. Application of bioluminescence imaging (BLI) to the study of the animal models of human infectious diseases. Chapter in BR Moyer et al. Pharmaco-imaging in drug and biologics development. American Association of Pharmaceutical Sciences, 2014.

Griffin JFT. A strategic approach to vaccine development: animal models, monitoring vaccine efficacy, formulation, and delivery. Advanced Drug Delivery Reviews 2002; 51: 851–861.

Hatziioannou T, Evans DT. Animal models for HIV/AIDS research. Nature Reviews Microbiology 2012; 10: 852–867.

Louz D, et al. Animal models in virus research: their utility and limitations. Critical Reviews in Microbiology 2013; 39: 325–361.

Tisoncik JR, et al. Into the eye of the cytokine storm. Microbiology and Molecular Biology Reviews 2012; 76: 16–32.

Original Contributions

Belser JA, Maines TR, Gustin KM, Katz JM, Tumpey TM. Kinetics of viral replication and induction of host responses in ferrets differs between ocular and intranasal routes of inoculation. Virology. 2013 Apr 10;438(2):56–60. doi: 10.1016/j.virol.2013.01.012. Epub 2013 Feb 13.

Bull KR, Rimmer AJ, Siggs OM, Miosge LA, Roots CM, Enders A, Bertram EM, Crockford TL, Whittle B, Potter PK, Simon MM, Mallon AM, Brown SD, Beutler B, Goodnow CC, Lunter G, Cornall RJ. Unlocking the bottleneck in forward genetics using whole-genome sequencing and identity by descent to isolate causative mutations. PLoS Genetics. 2013; 9(1):e1003219.

Cook SH, Griffin DE. Luciferase imaging of a neurotropic viral infection in intact animals. J Virology 2003, 77: 5333–5338.

Doudna JA, Charpentier E. The new frontier of genome editing with CRISPR-Cas9. Science 2014, 346: 125096.1-125096.9.

Gerdts V, et al. Use of animal models in the development of human vaccines. Future Microbiology 2007; 2: 667–675.

Gitlin AD, Nussenzweig MC. Fifty years of B lymphocytes. Nature 2015, 517: 139–141.

Go, JT et al., 2009 pandemic H1N1 influenza virus elicits similar clinical course but differential host transcriptional response in mouse, macaque, and swine infection models. BMC Genomics. 2012. 13: 627. doi: 10.1186/1471-2164-13-627.

Gowen BB, Holbrook MR. Animal models of highly pathogenic RNA viral infections: hemorrhagic fever viruses. Antiviral Research 2008; 78: 79–90.

Griffin DE. Role of the immune response in age-dependent resistance of mice to encephalitis due to Sindbis virus. J Infectious Diseases 1976; 133:456–464.

Herfst S, et al. Airborne transmission of influenza /H5N1 virus between ferrets. Science 2011: 336, 1534–1541.

Harrison MM, Jenkins BV, O'Connor-Giles KM, Wildonger J. A CRISPR view of development. Genes and development 2014; 28: 1859–1872.

Holbrook MR, Gowen BB. Animal models of highly pathogenic RNA viral infections: encephalitis viruses. Antiviral Research 2008; 78: 69–78.

Jax Mice Database. The Jackson Laboratory. Accessed June 5, 2014.

Jinek M, et al. A programmable dual-RNA-guided DNA endonuclease in adaptive bacterial immunity. Science 2012, 337: 816–821.

Lorenzo ME, Hodgson A, Robinson DP, Kaplan JB, Pekosz A, Klein SL. Antibody responses and cross protection against lethal influenza A viruses differ between the sexes in C57BL/6 mice. Vaccine 29 (2011) 9246– 9255.

Menke DB. Engineering subtle targeted mutations into the mouse genome. Genesis 2013; 51: 605–618.

Moresco EM, Beutler B. Going forward with genetics: recent technical advances and forward genetics in mice. American J of Pathology 2013; 182: 1462–1473.

Mouse Genomes Project. Sanger Institute. Accessed June 5, 2014.

Nischang M, et al. Modeling HIV infection and therapies in humanized mice. Swiss Medicine Weekly 2012; doi: 142:w13618.

Palermo RE, Tisoncik-Go J, et al. (2013). "Old World Monkeys and New Age Science: The Evolution of Nonhuman Primate Systems Virology." ILAR Journal 54(2): 166–180. http://ilarjournal.oxfordjournals.org/content/54/2/166.long.

Peng X, Alföldi J, Gori K, Eisfeld AJ, et al. The draft genome sequence of the ferret (Mustela putorius furo) facilitates study of human respiratory disease. Nat Biotechnol. 2014 Dec;32(12):1250–5. doi: 10.1038/nbt. 3079. Epub 2014 Nov 17.

Sellers RS, et al. Immunological variation between inbred laboratory mouse strains: points to consider in phenotyping genetically immunomodified mice. Veterinary Pathology 2012; 49: 32–43.

Part II

Systems-Level Approaches to Viral Pathogenesis

How does the transition from reductionism to systems biology impact our concept of viral pathogenesis? It embodies a shift from a focus on the properties of the virus (virulence, attenuation, acute and persistent infections) or the host (susceptible or resistant) to a holistic view of the virus–host combination. This interaction must be considered in a dynamic context, since the innate response is initiated very soon after viral invasion and this—in turn—triggers a virus response with selection of the fittest virions within the invading swarm. The evolving methods of systems and computational biology can capture and analyze this dynamic interaction between virus and host.

Part II—*Systems-level approaches to viral pathogenesis*—explores the wide variety of "omics" technologies and computational methods that are driving innovations in viral pathogenesis. In addition to presenting the new methods, this section demonstrates their application to representative viral infections, aiming for a synthesis with established knowledge described in Part I. Finally, Part II prepares the reader for the applications of systems biology to practical challenges in infectious disease, which are discussed in Part III.

Part II opens with a chapter on *Systems virology* that spotlights the host transcriptional response to infection, including the dynamics of innate and acquired immunity, and diagnostic signatures. Host response networks are described, along with their potential utility for drug design. The application of the systems approach is illustrated by data for influenza, HIV, and hepatitis viruses.

The next chapter on *the virus–host interactome* describes the use of large-scale genetic and protein–protein interaction screens to identify host proteins that promote or inhibit viral survival. This chapter delves into genetic screens, including loss-of-function, gain-of-function, and transcriptional methods. The use of protein screens is explained, including two-hybrid screening, microarray screening, and affinity purification and mass spectrometry assays. The chapter concludes with a discussion of data analysis and integration of information from different methodologies.

Host genetics reviews classical technologies to identify allelic variation in host genes that regulate the outcome of viral infections, with specific examples. This chapter then describes next-generation computational and molecular technologies and platforms, including the manipulation of host genes. The dissection of complex genetic determinants of host susceptibility is discussed, with attention to challenges and analytic approaches. These methodological and computational advances are poised to transform our understanding of the impact of host genetics for both experimental models and viral diseases in human populations.

Metabolomics and lipidomics is the subject of a chapter that describes the effects of infection on a wide variety of essential host metabolic processes, the full extent of which can only be seen by using a systems approach. Infection impacts catabolic reactions that convert nutrients to energy, and anabolic reactions that lead to the synthesis of larger biomolecules. In addition, there is a description of the role of metabolites in viral processes, and the use of metabolomic profiling to investigate the interaction of viral pathogens with the infected host.

Part II concludes with *Mathematical modeling*, in which a quantitative approach to viral dynamic modeling is reviewed. Using a few simple differential equations, data from human patients infected with HIV or HCV are analyzed. This approach estimates the daily rates of viral production and clearance and the life spans of infected cells. In the case of HIV, these methods have revealed the existence of populations of both long-lived productively infected cells and latently infected cells. Modeling is also used to predict the impact of HIV treatment and the need for combination therapy to prevent drug resistance. In the case of HCV, modeling has quantified the effectiveness of new antivirals and predicted the duration of treatment required to cure HCV infection.

Chapter 11

Systems Virology

Why everybody wants to measure everything

Marcus J. Korth[1,2], G. Lynn Law[1,2]

[1]*Department of Microbiology, University of Washington, Seattle, WA, USA;* [2]*Washington National Primate Research Center, University of Washington, Seattle, WA, USA*

Chapter Outline

1. INTRODUCTION

Virologists have long known that viral pathogenesis must be studied from the standpoint of both the virus and the host. Nevertheless, given its relative simplicity, studying the virus has always been more tractable. As outlined in the previous chapters, virus-centric approaches have yielded a tremendous amount of information about viral genetics, viral replication cycles, and host and tissue tropisms. Along the way have come insights into host innate and adaptive immune responses and the many ways in which viruses antagonize these responses while exploiting other cellular processes to their advantage. In the last decade, however, new opportunities to study the host response have emerged. In 1990, the National Institutes of Health and the Department of Energy announced a plan to map and sequence the human genome. Eleven years later, the first draft of the genome was released, and in 2003, the project was declared complete.

The sequencing of the human genome dramatically changed the field of viral pathogenesis. Virologists were now able to move beyond virus-centric or single-gene approaches and instead investigate the host response to infection on a genome-wide scale. With the human genome sequence in hand, it became possible to predict the complete constellation of human genes, their corresponding mRNA transcripts, and encoded protein products. This information spurred the development of methods to measure global gene expression and protein abundance, which in turn mandated the development of computational methods to interpret the resulting avalanche of data (Sidebar 1).

In this chapter, we focus on the insights into viral pathogenesis that are provided by examining the host transcriptional response, including the dynamics of innate and acquired immunity, diagnostic signatures, and the identification of targets for antiviral drugs. We also touch briefly on data interpretation and on protein and metabolite profiling. Although the methods used to measure protein and metabolite abundance (i.e., chromatography and mass spectrometry) differ from those used to measure gene expression, downstream data analysis approaches are similar, and the integration of gene expression and protein and metabolite abundance data brings us closer to a true systems level understanding of virus–host interactions. The use of large-scale genetic and protein–protein interaction screens to identify host proteins that promote

Viral Pathogenesis. http://dx.doi.org/10.1016/B978-0-12-800964-2.00011-2

Sidebar 1 The evolution of systems biology

The sequencing of the human genome and the advent of high-throughput molecular profiling are widely credited with giving rise to systems biology. In one sense this may be true. The convergence of genome sequence information, profiling technologies, and computational advances made it possible to examine biological systems on a scale never before possible. However, the concepts underlying systems biology, and an understanding of the need to comprehend complete systems, have deeper roots. The notion of emergent properties–properties or outcomes that cannot be predicted by an understanding of the individual parts of a system alone—dates back at least to the time of Aristotle (384–322 BC), who stated: "the whole is something over and above its parts and not just a sum of them all." Nevertheless, reductionism—the idea that complex systems can be analyzed and understood by reducing them to manageable pieces—held sway through much of modern history. In the early 1900s, views began to change. It became apparent, for example, that biological systems have hierarchies of organization, and that components of a system behave differently in isolation than when in the intact system (Trewavas, 2006).

The first known use of the term systems biology, and its proposal as a distinct discipline, is attributed to Mihajlo Mesarovic in his 1968 book *Systems Theory and Biology*. Then, as now, systems biology was met with some skepticism. A reviewer of the book noted: "There is no doubt that system-theoretic ideas seem somewhat strange, and perhaps just a little frightening, to the present generation of structurally-oriented biologists" (Rosen, 1968). The current concept of systems biology—driven by genome-based technologies and mathematics—is most often associated with Leroy Hood, founder of the Institute for Systems Biology. Hood proposed that biological systems are composed of two types of information, genes and networks of regulatory interactions, and that biology be viewed as an informational science (Ideker et al., 2001). In this view, studying biological systems requires detailed knowledge of the components of the system, systematic perturbation of the system, monitoring of gene, protein, and pathway responses, and the formulation of mathematical models to describe the system and its response to perturbation (viral infection being an example of such a perturbation). Today, systems biology has become a driving force in biology and medicine, although it is still often criticized for being too focused on data acquisition. As put by Nobel Laureate Sidney Brenner: "Everybody wants to measure everything, you'll never get anything out of it." Such criticisms may eventually be muted, however, as systems biology continues to evolve, driven currently by rapid advances in computing, mathematics, and network modeling, which will be essential to making complex systems comprehensible.

or inhibit viral survival is the subject of Chapter 12, The Virus–Host Interactome.

2. USING GENOMICS TO STUDY VIRAL PATHOGENESIS

On the most basic level, profiling the host transcriptional response to viral infection entails measuring changes in the level of mRNA transcripts present in a cell population in the presence or absence of virus. Indeed, the first published application of large-scale genomic profiling in virology was essentially this simple; the study examined the transcriptional response of primary human fibroblasts at three time points following infection with human cytomegalovirus (Zhu et al., 1998). Yet even this simple experimental plan, which monitored the expression of approximately 6000 genes, revealed not only the complexity of the host response, but also the complexity of data interpretation. Changes in transcript abundance can be due to changes in synthesis, stability, or degradation and such changes may or may not correspond to changes in protein abundance or activity. The transcriptional response is dependent upon time, cell type, virus, and other parameters. Even today, the biological function of many genes is unknown. And with the number of transcripts profiled by current technologies numbering in the tens of thousands, the computational requirements for data analysis are considerable.

The initial study of the host response to human cytomegalovirus identified 258 mRNA transcripts that changed by a factor of four or more, including transcripts encoding major histocompatibility complex I surface receptors and multiple components of the pathway that produces prostaglandin E2 (an inflammatory mediator). The authors concluded: "The global analysis of changes in mRNA levels provides a catalog of genes that are modulated as a result of the host–pathogen interaction and therefore deserve further scrutiny." To a large extent, this forward-looking statement also sums up what is perhaps the main challenge associated with genomic profiling. The global analysis of transcription produces a *catalog* of differentially expressed genes, and oftentimes this catalog is extremely large. Investigators must devise strategies to sift through these catalogs and determine which genes deserve further scrutiny. Ideally, genomic profiling should also do more than produce lists; it should reveal interrelationships between genes, the structure and activity of gene networks, and the function of genes for which no role has been previously ascribed. There is also meaning to be gained from patterns within the data. As described later, this information can be used for predictive or diagnostic purposes or for computational screens for new antiviral drugs.

2.1 Transcriptional Profiling

The earliest assays for large-scale transcriptional profiling consisted of cDNAs or oligonucleotides spotted onto

Sidebar 2 Microarrays

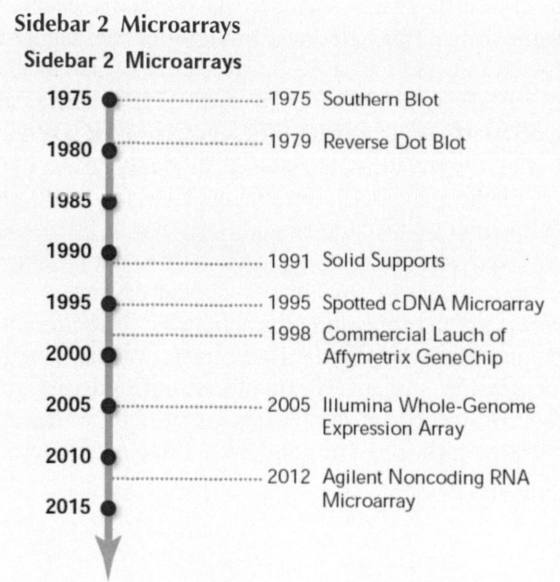

DNA microarrays can trace their ancestry to the Southern blot, in which genomic DNA, digested into fragments by restriction enzymes, is immobilized on a permeable membrane filter for subsequent detection by labeled DNA hybridization. The Southern blot gave rise to the "reverse" dot blot, in which synthetic oligonucleotides of known sequence (called probes) were immobilized on permeable membrane supports. The collection of DNA sequences to be analyzed (called targets) were then labeled and applied to the membrane under hybridization conditions. The use of permeable membranes, however, made it difficult to control the size and shapes of the spotted DNA and limited miniaturization. These limitations were overcome by the introduction of solid supports, which provided the ability to accurately control the size, shape, and location of the spots. The robotic spotting of cDNAs onto glass slides—the first of what we now know as microarrays—was pioneered by Patrick Brown at Stanford University (Schena et al., 1995). In an interview with *Discover* magazine, Brown explained that when he applied for a grant to develop the technology: "The microarray part was thoroughly rejected…but I just decided that I would make one anyway." Importantly, the use of solid supports also facilitated the development of methods for the in situ synthesis of nucleic acids on the surface using approaches such as ink jet fabrication (spotting droplets of nucleotide reagents instead of droplets of ink) or photolithographic methods (similar to those used in the semiconductor industry). Today, major commercial providers of microarrays include Affymetrix, Agilent Technologies, Illumina, and NimbleGen, with platforms varying in the length of oligonucleotide used, the number of oligonucleotides representing each gene, and the methods used for oligonucleotide synthesis and attachment to solid supports. Microarray technology has also been adapted for the profiling of noncoding RNA expression, DNA methylation, and single nucleotide polymorphisms, as well as for promoter analysis and the detection of genome-wide DNA copy number variation.

nitrocellulose membranes. These first "arrays" have since been replaced by commercially available platforms in which tens of thousands of oligonucleotides are arrayed at high density on glass slides or other solid supports. (Protein microarrays and protein–protein interaction profiling are discussed in Chapter 12, The Virus–Host Interactome.) Yet even the wealth of information available through the use of microarrays cannot compare to the level of information obtained through the direct sequencing of RNA transcripts (RNA-seq). Next-generation sequencing, a sequencing-by-synthesis approach that has replaced the first-generation Sanger sequencing method, is capable of yielding a truly comprehensive view of the transcriptome (discussed in more detail in Section 6 of this chapter; Sidebar 2).

Whether generated by microarray analysis or RNA-seq, a transcriptional profile is the pattern of gene expression that is observed in a biological system. Most often, a comparison is made between a normal or resting state and one or more time points following perturbation of the system. The simpler the biological system, the easier it is to manipulate and to generate and interpret transcriptional data. For virology, a simple system means infecting cultured cells. Cell culture experiments facilitate the rapid examination of multiple time points or viruses, and they are useful for examining and modeling intracellular signaling in response to infection. However, the limited phenotypic parameters that can be studied using cultured cells—viral replication or cytopathic effect—makes it difficult to study viral pathogenesis. In contrast, with an animal model, it is possible to measure a variety of disease parameters—e.g., virus yields, clinical signs, gross pathology, histopathology, and time to recovery or death—and the cells or tissues that are examined come from their natural environment. Of course, these benefits are balanced by the cost, ethics, and complexity of animal models.

2.2 Interpreting Transcriptional Profiles

The interpretation of the transcriptional profiles associated with viral infection, or other perturbation, can take a variety of forms. In all cases, it begins with the computational determination of genes that are differentially expressed between conditions (or time points) to a statistically significant degree. This information alone can yield important scientific insights. For example, transcriptional profiling of interferon-treated cultured cells, or tissues from whole animals, has been used to define the hundreds of genes that are induced by this innate immune cytokine, many of which are important players in the innate antiviral response. With an understanding of interferon-induced gene expression in hand, it is then possible to evaluate the extent to which a virus may induce interferon signaling, or alternatively, encode mechanisms to counteract the interferon response.

Similarly, transcriptional profiling of gene knockout cells or animals can be used to evaluate the role of specific genes in the host response to infection. These studies reveal both the changes in gene expression that occur as a result of the gene knockout as well as the effect of the knockout on the host response. Often, these studies also reveal the existence of compensatory signaling mechanisms, which can complicate data interpretation. A slight twist on this approach has been used to study *Mx1*, an interferon-induced gene that confers resistance to influenza virus infection. *Mx1* is lacking in most laboratory strains of mice, and these animals are highly susceptible to many strains of influenza virus, including the reconstructed 1918 pandemic virus. Transcriptional profiling of the lungs of 1918 virus-infected wild-type BALB/c mice (which lack a functional *Mx1*), and *knock-in* mice carrying a functional *Mx1*, demonstrates profound differences in gene expression, which correlate with reduced mortality in $Mx1^{+/+}$ mice (Cilloniz et al., 2012). Treatment of $Mx1^{+/+}$ mice with interferon prior to infection increases survival to 100% and is associated with an increase in the expression of genes related to intercellular signaling or cellular movement and a decrease in the expression of inflammatory cytokine and chemokine genes (Figure 1). Here, transcriptional profiling, attached to a relatively straightforward study design, yielded information on the function of *Mx1* in mediating resistance to influenza virus and on the role of interferon signaling in this process.

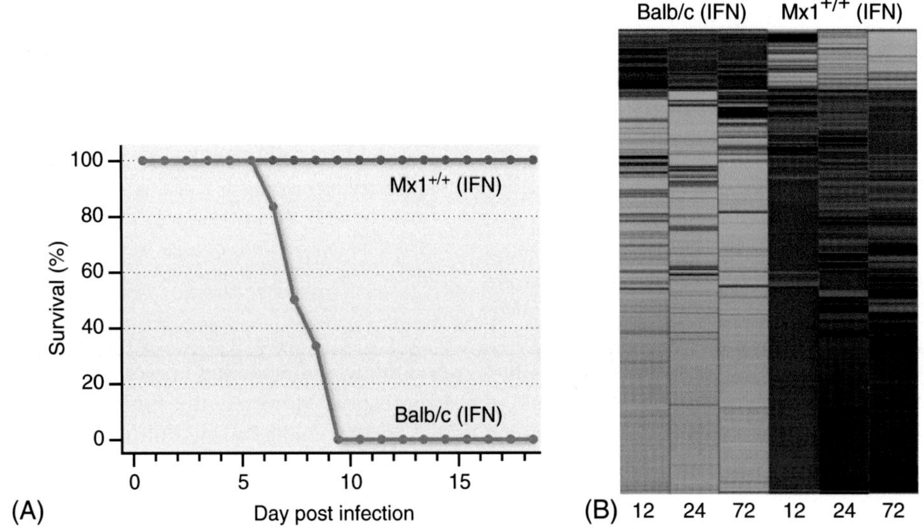

Biological function	P-value
Cell-to-cell signaling and interaction	1.91E-06 – 7.17E-02
Cellular movement	1.17E-05 – 7.17E-02
Antigen presentation	3.5E-04 – 7.09E-02
Cellular growth and proliferation	2.90E-03 – 7.17E-02
Cellular development	5.56E-03 –7.06E-02
Cell signaling	5.79E-03 – 6.09E-02
Molecular transport	5.79E-03 – 6.67E-02
Vitamin and mineral metabolism	5.79E-03 – 6.52E-02
Amino acid metabolism	1.05E-02 – 3.42E-02
Gono oxproccion	1.05E 02 6.52E 02

(C)

FIGURE 1 Interferon-treated $Mx1^{+/+}$ mice are resistant to lethal 1918 pandemic influenza virus infection. (A) Survival plot of a total of 18 mice (9 animals/mouse strain; BALB/c mice lack a functional *Mx1* gene). (B) Heat map illustrating 2071 differentially expressed genes associated with complete protection of $Mx1^{+/+}$ animals compared with the wild-type animals during interferon treatment; cutoff values were ≥twofold change and $P \leq 0.01$ (ANOVA), false discovery rate corrected. Red indicates up-regulation and blue indicates down-regulation. (C) Top 10 biological functions, as determined by using Ingenuity Pathway Analysis, assigned to the 2071 differentially expressed genes. *Adapted from Cilloniz et al. (2012), with permission.*

In the example above, differentially expressed genes are viewed as a heat map, where relative increases or decreases in gene expression are depicted using color gradations, and genes that exhibit similar patterns of expression are grouped together (through the use of various clustering algorithms). Once differentially expressed genes are identified, it is important to know the biological or biochemical function of their encoded products. For many genes, this information can be obtained from a variety of public (e.g., DAVID, maintained by the National Institute of Allergy and Infectious Diseases) or proprietary (e.g., Ingenuity Pathway Analysis) databases. These databases bring together information about individual genes from multiple sources, including the scientific literature and other databases. DAVID, for example, includes over 40 annotation categories, including gene ontology, protein–protein interactions, protein functional domains, disease associations, and tissue expression.

Bioinformatics tools are then used to sift through this information to deconstruct large lists of differentially expressed genes into functionally related groups. Again as seen in the example above, grouping genes in this way helps to interpret and understand the biological information available in high-throughput data. It is important to keep in mind, however, that many genes lack functional annotation or they are only poorly annotated. This means that considerable information may be lost if data analyses do not move beyond a survey of genes with already known function. Fortunately, one of the broader outcomes of transcriptional profiling studies should be an increase in functional annotation. This will come in part as investigators use network modeling approaches (described later) to piece together the interactions or common expression patterns that link various genes and which can be used to infer gene function.

2.3 Dynamics of the Host Response to Influenza Virus

Evaluating how transcriptional profiles change over time provides insights into how the timing and magnitude of the host response impacts viral pathogenesis and disease outcome. This has been particularly well studied in the case of influenza virus. This virus is renowned for its ability to continually generate new variants, and although most variants cause a relatively mild respiratory disease, others can cause severe and even fatal infections. The 1918 pandemic virus, for example, caused over 50 million deaths worldwide, making it responsible for one of the deadliest infectious disease outbreaks in human history. Since the late 1990s, highly pathogenic avian H5N1 influenza viruses have caused sporadic infections in humans (popularized in the news media as bird flu) with a mortality rate estimated at 60%.

Transcriptional profiling of lung tissue from mice or macaques experimentally infected with highly pathogenic influenza viruses has revealed that these viruses elicit a rapid induction of proinflammatory cytokine and chemokine genes, an event often referred to as a cytokine storm. These genes remain highly expressed until the death of the animal. In contrast, animals infected with less-pathogenic viruses exhibit a rapid induction of interferon and innate immune response genes, a response that resolves over time as the animals recover. In macaques, the 1918 virus also induces a disproportionate induction of genes associated with the inflammasome, a group of genes that are thought to be part of a protective innate immune response to influenza viruses. Excessive activation of the inflammasome, however, appears to be detrimental rather than protective. In mice, highly pathogenic avian H5N1 viruses also down-regulate anti-inflammatory genes, including *Alox5*, responsible for the biogenesis of lipoxins, and *Socs2*, encoding a suppressor of cytokine signaling that can be induced by lipoxins to control the inflammatory response. This excessive and sustained inflammatory response appears to promote immunopathology.

An innovative study using automated image analysis, gene expression profiling, and flow cytometry has shed additional insight into the relative contributions of direct viral damage or immunopathology to lethal influenza viral infection (Brandes et al., 2013). Using what is referred to as a top-down systems analysis approach (the breaking down of a complex system into finer details), transcriptional profiles were first obtained from whole lung and then from individual immune cell populations isolated from the lungs of mice infected with lethal or nonlethal influenza virus. Rather than focusing on specific gene expression changes, transcriptional data were organized into 50 modules consisting of genes with coordinate patterns of expression (Figure 2). A module of genes annotated as highly proinflammatory was uniquely associated with lethal infection. This inflammatory signature was shown to largely originate from neutrophils that rapidly migrate to the site of infection (which is poorly contained in the case of the lethal virus). The authors suggest that this initiates a chemokine-driven feed-forward pathway in which the first neutrophils at the scene release chemokines that attract additional neutrophils, resulting in a rapidly escalating inflammatory response and tissue damage. In support of this finding, attenuation of this self-amplifying process by experimental reduction (but not elimination) of neutrophil numbers increased survival without changing viral spread.

An alternative computational methodology incorporating a geometric representation method (singular value decomposition-multidimensional scaling) has been used to visualize and quantify the kinetics of the host transcriptional response to wild-type and attenuated variants of highly pathogenic avian H5N1 viruses (Tchitchek et al., 2013). This approach was used to analyze 230 transcriptomic and 198 proteomic profiles derived from the lungs

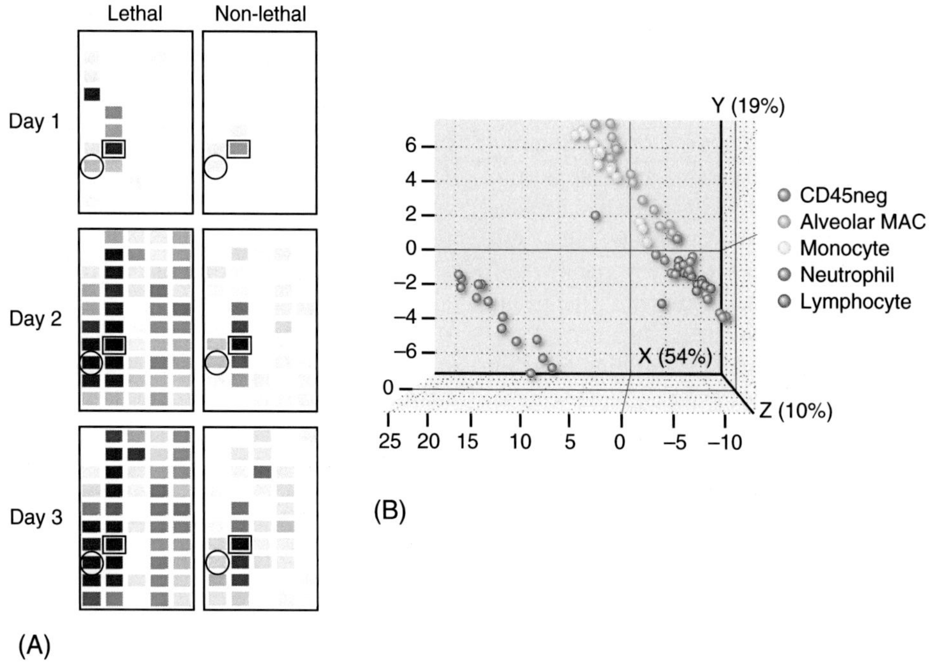

FIGURE 2 Shared and virus-specific innate inflammatory signatures characterize the response to influenza virus infection. (A) Modular map analysis reveals early shared antiviral signatures (module bounded by a rectangle) and virus-specific inflammatory signatures (module bounded by a circle). (B) Neutrophil infiltrates largely account for the lethality-associated inflammatory signature. Shown is principal component analysis of sorted cell microarrays based on genes in the inflammatory module. Amplitude in X largely depends upon whether samples were obtained from neutrophils. *Adapted from Brandes et al. (2013), with permission.*

of infected mice. The analysis revealed that the wild-type and mutant viruses elicit the differential expression of many of the same genes, but that it is the magnitude (the degree to which a gene is differentially expressed) and velocity of the initial host response (the speed at which these changes occur) that best correlates with pathogenic outcome.

2.4 Dynamics of Innate Immunity and AIDS

Transcriptional profiling studies have also shed light on the importance of the dynamics of immune activation in AIDS. This has come primarily from studies of various nonhuman primate species infected with simian immunodeficiency virus (SIV). Natural hosts for SIV, such as sooty mangabeys and African green monkeys, do not develop AIDS when naturally or experimentally infected with the virus. In contrast, Asian monkeys, such as rhesus or pig-tail macaques, develop AIDS following SIV infection. Several independent transcriptional profiling studies have shown that natural hosts for SIV exhibit an innate immune response to the virus that is comparable to that exhibited by Asian macaques. Initial levels of viral replication are also comparable. However, in natural hosts the innate immune response is of limited duration, whereas in macaques the response is sustained (Figure 3) (Jacquelin et al., 2009). This leads to chronic immune activation, which is associated with progression to AIDS in SIV-infected macaques and in HIV-infected individuals.

Because innate antiviral responses are often accompanied by inflammatory reactions, negative regulatory mechanisms, such as the induction of *IL-10*, are necessary to return antiviral responses to baseline to prevent the harmful effects of immune activation. The question remains as to why natural host species are able to resolve the initial response to SIV, whereas macaques (or most HIV-infected individuals) cannot. One hypothesis, proposed as the West Coast Model of immune activation, suggests that the kinetics of activation holds the key (Benecke et al., 2012). The West Coast Model likens virus infection to a wave on the beach, and immune activation as the mounting of a surf board. Mount too early or too late and bad things happen. If correct, the model has important implications for AIDS vaccine design as it suggests that the question of whether an adaptive immune response can be mounted against SIV (or HIV) could be irrelevant. Rather, it may be beneficial to devise strategies to modulate the timing or duration of the response, or to prevent chronic immune recognition of the virus. (Additional information on the kinetics of HIV and SIV infection and lessons learned from natural host species can be found in Chapter 9, HIV and AIDS)

3. DIAGNOSTICS AND PROGNOSTICS

The concept of using gene expression profiles for diagnostic purposes was first proposed in cancer biology, when it was

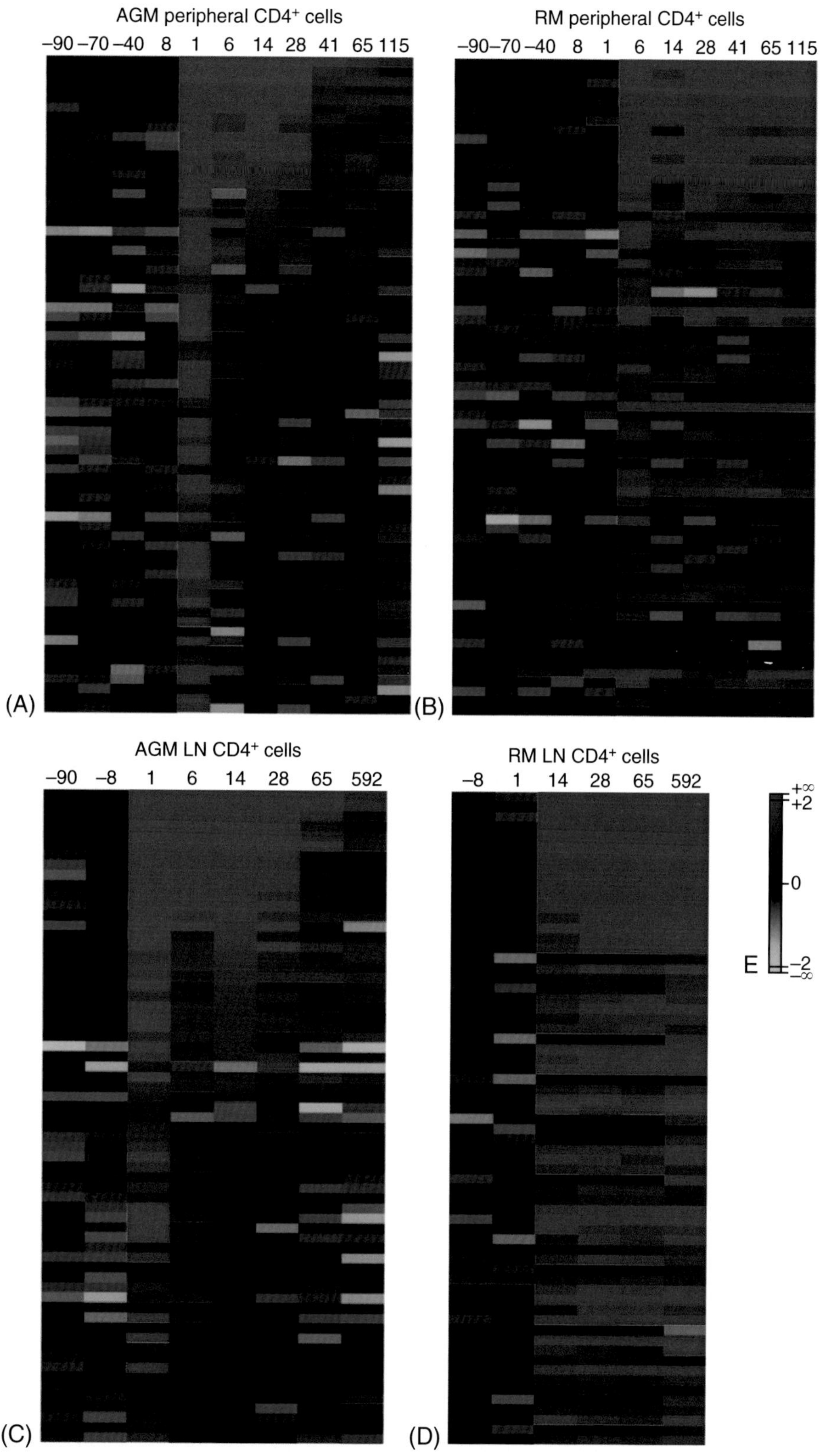

FIGURE 3 Type I interferon-stimulated gene (ISG) expression in blood and lymph node CD4+ cells obtained from monkeys infected with simian immunodeficiency virus. The genes of the ISG cluster that were significantly regulated ($P < 0.05$) in at least one of the two species are represented as heat maps. Mean values of the log2Q of type I ISG expression in peripheral (A and B, respectively) and lymph node (LN) CD4+ cells (C and D, respectively) from six African green monkeys (AGM) and six rhesus macaques (RM) are shown. Many of these genes can also be induced by type II interferon. (E) The color scheme indicates the log2Q. Red indicates up-regulation and green indicates down-regulation. *Adapted from Jacquelin et al. (2009), with permission.*

Sidebar 3 Systems biology in cancer research

Many of the systems approaches that are being used in virology were pioneered in the field of cancer research. Cancer biologists were quick to embrace systems biology as a way to understand how genetic and epigenetic aberrations perturb intracellular signaling networks thereby leading to carcinogenesis. Systems approaches have already yielded clinical benefits to patients, for example, by giving rise to new classification schemes for breast and pancreatic cancer, which in turn allow clinicians to identify patients most likely to benefit from a particular therapy. The cancer research field has been aggressive in developing large-scale systems biology resources, such as The Cancer Genome Atlas (TCGA), the International Cancer Genome Consortium, and The Cancer Proteome Atlas. The field is also driving the development of new mathematical and modeling methods for studying the relationship between intracellular signaling and the behavior of cells at the tissue level, and for the imaging of cells to determine how their spatial orientation and interactions with the environment interplay with gene expression patterns and tumor behavior. Many of these new approaches will no doubt make their way into virology research.

It should be noted, however, that cancer research also draws on virology. A good example is the use of DNA tumor virus proteins for cancer gene discovery (Rozenblatt-Rosen et al., 2012). In this study, proteins from four types of tumor viruses, papillomavirus, Epstein–Barr virus, adenovirus, and polyomavirus, were tested for their ability to interact with 13,000 human gene products. The effect of expressing individual viral genes in cell culture was also assessed by microarray analysis. These data were used to build a virus–host perturbation network that reveals genes and pathways commonly affected by the tumor virus proteins and that are likely to contribute to cancer.

discovered that breast tumor transcriptional profiles could be used for tumor classification. Because different viruses elicit different host transcriptional responses, it might also be expected that gene expression profiles can be used for diagnostic purposes or for making predictions about infection or therapeutic outcome. In clinical virology, most diagnostic tests rely on the ability to detect a particular virus in the bloodstream, either through detection of viral antigens or viral genome sequences. In contrast, host genomic signatures can be used to detect unknown as well as known viruses, or as in cancer, be used as guides for choosing therapy or as predictors of therapeutic or disease outcome (Sidebar 3).

3.1 Genomic Markers of Virus-Induced Liver Disease

One area in which the search for host diagnostic markers has been particularly intense is in hepatitis virus-induced liver disease and hepatocellular carcinoma. Liver disease caused by hepatitis C virus (HCV), for example, can take decades to develop and is often asymptomatic until disease has progressed to cirrhosis or hepatocellular carcinoma. At that point, transplantation may be the only remaining treatment option. Diagnosis is additionally hampered by the lack of reliable noninvasive detection methods as an alternative to percutaneous liver biopsy. A biomarker or genomic profile that could be detected in the blood is therefore highly desirable for early diagnosis of liver disease.

A variety of individual candidate blood biomarkers have been identified, such as aspartate and alanine aminotransferases, albumin, and alkaline phosphatase. These biomarkers provide information about liver function, but values can sometimes be normal in people with liver disease or damage, and these tests do not provide information about disease etiology. Unfortunately, diagnostic genomic signatures in blood have been hard to come by, and studies have therefore focused on the identification of such markers in liver tissue. Although profiling liver gene expression does not get around the need for a liver biopsy, the hope is that genomic profiles will be predictive in advance of tissue injury. There have been a number of reports of liver gene expression signatures, ranging from a half dozen to several hundred genes, which may be prognostic for hepatocellular carcinoma and that may be useful for risk-adjusted surveillance approaches (Hannivoort et al., 2012).

Progress in identifying genomic markers for the diagnosis and prognosis of HCV-induced liver disease has been hampered by the length of time from infection to disease, which makes prospective studies difficult. There is a group of patients, however, in which this time period is compressed. When individuals receive a liver transplant because of end-stage liver disease caused by HCV infection, the transplanted liver quickly becomes infected. While many transplant recipients show no biochemical or histological evidence of liver injury in the first 10 years after transplantation, approximately one-third develop rapidly progressive fibrosis, with the onset of cirrhosis occurring in as little as 5 years after transplantation.

Liver transplant recipients therefore represent a unique study population for discovering gene expression changes that may be predictive of disease progression. In what is perhaps the largest study focused on this population, advanced computational approaches, such as singular value decomposition-multidimensional scaling, were used to analyze transcriptional data obtained from serial liver biopsies from 57 patients (Rasmussen et al., 2012). This study revealed that within 3 months of transplantation over 400 genes were differentially expressed between progressors (who developed adverse clinical outcomes 4–7 years after transplantation) and nonprogressors. This included the down-regulation of genes associated with immune and inflammatory responses, cell cycle progression, and metabolic functions in patients who progressed to severe liver disease. A companion study identified a proteomic signature, indicative of oxidative stress, which could also distinguish progressors from non-progressors (Diamond et al., 2012). Significantly, these studies showed that transcriptional or proteomic markers of disease progression can be detected prior to histological evidence of severe liver injury. Such markers may therefore eventually be the basis for a diagnostic test that can identify patients at high risk of disease progression after transplantation, and perhaps more broadly outside of the transplant setting.

3.2 Discrimination between Viral and Bacterial Respiratory Infection

The etiologic diagnosis of respiratory infections is challenging. Moreover, because many of these infections are caused by bacterial pathogens, physicians will often treat patients with antibiotics even without a confirmed diagnosis in an attempt to provide speedy resolution of symptoms. It is therefore important to develop a means to rapidly distinguish between viral and bacterial infections as well as to identify the specific etiologic agent. Even though respiratory pathogens are typically confined to the respiratory tract, there is growing evidence that different immune cell types induce gene expression signatures in the blood that may be used to accurately diagnose acute respiratory viral infection.

In an initial study aimed at identifying such signatures, volunteers were experimentally infected with rhinovirus, respiratory syncytial virus, or influenza virus and blood samples were taken at set intervals following challenge (Zaas et al., 2009). Microarray analysis of blood gene expression patterns identified a 30-gene "acute respiratory viral" signature that was common to symptomatic individuals from all three viral challenges and which could distinguish between symptomatic individuals and uninfected controls. The signature could also accurately distinguish persons with influenza A virus infection from healthy controls in an independent community-based cohort. In addition, when used to analyze publicly available gene expression data from the blood of patients with bacterial respiratory infection, the signature can accurately distinguish viral from bacterial infection.

More recently, this same 30-gene set has been incorporated into a reverse transcription polymerase chain reaction (RT-PCR) assay (Zaas et al., 2013). RT-PCR is an established diagnostic platform, and moving the acute respiratory signature to this platform represents an important step toward eventual clinical use. The assay was tested in a cohort of 102 individuals arriving at an emergency room

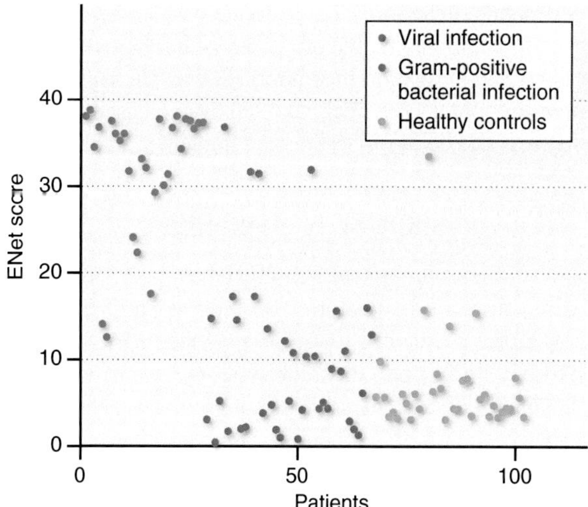

FIGURE 4 A reverse transcription polymerase chain reaction gene expression classifier accurately classifies individuals presenting to the emergency department with viral infection (blue) and distinguishes them from those presenting with Gram-positive bacterial infection (red) and healthy controls (green). ENet score (*y* axis) is a measure of the probability of having viral infection, with a score of 20 indicative of 50% probability of detection. *Adapted from Zass et al. (2013), with permission.*

with fever, and who were confirmed by standard microbiological assays to have a viral or bacterial infection. The RT-PCR assay showed 94% accuracy in distinguishing viral from bacterial infections, suggesting that measuring the expression of a small set of genes in blood samples can be used to classify viral respiratory illness in a real-world setting (Figure 4).

The stage is therefore set for using patient gene expression signatures in viral diagnostics and prognostics. However, there is still work to be done in terms of identifying the most appropriate and minimal set of signatures for these assays, and for improving specificity to provide diagnosis of specific viral agents. Nevertheless, because these approaches provide additional and complementary information to that provided by microbiological assays, gene expression profiling does not need to be considered as a substitute for current diagnostic methods. Instead, a combined approach is likely to yield benefits in terms of rapid triage, the evaluation of febrile illnesses without clear etiology, and for understanding disease pathogenesis.

Unfortunately, in addition to scientific challenges, the use of genomic profiles in clinical settings is facing increased regulatory hurdles as well. This comes primarily in response to the use of faulty (possibly fraudulent) genomic markers to select therapy for patients enrolled in a clinical trial to test alternative chemotherapy approaches to treat nonsmall-cell lung cancer (Kurzrock et al., 2014). In the aftermath to this trial, the Institute of Medicine (an arm of the National Academy of Sciences that provides advice to policy makers) has recommended that diagnostic tests that use genomic data be viewed as devices rather than as laboratory-based tests.

This designation requires that genomic-based diagnostics be subject to additional regulatory controls and be overseen by the Food and Drug Administration (FDA). These increased regulatory burdens are likely to result in increased development costs and delays in testing and implementation.

4. HOST-RESPONSE NETWORKS

Diagnostics and prognostics can be developed on the basis of gene expression signatures. So long as such signatures are accurate predictors, they do not necessarily have to impart any insight into the underlying mechanisms of viral pathogenesis (though often they do). Similarly, an examination of heat maps and functional annotations can on their own yield considerable information about the host response to infection. To use large-scale transcriptional information to gain insight into how things work at a molecular level, or to identify regulatory or drug targets, it is necessary to organize this information in different ways to look for interrelationships among the data. One of the most useful ways to visualize these relationships is in the form of biological networks. Such networks can be built using gene coexpression, the direct interaction of encoded proteins, or shared regulatory mechanisms, such as the binding of transcription factors to target genes.

4.1 Network Hubs and Bottlenecks

Biological networks are typically represented by graphs that contain nodes (genes or proteins) and lines (referred to as edges) connecting the nodes. In a gene coexpression network, for example, the edges are determined using statistical measures of expression correlation. The resulting network can then be analyzed using various computational methods to link network topology—the arrangement and connections of the components of a network—with biological properties.

One of the most common methods used to identify important elements of a network is centrality analysis. Many of the concepts of centrality analysis were first developed for analyzing social networks, and indeed, biological and social networks have many similarities. When analyzing a social network, centrality analysis might be used to determine the most influential person in the network. When analyzing a biological network, the same types of analyses can be used to identify key players in biological processes.

Centrality analysis is used to look for nodes in the network that are highly interconnected or that regulate (or restrict) the flow of information. In biological networks, genes that are highly interconnected—referred to as hubs—are often functionally important. Similarly, genes that connect or bridge multiple subnetworks—referred to as bottleneck genes—are positioned to play powerful roles in regulating network signaling even though they may have fewer connections than hub genes. Hub and bottleneck genes have both been shown to be significantly more likely to be essential for microbial virulence than their nonbottleneck or nonhub counterparts.

These same methods can be used to analyze networks derived from proteomic and lipidomic profiling. In such an analysis of HCV-infected hepatoma cells, two mitochondrial fatty acid oxidation enzymes, DCI (Enoyl-CoA Delta Isomerase 1)and HADHB (Hydroxyacyl-CoA Dehydrogenase/3-Ketoacyl-CoA Thiolase/Enoyl-CoA Hydratase), were identified as network bottlenecks and possible points of control through which HCV disrupts cellular metabolic homeostasis (Figure 5) (Diamond et al., 2010). Targeting DCI for knockdown by siRNA techniques subsequently showed that DCI is required for productive HCV infection in cultured cells. In this case, network analysis provided the information needed to sort through large numbers of proteins and to focus in on a specific target for further scrutiny. Changes in fatty acid oxidation have also been more broadly linked to inflammatory processes, dendritic cell maturation, and regulation of the immune response, suggesting that modulation of fatty acid oxidation may represent a target for antiviral therapy.

4.2 Targets for Host-Directed Antiviral Therapies

An understanding of network topology therefore provides the opportunity to identify potential host targets for therapeutic intervention. Most current antiviral drugs are directed against specific viral protein targets. Such drugs are narrow in spectrum, meaning they are effective only against a specific virus, and they are vulnerable to the emergence of viral resistance (through mutation of the viral genome). Moreover, most medically important viruses have small genomes, so the number of potential targets is limited. In contrast, treating viral infection by targeting the host increases the number of targets and decreases the likelihood that resistant viruses will emerge. Targeting host factors may also increase the likelihood that the drug will be effective against a wider spectrum of viral pathogens.

The analysis of coexpression networks using approaches such as centrality analysis is one way to identify potential host targets (such as DCI in the example above). However, combining network analysis with other information, such as protein–protein interaction data, may improve target identification. This was demonstrated using proteomic data derived from HCV-infected hepatoma cells and from liver tissue obtained from HCV-infected patients (McDermott et al., 2012). When protein–protein interaction data (e.g., host proteins known to interact with HCV proteins) were integrated into protein coabundance networks, network topology analysis of the integrated network provided improved discrimination of bottleneck and hub proteins.

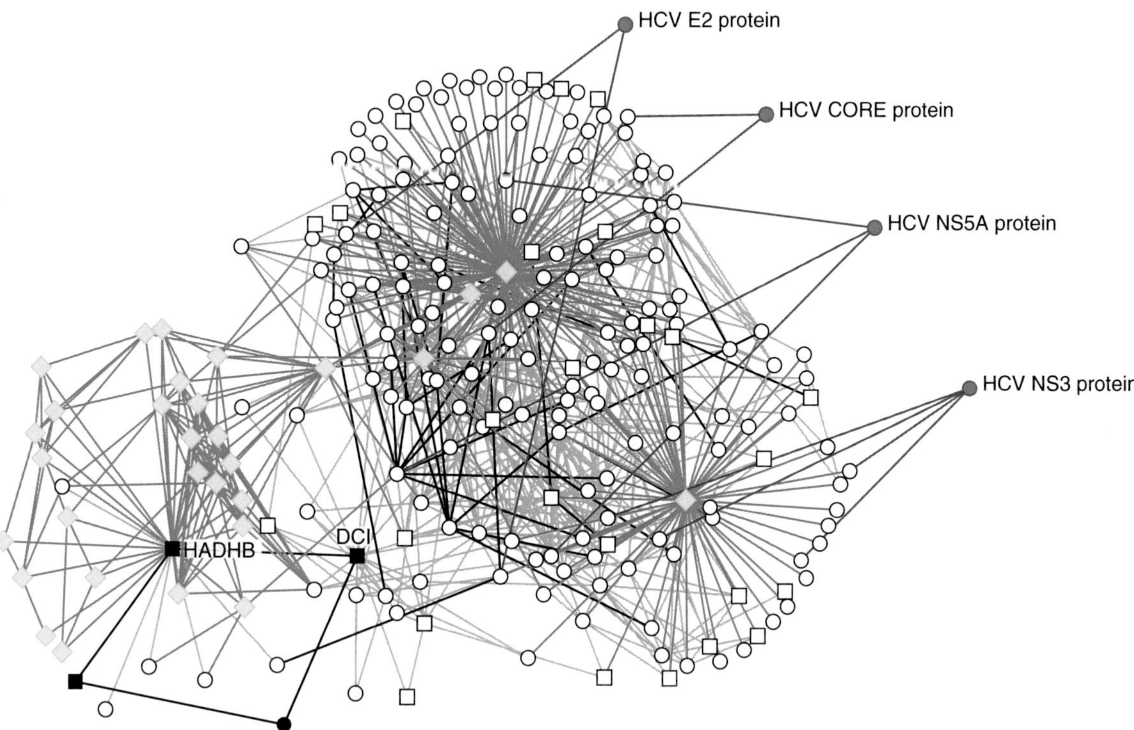

FIGURE 5 The integrated network surrounding several key bottlenecks identified by computational modeling. The neighbors of bottlenecks in the integrated network are shown. Relationships between the proteins and lipid species are gray for proteomics correlation, purple for lipidomics–proteomics correlation, black for protein–protein interactions, and red for interactions with viral proteins. Lipid species are indicated as yellow diamonds, HCV proteins are red, mitochondrial proteins are squares, and proteins involved in fatty acid β-oxidation are in black. *Adapted from Diamond et al. (2010), with permission.*

A similar analysis looked directly at host factors that interact with the HCV-encoded NS5A protein (Tripathi et al., 2013). Through its interaction with other HCV proteins and host factors, NS5A plays an important role in HCV infection, including regulating viral replication, the production of viral particles, and interferon resistance. NS5A has therefore become an attractive target for antiviral therapy. However, rather than targeting NS5A directly, it may be possible to target one or more host factors that interact with NS5A. To identify such host targets, NS5A-interacting proteins were identified through literature mining and by using a yeast two-hybrid approach (a molecular biology technique used to discover protein–protein interactions). The resulting 132 host proteins were used to build an interaction network, which was further expanded by incorporating protein–protein interactions for the proteins targeted by NS5A (resulting in 1442 proteins with 6263 interactions between them). Topological analysis was then used to identify bottleneck and hub proteins.

Intriguingly, these analyses suggest that NS5A preferentially interacts with highly central proteins in the host protein interaction network. These proteins have functions in a variety of cellular processes, including innate immunity, chemokine signaling, cell-to-cell communication, and cellular transport. Among the bottleneck proteins identified were two endoplasmic reticulum proteins, RTN1 and RTN3. These proteins are present in very low density lipoprotein transport vesicles, which have been reported to play a role in the production and release of infectious HCV. In cell culture assays, knockdown of RTN1 and RTN3 using siRNA has no effect on viral RNA levels, but significantly reduces viral titer. As regulators of viral propagation, RTN1 and RTN3 may be novel targets for anti-HCV therapy and perhaps more broadly as therapeutic targets for other viruses that depend upon lipoprotein vesicles for the release of infectious virus.

5. DRUG REPURPOSING

Developing new drugs takes an enormous amount of time, averaging 14 years from target discovery to FDA approval. Failure rates and costs are also extraordinarily high. So even once targets are identified and validated, a long road to drug development remains. One strategy to reduce the time and expense of drug development is to determine whether a drug approved to treat one disease might be repurposed to treat another. Similarly, many partially developed drug candidates could potentially be repurposed for new indications. Global transcriptomic methods have become central to drug repurposing efforts (Hurle et al., 2013), and the paradigm is finding its way into antiviral drug research.

5.1 Inverse Genomic Signatures

The inverse genomic signature approach is based on the proposition that a drug should have therapeutic benefit if it generates a gene expression profile that is the inverse of the signature associated with the disease (Figure 6; Peng et al., 2014). The approach therefore requires knowledge of the gene expression profiles induced by large numbers of drugs. Such information is being generated by the Connectivity Map project, which seeks to find connections between human diseases, gene expression profiles, and drug action. The Connectivity Map database contains over 7000 transcriptional profiles generated by treating cultured human cells with over 1300 compounds, many of which are FDA-approved drugs. Analytical tools can then be used to calculate a connectivity score, a measure of the similarity—or inverse similarity—between query gene expression signatures and profiles in the database.

When this approach was first used to identify drugs that inhibit influenza virus replication, a common 20-gene expression signature was identified from cultured human lung epithelial cells infected with different strains of human or avian influenza virus. This gene expression signature was then used to screen drug-associated profiles in the Connectivity Map database, and candidate antivirals were identified by their inverse correlation to the common signature. Eight potential antivirals were identified, six of which were subsequently determined to inhibit influenza virus replication, including the 2009 H1N1 pandemic influenza virus, which was not used to generate the 20-gene signature (Josset et al., 2010).

Because the use of genomic signatures for drug screening has the potential to dramatically reduce the time needed for drug development, the approach may be particularly beneficial when applied to emerging viral infections. Several recent studies have used the approach to screen for drugs against emerging influenza virus strains (e.g., H7N9 viruses), and Connectivity Map has also been used to identify drugs that may be effective against Middle East respiratory syndrome coronavirus (Josset et al., 2013). In all of these studies, follow-up validation using cell culture has demonstrated reductions in viral replication. However, it remains to be seen whether and how rapidly drugs (or classes of drugs) identified by this method actually make their way into clinical use. The same can be said for the cancer field, in which drug repurposing approaches using Connectivity Map have identified drug candidates against a variety of cancers. Many of these drugs have been validated in cell culture and rodent models, but have been slow to move into clinical trials.

5.2 Network-Based Approaches

As discussed earlier, networks can be used to represent regulatory and functional interactions between genes or proteins, or between combinations of genes, proteins, or metabolites. One goal of generating such networks is to identify *targets* for antiviral drugs. In contrast, network-based drug repurposing strategies aim to harness the information contained in networks to identify candidate *drugs*. Although network-based approaches can take a variety of forms, many are focused on understanding how diseases are connected to one another—through similar transcriptional or protein–protein interaction networks, for example—or

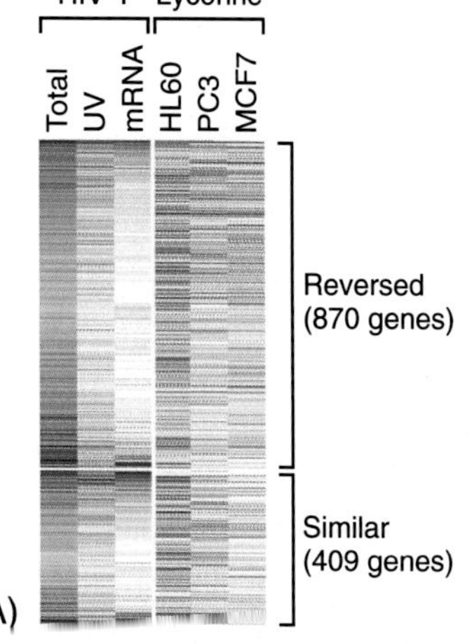

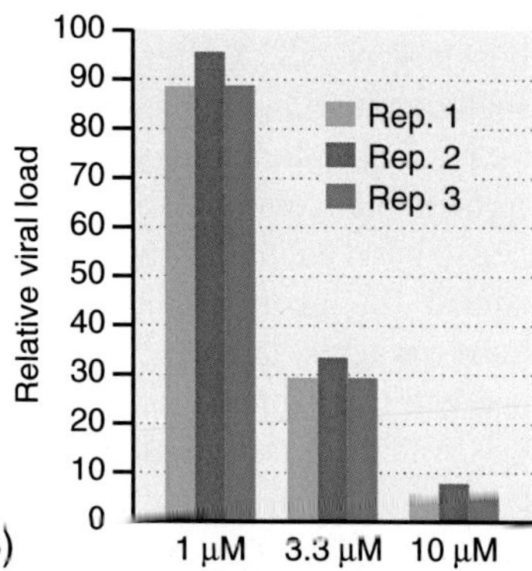

FIGURE 6 Side by side comparison of gene expression changes induced by HIV-1 infection of a human CD4+ T cell line (SUP-T1) and by treatment of three different cell lines with the drug lycorine. Gene expression-based antiviral drug repurposing predicts that a drug should have therapeutic benefit if it generates a gene expression profile that is the inverse of the signature associated with viral infection. (A) Columns represent individual conditions: Total RNA-seq of HIV-1 infection (Total) or UV-inactivated virion infection (UV), mRNA-seq of HIV-1 infection (mRNA), and lycorine treatment of three different cell lines. Lycorine-induced expression profiles were obtained from Connectivity Map. (B) Relative viral loads (quantified by qPCR) in HIV-1-infected SUP-T1 cells treated with lycorine showing inhibition of HIV-1 replication. *Adapted from Peng et al. (2014), with permission.*

how drugs are connected to one another through their mechanism of action. For example, Connectivity Map can also be used to construct "drug networks" to identify connections between drugs on the basis of shared transcriptional effects. In this case, different drugs (rather than genes or proteins) form the nodes of the network. Drugs can also be connected by side-effect similarity, which can be used to infer whether two drugs share a target.

Various networks—transcriptional, protein–protein interaction, disease, or drug—can also be integrated into multilayer networks that can be probed for drug–disease relationships. For example, an integrated analysis of gene expression data from 54 diseases (including several viral diseases) and human protein–protein interaction data yielded a network of 138 disease relationships (Suthram et al., 2010). Within this integrated network, a set of common pathways was identified, and many of the proteins in those pathways were found to be targets of existing drugs. By identifying disease relationships and shared drugs, it was possible to make predictions about the repurposing of drugs from one disease to another. With the ever-increasing amounts of genomic data being generated in infectious disease research, network-based approaches are sure to be increasingly exploited for the repurposing of drugs to fight viral infections.

6. NEW VIEWS OF THE TRANSCRIPTIONAL LANDSCAPE: LONG NONCODING RNAs AND VIRAL INFECTION

The direct sequencing of RNA transcripts is yielding exciting new views of the transcriptional landscape. The staggering amount of information available through sequencing is exemplified by the Encyclopedia of DNA Elements (ENCODE) project, funded by the National Human Genome Research Institute. The overall goal of the project is to identify and characterize all functional elements in the human genome, including cataloging of the complete repertoire of RNAs produced by human cells. The project has revealed that as much as three quarters of the human genome is capable of being transcribed and that cells contain many varieties of RNA transcripts (Djebali et al., 2012). These include polyadenylated and nonpolyadenylated transcripts, known and unannotated protein-coding transcripts, and long noncoding RNAs (>200 nucleotides) such as intergenic transcripts. Small noncoding RNAs (<200 nucleotides) are also abundant, including microRNAs (miRNAs), piwi-interacting RNAs (piRNAs), small nuclear RNAs (snRNAs), small nucleolar RNAs (snoRNAs), and transfer RNAs. Indeed the transcriptional landscape is so complex that it has called into question the very definition of a gene!

Virologists are beginning to explore the ramifications of this newfound transcriptional complexity on viral pathogenesis. In one of the first studies to do so, RNA-seq was used to profile the host response to infection with severe acute respiratory syndrome coronavirus (SARS-CoV) (Peng et al., 2010). In the lungs of mice infected with this virus, over 10,000-long noncoding RNAs were identified, and nearly 1500 were differentially expressed in response to infection. Comparable expression profiles were observed in cell lines infected with influenza virus or treated with type I interferon, suggesting that these RNAs may be involved in the innate response to a variety of viruses. Similar analyses that focused on the sequencing of small RNAs revealed that SARS-CoV infection also induces the differential expression of different classes of small noncoding RNAs.

Distinct patterns of noncoding RNA expression have now been observed in the response to many different RNA and DNA viruses. Moreover, long noncoding RNAs are not limited to cellular transcription. For example, RNA-seq has revealed that human cytomegalovirus (HCMV), a 240-kb DNA virus, produces hundreds of previously unidentified transcripts, including alternatively spliced transcripts and long noncoding RNAs (Gatherer et al., 2011). HCMV is also one of several viruses known to encode its own miRNAs, and RNA-seq has further revealed that HCMV encodes additional novel forms of small RNAs (Stark et al., 2012). Even viral genomes are capable of producing complex transcriptional profiles.

The next important step, of course, is to determine the biological functions of these newly found long noncoding RNAs. There is already evidence that long noncoding RNAs play roles in transcriptional and epigenetic gene regulation, developmental processes, and in a variety of diseases, including neurological and immune disorders and cancer. These RNAs may function through direct interaction with specific genome sequences (thereby affecting chromatin remodeling and gene transcription), transcription factors, or other components of the transcriptional machinery. For example, the long noncoding RNA, lincRNA-Cox2, has been demonstrated to interact with heterogeneous nuclear ribonucleoproteins to mediate both the activation and repression of multiple immune response genes (Carpenter et al., 2013). Similarly, lncRNA-CMPK2 has been reported to be a negative regulator of the interferon response and is itself among over 200 long noncoding RNAs induced by interferon (Figure 7; Kambara et al., 2014). Knockdown of lncRNA-CMPK2 expression reduces HCV replication in interferon-treated hepatocytes, and lncRNA-CMPK2 is up-regulated in liver samples from HCV-infected patients, suggesting it may also play a role in modulating the interferon response in these individuals.

Unfortunately, determining the biological functions of long noncoding RNAs has so far proven challenging.

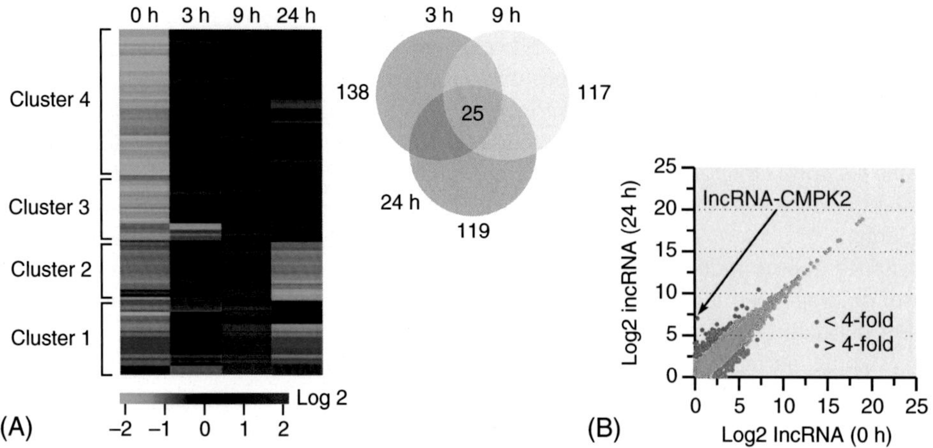

FIGURE 7 Interferon-α induces hundreds of long noncoding RNAs in human hepatocytes. (A) Heat map and Venn diagram of lncRNAs showing fourfold or greater change in expression following interferon-α treatment. (B) Scatter plot depicting the annotated lncRNAs that show a statistically significant change of fourfold or more after 24 h of interferon stimulation. The gray dots mark lncRNAs that did not show a significant change in expression. Red and blue dots correspond to up-regulated and down-regulated lncRNAs, respectively. The location of lncRNA-CMPK2, a negative regulator of the interferon response, is shown. *Adapted from Kambara et al. (2014), with permission.*

Unlike proteins, the function of long noncoding RNAs cannot presently be predicted from primary sequence or secondary structure. And with thousands of such RNAs to choose from, it is difficult to know which to pursue. In the example above, lncRNA-CMPK2 was chosen for study because of its over 100-fold induction in response to interferon stimulation. However, long noncoding RNA expression can also be integrated into network models, which may help to provide additional information as to which RNAs should be the focus for follow-up studies. With the advent of new genome editing techniques, such as CRISPR-Cas9, it will also be possible to design large-scale screens to identify long noncoding RNAs that may be required for viral replication or virus-induced cytopathic effects (for details on CRISPR-Cas9 and large-scale interaction screening, see Chapter 12, The Virus–Host Interactome).

Unquestionably, RNA-seq and the discovery of long noncoding RNAs have ushered in a new era in the study of viral pathogenesis. It will be essential to gain an understanding of the role of these RNAs during viral infection to fully understand the mechanisms by which viruses cause disease. It is also likely that a better understanding of long noncoding RNA function will lead to new therapeutic options. RNA-based therapeutics—most of which are focused on the use of antisense oligonucleotides to degrade specific mRNAs—are already in development. Indeed, an antiviral drug, fomivirsen (an antisense oligonucleotide that blocks the synthesis of a key cytomegalovirus protein), was the first drug of this type to be approved by the FDA. Antisense transcripts could be similarly used to deplete specific long noncoding RNAs, or small molecules could be used to disrupt the interaction of long noncoding RNAs with their protein or DNA partners.

7. CONCLUDING REMARKS

Ongoing advances in technology will continue to spur new approaches for studying the host transcriptional response to viral infection. While most current studies using animal models still examine the RNA profiles of whole tissues, it is becoming increasingly common to augment this approach with the analysis of isolated cell populations. But even the analysis of isolated cell types results in averaging the transcriptomes of millions of cells. RNA-seq performed on single cells is therefore emerging as a new frontier in transcriptional profiling. The ability to perform single-cell analyses has become possible through improvements to methods for cell isolation and the conversion of miniscule amounts of cellular RNA into cDNA for sequencing. Although single-cell analyses are not necessary (or appropriate) for all types of studies, the capability provides unique opportunities, such as profiling cell-to-cell variability. Such analyses have revealed surprising variability in the expression of hundreds of immune genes across single cells, as well as variation in splicing patterns (Shalek et al., 2013).

Efforts are also underway to incorporate an understanding of how the epigenome—the heritable, and potentially reversible, genome-wide chemical changes to the DNA and histone proteins of an organism—impact the host response to viral infection. Epigenetic modifications, such as DNA or histone methylation, result in changes in gene expression through alterations in chromatin structure. These modifications can occur in response to a multitude of environmental changes, including viral infection. In addition, several viruses, such as human cytomegalovirus and Epstein–Barr virus, use epigenetic mechanisms in part for switching between latent and active infection (Ernberg et al., 2012). Although studying the involvement of epigenetic mechanisms in viral infection is still in its infancy, such studies are likely to offer new clues into disease mechanisms.

Clearly, high-throughput molecular profiling has revealed that the host response to viral infection is more complicated than ever before thought possible. Moreover, gene expression, protein abundance, and all of their attendant regulatory mechanisms are only part of the picture. As discussed in other chapters, consideration must also be given to protein–protein interactions, host metabolism, host genetics, and even the myriad microorganisms that form the human microbiome. With each new technological advance comes a new avalanche of data to a field that some argue is already suffering from information overload. Perhaps what is needed most are improved computational methods for integrating diverse types of data and for identifying new interrelationships, including cooperative or synergistic interactions. These may arrive in the form of new geometric approaches and links between geometry, information theory, and probability theory (Law et al., 2013). In the meantime, even if the complete picture will have to wait to be assembled, there are plenty of discoveries to come to keep virologists energized well into the future.

REFERENCES

Benecke A, Gale Jr M, Katze MG. Dynamics of innate immunity are key to chronic immune activation in AIDS. Curr Opin HIV AIDS 2012;7: 79–85.

Brandes M, Klauschen F, Kuchen S, et al. A systems analysis identifies a feedforward inflammatory circuit leading to lethal influenza infection. Cell 2013;154:197–212.

Carpenter S, Aiello D, Atianand MK, et al. A long noncoding RNA mediates both activation and repression of immune response genes. Science 2013;341:789–92.

Cilloniz C, Pantin-Jackwood MJ, Ni C, et al. Molecular signatures associated with Mx1-mediated resistance to highly pathogenic influenza virus infection: mechanisms of survival. J Virol 2012;86:2437–46.

Diamond DL, Krasnoselsky AL, Burnum KE, et al. Proteome and computational analyses reveal new insights into the mechanisms of hepatitis C virus-mediated liver disease posttransplantation. Hepatology 2012;56:28–38.

Diamond DL, Syder AJ, Jacobs JM, et al. Temporal proteome and lipidome profiles reveal hepatitis C virus-associated reprogramming of hepatocellular metabolism and bioenergetics. PLoS Pathog 2010;6:e1000719.

Djebali S, Davis CA, Merkel A, et al. Landscape of transcription in human cells. Nature 2012;489:101–8.

Ernberg I, Karimi M, Ekstrom TJ. Epigenetic mechanisms as targets and companions of viral assaults. Ann N Y Acad Sci 2012;1230:E29–36.

Gatherer D, Seirafian S, Cunningham C, et al. High-resolution human cytomegalovirus transcriptome. Proc Natl Acad Sci USA 2011;108: 19755–60.

Hannivoort RA, Hernandez-Gea V, Friedman SL. Genomics and proteomics in liver fibrosis and cirrhosis. Fibrogenesis Tissue Repair 2012;5:1.

Hurle MR, Yang L, Xie Q, et al. Computational drug repositioning: from data to therapeutics. Clin Pharmacol Ther 2013;93:335–41.

Ideker T, Galitski T, Hood L. A new approach to decoding life: systems biology. Annu Rev Genomics Hum Genet 2001;2:343–72.

Jacquelin B, Mayau V, Targat B, et al. Nonpathogenic SIV infection of African green monkeys induces a strong but rapidly controlled type I IFN response. J Clin Invest 2009;119:3544–55.

Josset L, Menachery VD, Gralinski LE, et al. Cell host response to infection with novel human coronavirus EMC predicts potential antivirals and important differences with SARS coronavirus. mBio 2013;4:e00165–13.

Josset L, Textoris J, Loriod B, et al. Gene expression signature-based screening identifies new broadly effective influenza A antivirals. PLoS One 2010;5.

Kambara H, Niazi F, Kostadinova L, et al. Negative regulation of the interferon response by an interferon-induced long non-coding RNA. Nucleic Acids Res 2014;42(16):10668–80.

Kurzrock R, Kantarjian H, Stewart DJ. A cancer trial scandal and its regulatory backlash. Nat Biotechnol 2014;32:27–31.

Law GL, Korth MJ, Benecke AG, et al. Systems virology: host-directed approaches to viral pathogenesis and drug targeting. Nat Rev Micro 2013;11:455–66.

McDermott JE, Diamond DL, Corley C, et al. Topological analysis of protein co-abundance networks identifies novel host targets important for HCV infection and pathogenesis. BMC Syst Biol 2012;6:28.

Peng X, Gralinski L, Armour CD, et al. Unique signatures of long noncoding RNA expression in response to virus infection and altered innate immune signaling. mBio 2010;1:e00206–10.

Peng X, Sova P, Green RR, et al. Deep sequencing of HIV infected cells: insights into nascent transcription and host-directed therapy. J Virol 2014;88:8768–82.

Rasmussen AL, Tchitchek N, Susnow NJ, et al. Early transcriptional programming links progression to hepatitis C virus-induced severe liver disease in transplant patients. Hepatology 2012;56:17–27.

Rosen R. A means toward a new holism. Science 1968;161:34–5.

Rozenblatt-Rosen O, Deo RC, Padi M, et al. Interpreting cancer genomes using systematic host network perturbations by tumour virus proteins. Nature 2012;487:491–5.

Schena M, Shalon D, Davis RW, et al. Quantitative monitoring of gene expression patterns with a complementary DNA microarray. Science 1995;270:467–70.

Shalek AK, Satija R, Adiconis X, et al. Single-cell transcriptomics reveals bimodality in expression and splicing in immune cells. Nature 2013;498:236–40.

Stark TJ, Arnold JD, Spector DH, et al. High-resolution profiling and analysis of viral and host small RNAs during human cytomegalovirus infection. J Virol 2012;86:226–35.

Suthram S, Dudley JT, Chiang AP, et al. Network-based elucidation of human disease similarities reveals common functional modules enriched for pluripotent drug targets. PLoS Comput Biol 2010;6:e1000662.

Tchitchek N, Eisfeld AJ, Tisoncik-Go J, et al. Specific mutations in H5N1 mainly impact the magnitude and velocity of the host response in mice. BMC Syst Biol 2013;7:69.

Trewavas A. A brief history of systems biology. Plant Cell 2006;18:2420–30.

Tripathi LP, Kambara H, Chen YA, et al. Understanding the biological context of NS5A-host interactions in HCV infection: a network-based approach. J Proteome Res 2013;12:2537–51.

Zaas AK, Burke T, Chen M, et al. A host-based RT-PCR gene expression signature to identify acute respiratory viral infection. Sci Transl Med 2013;5:203ra126.

Zaas AK, Chen M, Varkey J, et al. Gene expression signatures diagnose influenza and other symptomatic respiratory viral infections in humans. Cell Host Microbe 2009;6:207–17.

Zhu H, Cong JP, Mamtora G, et al. Cellular gene expression altered by human cytomegalovirus: global monitoring with oligonucleotide arrays. Proc Natl Acad Sci USA 1998;95:14470–5.

FURTHER READING

Csermely P, Korcsmaros T, Kiss HJ, et al. Structure and dynamics of molecular networks: a novel paradigm of drug discovery: a comprehensive review. Pharmacol Ther 2013; 138: 333–408.

Korth MJ, Tchitchek N, Benecke AG, et al. Systems approaches to influenza-virus host interactions and the pathogenesis of highly virulent and pandemic viruses. Semin Immunol 2013; 25: 228–39.

Menachery VD, Baric RS. Bugs in the system. Immunol Rev 2013; 255(1): 256–74.

Tisoncik JR, Korth MJ, Simmons CP, et al. Into the eye of the cytokine storm. Microbiol Molec Biol Rev 2012; 76(1): 16–32.

Chapter 12

The Virus–Host Interactome

Knowing the Players to Understand the Game

Monika Schneider[1], Jeffery R. Johnson[2], Nevan J. Krogan[2], Sumit K. Chanda[1]
[1]*Sanford-Burnham Medical Research Institute, La Jolla, CA, USA;* [2]*University of California, San Francisco, CA, USA*

Chapter Outline

1. INTRODUCTION

The interactions between viruses and human host cells encompass the activation of immune defenses, viral countermeasures, and viral hijacking of cellular proteins. As discussed in previous chapters, mammalian cells have a robust arsenal of antiviral and innate immune protective mechanisms, and viruses have evolved proteins that specifically target those host responses. It is the molecular interactions between virus and host that ultimately determine pathogenic outcome. Just as knowing the positions and roles of players is essential to understanding a football or basketball game, knowing the complete roster of relevant genes and proteins, and the roles that they play, is essential to understanding viral pathogenesis. This chapter describes the high-throughput screening approaches being used to identify the molecular "players" in viral pathogenesis, how they interact with one another, and the roles that they play in determining pathogenic outcome.

2. GENETIC FUNCTIONAL SCREENS

In the context of cellular infection, not only do viruses depend on their own genes for successful replication, but they also use genes expressed by the host. The cellular processes that are hijacked vary with the stage of infection, and conversely, viral infection induces a cohort of cellular genes that help the cell fend off the infecting virus. Viruses frequently mutate to adapt to the environment or to improve infection efficiency. In contrast, host proteins mutate at a much slower rate, if at all, and therapeutically targeting host proteins that are important for viral replication may be more successful at thwarting viral escape. A variety of high-throughput screens have therefore been developed to identify host proteins that are important for viral replication (Panda and Cherry, 2012).

Host proteins that promote the survival or replication of a virus are referred to as *host factors,* and host proteins that limit viral survival or replication are termed *restriction factors* or antiviral factors. The loss of host restriction factors results in increased viral replication, whereas overexpression of restriction factors reduces viral proliferation. Examples of broad-acting host restriction factors include the products of many interferon-stimulated genes (see Chapter 4, Innate Immunity).

There are several genetic tools available to uncover host and restriction factors (Table 1). Loss-of-function and

Viral Pathogenesis. http://dx.doi.org/10.1016/B978-0-12-800964-2.00012-4

TABLE 1 Methods Used to Probe the Virus–Host Interactome

Method	How It Works	Advantages	Disadvantages
Loss of function	Transient or stable siRNA or shRNA expression to knockdown expression of target genes	Can be used in high-throughput screens; can reveal genes that play a role in viral infection	False negatives and positives due to off-target activity of mismatched siRNAs or irrelevant immune activation
Gain of function	Overexpression of cDNAs	Can be used in high-throughput screens	False positives due to nonphysiological levels of the target gene
RNA-seq	Whole-genome sequencing of mRNA or small RNA at a given time point	Quantifies all transcripts in the cell	Only captures a snapshot of given infection
Yeast two-hybrid	Pairwise introduction of proteins that are fused to complementary fragments of a readout system	Low-cost and scalable	High false-positive and false-negative rates
Protein microarray	Whole cell lysate incubated with proteins or peptides immobilized on a solid surface	Can identify otherwise transient enzyme–substrate interactions	Protein complexes cannot be identified if the bait protein is not a direct interactor; high false-negative rate
Affinity purification and mass spectrometry	Protein complexes co-purified with tagged proteins then identified through mass spectrometry	Identifies indirect interactions; complexes are formed in vivo, so more physiologically relevant	Direct interactions are difficult to identify

gain-of-function screens probe individual genes to determine how each affects virus viability, whereas RNA-seq technologies measure the quantity of all transcripts induced or reduced by infection at a given time. The best tool to use depends upon the desired information, and each assay can be adjusted for optimal results. The details of genetic functional screens are discussed in the remainder of this section.

2.1 Loss-of-Function Screening Using RNA Interference

RNA interference (RNAi) screens are commonly used to globally identify the proteins that are involved in a cellular phenotype of interest. This method makes use of the cellular RNAi pathway, which naturally produces regulatory microRNAs (miRNAs) and small interfering RNAs (siRNAs) to specifically silence target genes. In the RNAi pathway, noncoding RNAs are processed into shorter stem-loop structures called pre-miRNA, which are then further cleaved into 20–22 base-pair siRNAs. One strand of the double-stranded siRNA, termed the guide strand, is then incorporated into the RNA-induced silencing complex for targeting to the corresponding mRNA. siRNAs and miRNAs are both processed through this pathway and bind to target mRNA through an eight-nucleotide seed region, but with different end results.

siRNAs bind to their target mRNA and induce degradation of the transcript. In contrast, miRNAs bind to their target mRNA and inhibit translation without leading to transcript degradation, in part due to their ability to bind to transcripts without perfect complementarity. The RNAi pathway can be experimentally induced through the introduction of synthetic siRNAs or short hairpin RNAs (shRNAs) in a transient or stable manner. Algorithms and mathematical models have been developed to design and predict targeting efficiency on the basis of sequence characteristics, such as GC content and whether hairpin structures form within the siRNA.

shRNAs intersect the RNAi pathway at an earlier step than siRNAs. shRNAs are approximately 70 nucleotides in length and are engineered to form a hairpin structure, similar to pre-miRNAs. The hairpin structure is recognized by endogenous RNAi proteins, which then process the shRNA into functional siRNAs. shRNAs can be delivered to the cell using an exogenous vector, or they can be incorporated into the genome for stable silencing via a lentiviral vector.

Although siRNA and shRNA are effective means of reducing the expression of target genes, it is expensive to generate whole-genome siRNA or shRNA libraries. This hurdle can be bypassed through the use of in vitro-generated siRNAs, termed endoribonuclease-prepared siRNAs (esiRNAs) (Yang et al., 2002). This method uses a cDNA library to transcribe each gene; the long dsRNA is then digested by bacterial RNase III. The resulting esiRNAs effectively knock down target gene expression. All gene-targeting strategies using RNAi have their limitations, and as whole-genome screens are more widely used, reproducibility, consistency between assays, and off-target activities remain a concern (Jackson and Linsley, 2010).

2.2 CRISPR Genome Editing

A new system for gene knockdown, termed CRISPR (for clustered regularly interspaced short palindromic repeats), has recently been developed. CRISPR was originally discovered in bacteria, where it provides immunity against bacterial viruses by disrupting viral transcription (Barrangou et al., 2007). Genome-editing strategies using CRISPR take advantage of bacterial Cas9, an enzyme that catalyzes double-stranded DNA breaks, to specifically target and create deletions in the exons of a desired gene. CRISPR functions through the base pairing of a guide RNA (gRNA) to a specified genomic location. The gRNA directs Cas9 to the target sequence where it induces a double-strand break, leading to a sequence deletion or insertion. Unlike siRNA and shRNA, which can result in an incomplete knockdown of the target gene, CRISPR permanently edits the genome and can result in the complete knockout of a gene.

CRISPR has been used to achieve individual gene knockdown on a genome-wide scale by stably expressing Cas9 in mouse embryonic stem cells or in HeLa cells (Koike-Yusa et al., 2014; Zhou et al., 2014). A genome-wide library of pooled gRNAs is then generated and gRNAs are individually expressed using lentiviral vectors. There are few false positives, and the phenotype is stronger than that observed when the same genes are knocked down with a shRNA.

Although CRISPR offers several advantages to traditional genome editing tools, some studies have indicated that gRNAs can cause off-target gene mutation. In addition, genes that have multiple alleles may require multiple rounds of CRISPR treatment to completely knock out the target gene, and single-cell sorting is required to identify which cells harbor a homozygous or heterozygous knockout. Finally, as with RNAi, gene knockout using CRISPR is dependent on the ability of delivery vectors to enter the cell.

2.3 Gain-of-Function Screening

Genetic gain-of-function screens are the converse of loss-of-function screens since they ectopically express genes, sometimes in excess of physiological expression. There are several whole-genome libraries that can be used for this type of screening depending upon the readout and type of cell used. These include the Mammalian Gene Collection, which is curated by the National Institutes of Health, and the human ORFeome collection, which is curated by the Center for Cancer Systems Biology. Gain-of-function screens can also be run using either transient transfection or through the generation of stable expression cell lines.

There are scenarios in which gain-of-function screens are advantageous. For example, viral infection may cause the down-regulation of particular genes that have antiviral effects. Knocking down repressed or lowly expressed genes is not likely to result in a measureable phenotype. However, overexpression of these genes may overcome viral down-regulation, resulting in reduced viral proliferation. A disadvantage of overexpression is that it can generate more false positives than would be found in a loss-of-function screen because of nonspecific activities due to nonphysiological levels of gene dosage.

2.3.1 Application of Gain-of-Function Screens to Viral Pathogenesis

An example of using a gain-of-function screen to identify host antiviral proteins is shown in Figure 1. In this screen, interferon-stimulated genes (ISGs) were identified through published microarray and RNAi screening sets, and these data were used to curate a library of nearly 400 ISGs (Schoggins et al., 2011). The ISGs were then individually overexpressed to test which would have the ability to inhibit viral replication. Multiple viruses were used in the screen, including hepatitis C virus, HIV-1, yellow fever virus, West Nile virus, Venezuelan equine encephalitis virus, and Chikungunya virus. This approach revealed ISGs that were broadly antiviral as well as those that specifically inhibited only one or two viruses.

2.4 Vector Delivery Methods

The efficiency of an RNAi vector is dependent on the cell type and delivery method. There are several well-characterized methods for RNAi delivery, but desired throughput may limit the options available for a particular assay (Table 2). Lipid delivery methods use a phospholipid bilayer to form a vesicle around the siRNA or shRNA, which fuses with the cell membrane. Lipofection typically has a low level of toxicity, but sensitivity is dependent on the type of reagent used and the cell type being transfected. This method can be used for high-throughput screens. Electroporation delivers an electric pulse to the cells, which results in membrane pores, and the charge of the electric pulse helps to move the nucleotide vector into the cell. Electroporation can cause considerable cell death and requires transfer of the cells between the electroporation device and the plating well. Because of this extra step, this delivery method has limited throughput. Finally, viral vectors can be used for RNAi delivery. Retroviral vectors become integrated into the cellular genome, allowing for stable expression of the delivered shRNA. In contrast, adenoviruses do not incorporate into the genome and are therefore only useful for transient expression.

Due to their effectiveness at specifically reducing the expression of target genes, RNAi technologies also have

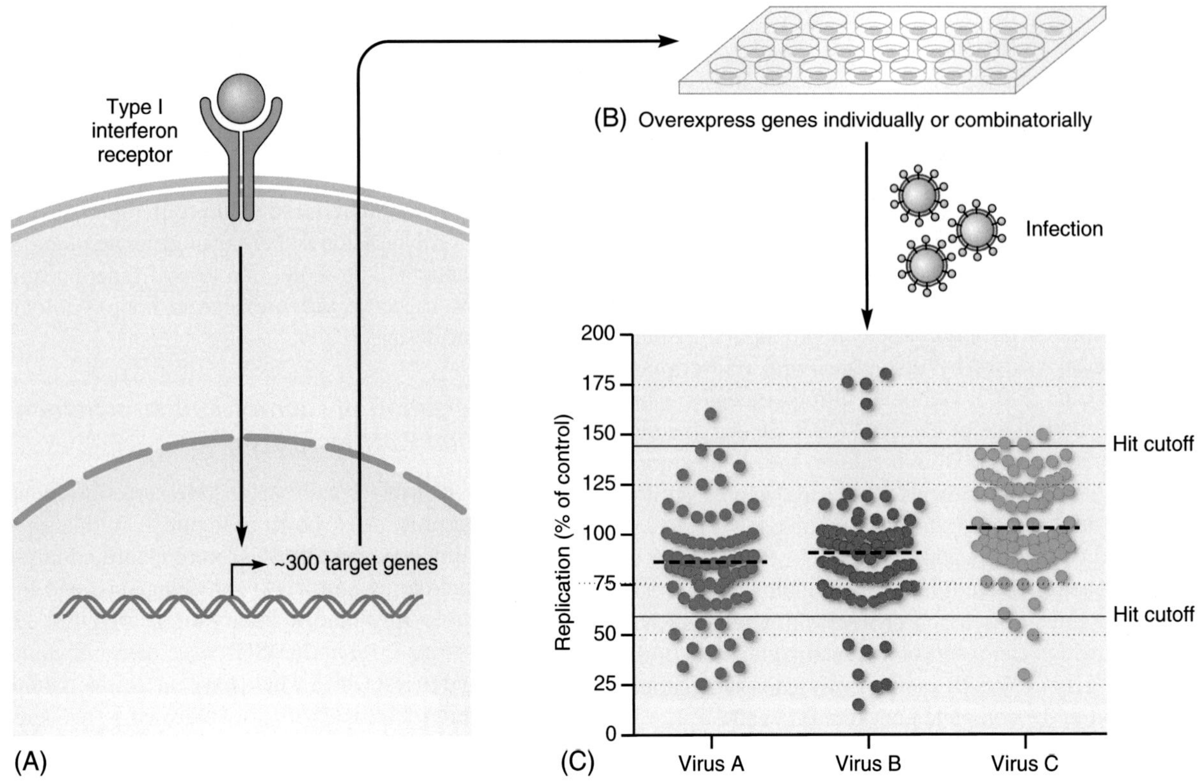

FIGURE 1 Gain-of-function screen to identify host antiviral proteins. (A) A library of genes that are induced after type I IFN treatment (ISGs) are subcloned into a lentiviral expression vector. (B) The genes are overexpressed and individual sets of cells are infected with a panel of different viruses. (C) Replication efficiency is measured, and those genes that significantly inhibit viral replication (lower cutoff) are further investigated for antiviral effects. Hits above the upper cutoff represent ISGs that enhance viral replication.

TABLE 2 Methods to Deliver Interfering RNA to Cells

Entry Method	How It Works	Potential Off-target Effects	Throughput
Lipid vectors	Lipid vesicle merges with the cell and releases through endosome	Cell death; irrelevant immune activation	High
Electroporation	Electric charge forms pores in the cell membrane, RNAi vectors enter into cytoplasm	Cell death	Low
Viral vectors	Fusion with cell membrane	Irrelevant immune activation; targeting wrong gene	Medium

great potential for therapeutic use. However, the delivery of RNAi vectors in vivo is a much greater challenge than delivery to cell lines. Uptake of siRNA or shRNA by target cells in vivo is poor, and the RNAi vectors themselves are often rapidly degraded once they enter the bloodstream. There has been some success in targeting specific organs, such as the liver, but systemic delivery poses additional challenges. There are, however, promising avenues for this type of delivery, including chemical modification of the siRNA or shRNA, cholesterol conjugation, viral vectors, and various polymers (Kanasty et al., 2013).

2.4.1 Applications of RNAi Vector Delivery to Viral Pathogenesis

Despite therapeutic delivery challenges, methods have been developed for screening siRNAs in vivo. For example, tumor suppressors have been identified by introducing shRNAs—targeting the mouse orthologs of genes deleted in human liver cancer—into premalignant cancer cells and then transplanting these cells into mice and testing for their ability to promote the formation of tumors (Zender et al., 2008). Virus–host interactions have also been probed by incorporating miR-30-based hairpins into Sindbis virus. The shRNAs that conferred a survival advantage for the virus were then identified by using mRNA-seq. This method identified novel host factors required for antiviral gene transcription as well as many known ISGs (Varble et al., 2013).

2.5 Screening Formats and Readouts

Screening "hits" are defined as those genes that fall above or below a certain threshold, which is set by the control values. There is no uniform way to determine hits, and each research group has unique ways of identifying genes of value. Generally, the quality of a screen is determined through the calculation of a Z prime, which uses the means and standard deviations of the control samples to estimate the intrinsic value of the experimental samples. The Z prime cannot exceed 1, which indicates a perfect assay, and a good separation between negative and positive controls yields a Z prime between 0.5 and 1. Different methods of determining hits may account for some of the variability seen in screens that have been run using similar parameters.

Loss-of-function assays can be run as whole-genomic or targeted subgenomic screens, and siRNAs or shRNAs can be delivered in gene-based pooled or in arrayed formats. In pooled libraries, two to four siRNAs for the same gene are delivered simultaneously, which increases the likelihood of achieving significant gene knockdown. However, if one of the siRNAs is toxic to the cell, then that gene will be eliminated from the screen regardless of the effects of the other siRNAs. Following screening using a pooled library, the pools need to be deconvoluted through confirmation of phenotype and knockdown using each siRNA individually.

In an arrayed format, each siRNA or shRNA is plated in an individual well. Because multiple siRNAs or shRNAs may be used for each gene, this method greatly increases the size of the initial screen. However, screening in this way allows for the identification of toxic siRNAs and eliminates the need for a deconvolution step. With this method, target genes that have two or more effective siRNAs are apparent immediately, and thus off-target effects can be minimized.

2.5.1 *Viral Replication Readouts and Their Applications to Viral Pathogenesis*

There are several types of readouts that can be used to measure viral replication. A luciferase reporter can be incorporated into the virus, such that is it activated when the virus replicates. Lower values will be reported if the virus fails to enter the cell or if it is unable to subsequently propagate. This method was used in a screen to identify host factors involved in influenza virus replication (Konig et al., 2010). The influenza virus used had the viral HA gene replaced with a Renilla luciferase gene, which allowed luciferase activity to be measured following infection. Almost 295 genes were identified as being important for early replication steps, including those involved in endosomal processes, intracellular trafficking, and ubiquitination.

Another method to measure viral replication is through tracking the expression of a viral protein, or by replacing a viral protein with a tagged or fluorescent protein, such as green fluorescent protein. After infection, the cells can then be stained for the endogenous or tagged protein, and the amount of virus in each cell can be measured through high-content imaging, which uses robotics to image proteins or process cells through fluorescent microscopy. This method was also used to identify host factors required for influenza virus replication (Brass et al., 2009). Cells were transfected with siRNA pools, and after influenza virus infection, the cells were immunostained for the presence of HA, which was used as a surrogate for viral replication (Figure 2). Nearly 150 host antiviral genes were identified as potential targets after validation assays. This loss-of-function screen identified an important family of host restriction factors, call the IFITM (interferon-inducible transmembrane) proteins, which potently inhibit the replication of influenza virus, West Nile virus, and Dengue virus. IFITM3 was later confirmed as a broad-acting restriction factor through a gain-of-function screen using a curated list of ISGs (Schoggins et al., 2011).

Finally, spreading replication can be measured by quantifying the amount of new virus produced from the originally infected cells. Host factors necessary for late influenza virus replication were identified by either measuring the presence of a virus-specific protein 24h after infection, or by transferring the supernatant of the infected cells onto an uninfected reporter cell line. By using this approach, nearly 300 host genes were found to positively influence influenza virus proliferation (Karlas et al., 2010).

2.5.2 *Host Response Readouts and Their Applications to Viral Pathogenesis*

Measurement of the host response often focuses on the inflammatory pathways that are activated following viral infection. Generally, the focus is on the Toll-like receptor, RIG-I, and interferon (IFN) signaling pathways, as these are well-defined viral innate response pathways (see Chapter 4, Innate Immunity). These pathways converge on NF-κB, IRF3, and IFN-β transcription, and binding sites for these transcription factors can be attached to a luciferase reporter gene so that activation of the signaling pathway can be quantified. Additionally, the translocation of these transcription factors from the cytoplasm to the nucleus, which also signals activation of the pathway, can be measured through high-content imaging. Further, the host response can be measured by quantification of protein production, gene expression, or receptor up-regulation.

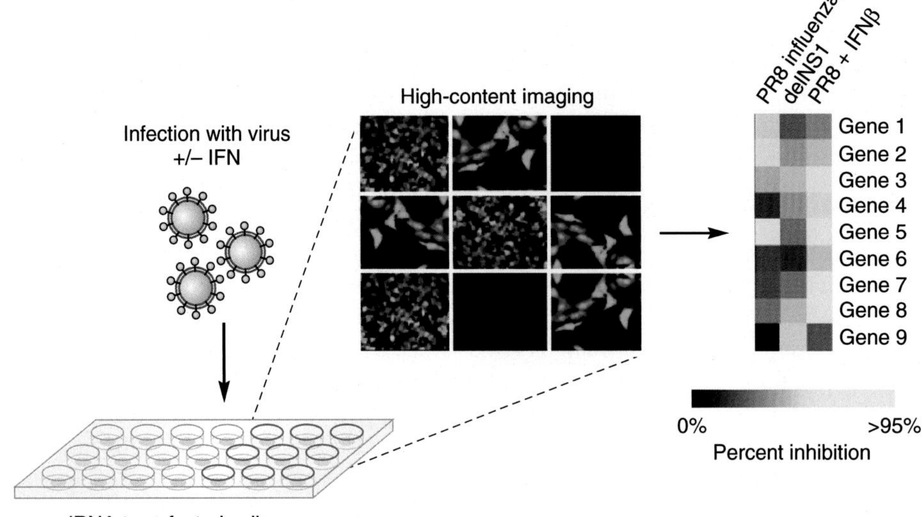

FIGURE 2 siRNA screen to measure the effects of ISGs on influenza virus replication. Cells are transfected with siRNA pools, and after infection with wild-type influenza virus (PR8), influenza virus with a deletion in the host immune inhibiting NS1 gene, or wild-type virus in the presence of IFN-β, cells are immunostained for hemagglutinin (HA). HA, which is an influenza-encoded protein that can be used as a surrogate measure of viral replication levels, is quantified using imaging technologies after RNAi treatments. Inhibition of viral replication after gene depletion reveals host factors that are required by the virus for replication. Prior stimulation with interferon enables an understanding of the virus–host interaction in cells that are in a preexisting antiviral state. The heat map reflects potential effects of knockdown of individual genes on the replication levels of wild-type influenza virus; influenza virus with a deletion in the NS1 gene; or wild-type influenza in cells pretreated with IFN-β.

As an example, in a search for genes needed for the induction of RIG-I, IFN-β signaling was used as a readout for the host response to Sendai virus (SeV) infection (Baril et al., 2013). This screen incorporated a genome-wide lentiviral-based shRNA library and cells expressing a luciferase reporter gene under the control of an IFN-β promoter. After SeV infection, those genes that either increased or decreased IFN-β activation were identified and were further mapped to the RIG-I signaling pathway using follow-up assays. In this way, WNT family members were identified as novel negative regulators of RIG-I signaling.

2.6 Integrating Data from Multiple Screens

RNAi screens usually generate an abundance of data and hits. Perhaps surprisingly, when similar screens are performed by different groups, there is often little overlap in the hits from each screen. It can therefore be difficult to determine genes that may be valuable for follow-up studies or therapeutic targeting. In the case of the three RNAi screens described above—which together identified over 700 genes that potentially impact influenza virus replication—there was only *one* common gene hit. The differences in the outcomes of these screens are likely due to variability in the type of cells used, the readout used, the type of virus, and the time points used for the assays.

Although these results may at first appear discouraging, a closer look at the data reveals that certain cell signaling pathways and protein complexes are overrepresented by the hits in these screens. A *meta*-analysis uncovered several common groups of host factors that are involved in processes crucial to influenza proliferation (Stertz and Shaw, 2011). Most of these overlaps occur at the gene pathway and protein complex level rather than at the level of individual genes.

2.7 Transcriptional Profiling

Viral infection has a dramatic effect on the expression, splicing, and turnover of cellular mRNAs. The full extent of these changes is beginning to be elucidated through RNA-seq, which is a method that builds upon the advances made by next-generation sequencing (see Chapter 11, Systems Virology). RNA-seq provides a snapshot of the global cellular transcriptome at the time of sample collection.

As described above, RNAi screens can illuminate the signaling pathways that are hijacked by a virus. In contrast, RNA-seq reveals the impact of infection on cellular transcription. Historically, DNA microarrays have served this purpose; however, the advantage of using RNA-seq is that all RNA within a cell population can be quantified. RNA-seq can therefore decipher splice variations and noncoding transcripts, which exponentially increases the information provided by each sample. However, sampling at multiple time points is usually necessary to provide the desired genomic information, and detailed kinetic studies can quickly become expensive.

RNA-seq can be used to simultaneously analyze both host and viral transcription, which is particularly informative for DNA viruses with large genomes. For example, this approach was used to analyze the expression patterns of both murine cytomegalovirus and infected mouse embryonic fibroblasts (Juranic Lisnic et al., 2013). The sequencing of samples from nine different time points after infection revealed many novel viral transcripts, including antisense transcripts, spliced transcripts, and transcripts that overlapped multiple annotated genes. In fact, the most highly abundant transcripts did not have known functions. On the host side, many of the most up-regulated genes were associated with transcription and cellular defense. Down-regulated genes not previously known to have a role in infection were also identified.

Because RNA-seq can be used to analyze noncoding RNAs, it is beginning to be used to investigate the roll of such RNAs in viral infection. Certain miRNAs have already been demonstrated to play roles in viral persistence or host defense; however, a comprehensive picture of the miRNAs that are present or induced during viral infection is lacking. Further, most miRNAs still have unknown significance in the context of infection. Deep sequencing of RNA isolated from the lungs of four different mouse strains that had been infected with SARS coronavirus or influenza virus revealed extensive differential expression of diverse classes of short noncoding RNAs (Peng et al., 2011). Such studies are providing a greater understanding of overall transcriptional changes due to viral infection and provide an important resource for determining the functional role of noncoding RNAs in the antiviral response.

3. PHYSICAL PROTEIN INTERACTION SCREENS

Proteins typically act in complexes that direct their specificity, activity, localization, and interactions with other protein complexes and cellular machinery. It is therefore desirable to comprehensively identify members of protein complexes in order to enhance the functional and mechanistic knowledge of individual proteins. For uncharacterized proteins lacking functional annotation, identifying interacting proteins can be used to assign functions through testable "guilt-by-association" hypotheses. For well-characterized proteins, an unbiased interactome characterization may uncover previously unknown functions.

Of particular relevance to viral pathogenesis, the identification of physical interactions between virus and host proteins provides putative targets for therapeutic intervention. Furthermore, host proteins that interact with viruses are frequently under strong selective pressure and thus may be unable to escape antiviral drugs, in contrast to rapidly evolving viral proteins. In any case, the identification of virus–host protein–protein interactions identifies viral vulnerabilities and dependencies on the host cell.

Screens to identify protein–protein interactions can be divided into two categories: those that identify direct, physical interactions, and those that identify components of a protein complex. Screens identifying direct, physical interactions frequently use yeast two-hybrid methods or other complementation assays. Protein arrays may also identify direct, physical interactions as well as enzyme–substrate relationships. Screens that identify components of a protein complex frequently rely on protein co-purification using co-immunoprecipitation or affinity purification. Each of these types of screens is discussed in the remainder of this chapter.

3.1 Two-Hybrid Screening

Two-hybrid screening approaches are high-throughput complementation assays that test for protein–protein or protein–DNA interactions. The assay is typically performed by introducing proteins of interest pairwise into yeast, with each protein fused to a transcription factor that has been split into two complementary fragments. Conventionally, the protein fused to the N-terminal DNA-binding domain of the transcription factor is referred to as the "bait" and the protein fused to the C-terminal activation domain as the "prey." When brought into close proximity to one another through interaction between the bait and prey proteins, the binding and activation domains of the transcription factor function to activate transcription of a reporter gene. The reporter gene may encode for antibiotic resistance, such that interacting clones can be selected by applying antibiotic pressure. Alternatively, the reporter gene may code for a lethal gene, such that a physical interaction results in a reduction in colony size. Although two-hybrid approaches are typically performed using yeast, these assays have also been adapted to bacterial and mammalian systems (Joung et al., 2000). The disadvantages of two-hybrid approaches are that they often have high false-positive and false-negative rates. Producing proteins at far higher abundance than is biologically relevant can lead to spurious interactions that elevate the false-positive rate. False negatives may occur if N- or C-terminal fusions disrupt interaction interfaces, or if proper protein folding, processing, or posttranslational modifications cannot be recapitulated.

Two-hybrid approaches have been used to comprehensively characterize interactions between host proteins and proteins derived from a variety of viruses, including Kaposi sarcoma-associated herpesvirus, varicella-zoster virus, murine γ-herpesvirus 68 (MHV-68), vaccinia virus, SARS coronavirus, influenza virus (Friedel and Haas, 2011). In the case of influenza virus, an integrated approach was used to identify and validate interactions between viral and human proteins by complementing a comprehensive yeast two-hybrid assay with additional large-scale experiments (Shapira et al., 2009). This included the measurement of cellular transcriptional responses following transfection with influenza viral RNA, IFN-β treatment, and infection with an influenza strain lacking the NS1 gene (responsible for inhibiting the innate immune sensing of viral RNA and downstream IFN production). A set of genes found to be regulated in either the two-hybrid screen or the gene expression screens were also tested in siRNA knockdown screens measuring influenza replication and IFN-β production. Integrating the data resulting from these various assays revealed that viral polymerase subunits were enriched for interactions resulting in the positive regulation of IFN production, suggesting that the viral polymerase, in addition to NS1, plays a role in inhibiting the IFN response.

A similar integrated approach was used to characterize virus–host interactions for murine γ-herpesvirus MHV-68 (Lee et al., 2011). Using a yeast two-hybrid approach, a library of 84 MHV-68 genes was screened against each other to identify 23 intraviral interactions. The library was also screened against a cDNA library derived from human liver cells to identify 243 virus–host interactions. An affinity purification approach validated 70% of the intraviral interactions, giving an estimate of the false-positive rate of the yeast two-hybrid screen. Network analyses indicated that cellular proteins targeted by MHV-68 had more partners in a cellular protein–protein interaction network than expected by chance. This integrated screening and validation approach therefore yielded viral–viral and viral–host protein interaction networks.

3.2 Protein Microarray Screening

Protein microarrays are constructed by immobilizing proteins at high density on a solid surface. The proteins may be individually purified, or they may be synthetic peptides generated using chemical peptide synthesis. The task of cloning and purifying thousands of native, full-length proteins to immobilize on an array may seem an insurmountable task, but arrays containing an impressive 17,000 full-length human proteins have been constructed (Hu et al., 2012). Proteins immobilized at addressable locations on a microarray can be used to identify not only protein–protein interactions, but also interactions with nucleic acids, antibodies, and small molecules.

The most cited limitation of protein microarrays relates to the comprehensiveness of the protein libraries available on the chip. In addition, proteins immobilized on microarrays are typically produced in bacteria or yeast and are therefore prone to false-negative interactions due to incorrect folding or lack of posttranslational modifications (similar to that observed using two-hybrid assays).

Protein microarrays have been used to identify conserved substrates for viral kinases that may represent targets for antiviral drugs. Herpes simplex virus, human cytomegalovirus, Epstein–Barr virus, and Kaposi's sarcoma-associated herpesvirus each encode a serine/threonine kinase that is necessary for viral replication and spread. To determine the extent to which substrates for these kinases are conserved, a human protein microarray was used to identify the cellular substrates of each kinase (Li et al., 2011). This approach resulted in the identification of 643 nonredundant substrates, 110 of which are shared by at least 3 kinases. The shared substrates were then mapped onto a network of existing data for protein–protein interactions, enzyme–substrate relationships, and gene ontology functional classes resulting in the identification of a highly connected cluster of DNA damage response proteins.

3.3 Affinity Purification and Mass Spectrometry Screening

Modern mass spectrometry platforms are increasingly sensitive and capable of characterizing complex protein mixtures at unprecedented depth. These platforms can be used to identify protein complexes that are formed in vivo and purified intact, thus the approach can identify protein interactions as they occur in a natural biological system. A major advantage of co-purification approaches is that they identify many proteins contained within a complex, making the process of identifying complexes and pathways associated with a protein of interest a much simpler task than with two-hybrid or protein microarray approaches (which rely on bioinformatics approaches and public databases to identify complexes and prioritize interactions). Affinity purification coupled with mass spectrometry, termed AP-MS, has been used to characterize a wide range of biological systems, including virus–host interactions.

Although many biological systems are amenable to AP-MS analysis, the use of affinity-tagged proteins introduces an inherently synthetic aspect to the assay. The most common method of introducing a tagged protein into a cell is by transfection, a method that does not work well for many types of primary cells. Furthermore, AP-MS requires a fairly large amount of starting material, which is not scalable for many cell types and primary cell systems, and AP-MS screens do not identify direct, physical interactions. Entire complexes are co-purified such that the specific interactions between members of a complex may be obscured. Cross-linking approaches combined with AP-MS can overcome this problem by providing distance constraints between proteins within a complex. Cross-linking AP-MS approaches are still highly specialized, particularly with respect to the bioinformatics interpretation of the spectra of cross-linked peptides.

AP-MS has been used to identify HIV–human protein interactions. For this approach, all the genes associated with the HIV genome, as well as unprocessed polyproteins, were cloned into a vector that contained a dual affinity tag fused

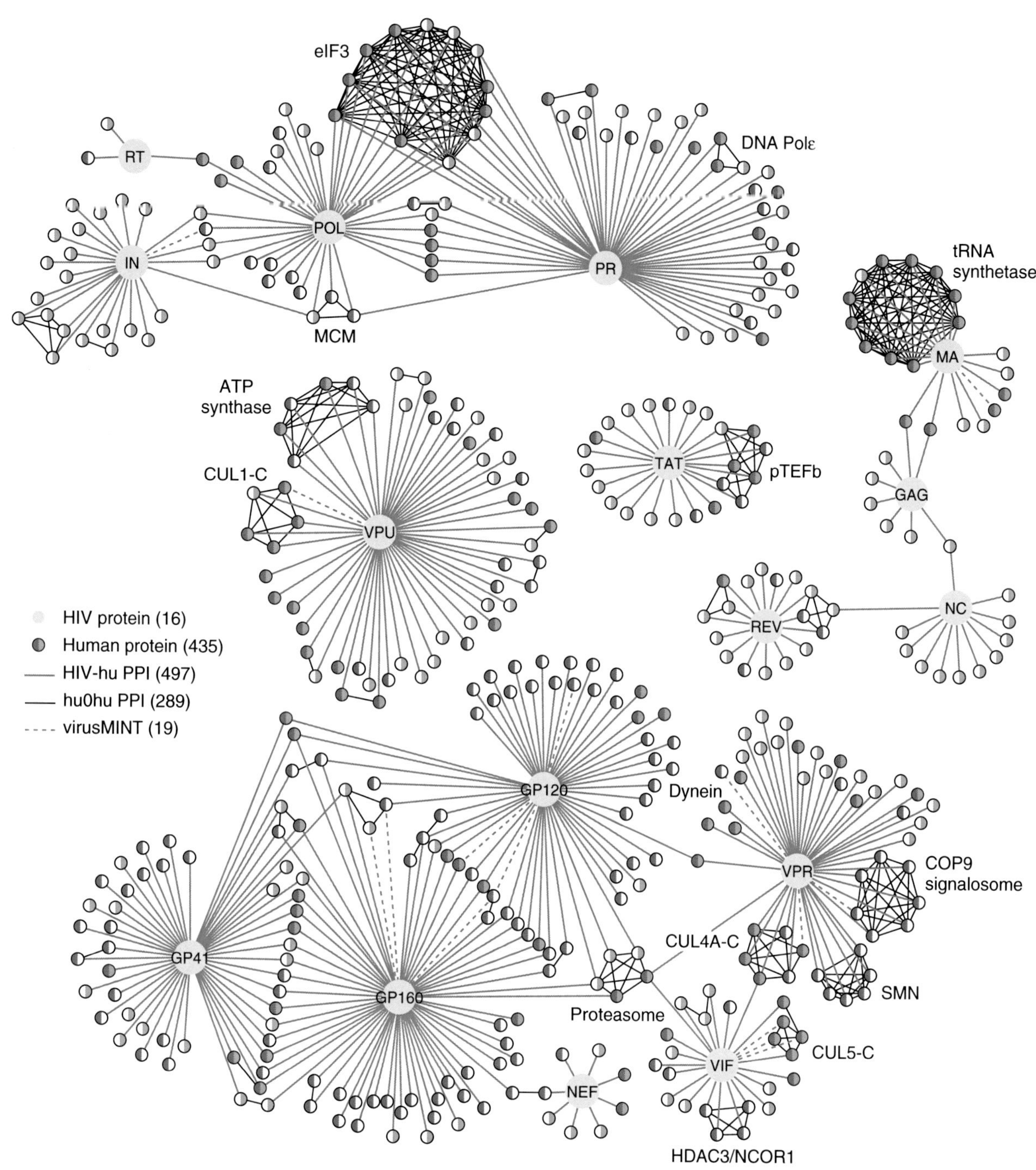

FIGURE 3 Network representations of HIV–human protein interactions. Yellow nodes correspond to viral proteins and all other nodes represent human factors (red indicates the interaction was identified in Jurkat cells and blue indicates the interaction was identified in HEK293 cells). Blue edges between nodes correspond to the HIV–human protein interactions that were identified whereas black edges represent connection between human proteins derived from several protein–protein interaction databases. *Adapted from Jäger et al. (2011), with permission.*

at the C-terminal end of the proteins (Jäger et al., 2011). Each clone was transiently transfected into HEK293 cells and was also used to generate stably expressed, tetracycline-inducible versions in Jurkat cells. HIV–human protein complexes were purified by affinity purification and the resulting complexes were analyzed by mass spectrometry. An unsupervised scoring system was then used to identify host–pathogen protein–protein interactions and to separate nonspecific from specific interactions. Using this score, a high-confidence interaction map was generated and overlaid with human–human protein interactions and HIV–human genomics data (Figure 3).

4. DATA INTEGRATION AND NETWORK ANALYSIS

Data integration is crucial to recognizing signaling pathways and nodes of activity that mediate the interactions between virus and host. This integration can be performed through side-by-side analysis of multiple loss-of-function screens, or it can be done by integrating orthogonal datasets. Protein–protein interaction studies provide clear evidence of which individual proteins interact with each other. Network representations then allow host–pathogen interactions to be visualized.

Protein–protein interaction data can also be applied to genetic screens to better understand the functional significance of each gene (Shapira et al., 2009). However, individual gene arrays make it more difficult to determine genetic interactions, or how the absence of two or more genes will affect the virus–host interactome. Genetic interactions can further reveal whether two genes function independently or in conjunction with each other. For example, if knocking down two genes independently results in reduced viral replication, there is no easy way to determine whether it is because both genes are in the same pathway or because the genes are in pathways that act in parallel. If both genes are knocked down at the same time, and the reduction in viral replication is additive, then it is likely that these genes are in independent signaling pathways. If viral replication is reduced to a similar level as either gene alone, then the genes are likely in the same pathway. Predictions can be made based on protein–protein interactions, but designing studies that systematically test combinations of genes is the most effective method to determine the genetic and functional relationships between these target genes.

Although data integration and analysis is challenging, there have been many efforts to make high-throughput data more accessible and universal (Masseroli et al., 2014). Close interactions between virologists and biostatisticians are necessary to develop proper analysis tools. The quest to elucidate the virus–host interactome has already benefited greatly from open-source databases. For example, Gene Ontology (GO) is a bioinformatics database that allows researchers to categorize their gene lists into functional groups, and GO data can be complemented with commercial (e.g., Ingenuity Pathway Analysis) or open-source (e.g., Cytoscape) network analysis programs. These programs are used to visualize clusters of activities and provide an idea about how the genes are interrelated. Using screening data, GO, and known protein interactions, a predicted interaction network can be generated. Importantly, the cumulative results from genome-wide screens provide a valuable resource for the field and are being used to create a clearer picture of virus–host interactions.

5. REPRISE

Identifying all of the viral and cellular factors that impact viral infection, replication, and pathogenesis is a monumental task. Although a truly comprehensive determination of every such factor is currently beyond reach, a variety of experimental approaches are being used to work toward this goal. Gene-based approaches include loss-of-function screening using siRNAs or shRNAs, and screens of this type have identified hundreds of human host factors required for influenza virus replication. Newer CRISPR-based approaches, which provide improved targeting and stable and complete gene knockouts, are also being developed for use in high-throughput screens. Gain-of-function screens complement these approaches by providing the ability to identify host antiviral genes that may be down-regulated by the infecting virus. RNA-seq, which can identify and quantify entire viral and cellular transcriptomes, is yielding new views into the complexity of transcription and the incredible diversity of coding and noncoding RNA transcripts. Many RNAs that no one had ever before thought to look for are likely to have roles in the virus–host interactome.

Protein-based approaches provide information on how viral and host factors interact with one another and how such interactions contribute to infection outcome. Two-hybrid screens and protein microarrays are used to identify direct, physical interactions between proteins, or between proteins and nucleic acids, or other small molecules. Affinity purification and mass spectrometry complement these approaches by identifying the members of protein complexes. Together, these approaches are being used to identify cellular substrates for viral enzymes and to construct detailed network models of the interactions between viral and host proteins.

In this chapter, the focus has been on identifying the genes and proteins, and the protein–protein interactions, which contribute to the virus–host interactome. As we will see in the following chapters, even the most comprehensive lists of genes and proteins will be only part of the overall roster of factors involved in viral pathogenesis. Viruses also impact host metabolic processes and interact with host metabolites. In addition, players beyond the virus and host—the components of the host microbiome—contribute to infection outcome. In the following chapter, we take up the contribution of host genetics to virus–host interactions and see how components of the virus–host interactome may therefore differ between individuals, with sometimes dramatic consequences.

REFERENCES

Baril M, Es-Saad S, Chatel-Chaix L, et al. Genome-wide RNAi screen reveals a new role of a WNT/CTNNB1 signaling pathway as negative regulator of virus-induced innate immune responses. PLoS Pathog 2013;9:e1003416.

Barrangou R, Fremaux C, Deveau H, et al. CRISPR provides acquired resistance against viruses in prokaryotes. Science 2007;315:1709–12.

Brass AL, Huang IC, Benita Y, et al. The IFITM proteins mediate cellular resistance to influenza A H1N1 virus, West Nile virus, and dengue virus. Cell 2009;139:1243–54.

Friedel CC, Haas J. Virus-host interactomes and global models of virus-infected cells. Trends Microbiol 2011;19:501–8.

Hu CJ, Song G, Huang W, et al. Identification of new autoantigens for primary biliary cirrhosis using human proteome microarrays. Mol Cell Proteomics 2012;11:669–80.

Jackson AL, Linsley PS. Recognizing and avoiding siRNA off-target effects for target identification and therapeutic application. Nat Rev Drug Discov 2010;9:57–67.

Jäger S, Cimermancic P, Gulbahce N, et al. Global landscape of HIV-human protein complexes. Nature 2011;481:365–70.

Joung JK, Ramm EI, Pabo CO. A bacterial two-hybrid selection system for studying protein-DNA and protein-protein interactions. Proc Natl Acad Sci USA 2000;97:7382–7.

Juranic Lisnic V, Babic Cac M, Lisnic B, et al. Dual analysis of the murine cytomegalovirus and host cell transcriptomes reveal new aspects of the virus-host cell interface. PLoS Pathog 2013;9:e1003611.

Kanasty R, Dorkin JR, Vegas A, et al. Delivery materials for siRNA therapeutics. Nat Mater 2013;12:967–77.

Karlas A, Machuy N, Shin Y, et al. Genome-wide RNAi screen identifies human host factors crucial for influenza virus replication. Nature 2010;463:818–22.

Koike-Yusa H, Li Y, Tan EP, et al. Genome-wide recessive genetic screening in mammalian cells with a lentiviral CRISPR-guide RNA library. Nat Biotechnol 2014;32:267–73.

Konig R, Stertz S, Zhou Y, et al. Human host factors required for influenza virus replication. Nature 2010;463:813–7.

Lee S, Salwinski L, Zhang C, et al. An integrated approach to elucidate the intra-viral and viral-cellular protein interaction networks of a gamma-herpesvirus. PLoS Pathog 2011;7:e1002297.

Li R, Zhu J, Xie Z, et al. Conserved herpesvirus kinases target the DNA damage response pathway and TIP60 histone acetyltransferase to promote virus replication. Cell Host Microbe 2011;10:390–400.

Masseroli M, Mons B, Bongcam-Rudloff E, et al. Integrated bio-search: challenges and trends for the integration, search and comprehensive processing of biological information. BMC Bioinformatics 2014;15(Suppl. 1):S2.

Panda D, Cherry S. Cell-based genomic screening: elucidating virus-host interactions. Curr Opin Virol 2012;2:784–92.

Peng X, Gralinski L, Ferris MT, et al. Integrative deep sequencing of the mouse lung transcriptome reveals differential expression of diverse classes of small RNAs in response to respiratory virus infection. mBio 2011;2:00198–211.

Schoggins JW, Wilson SJ, Panis M, et al. A diverse range of gene products are effectors of the type I interferon antiviral response. Nature 2011;472:481–5.

Shapira SD, Gat-Viks I, Shum BO, et al. A physical and regulatory map of host-influenza interactions reveals pathways in H1N1 infection. Cell 2009;139:1255–67.

Stertz S, Shaw ML. Uncovering the global host cell requirements for influenza virus replication via RNAi screening. Microbes Infect 2011;13:516–25.

Varble A, Benitez AA, Schmid S, et al. An in vivo RNAi screening approach to identify host determinants of virus replication. Cell Host Microbe 2013;14:346–56.

Yang D, Buchholz F, Huang Z, et al. Short RNA duplexes produced by hydrolysis with *Escherichia coli* RNase III mediate effective RNA interference in mammalian cells. Proc Natl Acad Sci USA 2002;99:9942–7.

Zender L, Xue W, Zuber J, et al. An oncogenomics-based in vivo RNAi screen identifies tumor suppressors in liver cancer. Cell 2008;135:852–64.

Zhou Y, Zhu S, Cai C, et al. High-throughput screening of a CRISPR/Cas9 library for functional genomics in human cells. Nature 2014;509:487–91.

Chapter 13

Host Genetics

It Is Not Just the Virus, Stupid

Martin T. Ferris[1], Mark T. Heise[1,2], Ralph S. Baric[2]

[1]*Department of Genetics, University of North Carolina, Chapel Hill, NC, USA;* [2]*Department of Microbiology and Immunology, University of North Carolina, Chapel Hill, NC, USA*

Chapter Outline

1. INTRODUCTION

Much of the focus of viral pathogenesis is on viral virulence factors; however, virulence can only be defined in the context of a virus–host combination. Furthermore, there is abundant evidence that highly pathogenic viruses have been selected for specific host genetic variants within the target population. The identification and mechanistic dissection of host genetic polymorphisms therefore has powerful implications for our understanding of viral diseases, including the identification of novel therapeutic targets. In this chapter, we review the classical approaches used to identify host allelic variations that regulate disease susceptibility. We also discuss next-generation molecular technologies and computational approaches that are poised to revolutionize our understanding of the impact of host genetic variation on viral diseases at the population level.

Viruses require efficient pathways for transcription, translation, assembly, and release, while simultaneously avoiding detection and clearance by the host innate immune system. As discussed in Chapter 12 (The Virus–Host Interactome), host proteins that promote the survival or replication of a virus are referred to as host factors. With the advent of high-throughput screening technologies, it has become feasible to identify in a single experiment large numbers of host factors that are required to promote viral replication. In addition, high-throughput screens are used to identify constitutive or inducible cellular factors that exhibit antiviral activity (referred to as restriction factors). These high-throughput screens often use genetically deficient yeast or siRNA knockdown technologies. Although in vitro screens allow for the rapid identification of candidate host or restriction factors, an understanding of the role of these factors in viral pathogenesis requires in vivo studies.

The use of mice carrying genetically defined deletions of specific genes (gene knockouts) has identified a number of host genes that play a critical role in limiting viral disease. Many of these genes are related to immune responses, such as those involved within the complement pathway or in the development of adaptive immune responses. However, many other host genes influence intrinsic biological activities such as transcription, translation, and intracellular transport. Removal of many of these genes has divergent effects that are dependent upon the virus. A given host gene may play a protective, pathologic, or neutral role during infection, and this will vary between viruses. Individual viruses co-opt divergent host mechanisms, and the specific host response can have radically different results depending upon that interaction. Such approaches have revealed the complex interactions between viral pathogens and the host immune system and highlight the need to study viral infections through integrated systems.

The use of classical genetic approaches has also provided significant insight into the role that specific host genes or pathways play in modulating disease severity. Much of this progress has come from studies comparing mouse strains with differing susceptibility to specific viral pathogens and the subsequent mapping of the

Viral Pathogenesis. http://dx.doi.org/10.1016/B978-0-12-800964-2.00013-6

polymorphic genes responsible for strain variation. For example, studies in the 1920s identified mouse strains that differed in their susceptibility to infection by a number of flaviviruses, including West Nile virus and Japanese encephalitis virus, revealing that differential disease responses are driven by a single autosomal-dominant genetic locus. Follow-up studies using positional cloning showed that animals carrying a functional allele of the *OAS1b* gene are highly resistant to flavirvirus replication (Brinton and Perelygin, 2003). Similar studies identified mouse genes associated with resistance to influenza virus (*Mx1*) (Staeheli et al., 1988) and murine cytomegalovirus (*Ly49H*) (Lee et al., 2001).

Such Mendelian traits are also prevalent in humans and reveal an interesting complexity in disease responses. For example, although persons who are homozygous for defective alleles of the chemokine receptor *CCR5* (*CCR5Δ32*) are highly resistant to HIV infection (Huang et al., 1996), these individuals may be at increased risk for severe neurologic infection with West Nile virus (Glass et al., 2006). Similarly, the ABH fucosyltransferase 2 (*FUT2*) allele contributes to resistance against norovirus infection, whereas other alleles of FUT2 encode a necessary receptor for these viruses (Lindesmith et al., 2003). In fact, virus evolution to altered forms of *FUT2*-regulated blood group carbohydrate receptor ligands (e.g., A, B, H) results in new pandemic outbreak strains that target distinct human subpopulations depending on the type of *FUT2* allele (Lindesmith et al., 2008). These findings highlight the importance of studying polymorphisms that contribute to specific viral responses.

2. COMPLEX AND POLYMORPHIC GENETIC INTERACTIONS

Although there are cases where viral disease outcomes are governed largely as a Mendelian trait, the majority of viral disease traits are much more complex. That is, multiple genetic, environmental, and demographic effects interact to determine an individual's disease response. Furthermore, many of the genetic variants contributing to trait differences are due to polymorphisms that have more moderate effects rather than the extreme phenotypes typically seen with gene-knockout approaches (Figure 1). Within a polymorphic population, there can be multiple genetic variants affecting the expression and function of a single gene. In this way, three different alleles of the translation factor *Eif4E* all confer bean plants with some level of resistance to clover yellow vein virus. Importantly, each of these alleles is effective against a different subset of virus strains, showing the complexity of disease responses that can be driven by multiple segregating natural polymorphisms (Hart and Griffiths, 2013).

A statistically complex framework, developed from elegant agricultural experiments over 80 years ago, posited that even if traits are not under strict Mendelian control, those under some form of genetic control must be heritable. That is, individuals who have some resistance to a viral pathogen tend to have offspring that are also somewhat resistant to the same viruses. Assessment of such heritability was initially used to measure the overall impact of host genetic variants on phenotypic traits, then for the selection of favorable traits (more below), and eventually as the basis of genetic mapping approaches.

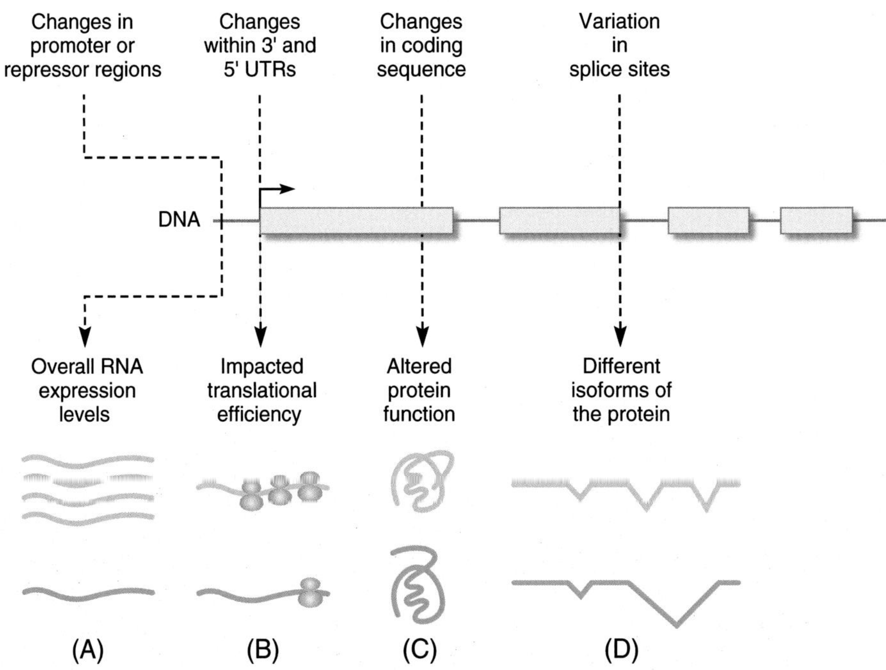

FIGURE 1 Effects of natural genetic variation. In contrast to gene knockouts, natural polymorphisms can impact genes in a variety of ways. These include (A) changes in promoter or repressor regions that impact overall RNA expression levels; (B) changes within 3′ and 5′ UTRs that impact translational efficiency; (C) changes in coding sequence that alter protein function, and (D) variation in splice sites that create different isoforms of the protein. Importantly, multiple changes can segregate within the same gene in a population, driving multiple expression and functional differences in disease responses.

These types of analyses were enthusiastically applied within the agricultural community and are still heavily used today. Controlled agricultural breeding allows for the explicit choice of breeding pairs, such that the next generation of animals will be enriched for a desired phenotypic trait. By iterating this process over generations, eventually the breeders fix large numbers of genetic variants within this population, and all animals have the advantageous traits of interest. Indeed, this is how milk and meat production, as well as crop yields, are regularly maximized. Also, it was quickly realized that these methods could be applied to disease resistance, such as the breeding of resistance to potato viruses. Furthermore, with the development of a variety of molecular markers, the classical need to use phenotypic markers to aid in the selection of resistant breeders has been reduced, allowing for more robust and rapid approaches. However, such approaches had only limited feasibility within human populations (and were used solely in the assessment of heritability) until mapping approaches were developed.

3. BIOMEDICAL ADVANCES IN VIRAL DISEASES

A key insight from genetic studies is that virus–host interactions and outcomes vary depending upon the virus. For example, the complement pathway is a complex pathway involved in the control of pathogens through several recognition and response arms. Depending on the recognition or response arms activated, differential disease outcomes can result (Stoermer and Morrison, 2011). As an example, the mannose binding lectin (MBL) complex, which binds sugar moieties on pathogens to activate the complement cascade, exhibits a direct antiviral activity against flaviviruses in part by blocking viral fusion to cells. In contrast, MBL not only fails to protect against Ross River virus (RRV) infection, but MBL deposition within RRV-infected joint and muscle tissue promotes complement activation and subsequent inflammatory tissue destruction (Gunn et al., 2012). Since the MBL genes are highly polymorphic in humans, and MBL levels correlate with RRV-induced disease severity, genetic variation in human MBL genes may be associated with susceptibility to RRV-induced disease. These types of studies also have important implications for the development of antiviral therapies targeting the complement cascade. Although therapies designed to block complement activation might be beneficial in the case of RRV, this type of intervention might have adverse consequences for individuals infected with West Nile virus or dengue virus.

The development of technologies to create gene-specific knockouts, especially in mice, dramatically increased the ability to determine how specific host genes affect viral pathogenesis. Studies using knockout mice have contributed to the identification and characterization of a wide array of host genes that play either protective or pathologic roles during viral infection (Table 1). As noted above, components of the host complement cascade play a protective role during West Nile virus infection, but exacerbate RRV-induced arthritis. Similarly, a deficiency in the chemokine receptor *CCR2* reduces early stage lung pathology during influenza virus infection, but exacerbates disease following Chikungunya virus infection by altering the inflammatory infiltrate and resultant arthritic disease. Importantly, in addition to providing insights into the role that a specific molecule plays in the pathogenesis of specific viral diseases, these types of studies have revealed important principles about the role of host genes in viral pathogenesis (Sidebar 1).

Although gene-specific knockouts are a powerful tool for investigating the role of specific genes or host pathways in viral pathogenesis, targeted knockout mice are not generally used as a screening platform for identifying novel genes that regulate the response to viral infection. Many of the advances in innate and intracellular responses to viruses have been derived either from high-throughput in vitro screens, or manipulation of the more tractable *Caenorhabditis elegans* or *Drosophila* systems. For example, such studies identified the highly evolutionarily conserved Toll-like receptor (TLR) pathway. However, it is possible to conduct unbiased screens using mice, such as by

TABLE 1 Gene Removal and Discordant Pathologic Responses

Gene	Protective Role	Pathologic Role
MBL (*MBL-1*/ *MBL-2*)	West nile virus (Fuchs et al., 2010)	Ross river virus (Gunn et al., 2012)
C3	West nile virus (Mehlhop et al., 2005), influenza virus (Kopf et al., 2002)	Ross river virus (Morrison et al., 2008)
CCR5	West nile virus (Glass et al., 2006), influenza virus (Dawson et al., 2000)	Vaccinia virus (Rahbar et al., 2009)
Ifn-γ	Mouse hepatitis virus (Pewe and Perlman, 2002)	Herpesvirus (Schijns et al., 1994)
CCR2	Chikungunya virus (Poo et al., 2014)	Influenza A virus (Dawson et al., 2000)

Sidebar 1 Key findings from genetic manipulation studies

1. A large number of immune-related genes influence viral pathogenesis either by providing protection or enhancing pathology (Table 1). In addition, many host genes with no obvious immune-related functions can also impact viral disease. These can include viral receptors, such as the Sindbis virus receptor, NRAMP, but also host genes with effects at other stages of viral infection, such as transcription and translation, as with the translation factor Eif4E and its role in clover yellow vein virus pathogenesis (Hart and Griffiths, 2013).
2. Genes within host response pathways are often redundant. Many host immune pathways, such as the complement pathway (Stoermer and Morrison, 2011) or the TLR pathway (Hoebe, 2009), have multiple-sensing molecules, as well as a conserved set of downstream signaling molecules and response pathways. In this way, a specific virus might interact with multiple sensors within the same pathway. Thus, dissection of such pathways can often require complex experiments and the generation of double (or even triple) knockouts in the same animals
3. Many genes play different roles when interacting with different pathogens. Just as there can be redundant-sensing pathways for some pathogens, the specific molecules that a virus interacts with can modulate disease progression. In this way, abrogation of a specific pathway member can either enhance or ameliorate disease (Table 1). In more extreme examples, such as the role of *Hc* in influenza virus infection, a molecule can be protective for one virus, but have no role in the pathogenesis of a closely related virus. Such results illustrate the difficulty in understanding the general role of immune pathways when describing viral immunity and pathogenesis.

introducing mutations into the mouse genome through the use of *N*-ethyl-*N*-nitrosourea (ENU). ENU is an alkylating agent that induces point mutations in sperm progenitor cells at a predicable frequency. When male mice are treated with ENU and then bred to wild-type female mice (G_1 mice), their offspring carry a subset of the mutations derived from ENU mutagenesis. Through subsequent rounds of breeding, ENU-induced mutations affecting immune phenotypes that drive differential viral disease responses can be identified (Figure 2).

Historically, the identification of causal ENU-induced mutations has required a time- and labor-intensive mapping effort. However, with the advent of high-resolution whole-exome or genome sequencing, the efficiency of identifying causal ENU mutations has been significantly enhanced, thereby increasing the ability of this approach to identify genes impacting specific phenotypes. The power of this system is illustrated by the identification of a large number of innate immune genes that modulate the host response to herpesvirus infection, as well as the identification of several nonobvious host factors that regulate homeostasis and protect against viral disease. For example, ENU mutagenesis initially identified *UNC93B1* as a mediator of herpesvirus resistance in mice that is essential for signaling via multiple TLRs. Subsequent studies also demonstrated that UNC-93B deficiency in humans is associated with enhanced susceptibility to herpesvirus infection. Multiple ENU-induced mutations within the same gene can have strikingly different effects on phenotype, such as variant mutations affecting TLR signaling (Hoebe, 2009).

These types of studies, often termed reductionist, have provided new avenues for combating viral diseases, and have uncovered a variety of complex mechanisms and pathways involved in host immune responses and cellular function. Since the majority of these studies have been conducted within only one or a few genetic backgrounds, more complex genetic systems are needed to fully disentangle host responses to viral pathogens.

4. GENETIC MAPPING TO IDENTIFY VARIANTS AFFECTING COMPLEX TRAITS

Regardless of the number and type of markers used, genetic mapping follows the same principles (Figure 3). Individuals are typed at a series of markers (whose positions relative to each other are known), and these individuals are also phenotyped for traits of interest. Whether these are closely related populations (e.g., a large family pedigree of humans or an F2 cross between two inbred strains of animals) or larger population-level association studies, all such studies rely on recombination within gametogenesis to break apart genome structure. As a result, only markers physically close to the polymorphic loci of interest will remain associated or linked throughout the study population. Therefore, at each marker, the significance of the association between the trait values and the two alleles at that marker is determined. When polymorphisms controlling a trait are unlinked to a marker (e.g., they are on different chromosomes), then there will be completely random association between the two variants at the marker and the trait of interest. However, when tested markers exist on the same chromosome as a polymorphism controlling a trait, there will be an increased association of one of the two marker variants and an increase in the trait of interest. Furthermore, for the markers near the trait-controlling polymorphism, there will be a tighter and tighter association between the marker variant and an increase in the trait of interest. Indeed, with sufficient numbers of individuals, it is possible to identify polymorphisms driving subtle effects.

Genetic markers have undergone development from restriction fragment length polymorphisms or micro/minisatellite markers, to well-annotated single nucleotide

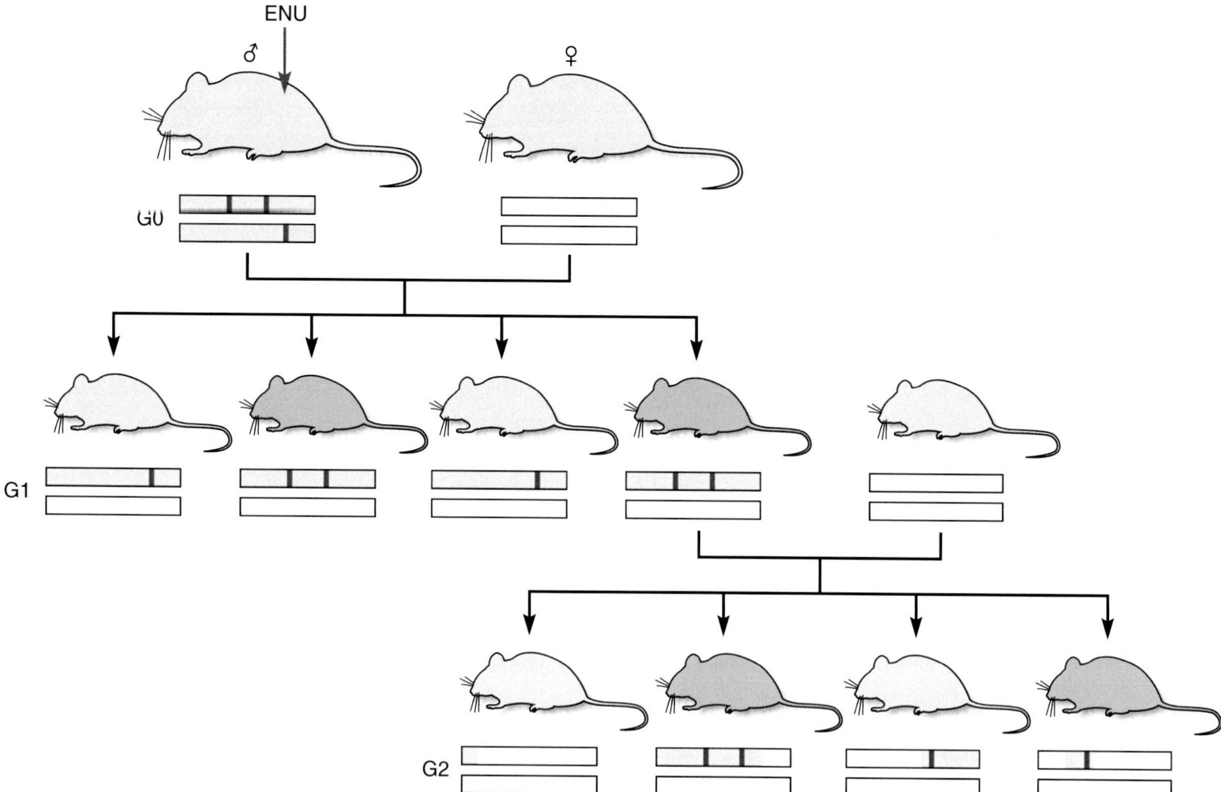

FIGURE 2 *N*-ethyl-*N*-nitrosourea (ENU) mutagenesis and differential disease responses. Following ENU mutagenesis of the testes of a male mouse, a large number of mutations are accumulated within the sperm of the animal (red bars on black chromosomes). This male is then bred to a female of a closely related strain (denoted by white chromosomes). G1 offspring of this mating receive a variety of mutations from the ENU-mutagenized sperm, some of which can confer novel phenotypes, such as susceptibility to a viral pathogen. Here, purple animals have acquired a common mutation showing enhanced pathology following viral infection, as compared to wild-type animals. Depending on the mode of action of the mutation (e.g., dominant in this case), these phenotypic variants can be seen in the G1 generation, or in later generations through backcrossing. Furthermore, by conducting further crosses, such as to a wild-type female mouse, a new mapping population (G2) can be used to identify the underlying causal mutations driving these phenotypes.

polymorphisms (SNPs) or even haplotype reconstructions (a set of tightly clustered heritable polymorphisms). Using current methods, it is possible to more finely identify both the chromosomal locations and the relative linkages between genetic factors. Careful analysis is needed to differentiate multiple genetic variants affecting many traits in close proximity to each other from pleiotropy (a single variant affecting multiple traits). However, analyses of host genetic contributions at specific loci suggest that there are often multiple variants close together that commonly affect a given trait, or set of related traits.

For example, many early mapping studies showed that the major histocompatibility complex (MHC) locus affects a wide variety of immune responses including complement function, B-cell and T-cell responses, and allergic responses. This genome region contains a large number of genes involved in antigen presentation, the complement cascade, innate immune responses, and development all in close physical proximity. Furthermore, as the sequence and annotation of mammalian genomes has progressed, other loci with tightly linked genes of related function have been identified. Such tightly linked genes hint at co-evolutionary pressures, but more practically they create difficulty in identifying specific polymorphic genes influencing phenotypes.

The development of genome-wide association studies (GWAS) is based on the assumption that historical recombination events within a population have broken apart all but the tightest associations. (This is in contrast to linkage analysis, which seeks to maintain long-term linkages between markers and phenotypes.) Therefore, only associations with causative (or intimately proximal polymorphisms) will be identified. The GWAS approach is something of a double-edged sword. In order to identify these very tight associations, many markers are necessary. This necessitates incredibly large cohorts of individuals with which to attain statistical significance. However, when associations do exist within these cohorts, those associations can detect incredibly small effects (those influencing disease phenotypes by <1% of the overall trait variation). Indeed, the GWAS literature shows that large numbers of identified SNPs have small-to-moderate effects on disease phenotypes and traits.

GWAS requirements, such as large numbers of individuals and the often subtle nature of SNPs on phenotypic traits, have limited the ability to investigate the role of human

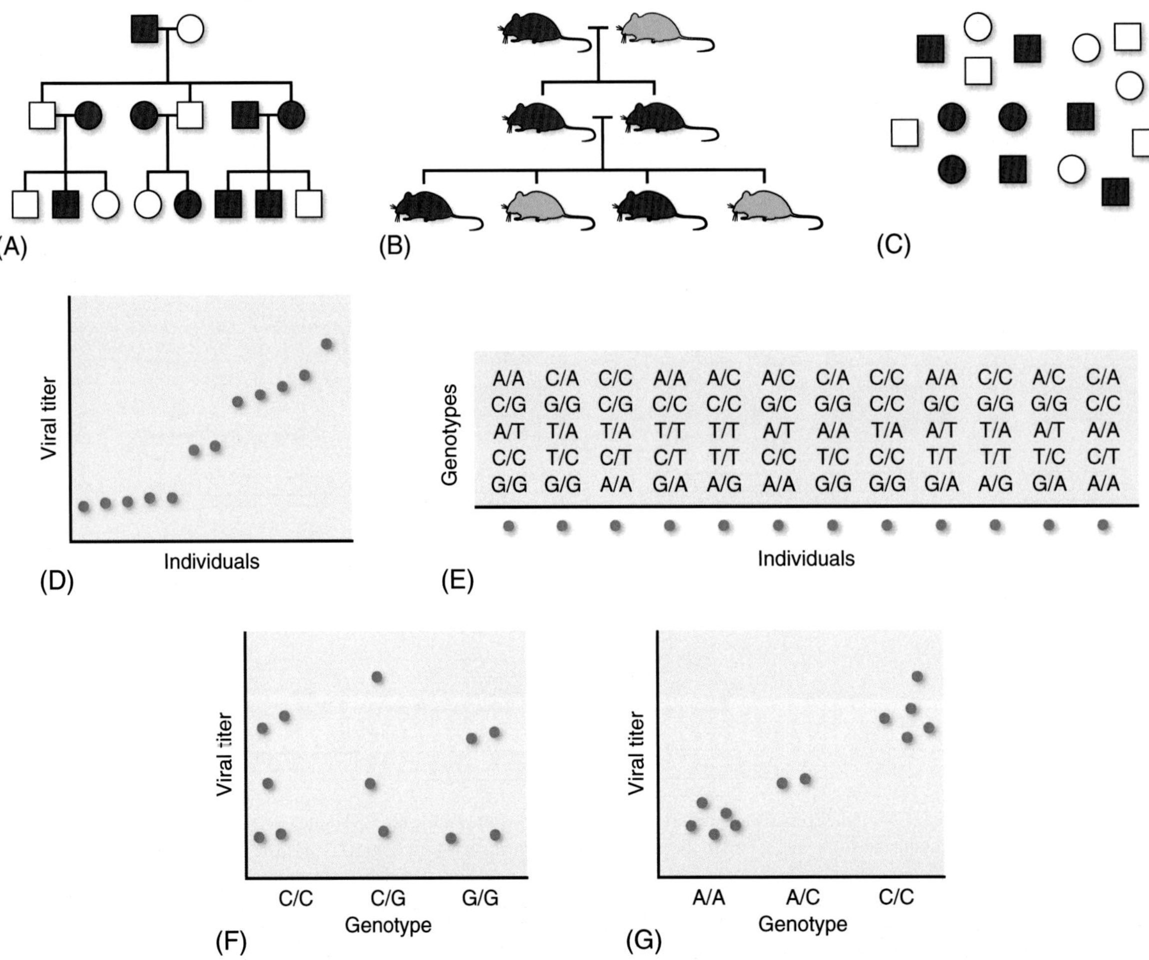

FIGURE 3 Genetic mapping involves the phenotyping and genotyping of a related population. (A) Family pedigrees or (B) crosses of experimental animals are used for linkage analysis. (C) Large cohorts of unrelated individuals are used for genome-wide association studies. Individuals are then (D) phenotyped and (E) genotyped at a number of genomic markers. Genetic mapping assesses the statistical strength of association between phenotypes and genotypes to distinguish (F) nonsignificant relationships from (G) those markers linked to genetic variants causing the phenotypic differences.

polymorphisms within the context of infection. Confounding factors—such as variation in viral doses and exposures, the effect of viral genetic variants on disease, the often narrow windows of symptomatic infection, and the numbers of individuals infected during an outbreak—can cloud subtle host genetic contributions to differential disease outcomes. Despite such limitations, for chronic viral infections, such as those caused by HIV or hepatitis C virus, GWAS studies have proven useful for identifying host genes that are associated with variation in viral control, disease progression, or treatment responses.

Perhaps most illustrative of the power of these approaches for chronic infections are studies of viral load and disease progression following HIV infection (see also Chapter 9, HIV and AIDS). GWAS studies of the response to HIV have identified a number of genes within the MHC that are associated with control of viral load. These include the HLA complex 5 gene (*HCP5*) and *HLA-C*, where a SNP in the 3′ UTR regulates *HLA-C* expression through microRNA interactions. Several GWAS studies have identified genes associated with long-term nonprogression or rapid progression to HIV disease, including *HCP5*, the *CXCR6* chemokine receptors, *PROX1* (a regulator of T cell IFN-γ production), and the SMAD family interacting protein PARD3. These studies provide new insights into the potential role of specific polymorphic genes in HIV pathogenesis, and they illustrate that HIV disease involves the complex interaction of multiple polymorphic genes that contribute to variation in disease progression (van Manen, van 't Wout, and Schuitemaker, 2012). In the case of hepatitis C virus, GWAS studies have identified a polymorphism upstream of *Il28B* that influences spontaneous virus clearance (Thomas et al., 2009). This SNP is also associated with improved responses to certain antiviral treatments.

Given the need to have extremely large cohorts of individuals at similar disease stages in order to study viral

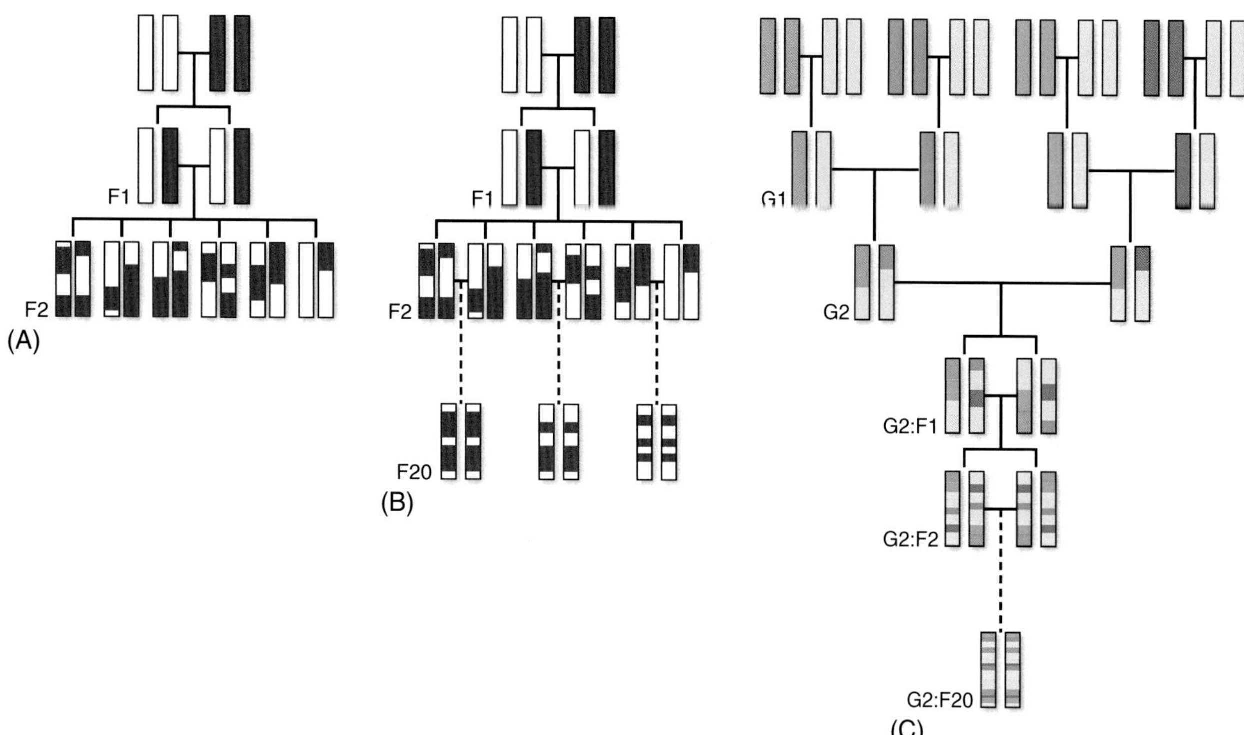

FIGURE 4 Recombinant Inbred models of genetic diversity. (A) Classical mapping populations, such as F2s, involve nonpermanent genotypes of individuals. By crossing F2 animals, followed by inbreeding (B) classical recombinant inbred panels can be generated. These panels are sets of lines all derived from common ancestors, but containing unique haplotype combinations of these founder lines, and allow for the assessment of genetic effects across time points, treatments, and conditions of viral infections. In order to better model the genetic diversity and complexity present within the human population, the Collaborative Cross panel (C), derived from eight classical inbred founder strains, was bred from a complex funnel design to incorporate genetic information from each founder line within each CC line. Following this funnel breeding, a large number of unique lines are generated and can be used for experimental analysis and genetic mapping of complex traits.

diseases on genome-wide scales, more targeted approaches have been used to carefully probe candidate genes, which are thought to be important in specific viral diseases. For example, initial candidate studies of the signaling lectin *L-SIGN/CLEC4M* identified a polymorphism associated with SARS-coronavirus resistance, and these findings appeared robust in an in vitro model (Chan et al., 2006). However, further analysis of patient cohorts failed to replicate this association, a cautionary note in interpreting genetic studies (Zhi et al., 2007).

5. RECOMBINANT INBRED PANELS

Animal models are critical for understanding the interaction between human pathogens and the host immune system (see Chapter 10, Animal Models). Mice offer the most tractable experimental system, in part because of well-developed molecular and genetic methods. Inbred mouse strains can be used to map genome locations contributing to differential disease responses between inbred strains. For certain large-effect genes, it is possible to identify causal genes contributing to disease differences in these populations. Genes such as *Mx1*, *Oas1b*, *CEACAM1a*, and *Cmv-1* (*Ly49H*) have been identified as major players in specific viral diseases. In addition, inbred mouse strains offer the ability to perform reproducible studies designed to elucidate the mechanisms by which a specific gene impacts disease. Researchers have designed recombinant inbred panels to extend studies using animals with different genetic variants.

Classical recombinant inbred (RI) panels (e.g., the C57BL/6J by DBA/2J (BxD) and A/J by C57BL/6J (AxB and BxA)) were derived from pairs of inbred lines. The generation of RI panels is initially identical to the generation of F2 animals for genetic mapping. However, instead of testing these F2 animals for viral disease, pairs of F2 animals are mated with a sibling, and the resultant animals undergo brother–sister mating until a set of new inbred lines are derived from this initial cross (Figure 4). Each new inbred line in this panel has an inbred genome composed of different segments from each of the two founder inbred lines. Furthermore, since each new inbred line is generated from two unique F2 animals, each one of the new RI lines has private recombination events. In this way, a panel of recombinant inbred lines provides a large number of inbred and reproducible mouse lines that are all related to each other. Each set of lines shares a subset of polymorphisms they received in common from the two founder lines, while at other loci they have different polymorphisms. Therefore, a whole RI panel can be used to conduct genetic mapping, or individual lines within a panel can be used to study specific aspects of viral disease.

A major advantage of RI panels is the ability to compare or contrast treatments across time points or viral infections. For example, use of the BxD panel to compare host responses to highly pathogenic H5N1 influenza virus and a less-pathogenic H1N1 influenza virus identified completely different polymorphic host loci controlling exacerbated disease. A major host genetic locus driving H5N1 susceptibility is a defective *Hc* (hemolytic complement) allele present within the DBA/2J strain. However, this defective *Hc* allele has no effect on disease responses following H1N1 infection.

Such comparative studies again highlight the complexity of infectious disease responses, relying on host genetic differences as well as virus–host genetic interactions. They also show the importance of understanding both the basal (uninfected) status of the immune system as well as how different viruses interact with specific variant host pathways to influence disease outcome. This was shown in a study of the whole-genome transcriptional responses of C57BL/6J and DBA/2J lines before and after H5N1 infection (Boon et al., 2009). Whereas many transcripts are commonly regulated in response to infection between these two lines, over one-third of transcripts have baseline differential expression between the two strains. Furthermore, approximately half of the genes are underneath quantitative trait loci (QTL) driving differential H5N1 responses. Therefore, high-priority candidates are primarily differentiated by basal expression differences versus differences in induction between strains. Such analyses significantly improve the identification of "core" immune or cellular responses that respond to viruses across genetic backgrounds. For instance, there are over 1800 transcripts that are commonly regulated following H5N1 infection of C57BL/6J and DBA/2J mice. The data also identify those transcriptional responses that are specific to individual RI lines (Figure 5).

RI panels and GWAS studies have demonstrated that genetic complexity, driven by multiple genetic variants, is a hallmark of many disease response phenotypes. In addition, the complex gene–gene interactions that exist between polymorphic loci can lead to radically different disease responses. GWAS studies have led to the recognition that complex gene–gene interactions (e.g., epistasis and dominance) contribute much of the genetic control of complex diseases. New methodologies, which require great increases in sample sizes and statistical power, are therefore needed to identify these interactions.

As a result, there is an increased interest in developing next-generation genetic reference populations that capture many of these attributes. Although such panels have been used in work on *Arabidopsis* and *Drosophila*, they have only recently been attempted for a mammalian model to study human biomedical traits. Specifically, the first such mammalian genetic reference population that incorporates more genetic complexity than traditional RI panels is the Collaborative Cross (CC) recombinant inbred panel (Collaborative Cross Consortium, 2012). This panel was derived from eight inbred mouse strains (A/J, C57BL/6J, 129S1/SvImJ, NOD/ShILtJ, NZO/HILtJ, CAST/EiJ, PWK/PhJ, and WSB/EiJ), which together contain high levels of genetic variants (approximately 40 million SNPs and four million small insertions/deletions, similar to levels of common human genetic variants), uniformly distributed across the genome. By including the wild-derived, inbred strains CAST/EiJ, PWK/PhJ, and WSB/EiJ, the CC contains input from the three major *Mus musculus* subspecies: *castaneus*, *musculus*, and *domesticus*; and these three strains contribute the majority of the genetic polymorphisms within the cross. Furthermore, as there are eight equal haplotype contributions at each locus, the minor allele frequencies are at roughly 12.5% (as opposed to under 1% in human populations), improving the ability to detect the role of rare variants and epistatic interactions on disease outcomes.

By breeding animals from these eight inbred strains together in a funnel design, and then inbreeding the resultant offspring together, new inbred lines are created. Due to the recombination events that occur throughout the funnel breeding and inbreeding processes, the genome of each new RI line is a mosaic of the eight founder strains. By repeating this process over and over, while shuffling the positions of the eight founder strains at the top of the funnel (Figure 4), multiple independent CC lines have been created.

A related population, Diversity Outbred mice, was started from incipient CC animals. Instead of defined inbred lines, this population is maintained fully outbred. Although not a classical genetic reference population (in that specific genotypes cannot be maintained and studied across treatments), Diversity Outbred mice can be used to study specific allele frequencies at a population level across different treatments. Furthermore, Diversity Outbred mice provide a useful experimental model to mimic outbred populations, such as humans.

High-resolution genotyping of CC and Diversity Outbred animals make it possible to accurately describe the genomic composition of the each population. These genetic details have two main advantages over other approaches. First, instead of genetic mapping based on individual SNPs, mapping can be conducted on founder haplotypes. In this way, false-positive associations between genome regions and phenotypes are reduced, true associations increase in significance, and it is possible to rapidly narrow QTL regions down to a smaller number of likely candidate genes

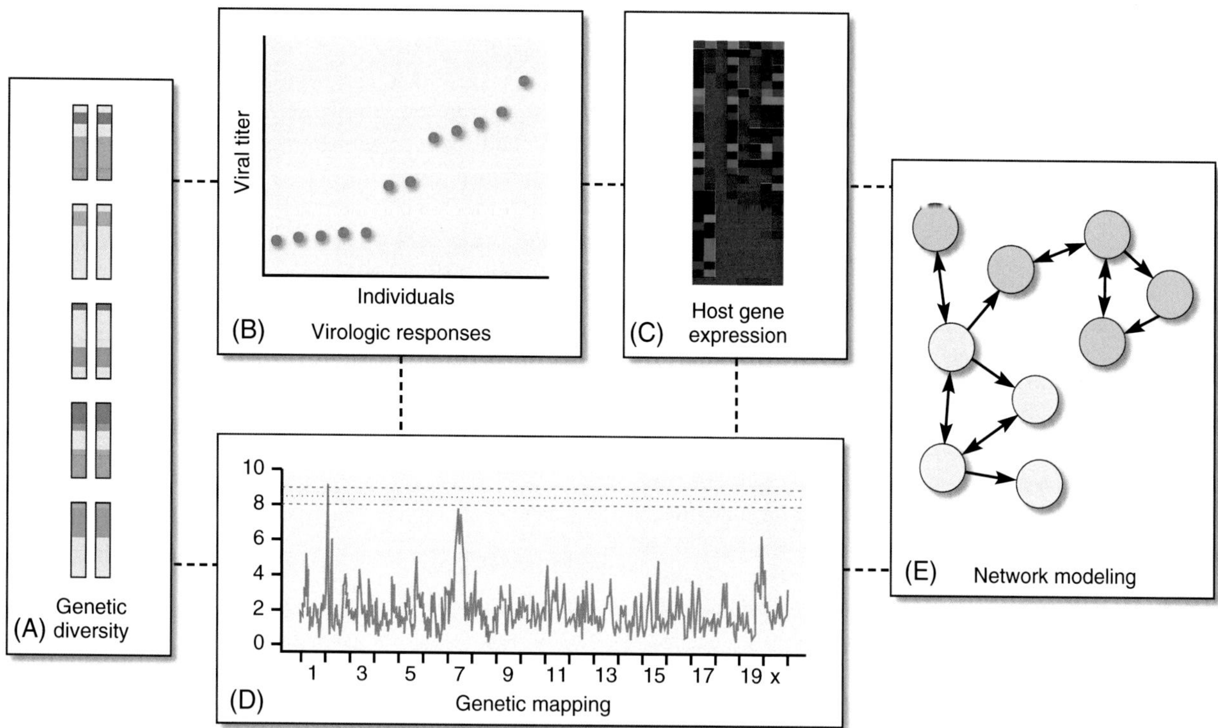

FIGURE 5 Systems genetics incorporates genetic complexity to explain differential disease responses. An advantage of genetically tractable, yet complex experimental systems such as RI panels, is the ability to explicitly integrate host genetic (A), virologic (B), and transcriptional responses (C) in order to identify polymorphic genetic loci (D) that contribute to differential virologic responses, and to develop transcriptional networks (E) that shed mechanistic insight into these polymorphic responses.

and features. In a study of H1N1 influenza virus infection of a progenitor population of CC animals, these approaches identified sets of high-priority candidate genes driving differential host responses, and showed that multiple antiviral *Mx1* alleles segregating within this population can differentiate antiviral and clinically protective disease responses (Ferris et al., 2013). Second, this genomic sequence and composition information is critical for correctly assessing microarray and RNA-seq data, avoiding misaligned sequences from RNA-seq, and correctly assessing microarray probes that are impacted by SNPs.

These genetic reference populations are powerful tools for identifying polymorphisms contributing to differential viral disease outcomes. Indeed, several studies have used these two populations to identify individual polymorphisms or polymorphic pathways that contribute to different immunological or disease responses (Phillippi et al., 2014). An additional use of such genetic reference populations is the detection of epistatic and other complex genetic interactions. The tractability of genetic reference populations can allow for direct assessment of these genetic interactions.

The shuffling of susceptibility alleles across the CC population has recently been used to identify a novel mouse model of Ebola virus hemorrhagic fever (Rasmussen et al., 2014). None of the founder lines of the CC, or other classical mouse strains, have provided an effective model of severe Ebola virus disease. However, several F1 crosses between CC lines develop a severe Ebola virus disease, including hepatic involvement, coagulation defects, and rapid mortality. These results further emphasize the utility of studying genetically variant models to understand disease processes. Furthermore, with the development of gene-editing approaches (discussed below), it is becoming possible to more fully study putative epistasis across a range of genetic backgrounds.

Elsewhere in this book, systems biology approaches to viral pathogenesis are described in detail (see Chapter 11, Systems Virology). These approaches integrate a variety of data types with whole-genome measures of transcription and translation to better understand the complex networks and pathways that are altered during infection, specifically targeting those that are causative for host-mediated protection or disease exacerbation. In addition to the genetic mapping approaches described above, genetic reference populations provide a novel and powerful overlay to systems biology approaches. Integrating variant disease pathologic, virologic, and transcriptional responses to virus infection, across a range of genetically distinct host backgrounds, allows for the direct assessment of the role of genetic variation on host response networks (Figure 5).

6. REPRISE

Throughout this chapter, we have described the approaches that have provided much of our understanding of host genetic determinants of viral disease. These studies have shown the need for sophisticated approaches to identify, manipulate, and characterize genetic variants contributing to complex responses. It is clear that many polymorphisms underlying complex diseases are pleiotropic (a single polymorphism impacts multiple aspects of the disease response), epistatic (some responses occur only when an individual has multiple interacting polymorphisms), and epigenetic (polymorphisms that either influence, or are activated or repressed by, genome modifications such as methylation).

Gene deletion approaches have allowed for the dissection of pleiotropic gene responses (e.g., many gene knockouts ameliorate weight loss and inflammation, but have no effect on viral titers; others have an effect on viral titers, but not on weight loss or inflammatory responses). However, it is only with the advent of the large next-generation genetic reference populations that epistatic interactions can reliably be detected. Such systems allow for the induction, characterization, and manipulation of epigenetic states, clarifying the role and mechanisms of epigenetic responses in complex viral diseases.

Molecular and experimental advances have provided additional approaches. One is the ability to use tissue-specific promoters to generate constitutive or tissue-specific gene knockouts. Inducible systems are often able to rescue embryonic lethal gene knockouts. Another is the ability to directly modify the genome via CRISPR/Cas9 or Talens gene-editing technologies, which enable the introduction of a single mutation, or combination of mutations, into a host genome (Niu et al., 2014; Yang et al., 2013). The influence of specific polymorphisms on disease responses can then be assessed across a variety of genetic backgrounds. By engineering mutations into specific backgrounds, or engineering all epistatic mutations into a naïve background, rare epistatic phenotypes can be more easily validated and their molecular underpinnings investigated.

Although there is concern that mouse models may not accurately recapitulate human disease, some notable cases have validated the power of using mouse models. For example, a polymorphism in human *Ifitm3* was identified through the sequencing of patients hospitalized for influenza infection, and a splice variant encoding a truncated Ifitm3 protein is enriched within these hospitalized individuals (Everitt et al., 2012). The anti-influenza effects of *Ifitm3* were then recapitulated in cell culture and with a knockout mouse model. That said, it remains unclear how often mouse knockouts will successfully translate to human polymorphisms. Furthermore, since epigenetic modifications can also impact the host response to environmental stimuli, it will be important to assess whether these types of effects can be accurately modeled in mice or other experimental systems.

Genetic methods have significantly contributed to our understanding of the influence of host genes on disease outcomes. It is clear that host-pathogen coevolution and selection has altered the population genetic structure of most species, and there is abundant evidence that host genetic variation has a major impact on viral pathogenesis. Much of the progress in this area has come through the use of gene-specific knockouts or forward genetic screens. Recent advances in systems genetics, including QTL analysis in humans, whole-genome sequencing, genetically complex mouse genetic reference populations, and genome-editing techniques, promise to significantly enhance our ability to study the polymorphic genes and pathways that influence disease outcome. These approaches are making a critical contribution to our understanding why different humans are either resistant or susceptible to specific viral infections.

REFERENCES

Boon AC, deBeauchamp J, Hollmann A, Luke J, Kotb M, Rowe S, et al. Host genetic variation affects resistance to infection with a highly pathogenic H5N1 influenza A virus in mice. J Virol October 2009;83(20):10417–26.

Brinton MA, Perelygin AA. Genetic resistance to flaviviruses. Adv Virus Res 2003;60:43–85.

Chan VS, Chan KY, Chen Y, Poon LL, Cheung AN, Zheng B, et al. Homozygous L-SIGN (CLEC4M) plays a protective role in SARS coronavirus infection. Nat Genet January 2006;38(1):38–46.

Collaborative Cross Consortium. The genome architecture of the collaborative cross mouse genetic reference population. Genetics February 2012;190(2):389–401.

Dawson TC, Beck MA, Kuziel WA, Henderson F, Maeda N. Contrasting effects of CCR5 and CCR2 deficiency in the pulmonary inflammatory response to influenza A virus. Am J Pathol June 2000;156(6):1951–9.

Everitt AR, Clare S, Pertel T, John SP, Wash RS, Smith SE, et al. IFITM3 restricts the morbidity and mortality associated with influenza. Nature March 25, 2012;484(7395):519–23.

Ferris MT, Aylor DL, Bottomly D, Whitmore AC, Aicher LD, Bell TA, et al. Modeling host genetic regulation of influenza pathogenesis in the collaborative cross. PLoS Pathog February 2013;9(2):e1003196.

Glass WG, McDermott DH, Lim JK, Lekhong S, Yu SF, Frank WA, et al. CCR5 deficiency increases risk of symptomatic West Nile virus infection. J Exp Med January 23, 2006;203(1):35–40.

Gunn BM, Morrison TE, Whitmore AC, Blevins LK, Hueston L, Fraser RJ, et al. Mannose binding lectin is required for alphavirus-induced arthritis/myositis. PLoS Pathog 2012;8(3):e1002586.

Hart JP, Griffiths PD. A series of eIF4E alleles at the bc-3 locus are associated with recessive resistance to clover yellow vein virus in common bean. Theor Appl Genet November 2013;126(11):2849–63.

Hoebe K. Genetic dissection of toll-like receptor signaling using ENU mutagenesis. Methods Mol Biol (Clifton, N.J.) 2009;517:239–51.

Huang Y, Paxton WA, Wolinsky SM, Neumann AU, Zhang L, He T, et al. The role of a mutant CCR5 allele in HIV-1 transmission and disease progression. Nat Med November 1996;2(11):1240–3.

Kopf M, Abel B, Gallimore A, Carroll M, Bachmann MF. Complement component C3 promotes T-cell priming and lung migration to control acute influenza virus infection. Nat Med April 2002;8(4):373–8.

Lee SH, Girard S, Macina D, Busa M, Zafer A, Belouchi A, et al. Susceptibility to mouse cytomegalovirus is associated with deletion of an activating natural killer cell receptor of the C-type lectin superfamily. Nat Genet May 2001;28(1):42–5.

Lindesmith L, Moe C, Marionneau S, Ruvoen N, Jiang X, Lindblad L, et al. Human susceptibility and resistance to norwalk virus infection. Nat Med May 2003;9(5):548–53.

Lindesmith LC, Donaldson EF, Lobue AD, Cannon JL, Zheng DP, Vinje J, et al. Mechanisms of GII.4 norovirus persistence in human populations. PLoS Med February 2008;5(2):e31.

van Manen D, van 't Wout AB, Schuitemaker H. Genome-wide association studies on HIV susceptibility, pathogenesis and pharmacogenomics. Retrovirology August 24, 2012;9(70):4690–9.

Mehlhop E, Whitby K, Oliphant T, Marri A, Engle M, Diamond MS. Complement activation is required for induction of a protective antibody response against West Nile virus infection. J Virol June 2005;79(12):7466–77.

Morrison TE, Simmons JD, Heise MT. Complement receptor 3 promotes severe ross river virus-induced disease. J Virol November 2008;82(22): 11263–72.

Niu Y, Shen B, Cui Y, et al. Generation of gene-modified cynomolgus monkey via Cas9/RNA-mediated gene targeting in one-cell embryos. Cell 2014;156:836–43.

Pewe L, Perlman S. Cutting edge: CD8 T cell-mediated demyelination is IFN-gamma dependent in mice infected with a neurotropic coronavirus. J Immunol (Baltimore, Md.: 1950) February 15, 2002;168(4):1547–51.

Phillippi J, Xie Y, Miller DR, Bell TA, Zhang Z, Lenarcic AB, et al. Using the emerging collaborative cross to probe the immune system. Genes Immun January 2014;15(1):38–46.

Poo YS, Nakaya H, Gardner J, Larcher T, Schroder WA, Le TT, et al. CCR2 deficiency promotes exacerbated chronic erosive neutrophil-dominated chikungunya virus arthritis. J Virol June 2014;88(12):6862–72.

Rahbar R, Murooka TT, Fish EN. Role for CCR5 in dissemination of vaccinia virus in vivo. J Virol March 2009;83(5):2226–36.

Rasmussen AL, Okumura A, Ferris MT, Green R, Feldmann F, Kelly SM, et al. Host genetic diversity enables ebola hemorrhagic fever pathogenesis and resistance. Science (New York, N.Y.) November 21, 2014;346(6212):987–91.

Schijns VE, Haagmans BL, Rijke EO, Huang S, Aguet M, Horzinek MC. IFN-gamma receptor-deficient mice generate antiviral Th1-characteristic cytokine profiles but altered antibody responses. J Immunol (Baltimore, Md.: 1950) September 1, 1994;153(5):2029–37.

Staeheli P, Grob R, Meier E, Sutcliffe JG, Haller O. Influenza virus-susceptible mice carry mx genes with a large deletion or a nonsense mutation. Mol Cell Biol October 1988;8(10):4518–23.

Stoermer KA, Morrison TE. Complement and viral pathogenesis. Virology March 15, 2011;411(2):362–73.

Thomas DL, Thio CL, Martin MP, Qi Y, Ge D, O'Huigin C, et al. Genetic variation in IL28B and spontaneous clearance of hepatitis C virus. Nature October 8, 2009;461(7265):798–801.

Yang H, Wang H, Shivalila CS, et al. One-step generation of mice carrying reporter and conditional alleles by CRISPR/Cas-mediated genome engineering. Cell 2013;154:1370–9.

Zhi L, Zhou G, Zhang H, Zhai Y, Yang H, Zhang F, et al. Lack of support for an association between CLEC4M homozygosity and protection against SARS coronavirus infection. Nat Genet June 2007;39(6): 692,4; author reply 694–6.

Chapter 14

Metabolomics and Lipidomics

Yet More Ways Your Health Is Influenced by Fat

Priscilla L. Yang
Department of Microbiology and Immunobiology, Harvard Medical School, Boston, MA, USA

Chapter Outline

1. INTRODUCTION

Cellular metabolism is comprised of the chemical reactions that occur in living cells. Broadly, these reactions can be divided into catabolic reactions that convert nutrients to energy and anabolic reactions that lead to the synthesis of larger biomolecules. The reactants and products of these chemical reactions are metabolites. The flow of genetic information from DNA (genome) to RNA (transcriptome) to protein (proteome) described by the Central Dogma leads to production of the metabolome. The composition of the metabolome is dynamic and reflects expression of the genome under specific conditions. Metabolomic changes induced by viral infection are the integration of virus-induced changes in both host gene expression and host protein function.

As early as the 1950s, Seymour S. Cohen advanced the idea of viral replication as a series of biochemical reactions whose reactants and products are amenable to dissection, writing: "Many of the most important biological questions have been rephrased as chemical problems. Questions now are being posed concerning the nature of the building blocks and the pathways of their biosynthesis. The time course of infection, duplication, and virus liberation is being dissected minute by minute in terms of the molecular transformations occurring in these systems." Cohen advocated that the tools of chemistry would enable understanding of viral processes at the molecular level. In the 60-odd years since Cohen first referred to this area of investigation as "chemical virology," advances in analytical methods have facilitated increasingly precise knowledge of the interaction of viruses with host metabolism.

The major classes of metabolites include amino acids, carbohydrates, nucleotides, lipids, coenzymes, and cofactors. These classes of compounds encompass an enormous diversity of molecular structures, physico-chemical properties, functions, and abundances. Due to the analytical challenges posed by this diversity, most studies require division of the metabolome into subsets of metabolites. The most common distinction is that between hydrophilic (polar) metabolites and hydrophobic

Viral Pathogenesis. http://dx.doi.org/10.1016/B978-0-12-800964-2.00014-8

TABLE 1 Classes and Functions of Lipids

Lipid Class	Examples	Function
Fatty Acids/Fatty Acyls		
Carboxylic acids and their derivatives	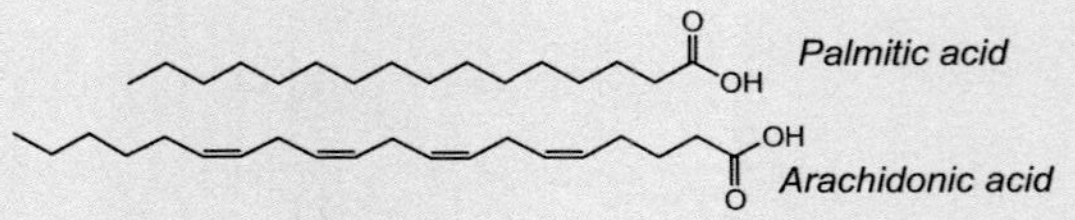 Palmitic acid	Energy storage
Hydrophobic tail varies in length and degrees of unsaturation	Arachidonic acid	Building block for other lipids
Glycerolipids		
Glycerol core (blue) with one, two, or three fatty acyl groups to form mono-, di-, and tri-acylglycerides (MAG, DAG, TAG)	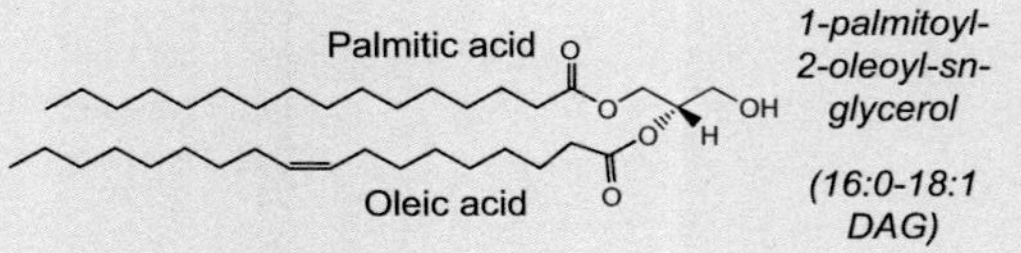 Palmitic acid, Oleic acid; 1-palmitoyl-2-oleoyl-sn-glycerol (16:0-18:1 DAG)	Energy storage
		DAG are also second messengers in signal transduction and pre-cursors for prostaglandin synthesis
		TAG are used for energy transport
Glycerophospholipids		
DAG backbone (blue) with phosphate at third alcohol (red)	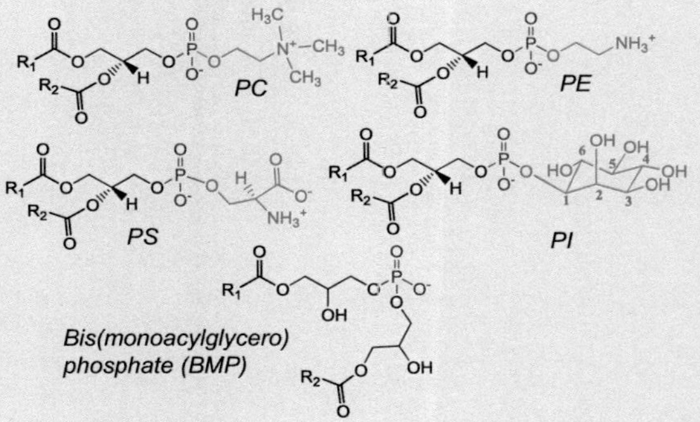 PC; PE	PC, PE, PS are structural components of membranes
Phosphate (red) esterified with choline, ethanolamine, serine, or inositol (green) results in phosphatidylcholines (PC), phosphatidylethanolamines (PE), phosphatidylserines (PS), and phosphatidylinositols (PI), respectively	PS; PI	PE and PS induce membrane curvature, as do MAGs like BMP
Acyl groups are represented as "R_1" and "R_2" and vary in length and degree of unsaturation	Bis(monoacylglycero) phosphate (BMP)	PS and PI serve as docking sites in membranes for signaling proteins
		PI are precursors of secondary messengers and form a "lipid code" signifying membrane identity
Sphingolipids		
Sphingosine base is the long-chain aliphatic amine backbone (R_1	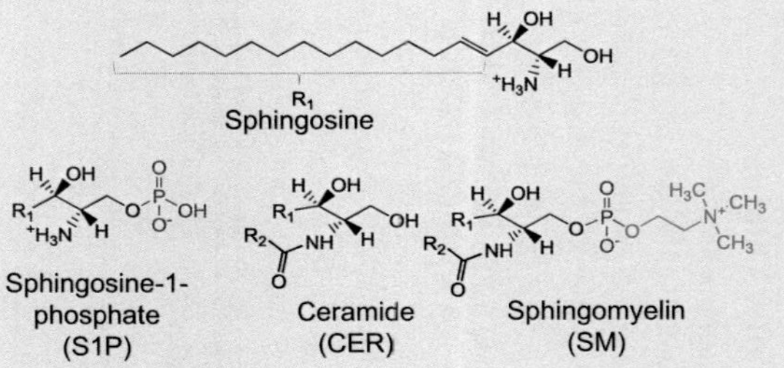 R_1 Sphingosine	S1P are extracellular signaling molecules
Ceramides (CER) are the sphingosine base acylated at the amine (R_2)	Sphingosine-1-phosphate (S1P)	CER and SM are structural components of membranes that can induce formation of lipid-ordered domains ("rafts") and membrane curvature
Sphingomyelin (SM) corresponds to a ceramide conjugated to phosphatidyl-choline (green) at the second alcohol	Ceramide (CER)	
R_2 varies in length and saturation	Sphingomyelin (SM)	

Sterols		
Sterols are defined by a four-ringed structure	Cholesterol	Cholesterol is the structural component of membranes
Rings are labeled A, B, C, D		Packing of cholesterol with SM induces lipid-ordered domains important for signal transduction and viral assembly
Sterol esters are formed by esterification with fatty acids	Desmosterol; 25-hydroxycholesterol	Sterols are also signaling molecules
		Sterol esters are used for energy storage
Bioactive Lipids		
Bioactive lipids are derived from arachidonic acid and related molecules	Arachidonic acid	Signal transduction via G-protein-coupled and peroxisome proliferator-activated receptors
These include prostaglandins, thromboxanes, lipoxins, leukotrienes, resolvins, protectins, and maresins	Prostaglandin E_2; Lipoxin A_4	Pro- or anti-inflammatory effects

(nonpolar) metabolites. Polar metabolites are soluble in aqueous solutions and include most sugars, purines and pyrimidines, nucleotides and nucleosides, acyl carnitines, organic acids, hydrophilic acids, amino acids, and phosphorylated compounds. These metabolites include most of the reactants and products involved in cellular respiration (e.g., glycolysis, the TCA cycle, or the pentose phosphate pathway) and in the production of building blocks for synthesis of large biopolymers such as DNA, RNA, proteins, and oligosaccharides. The nonpolar or hydrophobic metabolites are traditionally referred to as lipids. These metabolites function in energy storage, membrane structure, and signal transduction. Reflecting the diversity and complexity of lipid structures and functions (Table 1), the study of lipid metabolites has developed as a subfield within metabolomics. Thus, the complete metabolome formally includes both hydrophilic and hydrophobic metabolites, but the term "metabolome" is now frequently used to refer to the hydrophilic metabolome and the term "lipidome" is used to refer to hydrophobic metabolites. This chapter provides background and a framework for examining the role of lipid metabolites in viral processes and highlights general themes in our current understanding of the function of lipids in viral replication and pathogenesis.

Viruses require hydrophilic and lipid metabolites from the host. First, viruses are unable to manufacture the primary metabolites required for synthesis of new virions. The nucleotides, amino acids, and lipids that are structural components of the viral particle are all commandeered from the host cell. Likewise, cellular respiration provides the fuel required to drive the energetically costly process of viral replication. The effects of many viruses on metabolic pathways that produce the building blocks and energy needed for viral replication are well-documented. Although these perturbations are thought to support viral replication, the underlying molecular mechanisms are still the subject of much investigation. Second, many viruses use specialized membranes for viral entry, gene expression, genome replication, and assembly. The distinct morphology and functional properties of these membranes presumably reflects their specialized composition, which has been optimized for the viral process that they support. Analysis of the chemical composition and biophysical properties of these membranes, as well as the molecular mechanisms underlying their biogenesis, is therefore central to understanding their function in viral processes. Third, signal transduction by bioactive lipid metabolites plays a major role in the host response to viral infection and, in many cases, in pathogenesis. Identification of bioactive lipids associated with viral replication or the host inflammatory response holds promise for the identification of biomarkers of infection and disease progression as well as for identifying potential points for pharmacological intervention (Box 1).

Metabolomic profiling enables the systematic identification of perturbations of metabolism that arise from the interaction of viral pathogens with the host. Differences in metabolomic profiles reflect changes in protein function. This contrasts and complements the changes in gene expression and protein abundance detected by transcriptomic and

Box 1 Metabolomics for the discovery of antiviral targets

Metabolic enzymes and other metabolite-binding proteins are an attractive source of potential antiviral targets. First, metabolic reactions are essential to the infectious cycle of all viruses. Synthesis of new virions is itself comprised by the chemical conversion of nucleotides, amino acids, and lipids into progeny virions. Host-catalyzed chemical reactions that lead to the synthesis of these basic building blocks, and to the release of energy, have direct impacts on viral replication. Second, since the natural function of metabolite-binding sites on proteins is to bind to small molecules, these sites are generally reasonable molecular targets for drug discovery efforts. Indeed, many existing drugs mimic the interaction of naturally occurring metabolites with their respective enzymes and receptors. These "anti-metabolites" act by preventing utilization of the natural metabolite. Traditional antivirals targeting viral polymerase and protease activities are good examples of drugs that act by this mode. They inhibit the natural catalytic function of these enzymes by binding to the active site or to allosteric sites. These antiviral drugs are selective because they target metabolic reactions unique to the virus. Although this approach has produced some of the most successful drugs on the market today, it is inherently limited because viral genomes encode a very limited number of metabolite-binding proteins, and because resistance can develop through mutations that affect drug binding without affecting catalytic function. In addition, the selectivity of these drugs for their specific viral targets usually limits their use to a narrow spectrum of closely related viruses.

Targeting essential metabolic reactions catalyzed by host factors offers an alternative antiviral strategy. Viruses are fully dependent upon host metabolism and do not themselves encode the enzymes needed to produce the metabolites required for their own replication. Inhibition of a metabolic reaction critical for viral replication is therefore unlikely to be overcome by direct mutations of the viral genome. Likewise, the centrality of host metabolism to the viral infectious cycle makes it likely that phylogenetically related viruses have shared dependencies on specific metabolic pathways. Although the host proteome is replete with enzymes and proteins that bind to small-molecule metabolites, the challenge is to identify those that can mediate antiviral activity without affecting normal host cell function. As advances in analytical methods have enabled both broader and deeper surveys of virus-induced perturbations of host metabolism, identification of targets that mediate adequately selective antiviral effects has come closer to reality.

proteomic studies. The combination of these approaches provides the fullest picture of virus–host interactions occurring at both the molecular and pathway levels.

2. DEFINING THE ROLE OF THE LIPIDOME IN VIRUS–HOST INTERACTIONS

The goal of lipidomic analysis is the quantitative characterization of all lipids in a biological system, including the dynamic changes that occur in response to stimuli and the physiological consequences of these changes. This is technically challenging due to the structural diversity of lipid molecules and the inability to predict or infer lipid structure based on genomic, transcriptomic, or proteomic data. Lipidomic analysis is also challenging because the abundance of individual species can vary over a wide range. Phospholipids are so abundant that they constitute up to 50% of the total mass of lipids in the cell, whereas prostaglandins and other bioactive signaling lipids may be present at only picomolar concentrations. No single analytical method is sufficient to capture the full lipidome of a biological system, and as with other "omics" approaches, the methods used for quantitative analysis of large lipidomic data sets affect interpretation of the data. Strategies currently being used to examine the interaction of viruses with host lipids generally fall into one of three categories. First, there are classical reductionist approaches using metabolic labeling and biochemical reconstitution experiments to examine the potential function of a limited set of specific lipids in viral processes. Second, it is possible to infer roles for different classes of lipids through transcriptomic and proteomic studies that detect changes in the expression of host enzymes responsible for the synthesis, metabolism, and trafficking of these lipid classes. Third, global analytical chemistry methods allow direct detection and quantification of specific lipid species that change as a function of viral infection. This section highlights the types of biological questions that can be answered using these different approaches.

2.1 Reductionist Approaches

One of the first demonstrations that viruses affect host metabolism arose from the work in the 1950s of Hilton Levy and colleagues, who monitored the appearance of lactic acid in tissue culture supernatants and demonstrated that glycolysis is increased very early in poliovirus infection (Baron and Levy, 1956). Additional studies documented virus-induced changes in the uptake of isotopically labeled amino acids, phosphate, and other nutrients, suggesting perturbation of their respective metabolic pathways. These types of experiments were restricted by existing knowledge of specific metabolic reactions of interest. Efforts to "profile" virus-induced metabolomic changes in an unbiased manner were limited by the analytical methods available for detection and quantification of a finite set of metabolites (or metabolite classes).

Analogous reductionist approaches that focus on a particular metabolic pathway are useful today in corroborating discoveries made using newer, more discovery-based global methods (discussed below) and in examining the biochemical functions of individual lipid species in viral replication. Loss-of-function experiments that detect a viral phenotype when a specific lipid class is depleted from a host cell can be used to establish the essentiality of the depleted lipid for a viral process. Elegant studies of brome mosaic virus (BMV) replication in yeast led to the discovery that BMV RNA replication is blocked in yeast strains that have a reduced ratio of unsaturated to saturated fatty acids (UFAs and SFAs, respectively) due to mutations in the gene encoding Δ9 fatty acid (FA) desaturase, even when these mutations have no effect on cell growth (Lee et al., 2001). The antiviral effect of UFA depletion on BMV RNA replication is correlated with depletion of UFAs from replication membranes and decreases in membrane fluidity (Lee and Ahlquist, 2003). Rescue of BMV RNA replication by supplementation of growth medium with UFAs suggests a functional role of UFAs in this process. As another example, studies of Flock house virus (FHV) replication in a cell-free system allowed systematic examination of various glycerosphingolipids, sphingomyelin, and cholesterol in FHV RNA replication and the discovery that positive-strand RNA synthesis is promoted by glycerophospholipids. This effect varies depending on the length of the acyl chain and its degree of saturation (Wu et al., 1992). The major limitation of the approaches used to make these discoveries is that most viral processes cannot be recapitulated in genetically tractable organisms or in cell-free systems. A notable exception to this is the fusion step of viral entry, which has been examined for many enveloped viruses through the use of synthetic liposomes of defined composition and size.

2.2 Inferring Metabolomic Changes from Genomic or Proteomic Profiling

Examination of changes in host gene expression has become a useful tool for discovering how viruses affect metabolism. These studies implicitly assume that changes in enzyme abundance correspond to changes in the quantity or quality of metabolic output and that these changes are functionally relevant for viral replication, pathogenesis, or the host response. Systematic identification of virus-induced changes in steady-state host gene expression can now be routinely performed using global transcriptomic and proteomic methods (see Chapter 11). Conversely, RNA interference (RNAi) experiments that look for loss of replication upon depletion of a given host factor are also widely used to validate candidate pathways identified through transcriptomic and proteomic profiling studies and for the de novo discovery of metabolic pathways that are important for viral replication. Profiling and RNAi screening approaches have also been used to study viral pathogenesis with the goal of identifying pathways, and even specific host factors, that are critical for viral pathogenesis and that are candidate targets for antiviral intervention. For example, recent comparative transcriptomic analyses discovered that increased transcription of host lipid metabolic genes regulated by the liver X and retinoid X receptors (LXR/RXR) is characteristic of low pathogenicity influenza virus infections, whereas these changes are not observed during infection with H7N9 or other pathogenic avian influenza viruses or the highly pathogenic 1918H1N1 human influenza virus (Josset et al., 2012; Morrison et al., 2014).

Although these types of studies have been powerful tools in identifying virus-induced perturbations of the host cell, they provide an incomplete picture of a virus' effect on the host metabolome. Expression of many metabolic enzymes is regulated posttranscriptionally; moreover, the function of many enzymes is regulated posttranslationally through modifications that affect enzyme activity, oligomerization, or localization. Consequently, transcriptomic profiling experiments cannot predict all changes in protein abundance, and proteomic profiling experiments cannot detect changes in the activities and output of host metabolic enzymes. For related reasons, it is difficult to predict how RNAi-mediated depletion of a given metabolic enzyme affects the flux and molecular output of the pathways in which it functions. The potential limitations of relying upon only one type of profiling approach are illustrated by transcriptomic, proteomic, and lipidomic analyses of hepatitis C virus (HCV) infection. RNA profiling of HCV infection in vivo identified increased abundance of transcripts involved in glycosphingolipid synthesis and signaling by the serum response element binding protein (SREBP) pathway that are correlated with the onset of viremia (Su et al., 2002), suggesting that these pathways are important for HCV replication. Changes in the abundance of transcripts involved in lipid metabolism genes later in infection are correlated with steatosis and oxidative stress and thus are proposed to reflect a function of these gene products in pathogenesis (Smith et al., 2003; Bigger et al., 2004). A separate proteomic profiling of HCV infection in cell culture (Diamond et al., 2010) identified significant changes in enzyme abundance that were not predicted by the earlier transcriptomic studies. For example, changes in the abundance of proteins involved in energy production (i.e., glycolysis, the pentose phosphate pathway, and the citric acid cycle) early in infection were observed, suggesting that HCV increases energy production to meet the demands of viral replication. Parallel changes in the abundance of enzymes responsible for fatty acid oxidation have been interpreted to reflect a requirement for energy or increased acetyl-CoA for membrane synthesis. The perturbations detected by proteomic analysis likewise represent only a portion of the metabolomic changes induced by HCV since a parallel lipidomic profiling study detected HCV-induced enrichment of certain phospholipids and decreases in other phospholipid, sphingomyelin, and triacylglycerol species early in infection. Since not all lipids of these respective classes are uniformly affected, lipidomic analysis in this case enabled a higher resolution understanding of HCV's effects on lipid metabolism. This in turn provides an opportunity to identify particular molecules within a given pathway or lipid class that have a specific and optimized function in viral replication, pathogenesis, or the host response.

2.3 Global Analytical Methods

2.3.1 Nuclear Magnetic Resonance Spectroscopy

Nuclear magnetic resonance (NMR) spectroscopy monitors the absorption and emission of electromagnetic radiation by atomic nuclei. These occur at specific "resonance" frequencies affected by the local environment and thus provide data regarding structure and dynamics. Since NMR experiments do not require destruction or modification of the sample, the method is amenable to high-throughput analysis of biological samples and monitoring metabolic flux. Studies monitoring 50 or fewer species present at concentrations in the micromolar to millimolar range are the practical limit for metabolomic experiments due to resonance overlap and the limits of magnet strength. Although NMR-based metabolomic methods have not been widely used to study lipid function in the context of viral infection, they hold promise as a "fingerprinting" technique to discover biomarkers that are indicative of viral infection or disease state.

2.3.2 Mass Spectrometry

Mass spectrometry (MS) measures the molecular weight of a molecule by ionizing it and then measuring how its trajectory through a vacuum is affected by electric and magnetic fields. A mass spectrum plots the signal intensity (abundance) of a given ion as a function of its mass-to-charge (m/z) ratio, which enables measurement of molecular weight and also can provide structural information. Liquid or gas chromatography (LC or GC) is commonly used to separate molecules so that those with the same m/z ratio enter the spectrometer at different times. Electrospray ionization is the most commonly used method for lipidomic studies because it does not have a major structural bias and because it generates intact molecular ions and therefore enables ion abundance to be taken as a measure of the abundance of the parent metabolite. The combination of unique retention times and m/z ratios measured by quadrupole or Orbitrap mass spectrometers permits the routine quantitation of thousands of peaks in a single metabolomics profiling experiment with sensitivities as low as the femtomolar range.

MS-based profiling experiments can be performed in "targeted" or "untargeted" modes. Targeted studies focus on a limited set of metabolites for which retention times and m/z ratios have already been established using authentic standards. Untargeted profiling experiments monitor all detectable ions and use computational tools (e.g., XCMS or MetAlign) to identify differences in ion abundance between different samples. Assigning a structure to an ion of interest then requires tandem LC-MS/MS or other analytical experiments to compare the ion of interest with authentic standards. Targeted profiling provides a rapid and rigorous way to confirm the general effects of viral infection on a class of lipids or a lipid pathway, to test hypotheses generated from gene expression and proteomic analyses or untargeted metabolomic profiling, and to obtain higher resolution data regarding specific changes within pathways of interest. Untargeted analysis is more suited to the discovery of specific lipids that may be important in viral infection. As discussed above, targeted lipidomic profiling of HCV has been useful in interpreting and validating HCV-induced changes in lipid homeostasis inferred from transcriptomic and proteomic profiling studies. An independent untargeted study corroborated the targeted profile but in addition identified 26 distinct lipid species whose abundance changed by threefold or greater during HCV infection (Rodgers et al., 2012). One of these changes was a greater than 10-fold increase in desmosterol, a penultimate intermediate in cholesterol synthesis that is normally not abundant in the host cell. The discovery that HCV replication is inhibited under conditions in which desmosterol synthesis is blocked and rescued upon the addition of exogenous desmosterol (versus other sterols) suggests that this molecule has a specific function in HCV replication. *A priori*, there was little rationale for including desmosterol in the targeted study since this biosynthetic pathway is highly regulated and driven toward its end product, cholesterol. In another example, untargeted metabolomic profiling led to the discovery that *N*-acetyl-aspartate is the most up-regulated metabolite in cells infected with human cytomegalovirus (HCMV) (Rabinowitz et al., 2011). This change was unlikely to be detected in targeted studies because the synthetic route for this metabolite is unclear and it does not appear on standard metabolic maps or on lists of targeted metabolites.

An important distinction in MS-based studies is that between steady-state and flux analyses. Steady-state measurements quantify the abundance of metabolites at a given point in time and compare data collected from two or more different biological conditions (e.g., mock vs virus infected, high- vs low-pathogenicity strains, or early vs late infection). Any changes detected may reflect differences in synthesis or turnover. As illustrated by the aforementioned untargeted profiling studies of HCMV and HCV, this approach can be especially useful in detecting metabolites that are relatively rare in uninfected cells and that undergo large changes in abundance in the presence of a virus. Smaller fold changes in highly abundant molecules may also correspond to functionally significant changes in the total lipid composition of the cell.

Viruses may also affect reaction rates in a given biochemical pathway without altering steady-state metabolite abundance. For example, increased rates of cellular respiration during viral infection may not affect the steady-state abundance of ATP because it is consumed to fuel the energy demands of viral replication. Metabolic flux analysis directly measures changes in reaction rates for a given biochemical pathway. Pioneering work by Rabinowitz and colleagues has demonstrated the utility of LC-MS and isotopically labeled nutrients to profile kinetic flux through metabolic pathways (Rabinowitz et al., 2011). This allows discovery of specific pathways affected by viral infection and can also provide insights into underlying mechanisms. For example, comparative flux analysis of ^{13}C-labeled glucose in cells infected with HCMV or herpes simplex virus 1 (HSV-1) has revealed that both viruses induce the TCA cycle; however, HCMV-infected cells produce citrate with two ^{13}C atoms (reflecting synthesis via glycolytic flux) whereas HSV-1-infected cells produce citrate with three ^{13}C atoms (reflecting synthesis via anaplerotic flux) (Vastag et al., 2011) (Figure 1). The effect of each virus is distinct and their differences would not be detectable by steady-state measurements of transcript, enzyme, or even metabolite abundance. Analogous "fluxomic" analysis of lipid metabolic pathways may provide insights into the biogenesis of membranes utilized by viruses during replication.

3. LIPIDS IN VIRAL REPLICATION AND PATHOGENESIS

The membranes used in viral replication presumably have a specific composition that has been evolutionarily optimized to support the associated viral process; however, we know little about the molecular structures of specific lipids that function in viral replication and pathogenesis. Here, we discuss some of what is known about the function of lipids in different viral processes. Specific examples, primarily from studies of RNA viruses, are provided as examples but do not represent a comprehensive review of the current literature.

3.1 Lipids in Viral Entry

Lipids are central to the process of viral entry. First, glycosphingolipids and other lipids on the plasma membrane serve as entry factors mediating the initial attachment of many viruses to the host cell. Second, viral entry requires that the viral genome transit from the exterior of the host cell and past the plasma membrane to an interior cellular compartment. This requires fusion of the viral membrane with a cellular membrane (enveloped viruses) or penetration of a cellular membrane (nonenveloped viruses). Relatively more is known about fusion; however, both membrane fusion and

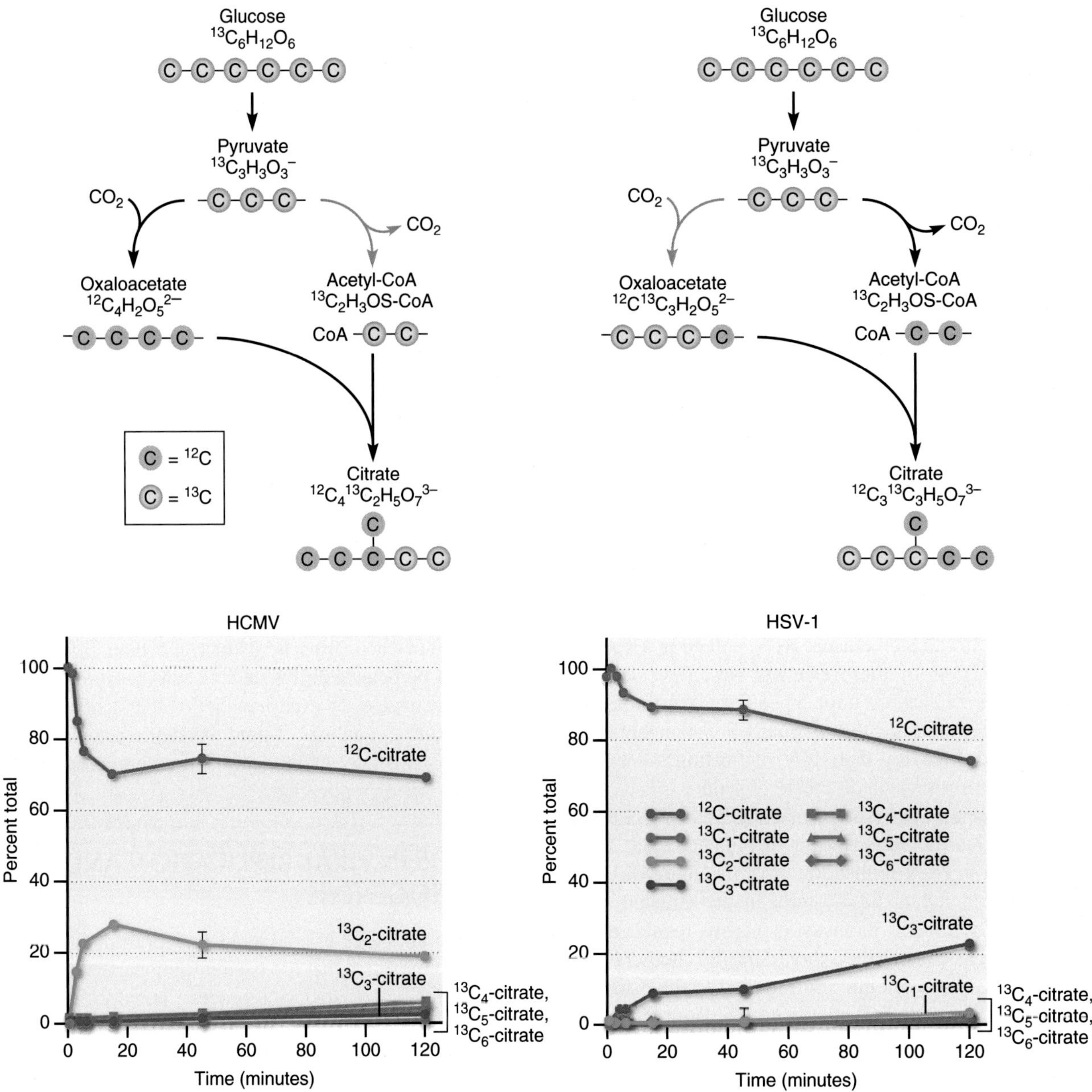

FIGURE 1 Flux analysis provides information on reaction rates and metabolic pathways affected by viruses. Herpes simplex virus 1 (HSV-1) and human cytomegalovirus (HCMV) infections both cause increases in the abundance of intermediates from glycolysis and the TCA cycle. Quantification of changes in metabolite abundance does not provide information on how these effects are produced (i.e., due to changes in influx, efflux, or both). By flux analysis monitoring conversion of ^{13}C-labeled glucose to citrate, Rabinowitz and colleagues determined that HSV-1 causes an increase in anaplerotic flux to the TCA cycle via pyruvate carboxylase (increased citrate with three ^{13}C-labeled carbons). In contrast, HCMV increases glycolytic flux, as evidenced by the accumulation of citrate labeled with two ^{13}C. *From Rabinowitz et al. (2011).*

penetration are affected by the lipid composition of target membranes.

3.1.1 Membrane Fusion

Fusion requires that the viral lipid bilayer transition from strongly positive to strongly negative curvature. Association of the viral envelope protein with both the viral and target membranes results in coupling of its structural rearrangements to membrane distortions that lower the energy barrier for reorganization of the two membranes as an initial hemifusion intermediate, and subsequently a fully fused product with a fusion pore that permits release of the viral genome (Figure 2). Physiological triggers, such as binding to a receptor or co-receptor at the plasma membrane (e.g., HIV) or exposure to acidic pH in the endosome (e.g., influenza virus), ensure that structural rearrangement of the envelope protein does not initiate prematurely.

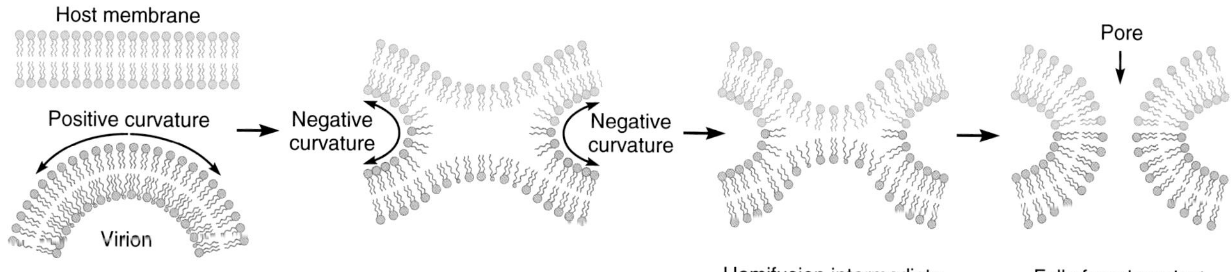

FIGURE 2 Membrane curvature during membrane fusion in viral entry. Fusion of viral and cellular membranes during enveloped virus entry requires the induction of membrane curvature to form, first, a hemifusion intermediate in which lipid mixing between outer leaflets has occurred and subsequently, a fully fused product. Lipid head group size and charge as well as interactions with proteins contribute to the induction of the necessary curvature (Text Box 2).

3.1.2 Membrane Composition

The functional properties of membranes, including membrane curvature, fluidity, elastic free energy, and lipid packing are determined by their lipid composition (Box 2). A long-standing hypothesis has been that the compositions of the viral lipid bilayer and the host target membrane are key determinants of the efficiency of fusion and other steps in viral entry.

Studies of specific viruses in biochemical reconstitution and cell-based models have revealed richness in the effects of lipid structure and function on membrane fusion during viral entry. Many viruses, including alphaviruses, HIV, HSV-1, HCMV, and influenza virus, require cholesterol (or other sterols) for efficient fusion (Box 3); however, there are differences in how and where cholesterol functions during entry (viral membrane, cellular membrane, or both). HIV entry requires cholesterol- and sphingolipid-rich lipid rafts in both the viral membrane, where they stabilize virion structure and infectivity, and in the plasma membrane, where they facilitate clustering of the CD4 receptor and CCR5/CXCR4 co-receptor. In contrast, Semliki Forest virus requires cholesterol because the conformational changes in the viral E1 protein that catalyze fusion cannot occur without its interaction with cholesterol (Umashankar et al., 2008). Influenza A virus entry requires cholesterol in the viral membrane but not in the target membrane, although the functional basis for this requirement is not well-understood.

Beyond requirements for cholesterol and sphingolipids, the fusion of many viruses may be regulated by the presence of compartment-specific lipids. Live-cell imaging studies have demonstrated that dengue virus enters via clathrin-mediated endocytosis and fuses in Rab7-positive late endosomes (van der Schaar et al., 2008), yet conformational changes in the dengue E protein can be triggered in vitro at the moderately acidic pH encountered by the virus in early endosomes. An additional requirement for negatively charged lipids in the late-endosomal membrane may explain the apparent delay between the initiation of fusion and release of the nucleocapsid. This is supported by the observation that negatively charged lipids, such as phosphatidylserine or bis(monoacylglycero)phosphate, increase the fusion of dengue virus with synthetic liposomes (Zaitseva et al., 2010; Nour et al., 2013). Negatively charged lipids may be required to trigger conversion of a "restricted hemifusion" intermediate to the fully fused product. Alternatively, the virus may initially fuse with small endosomal carrier vesicles in the early endosome and back-fusion with the late endosome requires negatively charged lipids.

3.2 Lipids in Viral Genome Replication

For many viruses, genome replication occurs on specialized membranes; however, the specific lipids in these membranes are generally not well characterized. Recent progress in studying the specialized replication compartments used by positive-sense RNA viruses highlights a growing appreciation for the function of membranes in this viral process.

3.2.1 Genome Replication of RNA Viruses

Localization of viral RNA replication to specialized replication compartments—essentially, virus-specific organelles—serves several functions. First, membrane localization promotes RNA replication by concentrating the reactants, catalysts, and cofactors required for RNA replication. Second, interaction of the macromolecular replication complex with the interior membrane of the compartment may scaffold factors in specific conformations and orientations necessary for activity. Third, sequestration of the viral genome within the compartment provides a mechanism for regulating the processes of replication, transcription, translation, and assembly that all compete for use of the genomic RNA. Spatial restriction of genome replication to the compartment also serves to shield the process from detection by the host response.

RNA viruses vary greatly in their use of different subcellular host membranes, including those of the rough endoplasmic reticulum (flaviviruses, picornaviruses, SARS-coronaviruses), mitochondria (nodaviruses), and plasma

Box 2 Lipid structure and function in membranes

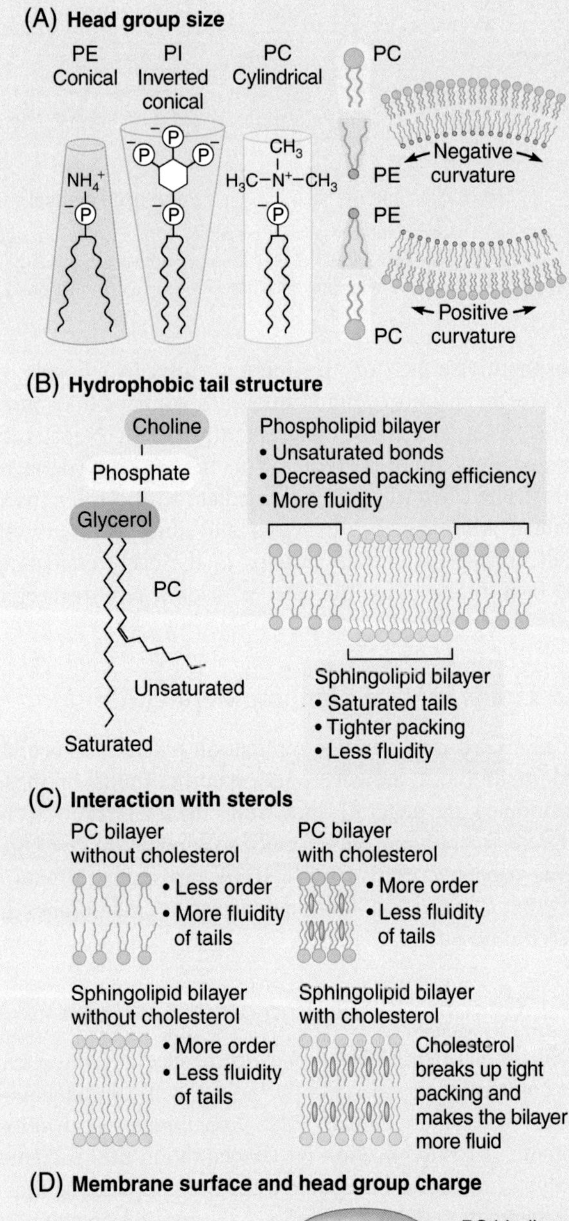

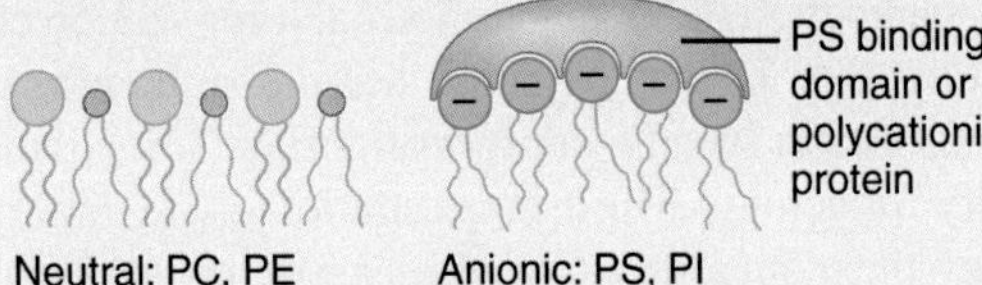

The physicochemical properties of membranes are dictated by their lipid composition, which is optimized for specific biological functions. This is evidenced by the striking differences in composition and function of different subcellular membranes (Holthuis and Menon, 2014). The plasma membrane is rich in saturated fatty acids, sphingolipids, and cholesterol, all of which contribute to the formation of liquid-ordered domains (also known as "lipid rafts") that have decreased fluidity relative to the endoplasmic reticulum (ER), whose mobility is associated with the low abundance of order-promoting lipids. Although the chemical composition of most membranes that support viral processes has not been characterized, analysis of the specific lipid molecules in viral membranes is expected to provide insight into their specialized function in viral replication.

(A) Head group size The relative sizes of head groups and hydrophobic tails of glycerophospholipids affect the overall shape of the lipid as well as the thickness and curvature of the bilayer. Phosphatidylcholines (PC) have relatively large head groups compared to phosphatidylethanolamines (PE), and these differences directly affect the shape of individual molecules and the propensity of these lipids to assemble into lipid bilayers and to induce membrane curvature. PC spontaneously form lipid bilayers that exist in a liquid crystalline state at physiological temperature. The smaller, pH-titratable head group of PE gives these lipids a cone shape that imposes lateral stress on the bilayer and that can promote negative membrane curvature.

(B) Hydrophobic tail structure The structures of the hydrophobic tails of glycerophospholipids and sphingolipids also influence the width and fluidity of membranes. Molecular features, such as *cis*-double bonds, reduce packing density and enhance membrane fluidity. Lipids with fully saturated tail groups pack more densely, resulting in decreased membrane fluidity. An important example of this is the tight packing of sphingolipids, which generally have fully saturated acyl chains or *trans*-double bonds. The taller, narrower shape and tighter packing of ceramides and other sphingolipids relative to their phospholipid counterparts promotes the formation of ordered domains (lipid rafts) that may be used by viruses at the plasma membrane and other sites.

(C) Interaction with sterols Cholesterol and other sterols in the lipid bilayer can alter packing density and membrane fluidity by packing against acyl chains in the hydrophobic tail groups. This can be critical in maintaining the fluidity of sphingolipid bilayers, which can otherwise form a solid–gel phase. Conversely, ordered packing of acyl chains around the fused four-ring structure of cholesterol reduces the entropy of phospholipid bilayers, leading to decreased mobility of the bilayer as well as increased membrane thickness and reduced membrane permeability.

(D) Membrane surface and head group charge The surface charge of membranes is important for interactions with proteins. The anionic head groups of phosphatidylserines and phosphatidylinositols interact with a variety of proteins. The head group of PC is zwitterionic over a wide range of pH resulting in bilayers that have no net charge. The head group of PE is zwitterionic, but can become anionic upon deprotonation of the ammonium ion; its charge and interactions with proteins can therefore be regulated by pH.

Box 3 Cholesterol in viral infections

The importance of sterols in viral infection is evident in the sensitivity of many viruses to perturbation of sterol homeostasis and the repressive effect of the antiviral interferon response on the sterol biosynthetic pathway. In addition to its importance in membrane structure and function, cholesterol is important as a biosynthetic precursor of hormones, vitamins, and other lipid-signaling molecules. Signal transduction by downstream metabolites of sterols, such as the oxysterols and vitamin D, has been increasingly implicated in both the innate and adaptive immune responses to viral pathogens. Sterols that occur upstream of cholesterol were previously thought to exist only as biosynthetic intermediates, but they are now known to have additional functions in signal transduction. For example, desmosterol, a penultimate intermediate of cholesterol, is an activator of the liver X receptor (LXR) and the dominant ligand of LXR in macrophage foam cells, where it activates and inhibits transcription of LXR and SREBP target genes, respectively, and suppresses expression of inflammatory genes (Spann et al., 2012).

Enteroviruses cause dynamic changes in cholesterol trafficking and metabolism (Ilnytska et al., 2013). Early in infection, cholesterol internalization is increased while cholesterol stored as esters in lipid droplets is concomitantly decreased. Ectopic expression of the viral 2BC protein alone is sufficient to trigger this increase in free intracellular cholesterol, much of which appears to be trafficked to sites of viral RNA replication via Rab11-positive recycling endosomes. Inhibition of viral replication when cholesterol trafficking is blocked suggests that these dynamic changes in trafficking are functionally significant for the virus. For example, cholesterol depletion appears to affect proteolytic processing of the viral polyprotein. Although the exact mechanisms underlying the effects of enteroviruses on cholesterol trafficking still need to be elucidated, it is known that virus-induced perturbations of PI4P and cholesterol metabolism and trafficking (Figure 4) are not mediated by changes in host gene expression but rather by posttranslational mechanisms affecting host protein function and localization.

Studies of cholesterol function in viral replication and cell biology have made widespread use of two sets of tools that bear comment. Proteins and natural products that bind to cholesterol, such as perfringolysin-O and fillipin, have been used to visualize the subcellular localization of cholesterol in microscopy studies. Methyl-beta-cyclodextrin has been used to deplete cells and membranes of cholesterol for loss-of-function studies. A limitation of these studies is that these reagents are not selective for cholesterol and bind to many other late-stage sterols. This selectivity (or lack thereof) should be recognized when designing and interpreting these types of experiments. For example, the loss of function caused by depletion of cholesterol with methyl-beta-cyclodextrin is commonly "rescued" by the addition of exogenous cholesterol without the inclusion of other sterols as controls. This overlooks the significant differences in the effects of cholesterol versus other sterols on the functional properties of membranes (Bloch, 1983). Likewise, the use of fluorophore-conjugated sterols in imaging experiments carries the risk that the fluorophore, which is generally comparable in size to the sterol moiety, affects the physical and/or chemical properties of the molecule.

or endosomal membranes (togaviruses). Regardless of the donor membrane, the replication compartments characterized to date can generally be categorized into one of two classes: invaginated vesicles or double-membrane vesicles (Paul and Bartenschlager, 2013) (Figure 3). Viruses such as the alphaviruses and flaviviruses replicate within single-membrane invaginations that are continuous with the donor membrane and connected to the cytosol through a "neck-like" structure. Viruses such as HCV, poliovirus, coxsackie B3 viruses, and SARS-coronavirus replicate their genomes within double-membrane vesicular structures that are connected to one another through their shared outer membrane. Membrane invagination requires negative membrane curvature, with bending of the membrane away from the cytoplasm, whereas double-membrane vesicles exhibit positive membrane curvature, with the membrane bending toward the cytoplasm. Membranes at the neck-like structure require membrane curvature of the opposite polarity. Formation of both types of replication compartments requires remodeling of the donor membrane through interaction with proteins or protein complexes as well as through alteration of the lipid content of the membrane.

3.2.2 Perturbation of Lipid Biosynthesis and Trafficking by RNA Viruses

Positive-strand RNA viruses perturb lipid biosynthesis and trafficking to facilitate formation and maintenance of the subcellular compartments where RNA replication occurs. The enteroviruses alter sterol trafficking and metabolism, and RNA replication for these viruses is sensitive to experimental conditions that inhibit cholesterol trafficking and/or deplete intracellular cholesterol (Box 3). The dengue virus NS3 protein directly recruits fatty acid synthase (FAS) to the replication compartment, and the resulting de novo synthesis of fatty acids is thought to enhance membrane fluidity and support membrane biogenesis at the site of RNA replication (Heaton et al., 2010). Although the molecular structures of the fatty acids produced have not been elucidated, activation of FAS enzymatic activity by NS3 reinforces the idea that lipids induced by the virus are directly involved in replication of the viral genome. Targeted profiling of the whole-cell lipidome of dengue virus-infected mosquito cells (Perera et al., 2012) has revealed increases in the steady-state abundance of unsaturated phospholipids, sphingomyelin,

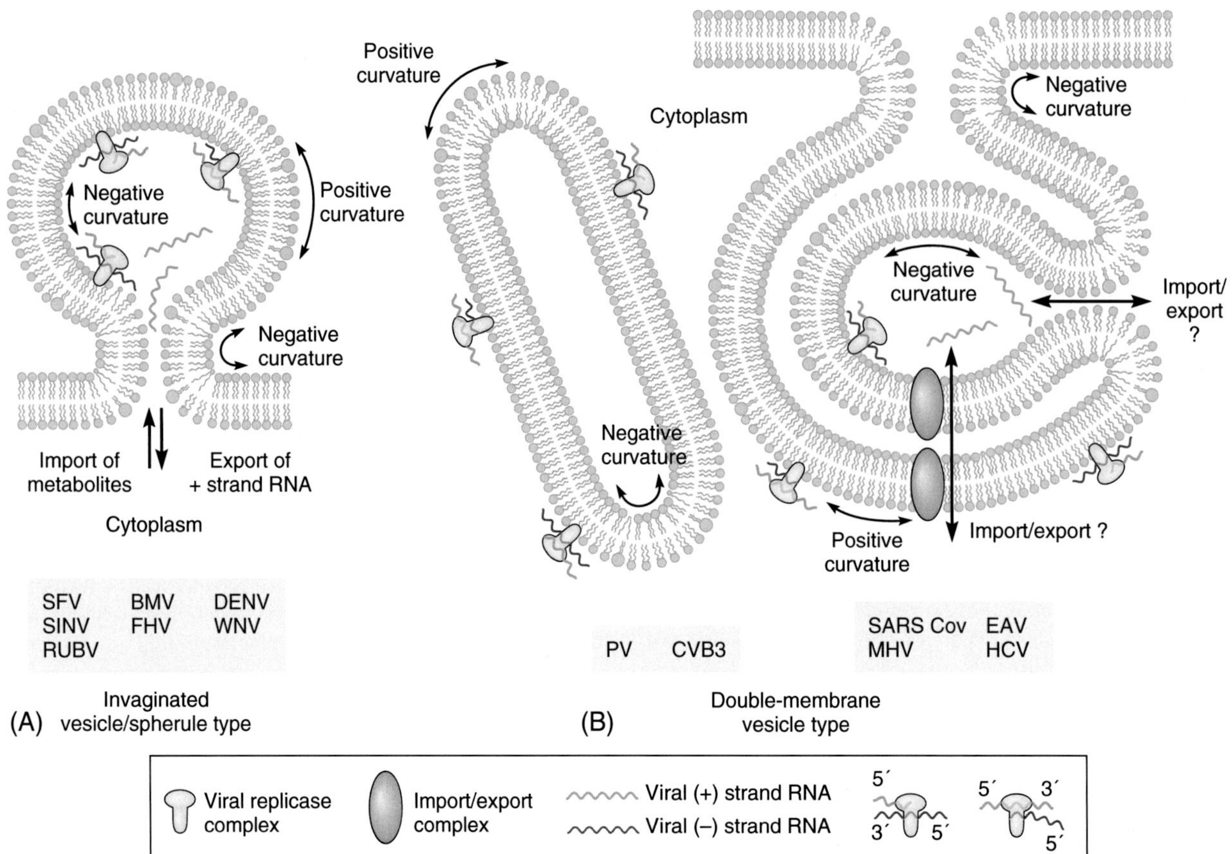

FIGURE 3 Membrane curvature in RNA virus genome replication membranes. (A) Membrane curvature is required to form invaginated vesicles. Negative curvature in the interior of the replication compartment, where RNA synthesis occurs, changes polarity at the neck structure that connects to the cytosol. Invaginated vesicle compartments are used for genome replication by Semliki Forrest virus (SFV), Sindbis virus (SINV), Rubella Virus (RUBV), Brome mosaic virus (BMV), Flock house virus (FHV), Dengue virus (DENV), and West Nile virus (WNV). (B) Double-membrane vesicle (DMV) RNA replication compartments have membrane topologies that also require the induction of negative and positive curvature although details of membrane topology and the site of RNA synthesis are still being defined. Two possibilities are illustrated. (*left*) The genome replication of enteroviruses, such as poliovirus (PV) and Coxsackievirus B3 (CVB3), is believed to occur on the outer membrane of single-membrane tubules and DMVs (inner membrane is not depicted for simplicity). Positive curvature on the exterior of the DMV is likely promoted by the presence of PI4P lipids, which have also been shown to recruit the $3D^{pol}$ of poliovirus (Figure 4). (*right*) Ultrastructural studies suggest that viruses such as hepatitis C virus (HCV), severe acute respiratory syndrome coronavirus (SARS-CoV), mouse hepatitis virus (MHV), and equine aterivirus (EAV) perform RNA replication in the lumen of the DMV, although it is less clear if an opening connects the interior of the DMV with the cytosol and/or if specific transporters facilitate import and export from this compartment. *From Paul and Bartenschlager (2013).*

and ceramide as well as diacylglycerol and phosphatidic acid. Biochemically isolated replication membranes exhibit enriched ceramide content, including specific enrichment of long-chain ceramide and dihydroceramide species that may induce negative curvature to facilitate invagination of the endoplasmic reticulum membrane and formation of the replication compartment. This is consistent with the known effects of ceramides on lipid bilayer structure, although it is also possible that these ceramides are generated as part of the host response or have some other function.

The replication membranes of enteroviruses, poliovirus and coxsackie B virus, and HCV are enriched in phosphatidylinositol-4-phosphate (PI4P) lipids, which in uninfected cells are synthesized by phosphatidylinositol 4-kinases in *trans*-Golgi-network membranes (Altan-Bonnet and Balla, 2012) (Figure 4). For the picornaviruses, the viral 3A protein recruits PI4KIIIβ (but not PI4KIIIα) to replication membranes by modulating the activity of the Arf1/GBF1 complex and also by either a direct or indirect interaction with the kinase. HCV enriches its replication membrane with PI4P through recruitment and activation of PI4KIIIα through an interaction with onstructural protein 5A (NS5A). In the absence of PI4KIIIα, HCV appears to use PI4KIIIβ, although the mechanism whereby HCV affects this enzyme is less clear.

What are the functional consequences of enrichment of PI4P lipids in viral replication compartment membranes? The presence of PI4P lipids appears to be required for viral RNA synthesis since steady-state RNA replication is inhibited in the presence of small-molecule inhibitors of PI4KIII activity or when the kinase is depleted through RNAi. Consistent

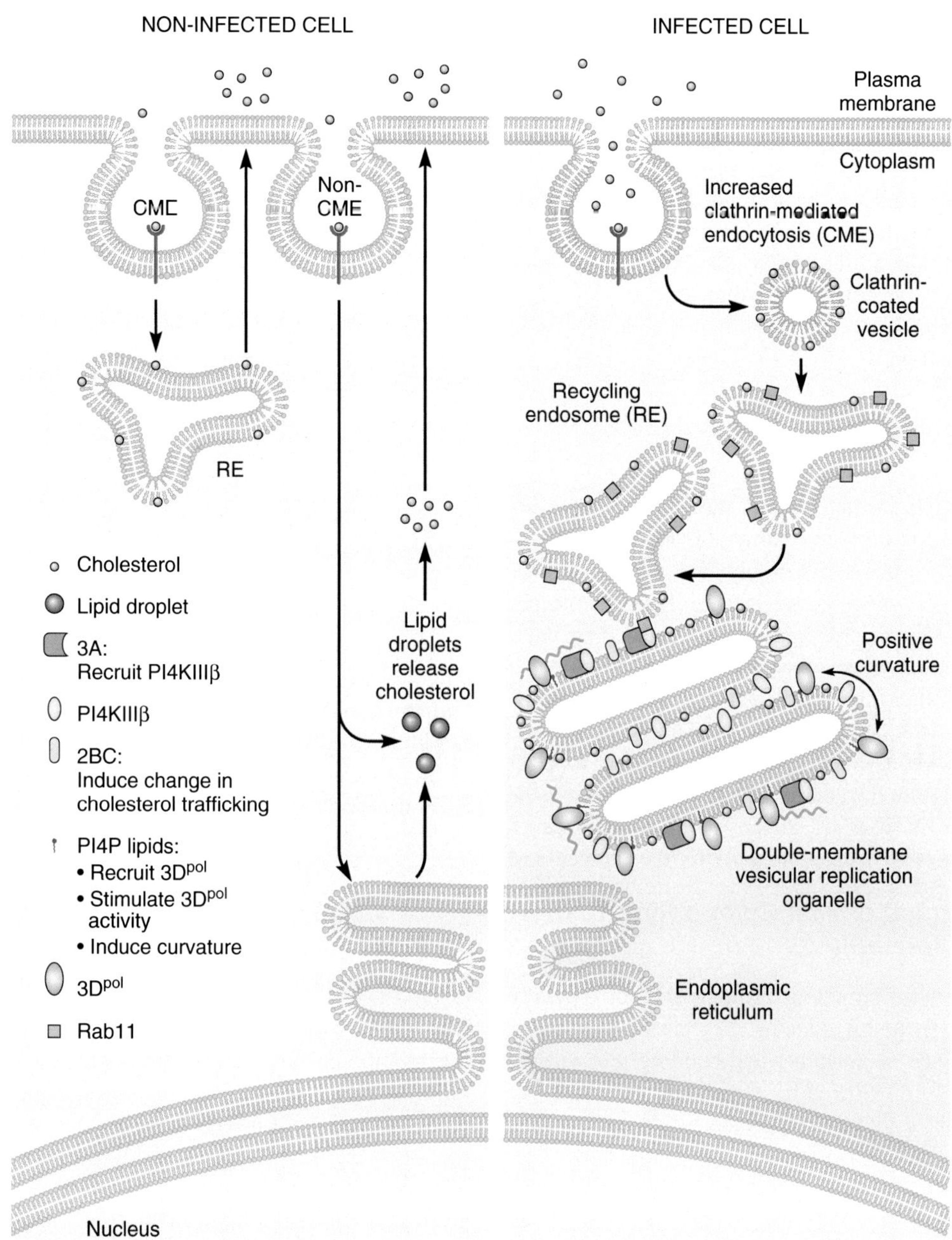

FIGURE 4 Perturbations of lipid metabolism by enteroviruses. Enteroviruses replicate on cholesterol-rich double-membrane vesicle-type (DMV-type) compartments formed from remodeling of the endoplasmic reticulum (ER), Golgi intermediate compartment (ERGIC), and Golgi apparatus. The inner membrane of the DMV is not depicted for visual simplicity. The viral 2BC protein enhances clathrin-mediated endocytosis (CME) and uptake of cholesterol, which is trafficked via recycling endosomes (RE) to the replication organelle. Direct or indirect recruitment of phosphatidylinositol-4-kinase IIIβ (PI4KIIIβ) by the viral 3A protein leads to synthesis of phosphatidyl-4-phosphate (PI4P) lipids in the replication membrane and also facilitates delivery of RE to the replication organelle via interaction of PI4KIIIβ with Rab11. A reduction in lipid droplets rich in sterol esters suggests that enteroviruses may also reduce formation of these lipid droplets and/or enhance release of sterols stored therein. *From Strating et al. (2013).*

with this, depletion of PI4P lipids reduces RNA synthesis in a cell-free model of poliovirus replication. Although the molecular basis for this effect on genome replication is still very much an open area of investigation, the cellular function of PI4P lipids is to recruit proteins to the Golgi membrane and other specific membranes by interaction with pleckstrin (PH) or epsin N-terminal homology domains. By analogy, the presence of PI4P lipids may recruit necessary components to the site of RNA replication. In support of this notion, the enterovirus RNA-dependent RNA polymerases have specific binding sites for PIP4, which may promote their recruitment from the cytosol to the site of replication.

A second possibility is that PI4P lipids induce the curvature needed to form the compartment where genome replication occurs. This could be a direct effect of PI4P structure on the membrane or an indirect effect resulting from recruitment of curvature-inducing PI4P-binding proteins, such as EpsinR and four-phosphate adaptor protein 2 (FAPP2). Although much remains to be learned about the function of PI4P lipids in positive-strand RNA virus replication, the requirement for this subset of phosphatidylinositol lipids over the other six phosphatidylinositol classes suggests a specific relationship between PI4P structure and function in viral RNA replication.

3.3 Lipids in Viral Assembly and Release

All enveloped viruses have viral membranes derived from the host cell, and the composition of the viral envelope has long been thought to provide information about the assembly and budding process. Assembly is initiated by trafficking of the viral genome and structural proteins to the host membrane where budding will occur. Budding itself requires the induction of positive membrane curvature as the nascent, enveloped virion is formed and negative membrane curvature at the neck connecting the nascent particle to the membrane from which it buds. Scission of the neck to sever the viral particle from the budding membrane completes the budding process. Viruses use diverse cellular membranes and mechanisms of varying complexity for these processes.

3.3.1 Recruitment of Viral Components to the Site of Assembly

Although inner viral structural proteins do not typically have membrane-spanning domains, they must somehow be recruited to the site of assembly and budding. For viruses such as HIV and influenza virus, which bud from raft-like microdomains at the plasma membrane, virion components undergo lipid-mediated co-sorting into the same membrane microdomain, which then serves as a platform for assembly and budding of new virions. In the case of HIV, Gag is initially recruited to the plasma membrane through its interaction with phosphatidylinositol (4,5) bisphosphate. Subsequent budding of virions requires both cholesterol and sphingolipids, consistent with a dependence on lipid-ordered domains to facilitate assembly of individual components into virions. Although other host factors contribute to these processes, the lipids involved appear to serve specific functions in spatial-temporal regulation of HIV assembly.

Viruses that bud at intracellular membranes must also concentrate virion components at sites of assembly. The host lipids and proteins used to direct this process are in some cases suspected but largely not known. HCV provides an apt illustration (Lindenbach and Rice, 2013; Filipe and McLauchlan, 2015). HCV core protein initially traffics to the surface of cytosolic lipid droplets (cLDs) where it interacts with the NS5A protein. Delivery of core and NS5A to cLDs early in infection is mediated by diacylglycerol O-acetyltransferase 1 (DGAT1), which catalyzes synthesis of triglycerides and formation of lipid droplets; Rab18, which resides on LDs and regulates vesicular trafficking; and tip-interacting protein 47 (TIP47), which regulates incorporation of triacylglycerols into LDs and LD maturation. In addition, trafficking of core to cLDs requires cytosolic phospholipase A2 (cPLA2), a key effector of arachidonic acid signaling.

Later in the replication cycle, budding of virions requires that cLD-associated core and viral genomic RNA traffic to sites where the viral E1 and E2 glycoproteins have accumulated on the lumenal side of the endoplasmic reticulum membrane. This is dependent on NS5A as well as interaction of NS2 with NS3/4A. Although the mechanisms underlying these trafficking events are still poorly understood, integration of HCV assembly with LD biogenesis is thought to enable coordination of viral translation, RNA replication, and assembly through regulated trafficking of viral components to or from the sites where these processes occur. Although our understanding of these processes is currently "protein-centric," the lipid molecules in these cellular pathways are unlikely to be passive bystanders. Investigation of the structure and function of specific lipid molecules in HCV assembly is anticipated to be necessary to understand the physical mechanisms regulating these dynamic processes.

3.3.2 Mechanisms for Inducing Membrane Curvature in Budding and Scission

Budding of viral particles requires the induction of positive membrane curvature leading to formation of a nascent virion attached to the budding membrane via a narrow-diameter neck region with negative curvature. Scission of this neck region completes budding of the viral particle. Viruses use protein–lipid and lipid–lipid interactions to achieve membrane curvature during virion budding and scission (Rossman and Lamb, 2013) (Figure 5). First, membrane curvature can be "scaffolded" by viral core proteins or proteins that bind to the head groups of the lipid bilayer. In this case, energetically favorable protein–lipid interactions are used to drive curvature of the membrane. Second, the insertion of proteins into the lipid bilayer can cause "hydrophobic displacement" leading to energetically unfavorable differences in the surface areas of the inner and outer membrane leaflets. Membrane curvature is induced in these cases to relieve the energy strain. Third, adjacent areas of lipid-ordered and lipid-disordered regions of a membrane can differ in thickness, resulting in energetically unfavorable interactions (referred to as "line tension") due to alignment of the polar head groups of the thinner lipid-disordered domain with the hydrophobic tails

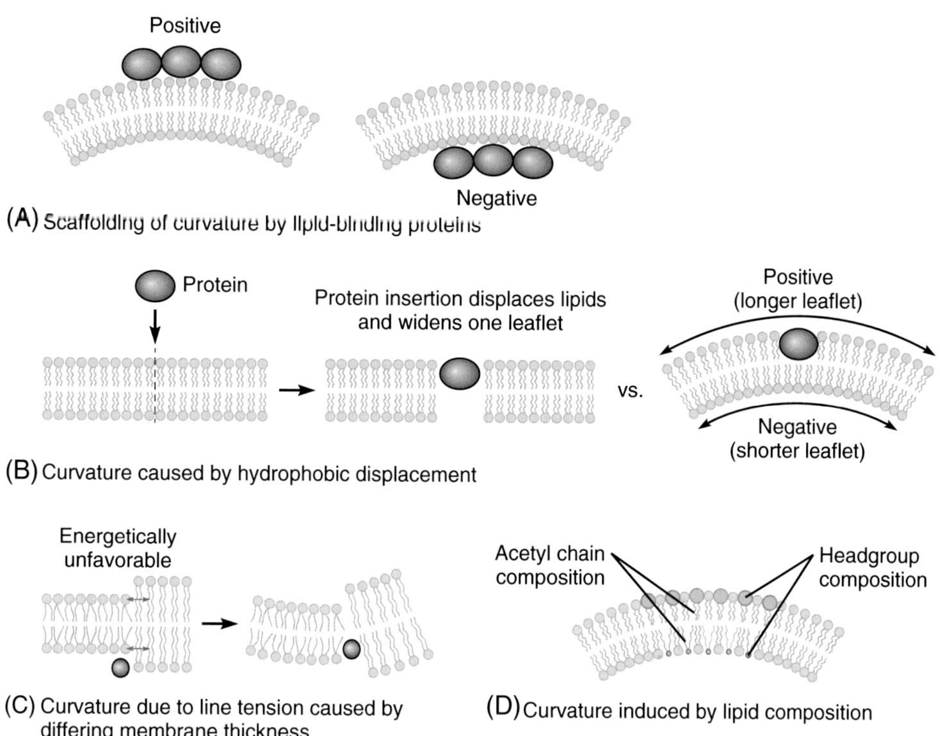

FIGURE 5 Mechanisms of inducing membrane curvature. (A) The interaction of membrane lipids with viral core proteins or other proteins that recognize lipid head groups can scaffold membrane curvature. (B) Insertion of a protein into the membrane causes asymmetric displacement of lipids. The induction of curvature is one physical mechanism for counteracting the resulting difference in surface areas of the inner and outer leaflets. (C) Adjacent regions of lipid-ordered and lipid-disordered membrane differ in thickness. This results in energetically unfavorable packing of hydrophilic head groups against hydrophobic tail groups. Membrane curvature can relieve this "line tension." (D) The structure and shape of individual lipids in the bilayer can affect membrane curvature (see Text Box 2). Cone-shaped lipids (PE) and inverse-cone-shaped lipids (PI) promote negative and positive membrane curvature, respectively.

of the thicker, lipid-ordered domain. Fourth, lipid structure can affect membrane curvature through intrinsic bending induced by the shape and charge of the component lipids (Box 2).

Perturbations of membrane curvature by viral glycoproteins on the outer leaflet of the budding membrane and matrix and capsid proteins on the inner leaflet are essential for the budding of many viruses. For viruses such as tick-borne encephalitis virus and hepatitis B virus, budding is driven by assembly of the viral glycoprotein on the surface of the nascent bud. This "pulling" force is sufficient to result in the formation and release of subviral particles that lack nucleocapsids. In contrast, for HIV and other retroviruses, the assembly of Gag exerts a "pushing force" that drives membrane curvature and is sufficient for assembly and release of virus-like particles. Membrane curvature during budding of influenza virus (see Section 3.3.3) as well as Semliki Forest virus and other alphaviruses appears to require the concerted action of pushing from the nucleocapsid and pulling by the viral glycoprotein.

Scission of the neck to release the viral particle from the budding membrane is a distinct step in viral replication, as evidenced by the detection of stalled intermediates that resemble "beads on a string" in the absence of viral proteins that recruit the scission machinery of the host cell or that catalyze membrane scission directly. Most viruses do not encode their own scission machinery, and the outer neck of the membrane requiring scission is topologically inaccessible to most cytoplasmic scission proteins (e.g., dynamin or Arf1). HIV and other retroviruses achieve scission by recruiting endosomal sorting complex required for transport (ESCRT) proteins to the site of viral assembly through late-domain motifs in HIV Gag and other retroviral capsid proteins. Other viruses (e.g., Poxviridae, Herpesviridae, Filoviridae, Rhabdoviridae, and Paramyxoviridae) that bud from a variety of subcellular membranes (nuclear, endoplasmic reticulum, or plasma membranes) also recruit the ESCRT machinery through interactions of viral late-domains with Tsg101, Alix, and Nedd4-like proteins, although there are undoubtedly variations in viral interaction with the ESCRT machinery. Viral scission can also occur via ESCRT-independent mechanisms. For example, respiratory syncytial virus has no known late-domain motifs and undergoes virus budding using a mechanism that requires Rab11-dependent pathway family-interacting protein 2 (Rab11-FIP2).

3.3.3 *Influenza Virus Assembly, Budding, and Scission*

Influenza virus assembly, budding, and scission present unique examples due to the extent to which these processes are mediated by viral proteins and also the extent to which they have been characterized (Martyna and Rossman, 2014). The two influenza virus glycoproteins, hemagglutinin (HA) and neuraminidase (NA), are targeted to the plasma membrane by specific sequences in the transmembrane domains of HA and NA as well as by palmitoylation of HA. Concentration of HA in the membrane induces curvature and is sufficient to permit budding of virus-like particles; however, budding of authentic virions requires other effectors, including the viral M1 and M2 proteins. M1 multimerizes to form a helical net that supports the viral membrane and connects the inner leaflet of the viral membrane to viral ribonucleoproteins (RNPs) on the virion interior. Pushing caused by multimerization of M1 beneath the budding membrane or by recruitment of RNPs initiates budding.

Scission of the neck structure is catalyzed by the viral M2 protein although the biophysical mechanism of scission is still not known. In infected cells, M2 localizes between ordered (raft) and disordered domains of the plasma membrane. As budding progresses, M2 concentrates in the neck region at the boundary between the ordered, cholesterol-rich viral membrane and the adjacent cholesterol-poor plasma membrane. Peptides corresponding to the amphipathic helix of M2 induce negative membrane curvature in a cholesterol-dependent manner; moreover, binding of M2 to lipid domains with high radii of curvature comparable to those in the neck of budding virions disrupts packing of lipid head groups in vitro. Insertion of the M2 amphipathic helix into the lipid-disordered phase may therefore cause membrane scission through the generation of line tension or packing defects. Cellular factors (e.g., Rab11) may also contribute to scission since mutations in the cytoplasmic tail of M2 can have significant effects on budding efficiency.

3.4 Lipids in the Inflammatory Response

The host response to viral infection in many cases contributes to viral pathogenesis through the overabundant production of cytokines (termed a "cytokine storm") and damage caused by inflammatory cells of the host response. Autocrine and paracrine signal transduction by bioactive lipids is central to both the activation and resolution of the host inflammatory response. These lipids are synthesized from arachidonic acid and other polyunsaturated fatty acids (e.g., eicosapentaenoic, docosahexaenoic, linoleic, and linolenic acids) via cyclooxygenase (COX), lipoxygenase (LOX), and cytochrome P450 pathway. Their release from membrane phospholipids is tightly regulated by a system of over 50 enzymes. COX-1 and COX-2 are responsible for synthesis of prostaglandins and thromboxanes, which generally exert a strong proinflammatory effect early in the host response. The three LOX enzymes (5-LOX, 12-LOX, and 15-LOX) synthesize leukotrienes and lipoxins as well as hepoxilins and hydroperoxy and hydroxyl fatty acids. Lipids produced by 5-LOX (LTB_4, LTC_4, LTE_4, 5-HETE) are proinflammatory, acting as chemoattractants for neutrophils and basophils and potently promoting bronchoconstriction, smooth muscle contraction, and increased vascular permeability. In contrast, lipids produced by 12-LOX and 15-LOX (LXA_4 and LXB_4) have anti-inflammatory activities that are important for resolution of the inflammatory response. Resolvins, protectins, and maresins produced from eicosapentaenoic and docosahexaenoic acids are also important in limiting infiltration of immune cells and tissue damage at the site of infection and are therefore important for resolution of the inflammatory response.

As illustrated by recent studies of influenza virus, analysis of the differential induction of bioactive lipid mediators of inflammation in high- versus low-pathogenicity viral infections has been informative for understanding pathogenesis and for suggesting new strategies to counteract or prevent misregulated host responses associated with severe disease. Parallel transcriptomic, proteomic, and targeted lipidomic profiling experiments have been used to identify differences in the induction of bioactive lipids mediating thc host inflammatory response to high and low pathogenicity strains of influenza virus. Targeted profiling of broncho-alveolar lavage samples from mice infected with the pathogenic mouse-adapted PR8/H1N1 virus, or lethal or sublethal doses of low pathogenicity X31/H3N2 virus revealed that the pathogenic phase of PR8 infection is associated with an elevation of 5-LOX proinflammatory metabolites (Tam et al., 2013). In contrast, the resolution phase of X31 infection is associated with elevated 12-LOX anti-inflammatory metabolites.

Building on these animal model studies, preliminary analysis of nasal wash samples from human influenza virus infections suggests that increased levels of 5-LOX-derived metabolites and decreased levels of 12-LOX-derived metabolites is also correlated with increased clinical symptoms and immune responses. Decreased production of metabolites from the 12/15-LOX pathway, DHA-derived protectin D1, and HDoHE in lung tissue is also correlated with high-pathogenicity viral infections (Morita et al., 2013). 12/15-LOX metabolites are thus markers of the resolution of inflammation during the host response to influenza virus, but whether they promote the resolution of the inflammatory response and protect from pathogenesis is unproven. Here, it is worth noting that Morita and colleagues identified protectin D1 as an inhibitor of influenza virus replication in human respiratory cells and found that its antiviral mechanism of action in cell culture is due to an effect on the nuclear export of viral transcripts. In the mouse model, administration of protectin D1 increases survival, but the extent to which this is due to an effect on viral replication versus mitigation of the host inflammatory response is less clear. These findings illustrate the reality that lipids and other small molecules can affect biological systems via multiple mechanisms, which may or may not be related. Detailed molecular characterization of lipid function in these systems is necessary to deconvolute and understand these effects.

4. CONCLUDING REMARKS: EXPLOITATION OF VIRAL DEPENDENCE ON HOST LIPIDS AS A SELECTIVE ANTIVIRAL STRATEGY

The study of lipid structure and function can be viewed as a challenge for systems biology. As the functional output of the expressed genome, the lipidome and metabolome represent the integration of all virus-induced changes in host transcription, RNA stability and processing, translation, posttranslational modifications, protein stability and localization, and enzymatic activity. A global view of these perturbations and their functional consequences is important for both basic and translational research efforts. As basic science, the interaction of viruses with host lipids is a fundamental aspect of viral replication and pathogenesis. We cannot understand the physical and molecular mechanisms driving viral processes without understanding the function of membranes associated with these processes and how, at a molecular level, lipid structure dictates function. Although our knowledge of the specific lipids required for viral replication and pathogenesis is arguably still too limited, the idea of antiviral strategies that exploit the interaction of viruses with host lipids is a compelling area for translational efforts. In part, this is because Nature already provides multiple examples in which the modulation of membrane composition is an effective defense against viral pathogens. For example, the interferon-inducible transmembrane proteins (IFITMs) exhibit broad-spectrum activity against many enveloped viruses and may do so by affecting the composition or packing of cellular membranes (Bailey et al., 2014). Consistent with this, the IFITMs affect intracellular lipid transport, and their presence in membranes enhances lipid packing and reduces membrane fluidity. Likewise, the effects of 25-hydroxysterol on sterol and sphingolipid synthesis are thought to directly affect the composition of host membranes targeted for viral fusion or used for other viral processes. Rational antiviral therapies that act through analogous mechanisms are therefore theoretically possible, although significant advances are needed in our understanding of how lipids function in viral infection to realize this potential.

We currently lack understanding of how lipid structure affects lipid function in vivo. This is a largely unexplored area of virus–host interactions despite clear significance for basic science and translational efforts. Ideally, tools analogous to site-directed mutagenesis studies of protein structure and function could be used to understand how seemingly subtle differences in structure (e.g., acyl chain length and degree and location of unsaturation, ratio of glycerophospholipids to sphingolipids, or head group charge and size) can lead to significant changes in function. In reality, this type of reductionist experiment is very challenging. Engineering advances that enable the synthesis of biomimetic-supported lipid bilayers from synthetic and cell-derived lipids in a medium-throughput manner may facilitate screens to identify specific lipids or combinations of lipids that are optimal for interaction with viral proteins. Viral processes, such as membrane fusion or penetration during entry or genome replication, may also be amenable to functional screens of this type that probe the relationship between lipid structure and membrane function in the associated virological process.

With respect to lipidomic profiling, current NMR and LC-MS-based approaches only permit the examination of lipid structure and abundance in biological samples derived from populations of cells. Determining the subcellular localization of lipids requires purification of the membrane of interest away from other cellular membranes or microscopy methods using fluorophore-conjugated lipid molecules or specific lipid-binding proteins to detect the lipid of interest. These approaches may be limited, respectively, because the membrane of interest cannot be biochemically isolated without perturbing it, because attachment of a fluorophore alters the physicochemical properties of the lipid molecule, and because the lipid-binding protein used for detection lacks sufficient specificity (Box 3). Alternative approaches that overcome these technical limitations are clearly needed, and emerging imaging methods may have a transformative effect on the study of lipids in viral processes in the cellular milieu. Raman-based microscopy methods rely on the vibrational frequency of chemical bonds and are especially well-suited to the imaging of lipids since they are rich in carbon-hydrogen bonds. Coherent Raman microscopy has been used to demonstrate variations in the degree of acyl chain saturation and order among droplets within the same cell and within the same droplet. This approach coupled with confocal fluorescence microscopy to detect viral proteins could be developed to study the chemical content of membranes associated with specific viral processes and to monitor the reactions that result in synthesis of these membranes in live cells. Here, isotopically labeled lipids of interest could be especially useful, since the imaging system can be tuned to detect only the isotopically labeled chemical bond and this could be used to distinguish a particular lipid species from other members of its class (e.g., visualization of deuterated desmosterol without background signal from endogenous cholesterol and other sterols). Imaging MS methods may also prove useful for monitoring lipid localization within the cell. Methods such as nanoscale secondary ion MS and desorption electrospray ionization MS can resolve the localization of a given metabolite in two-dimensional space and have been used to monitor lipid droplet formation in cells and tissues and may, with gains in spatial resolution and sensitivity, provide another tool for analysis of membrane-associated viral processes in cells.

As with all host-targeted antivirals, a crux issue is whether it is possible to inhibit viral replication or pathogenesis by targeting lipid metabolism without having undesired effects on the host. This concern arises especially because metabolism is central to host homeostasis and viability. The identification of lipid molecules that are required for viral replication but dispensable for the host, may present the best option for mitigating the risk of toxicity at the cellular or organismal level. Untargeted lipidomic profiling experiments and metabolomic flux studies are critical tools for this effort. The tools and approaches used to advance our understanding of how specific lipid molecules function in viral processes may also facilitate rational targeting of specific lipid molecules rather than wholesale blockade of entire biosynthetic pathways. This may enable minimization of toxicity and maximization of tolerability and safety. Rapid advances in methodology and understanding make this an exciting and rapidly evolving area of virology and systems biology.

REFERENCES

Altan-Bonnet N, Balla T. Phosphatidylinositol 4-kinases: hostages harnessed to build panviral replication platforms. Trends Biochem Sci 2012;37(7):293–302.

Bailey CC, Zhong G, Huang IC, Farzan M. IFITM-family proteins: the cell's first line of antiviral defense. Annu Rev Virol 2014;1:261–83.

Baron S, Levy HB. Some metabolic effects of poliomyelitis virus on tissue culture. Nature 1956;178(4544):1230–1.

Bigger CB, Guerra B, Brasky KM, Hubbard G, Beard MR, Luxon BA, et al. Intrahepatic gene expression during chronic hepatitis C virus infection in chimpanzees. J Virol 2004;78(24):13779–92.

Bloch KE. Sterol, structure and membrane function. Crit Rev Biochem Mol Biol 1983;14(1):47–92.

Diamond DL, Syder AJ, Jacobs JM, Sorensen CM, Walters KA, Proll SC, et al. Temporal proteome and lipidome profiles reveal hepatitis C virus-associated reprogramming of hepatocellular metabolism and bioenergetics. PLoS Pathog 2010;6(1):e1000719.

Filipe A, McLauchlan J. Hepatitis C virus and lipid droplets: finding a niche. Trends Mol Med 2015;21(1):34–42.

Heaton NS, Perera R, Berger KL, Khadka S, Lacount DJ, Kuhn RJ, et al. Dengue virus nonstructural protein 3 redistributes fatty acid synthase to sites of viral replication and increases cellular fatty acid synthesis. Proc Natl Acad Sci USA 2010;107(40):17345–50.

Holthuis JC, Menon AK. Lipid landscapes and pipelines in membrane homeostasis. Nature 2014;510(7503):48–57.

Ilnytska O, Santiana M, Hsu NY, Du WL, Chen YH, Viktorova EG, et al. Enteroviruses harness the cellular endocytic machinery to remodel the host cell cholesterol landscape for effective viral replication. Cell Host Microbe 2013;14(3):281–93.

Josset L, Belser JA, Pantin-Jackwood MJ, Chang JH, Chang ST, Belisle SE, et al. Implication of inflammatory macrophages, nuclear receptors, and interferon regulatory factors in increased virulence of pandemic 2009 H1N1 influenza A virus after host adaptation. J Virol 2012;86(13):7192–206.

Lee WM, Ahlquist P. Membrane synthesis, specific lipid requirements, and localized lipid composition changes associated with a positive-strand RNA virus RNA replication protein. J Virol 2003;77(23):12819–28.

Lee WM, Ishikawa M, Ahlquist P. Mutation of host delta9 fatty acid desaturase inhibits brome mosaic virus RNA replication between template recognition and RNA synthesis. J Virol 2001;75(5):2097–106.

Lindenbach BD, Rice CM. The ins and outs of hepatitis C virus entry and assembly. Nat Rev Microbiol 2013;11(10):688–700.

Martyna A, Rossman J. Alterations of membrane curvature during influenza virus budding. Biochem Soc Trans 2014;42(5):1425–8.

Morita M, Kuba K, Ichikawa A, Nakayama M, Katahira J, Iwamoto R, et al. The lipid mediator protectin D1 inhibits influenza virus replication and improves severe influenza. Cell 2013;153(1):112–25.

Morrison J, Josset L, Tchitchek N, Chang J, Belser JA, Swayne DE, et al. H7N9 and other pathogenic avian influenza viruses elicit a three-pronged transcriptomic signature that is reminiscent of 1918 influenza virus and is associated with lethal outcome in mice. J Virol 2014;88(18):10556–68.

Nour AM, Li Y, Wolenski J, Modis Y. Viral membrane fusion and nucleocapsid delivery into the cytoplasm are distinct events in some flaviviruses. PLoS Pathog 2013;9(9):e1003585.

Paul D, Bartenschlager R. Architecture and biogenesis of plus-strand RNA virus replication factories. World J Virol 2013;2(2):32–48.

Perera R, Riley C, Isaac G, Hopf-Jannasch AS, Moore RJ, Weitz KW, et al. Dengue virus infection perturbs lipid homeostasis in infected mosquito cells. PLoS Pathog 2012;8(3):e1002584.

Rabinowitz JD, Purdy JG, Vastag L, Shenk T, Koyuncu E. Metabolomics in drug target discovery. Cold Spring Harb Symp Quant Biol 2011;76:235–46.

Rodgers MA, Villareal VA, Schaefer EA, Peng LF, Corey KE, Chung RT, Yang PL. Lipid metabolite profiling identifies desmosterol metabolism as a new antiviral target for hepatitis C virus. J Am Chem Soc 2012;134(16):6896–99.

Rossman JS, Lamb RA. Viral membrane scission. Annu Rev Cell Dev Biol 2013;29:551–69.

Smith MW, Yue ZN, Korth MJ, Do HA, Boix L, Fausto N, et al. Hepatitis C virus and liver disease: global transcriptional profiling and identification of potential markers. Hepatology 2003;38(6): 1458–67.

Spann NJ, Garmire LX, McDonald JG, Myers DS, Milne SB, Shibata N, et al. Regulated accumulation of desmosterol integrates macrophage lipid metabolism and inflammatory responses. Cell 2012;151(1): 138–52.

Strating JRP, van der Schaar HM, van Kuppeveld FJM. Cholesterol: fa(s)t-food for enterovirus genome replication. Trends Microbiol 2013;21(11):560–1.

Su AI, Pezacki JP, Wodicka L, Brideau AD, Supekova L, Thimme R, et al. Genomic analysis of the host response to hepatitis C virus infection. Proc Natl Acad Sci USA 2002;99(24):15669–74.

Tam VC, Quehenberger O, Oshansky CM, Suen R, Armando AM, Treuting PM, et al. Lipidomic profiling of influenza infection identifies mediators that induce and resolve inflammation. Cell 2013;154(1):213–27.

Umashankar MC, Sanchez-San M, Liao M, Reilly B, Guo A, Taylor G, Kielian M. Differential cholesterol binding by class II fusion proteins determines membrane fusion properties. J Virol 2008;82(18):9245–53.

van der Schaar HM, Rust MJ, Chen C, van der Ende-Metselaar H, Wilschut J, Zhuang X, et al. Dissecting the cell entry pathway of dengue virus by single-particle tracking in living cells. PLoS Pathog 2008;4(12):e1000244.

Vastag L, Koyuncu E, Grady SL, Shenk TE, Rabinowitz JD. Divergent effects of human cytomegalovirus and herpes simplex virus-1 on cellular metabolism. PLoS Pathog 2011;7(7):e1002124.

Wu SX, Ahlquist P, Kaesberg P. Active complete in vitro replication of nodavirus RNA requires glycerophospholipid. Proc Natl Acad Sci USA 1992;89(23):11136–40.

Zaitseva E, Yang ST, Melikov K, Pourmal S, Chernomordik LV. Dengue virus ensures its fusion in late endosomes using compartment-specific lipids. PLoS Pathog 2010;6(10):e1001131.

Chapter 15

Mathematical Modeling

Solving Equations to Measure Viral Diseases—Math Rules

Alan S. Perelson
Theoretical Biology and Biophysics, Los Alamos National Laboratory, Los Alamos, NM, USA

Chapter Outline

1. HIV MODELING

1.1 The Simplest Model of Viral Dynamics

A very simple model of viral dynamics provided surprising insights into HIV infection. One critical element in modeling viral infections is keeping track of the change in viral load over time. For both HIV and hepatitis C virus (HCV), which generate chronic infections, one usually finds that after the acute phase of infection, the virus and host come into accommodation, such that the viral load when measured over periods of days, weeks, or months stays relatively constant. This constant level is called the viral set point. To gain information about the underlying processes that generate and clear the virus, one can perturb the system, for example, by drug therapy. When HIV chronically infected subjects were treated with the HIV protease inhibitor ritonavir, the level of plasma viremia decreased exponentially (Ho et al., 1995) (Figure 1). If one assumes that, before therapy, the viral load dynamics are due to viral production and clearance, one can construct the following differential equation to describe the time rate of change of the viral load, V:

$$\mathrm{d}V/\mathrm{d}t = P - cV, \tag{1}$$

where P is the rate of viral production and cV is the rate of clearance, with c being the rate of clearance per virus particle (virion) and V the virus concentration in plasma measured in terms of HIV RNA copies per milliliter. At the set point, $\mathrm{d}V/\mathrm{d}t=0$, the virus concentration does not change. When drug is given, virus production, P, is reduced. If we assume all production ceases, that is $P=0$, then this simple model predicts the viral level will change according to $\mathrm{d}V/\mathrm{d}t=-cV$, which has the solution $V(t)=V_0\exp(-ct)$, where V_0 is the baseline or set-point viral load. Therefore, this simple model predicts that the viral load will fall exponentially, as observed (Figure 1). Further, the rate of exponential decay, c, can be determined from the slope of the viral load decay curve when the natural logarithm of V, $\ln V(t)$, is plotted versus time. Thus, c can easily be determined from the data. In addition, at the pretherapy set point $\mathrm{d}V/\mathrm{d}t=0$ or $P=cV_0$, so the rate of viral production, before therapy, can be calculated from c and the baseline viral load, V_0. Because V_0 is measured in terms of HIV RNA per milliliter, one needs to multiply by the total fluid volume in which HIV can be found to determine the total body production rate of HIV. Assuming HIV is dispersed through the total extracellular body water, this amounts to 15 000 ml for a 70 kg person and yields a production rate of about 10^{10} virions per day (Perelson et al., 1996). This is an underestimate as the drug used, ritonavir, does not stop all viral production

Viral Pathogenesis. http://dx.doi.org/10.1016/B978-0-12-800964-2.00015-X

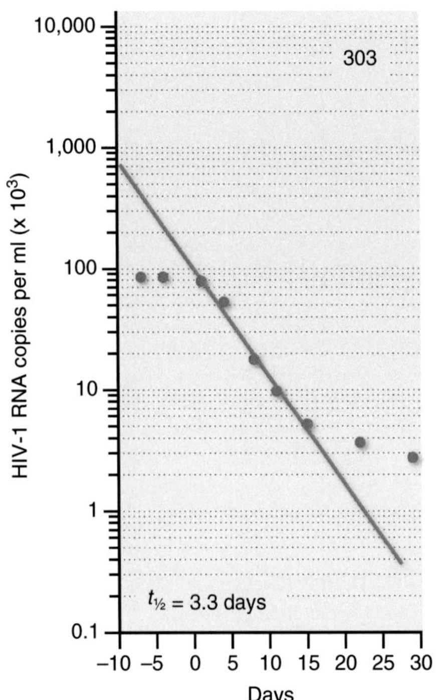

FIGURE 1 Decay of plasma viremia both before and after treatment was begun on day 1 with the protease inhibitor ritonavir. The solid line is the best-fit regression line and indicates that the initial viral decay in this patient is exponential. *Adapted with permission from Ho DD, Neumann AU, Perelson AS, Chen W, Leonard JM, Markowitz M. Rapid turnover of plasma virions and CD4 lymphocytes in HIV-1 infection. Nature;373:123–126.*

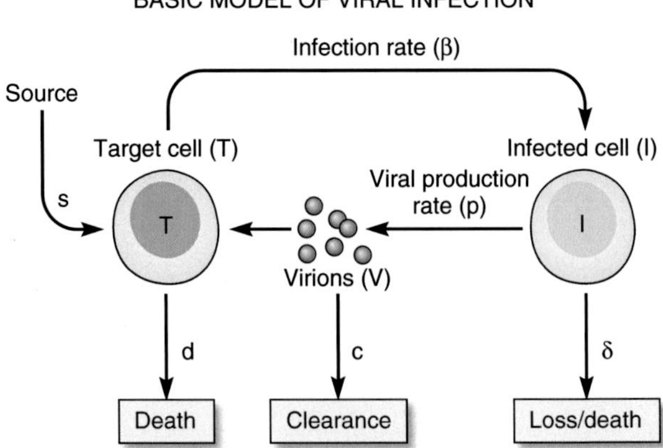

FIGURE 2 The basic viral dynamic model. Virions (red circles) infect target cells, T, and with infection rate constant β generate infected cells, I. Infected cells produce new virus particles, virions, at rate p per cell, and are lost by either death or cure at rate δ per cell. Virions are cleared from the circulation at rate c per virion. Target cells are created from a source at rate s and die at rate d per cell.

and hence the clearance rate must be higher than estimated from the data to account for the clearance of the continuing production. Conversely, this estimate may be inflated if the virus is not equally distributed throughout the total extracellular body water.

The remarkable thing about this calculation is that it shows how one can use data collected on therapy to determine how rapidly HIV is produced and cleared pretherapy. The original estimate of c using this approach suggested that HIV has a plasma half-life, $t_{1/2}\sim 2$ days, implying that every 2 days half of the body's HIV is produced (and cleared). Later estimates of the virion clearance rate using a method that does not rely on drug therapy (Ramratnam et al., 1999) showed that the plasma half-life is approximately 45 min and thus that the daily production of HIV is $\sim 10^{11}$ virions per day. These calculations totally changed the perception of HIV from a slow viral disease that played out over a decade to a picture of a highly dynamic, rapidly growing virus that was also rapidly being eliminated to produce a relatively constant set-point viral load.

The simple experiment of perturbing the set point with drug therapy in principle should be able to give additional insights. The assumption made above that ritonavir shuts off viral production is incorrect. In fact, protease inhibitors prevent viral maturation and cause noninfectious virus to be produced, as the polyprotein that encodes reverse transcriptase, integrase, and other viral proteins is not cleaved when protease is inhibited. These noninfectious viruses are still counted by HIV RNA assays. Thus, viral production is not directly influenced by this therapy and yet viral loads rapidly decrease when a protease inhibitor is given. To understand this apparent contradiction, let us look more closely at how virus is produced.

1.2 Basic Principles of Viral Dynamics

The simplest view of the events underlying viral infections are illustrated in Figure 2. Virus, V, interacts with cells susceptible to infection, so-called target cells, T, leading to the generation of infected cells, I, which then may produce a new generation of progeny virus particles. The diagram in Figure 2 can be converted into a set of equations, Eqn (B.1) given in Box 1, that describe the kinetics of the infection process. From the figure, one can immediately see that if a drug causes the production of noninfectious virus, then the infection process will be perturbed, but neither virus production from infected cells nor viral clearance will change. Therefore, to a first approximation, the viral level should not change unless the existing infected cells die rapidly and are not replaced because newly produced virus is noninfectious. Thus, Figure 2 suggests that the lifetime of productively infected cells should also be deducible from a careful analysis of the viral decline caused by administration of a protease inhibitor. Equation (B.2) in Box 1 can easily be solved if one assumes that during short-term therapy the number of target cells, T, remains at its baseline pretherapy steady-state value, $T_0=c\delta/\beta p$, and that therapy is with a 100% effective protease inhibitor ($\varepsilon_{PI}=1$, $\varepsilon_{RT}=0$). The solution is (Perelson et al., 1996)

Box 1 Models of viral infection

Models of viral infection are formulated as a system of ordinary differential equations. The basic model of viral infection contains the following three equations:

$$\begin{aligned} dT/dt &= s - d_T T - \beta VT \\ dI/dt &- \beta VT - \delta I \\ dV/dt &= pI - cV \end{aligned} \qquad (B.1)$$

where T denotes target cells, I denotes productively infected cells and V denotes virus. Target cells are generated at rate s, die at rate d_T per cell and become infected with rate constant β when virus, V, interacts with a target cell. Infected cells are generated at rate βVT and die at rate δ per infected cell. Last, virus, V, is produced by infected cells at rate p per cell and is cleared at rate c per virion.

Equation (B.1) been successfully used to model acute HIV infection if one assumes that initially all cells are uninfected and at time $t=0$ a small amount of virus is introduced into the body (Stafford et al., 2000). The model predicts that virus initially grows exponentially, reaches a peak, and then settles at a constant level, called the set point (Figure B1).

The model has also been used to analyze the effects of antiretroviral therapy (ART) by incorporating into the model the effects of antiretroviral drugs. For example, reverse transcriptase (RT) inhibitors block the ability of HIV to productively infect a cell. HIV protease inhibitors cause infected cells to produce immature noninfectious viral particles, V_{NI}. Thus, in the presence of these drugs, the model equations become,

$$\begin{aligned} dT/dt &= s - d_T T - (1 - \varepsilon_{RT})\beta VT \\ dI/dt &= (1 - \varepsilon_{RT})\beta VT - \delta I \\ dV_I/dt &= (1 - \varepsilon_{PI})pI - cV_I \\ dV_{NI}/dt &= \varepsilon_{PI}pI - cV_{NI}, \end{aligned} \qquad (B.2)$$

where ε_{RT} and ε_{PI} taken on values between 0 and 1 and represent the efficacies of RT and protease inhibitors ($\varepsilon=1$ being a 100% effective drug). Further, V_I and V_{NI} are the concentrations of "infectious" and noninfectious virus, respectively, and $V=V_I+V_{NI}$ is the total virus concentration. An HIV entry inhibitor can be modeled in the same way as an RT inhibitor. Modeling the effect of an HIV integrase inhibitor is more complicated and requires one to separate the infected cell population into pre- and postintegration subpopulations.

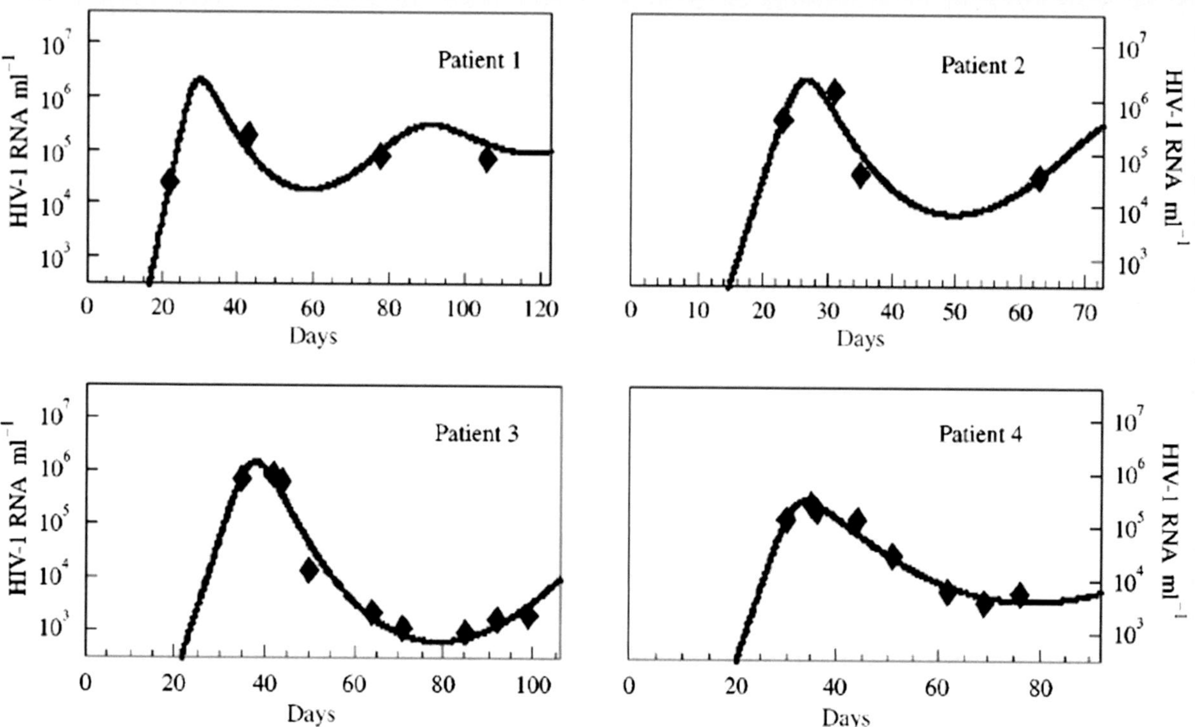

FIGURE B1 Fit of basic model, Eqn (B.1), to plasma viral load data from four acutely infected patients. *Adapted with permission from Stafford MA, Corey L, Cao Y, Daar ES, Ho DD, Perelson AS. Modeling plasma virus concentration during primary HIV infection. J Theor Biol 2000;203(3):285–301.*

$$V(t) = V_0 \exp(-ct) + \frac{cV_0}{c-\delta}\left(\begin{array}{l} \frac{c}{c-\delta}\left[\exp(-\delta t) - \exp(-ct)\right] \\ -\delta t \exp(-ct) \end{array}\right), \qquad (2)$$

where V_0 is the set-point viral load before initiation of therapy. This solution only depends on three parameters, V_0, c, and δ, where δ, the death rate of productively infected cells is the new parameter of interest. Allowing the target cell concentration, T, to vary necessitates using numerical methods to predict $V(t)$ but does not substantially alter the outcome of the analysis.

Fitting either the analytical solution (2) or the numerical solution of $V(t)$ to viral decline data allowed a minimal estimate of $\delta = 0.5$/day to be obtained (Figure 3) (Perelson et al., 1996). The estimate is minimal because, in reality, protease inhibitors are not 100% effective, and there will be small, continuing generation of new, productively infected cells (Perelson et al., 1996). Analysis using the same method but fitting to data obtained from patients treated with a very potent four-drug combination that is closer to being a 100% effective yielded an estimate of $\delta = 1.0$/day (Markowitz et al., 2003). This implies that productively infected cells have an average life span, $1/\delta$, of about 1 day while producing virus and decay with a half-life $t_{1/2} = \ln 2/\delta = 0.7$ days.

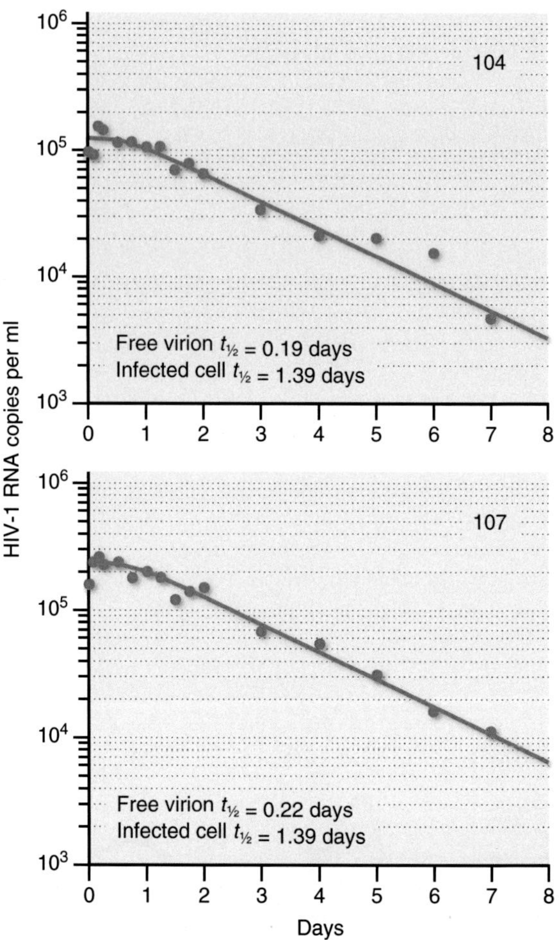

FIGURE 3 Fit of model given by Eqn (B.2) in Box 1 to viral load data obtained from two patients treated with the HIV protease inhibitor ritonavir. From the fit one can determine the parameters c and δ and hence the half-lives of virus in circulation and of infected cells. *Reproduced with permission from AAAS. Perelson AS, Neumann AU, Markowitz M, Leonard JM, Ho DD. HIV-1 dynamics in vivo: virion clearance rate, infected cell life span, and viral generation time. Science 1996;271:1582–1586.*

1.3 Long-Lived Infected Cells

When a potent combination of three antiretrovirals was given to HIV-infected patients, a rapid initial decline of HIV was again observed. But, after 1–2 weeks of therapy, the decline slowed, and a second phase of decline was uncovered (Figure 4). The exact origin of this second phase has been debated, but one possibility is that there exists a population of cells that live considerably longer than 1 day while productively infected. Such cells might be infected macrophages, which in cell culture can live for weeks. Or they may be infected resting T cells, which A. Haase and collaborators have shown to produce much less virus than activated CD4+ T cells (Reilly et al., 2007), the primary target cells for HIV, and hence may live longer while infected. If one constructs a generalization of the basic model with two types of productively infected cells, one short-lived and one long-lived, one can fit the long-term viral decays seen in patients treated with combination antiretroviral therapy (cART). Based on the rate of decay seen in the second phase,

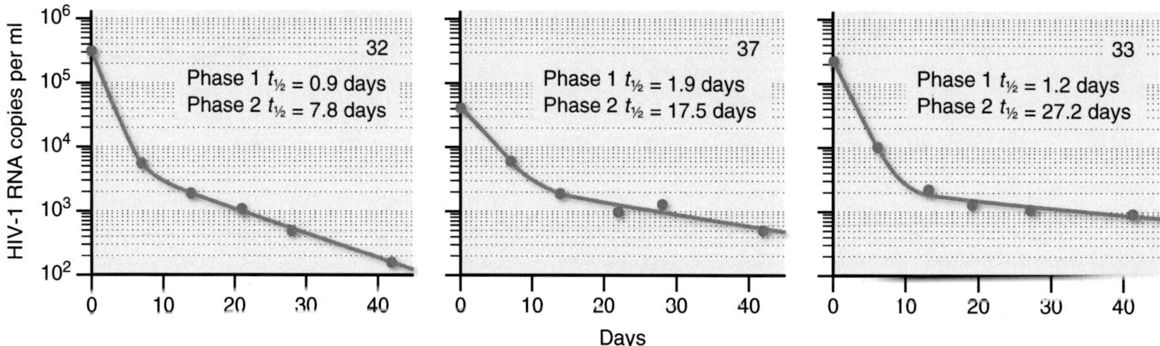

FIGURE 4 Two-phase viral decline kinetics. *Reproduced with permission from Perelson AS, Essunger P, Cao Y, Vesanen M, Hurley A, Saksela K, Markowitz M, Ho DD. Decay characteristics of HIV-1-infected compartments during combination therapy. Nature 1997;387:188–191.*

and an estimate of the number of infected long-lived cells at the time therapy is begun, one estimates that 2–3 years of 100% effective therapy could eliminate the long-lived infected cell population (Perelson et al., 1997). Therapy of this duration was shown not to eradicate all HIV-infected cells, likely due to the presence of even longer lived viral reservoirs, as stopping therapy leads to viral load rebounding back toward baseline.

1.4 Latently Infected Cells

When HIV infects a cell, its RNA genome is reverse transcribed into a DNA copy. This DNA is then transported into the nucleus where under the action of the HIV enzyme integrase the DNA is spliced into the infected cell's genome. This integrated copy of HIV's genome, called a provirus, is replicated along with the host cell's DNA if the cell divides, so that the cell's progeny will also carry the provirus. After infection, the provirus is usually transcribed into messenger RNA, new viral particles are produced and the cell is classified as productively infected. However, in rare cases, the provirus may remain transcriptionally silent, the infected cell producing no virus. Such cells are called latently infected. Latently infected cells are typically resting CD4+ memory T cells, and as such have long life spans.

Due to a change in a latently infected cell's environment, for example, encountering a cognate antigen that activates the resting T cell, a latently infected cell can transition into a productively infected cell. Thus, as long as latently infected cells are present, productive infection can be rekindled and HIV cannot be eradicated.

Using standard clinical assays, HIV viral loads can be measured down to 50 HIV RNA copies/ml. When the viral load falls below this threshold, the virus is said to be undetectable. In treated patients, HIV viral loads drop below the detection threshold during the second phase of decline, after a few months of treatment. However, using more sensitive "single copy" assays and larger volumes of blood, HIV RNA can be measured to levels below 1 copy/ml. Using such assays to follow HIV declines in patients on cART, whose viral loads have fallen below the 50 copy/ml standard threshold, has shown that HIV RNA continues to decay but at a yet slower rate. In one study, the half-life of this third phase was estimated to be 39 months (Palmer et al., 2008). In patients treated for long periods with no detectable viremia for many years, single copy assays may detect a few copies of HIV RNA per milliliter that persist and exhibit no noticeable decay. This has been called the fourth phase of decay (Palmer et al., 2008), but it might more appropriately be called a low-on-therapy set point.

Various models have been developed that include latently infected cells. Under potent cART, there may be very little ongoing replication suggesting that the activation of latently infected cells may be the source of the low-level residual viremia detected in long-term treated subjects. Models predicted that for low-viremia levels to be stable required that latently infected cells also attained a stable level. However, the latent cell activation rate required to sustain observed viral loads would rapidly deplete and eradicate the reservoir. Kim and Perelson (2006) therefore made the model-supported prediction that latently infected cells undergo homeostatic proliferation, which would maintain reservoir stability over long times. This was later shown to be the case by Chomont et al. (2009). Rong and Perelson (2009) proposed that, when a latently infected cell divides, both daughter cells need not remain latent; an asymmetric division could occur in which one daughter cell remained latent and the other became activated. Whether this occurs is not yet known.

One of the current clinical challenges in HIV research is to find methods of eliminating the latent reservoir. A number of pharmaceutical agents are being developed and tested for their ability to "wake up" latently infected cells and cause them to begin transcribing their latent provirus into HIV RNA. If such cells produce enough HIV RNA and HIV proteins, they might become targets for cytotoxic T cells or the cells might be killed by viral cytopathic effects. This approach has been called "shock and kill." Histone deacetylase inhibitors, such as vorinostat, have been shown to increase HIV RNA expression both in vitro and in vivo (Archin et al., 2012), thus applying the shock. Unfortunately, the shock induced by this agent has not been potent enough to lead to the kill, but research in this area continues. Mathematical models that include the action of latency-reversing agents are under development to help interpret clinical efforts to achieve viral eradication.

2. HEPATITIS C VIRUS

HCV is a positive strand RNA virus that primarily infects human hepatocytes. About 180 million people are infected worldwide and, as is the case for HIV, there is no vaccine. The standard therapy for HCV for over a decade has been a combination of interferon-α (IFN), or a longer acting form, pegylated interferon-α (PEG-IFN), with a nucleoside analog ribavirin (RBV). When given for 48 weeks, this combination was able to cure slightly less than 50% of patients with HCV genotype 1 infection, the most common form of infection in the United States. In May 2011, the first direct acting antiviral (DAA) for the treatment of HCV, the HCV protease inhibitor telaprevir, was approved. Subsequently, many other DAAs have been approved, including the protease inhibitors boceprevir and simeprevir, the NS5A inhibitors ledipasvir and ombitasvir, the nonnucleoside polymerase inhibitor dasabuvir, and the nucleotide HCV polymerase inhibitor sofosbuvir.

Mathematical modeling has played an important role in characterizing HCV viral kinetics, determining basic quantities such as the life span of infected cells and the clearance rate of HCV from circulation. Methods similar to those that we discussed in the context of HIV were used, but with some significant differences. For example, for HIV, we can only estimate the effectiveness of one therapy relative to another. But, as shown below, the absolute effectiveness of antivirals in blocking HCV production from infected cells can be determined by fitting appropriate models to short-term clinical data. Modeling has also played an essential role in quantifying the effectiveness and the mode of action of a variety of HCV antiviral agents, as well as the duration of therapy needed to cure the infection.

2.1 HCV Kinetics

As in the case of HIV, in a patient with chronic HCV infection the level of HCV RNA as measured in blood attains a set point, a viral load that remains approximately constant when measured over periods of days, weeks, and months. If this set point is perturbed by administering antiviral therapy, say with IFN-α, then the viral load declines in a biphasic manner with a rapid first phase followed by a second phase (Figure 5). Unlike HIV, where the first phase decline may involve a fall of one or two logs over the course of a week, for HCV the decline can be one or two logs over the first day of therapy with IFN and three logs or more with DAA therapy. In order to understand this, we can go back to the basic viral dynamic model shown in Figure 2. Interferon has been generally thought of as an agent that can put cells into an "antiviral state," where they are not susceptible to infection. However, as we discussed in the context of HIV therapy, if all a drug does is prevent new infection of uninfected cells, then to first approximation viral load should not fall, since infected cells still produce virus and virus is still being cleared. Using this argument, we then inferred that HIV infected cells should be short-lived and we estimated that the lifetime of productively infected cells is about 1 day. However, unlike HIV where the first-phase viral load decline continues for about a week, HCV viral loads fall multiple logs during the first day of therapy. To explain such a rapid viral load decline by a short half-life of infected cells would lead one to conclude that HCV-infected cells only live a few hours. Were that the case, then the liver would be rapidly destroyed by HCV infection, as liver cell proliferation could not keep up.

To understand the rapid decline in HCV, one must invoke a different mechanism. Neumann et al. (Neumann et al., 1998) suggested that IFN acts as direct antiviral and causes the production of HCV from already infected cells to decrease. Mathematically, they suggested that the basic

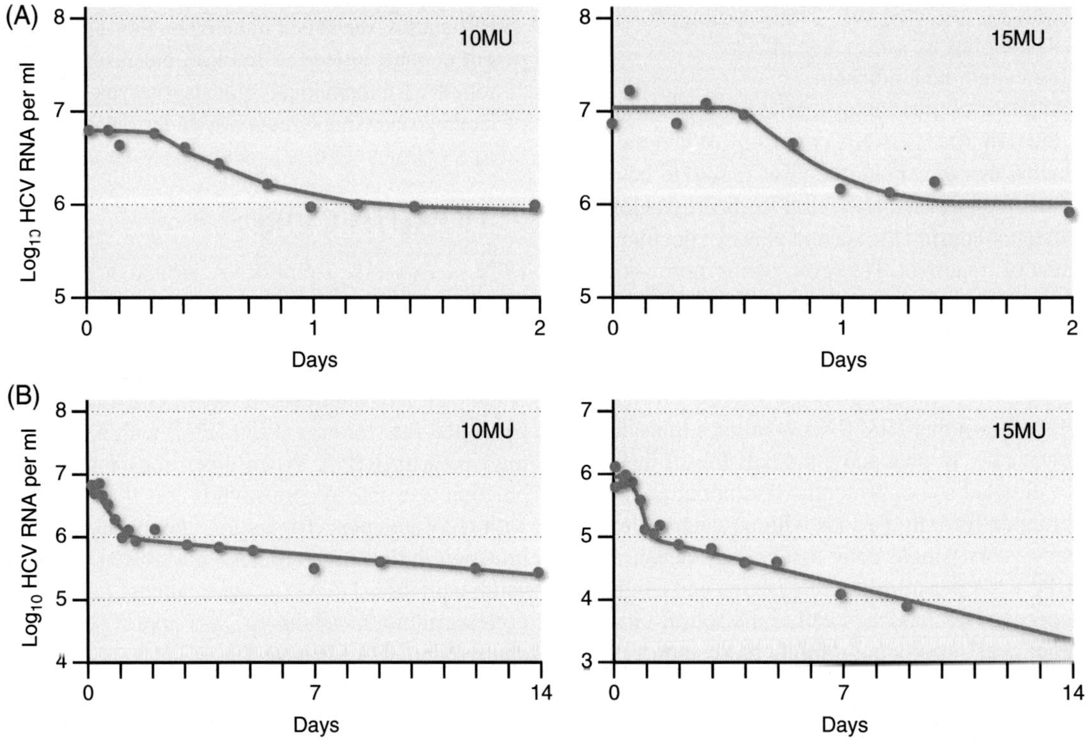

FIGURE 5 (A) Viral declines over the first two days of IFN therapy for a patient treated daily with 10 million units (MU) of IFN and a patient treated daily with 15 MU of IFN. (B) Continuation of these viral declines for 14 days shows the existence of a second phase of decline. Data (solid circles), best-fit model predictions (solid lines).

model Eqn (B.1) in Box 1, should be modified such that the viral equation becomes

$$dV/dt = (1-\varepsilon)pI - cV, \tag{3}$$

where ε is called the effectiveness of therapy in blocking viral production, with ε varying between 0 and 1, $\varepsilon=1$ being a 100% effective drug. Note that, at the pretherapy set point, $pI_0 = cV_0$, where the subscript 0 denotes the baseline value before therapy. If one assumes that over the first day of therapy the number of infected cells stays at I_0, then Eqn (3), with the initial condition $V(0)=V_0$, can be solved to yield

$$V = V_0\,,\ \text{for}\ \ t < t_0$$

$$V(t) = V_0\left[1-\varepsilon+\varepsilon\exp\left(-c\left(t-t_0\right)\right)\right],\ \text{for}\ t \geq t_0 \tag{4}$$

where t_0 is the pharmacological delay before IFN becomes effective. Interferon is a cytokine that binds IFN receptors and signals through these receptors, turning on hundreds of interferon-stimulated genes, whose gene products ultimately reduce viral production within an infected cell. The period of time, t_0, before any effect of IFN on viral load is seen is called the pharmacological delay. Equation (4) fits the viral load declines seen during the first two days of therapy well (Figure 5(A)) and allows one to estimate the effectiveness of therapy, ε, as well as the viral clearance rate, c. For IFN at doses of 10 or 15 MU daily the effectiveness was about 95%. Further, c was estimated as 6.2/day, which corresponds to an HCV $t_{1/2}=2.7$ h.

One can deduce from Eqn (4) that the initial viral load, V_0, will fall to a value equal to $V_0(1-\varepsilon)$ because the exponential term in Eqn (4), $\exp(-c(t-t_0))$, rapidly approaches 0. Note that this implies that, if a drug is 90% effective, that is if $\varepsilon=0.9$, then the viral load will rapidly decline by one log, whereas if it is 99% effective the decline will be two logs. Thus, from simply reading the magnitude of the first phase viral load decline one can get an estimate of the effectiveness of an HCV antiviral agent. Moreover, this estimate can be made in a one- or two-day clinical trial. For this reason, the first HCV protease inhibitor to go into human trials, BILN-2061, was given for only 2 days (Hinrichsen et al., 2004) and, using a viral kinetic (VK) model, the effectiveness ε was estimated to be above 99.5% in patients with mild disease (Herrmann et al., 2006).

At the end of the first phase, say with a 90% effective drug, the viral load has fallen 10-fold. Consequently, when an infected cell dies, it is less likely to be replaced with another infected cell as there is 90% less virus to infect new cells. We therefore expect the level of infected cells to continually decrease. Further, if there are fewer infected cells, less virus will be produced and the viral load will continue to fall. This fall corresponds to the second-decay phase. If we expand the mathematical model given by Eqn (3) to include the equation for infected cells given by Eqn (B.1) in Box 1, that is

$$dI/dt = \beta VT - \delta I, \tag{5}$$

and assume that over short treatment periods, for example, the 14 days in Figure 5, the number of target cells can be approximated by its constant, pretherapy, baseline level, ($T_0 = c\delta/p\beta$), and that pre-treatment the system is at steady-state, then Eqns (3) and (5) can be solved, yielding (Neumann et al., 1998)

$$V = V_0\,,\ \text{for}\ \ t < t_0$$

and

$$V = V_0\left[Ae^{-\lambda_1(t-t_0)} + (1-A)\,e^{-\lambda_2(t-t_0)}\right],\ \text{for}\ t \geq t_0, \tag{6}$$

where $\lambda_{1,2} = \frac{1}{2}\left[c+\delta\pm\sqrt{(c-\delta)^2+4(1-\varepsilon)c\delta}\right]$, $A = \frac{\varepsilon C-\lambda_2}{\lambda_1-\lambda_2}$ and t_0 is the pharmacological delay. One can show that when $c \gg \delta$, that is when the rate of virion clearance is much faster than the rate of infected cell loss, which is the case for HCV, $\lambda_1 \sim c$ and $\lambda_2 \sim \varepsilon\delta$. Thus, the rate of decline in the first phase reflects the rate of viral clearance, c, and for highly effective therapies (ε close to 1), the rate of second phase decline, δ, reflects the rate of infected cell loss. Fitting Eqn (6) to viral decline data obtained from patients treated with IFN, δ was found to be ~0.14/day.

This theory has been applied to fit data from scores of patients treated with IFN or PEG-IFN and RBV and provides excellent fits to the data. Two examples are given in Figure 5(B). For patients treated for longer than 2 weeks, changes in the number of target cells, T, may become important. Hepatocytes are known to proliferate in order to regenerate the liver if it is injured. Therefore, a generalization of the basic model includes the possibility of target cells proliferating or both target cells and infected cells proliferating (Dahari et al., 2007a,b). Such a model was used by Snoeck et al. (2010) to predict whether or not a patient treated with PEG-IFN alone or with PEG-IFN plus RBV for 48 weeks would be cured, that is cross a "cure boundary" of having less than one virus in the extracellular body water and less than one infected cell in the liver, based solely on that patient's early viral load decay data. Examining data from over 2000 patients they were able to correctly predict cure 99.3% of the time and correctly predict cure failure 97.1% of the time. This shows that a simple viral dynamic model can be used in a clinical setting to determine whether a patient should stop therapy, potentially avoiding side-effects and the high cost of therapy, or continue for the full 48 weeks of therapy.

2.2 Drug Pharmacokinetics and Pharmacodynamics

PEG-IFN is dosed once a week and the effectiveness of the drug might wane between doses. Models have therefore been developed in which the effectiveness of the drug is not taken as a constant, but rather is allowed either to vary as a function of time since the last dose, or as a function of the drug concentration as measured in plasma. Models in which the effectiveness is constant have been called constant effectiveness or CE models. Models in which the effectiveness is a direct function of drug concentration are called pharmacodynamic (PD) models. A commonly used PD model is the ε_{max} model in which the effectiveness, $\varepsilon = \frac{\varepsilon_{max} C(t)^n}{C(t)^n + EC_{50}^n}$, where $C(t)$ is the drug concentration, n is a Hill coefficient that determines the steepness of the dose–response curve, and EC_{50} is the drug concentration that corresponds to 50% of the maximum effectiveness, E_{max}. For HCV drugs, ε_{max} is generally taken to be 1. The concentration of the drug is then usually given by a pharmacokinetic (PK) model. For example, PEG-IFN is given once weekly by subcutaneous injection. The drug is absorbed into the blood and then eliminated. A PK model for a single dose of PEG-IFN is then given by (Powers et al., 2003)

$$C(t) = \frac{FD}{V_d} \frac{k_a}{k_e - k_a} \left[e^{-k_a t} - e^{-k_e t} \right] \quad (7)$$

where $C(t)$ is the PEG-IFN concentration in blood, k_a is the absorption rate constant and k_e is the elimination rate constant, D is the drug dose delivered, F is the bioavailability, that is the fraction of drug that is biologically active, and V_d is the volume through which the drug distributes. When multiple doses are given one needs to also account for any drug remaining from the prior dose. This formula predicts that the drug level increases after a dose is given, reaches a peak and then falls as shown in Figure 6. This PK model can then be coupled to a PD model, such as the E_{max} model to yield a drug effectiveness, and this effectiveness can then be used in a standard VK model thus yielding a PK/PD/VK model. When this was done for PEG-IFN, with measured drug concentrations fit to the PK model, it predicted that as drug levels fell the viral load would rebound, as occurs in patients (Powers et al., 2003; Talal et al., 2006) (Figure 6).

Typically, drug concentrations are not measured in every patient treated with antiviral drugs. In most circumstances viral kinetics but not drug pharmacokinetics are available for modeling purposes. In such circumstances, one can replace a detailed PK/PD model by an empirical time-varying effectiveness, $\varepsilon(t)$. For example, if drug is delivered frequently so that to a good approximation the drug concentration simply increases to a steady-state value, one can assume the effectiveness also increases and reaches a maximum steady-state value. The following varying effectiveness model (Shudo et al., 2008) has therefore proven useful.

$$\varepsilon(t) = \varepsilon_{max}(1 - \exp(-kt)) \quad (8)$$

where k is a constant controlling how rapidly the effectiveness increases.

2.3 HCV RNA Decline Kinetics with DAAs

The original VK theory of HCV was developed to explain the viral kinetics observed with IFN-based therapy. With DAAs, as with IFN-based therapies, the viral declines tend to be biphasic and can be fit with the standard viral dynamic model using either constant effectiveness or time-varying effectiveness depending on the drug. Interestingly, when the first HCV protease inhibitor to be approved by the FDA, telaprevir, was modeled, the biphasic model fit well but the estimates for the parameters c and δ, were significantly different than what had been found previously for IFN. The parameter c was estimated to about 12/day, which is twice the value previously estimated for patients treated with IFN, and δ was estimated as 0.56/day, about fourfold larger than the prior estimate of 0.14/day deduced with IFN-based therapies. As both c and δ, were thought to represent physiological processes of viral clearance and infected cell loss, it

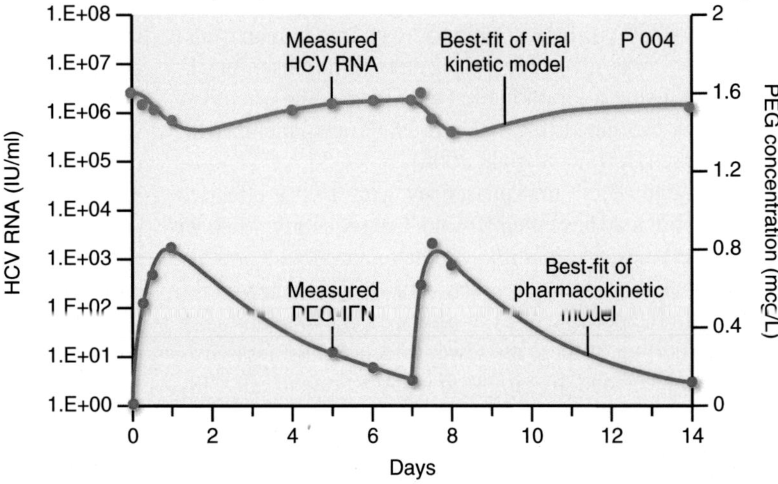

FIGURE 6 Patient given doses of PEG-IFN at day 0 and day 7 and the viral kinetic (VK) response. Blue symbols represent measured levels of HCV RNA and the blue solid line represents the best-fit of the VK model to the data. Red symbols represent the measured PEG-IFN concentration in blood and the red solid line represents the best-fit pharmacokinetic model.

was not clear why these discrepancies arose. Further, when VK data obtained with the HCV NS5A inhibitor daclatasvir was analyzed, the resulting estimate of c was 23/day. While both telaprevir and daclatasvir generated biphasic declines as predicted by the simple model, it was no longer clear that the basic viral dynamic model was properly accounting for the underlying biology.

For telaprevir, it was also shown that the estimated value of δ was positively correlated with $-\log_{10}(1-\varepsilon)$, that is the estimated magnitude of the first phase viral decline measured on a log scale. Thus, the more effective telaprevir was in blocking viral production, the faster the second phase decline. This correlation suggests a possible explanation for the high values of δ estimated for telaprevir. HCV is a +strand RNA virus that replicates in the cytoplasm. Unlike HIV, it does not integrate in the host cell's genome and therefore, if HCV RNA degrades, a cell can be cured. If a drug can effectively block HCV replication, the natural intracellular mechanisms of RNA degradation may be able to outpace HCV replication and lead to cure of the cell. Rather than solely representing the death rate of infected cells, δ may thus represent the loss of infected cells by the combined effects of cell cure and death.

2.4 Predicting the Duration of Therapy

Regardless of its origin, the very rapid second phase decline under DAAs had practical implications for the duration of therapy needed to cure HCV. Guedj and Perelson (2011) used the parameter estimates made by fitting a viral dynamic model to the HCV decline data observed in 44 patients treated with telaprevir to infer the duration of therapy required to clear the infection, assuming that the viral declines would continue at the same pace as when the viral load was detectable. To do this they constructed 10,000 in silico patients, each with a set of parameters picked from the observed distribution of parameter estimates, and simulated their viral declines. As in reality, each in silico patient had different viral decline kinetics, with some patients reaching the cure boundary faster than others. Overall they found that, by the end of 7 weeks of therapy 95% of the in silico patients were predicted to have eliminated all virus. The results of the first 6 weeks combination therapy trial for HCV, the SYNERGY trial, showed that 38 of 40 patients (95%) were cured with a three-drug DAA combination that was more potent than telaprevir monotherapy. The theoretical prediction of possible cure after 7 weeks with a less-potent regime may very well have provided the impetus for this trial.

2.5 Multiscale Models

By examining processes that occur intracellularly, we can gain insight into why estimates of the viral clearance rate, c, may depend on the drug being administered. At the time drug is administered, many infected cells will have already replicated HCV RNA. Some of this RNA will be in the cytoplasm waiting to be packaged and assembled into viral particles, while other HCV RNA molecules may already be in intracellular viral particles on their way to exiting the cell via its secretory pathway. Cells with this HCV RNA inside will continue to secrete virus even if new RNA replication is inhibited. Our earlier modeling assumption that viral production simply gets reduced by a factor $(1-\varepsilon)$ may therefore not be correct, and some drugs (e.g., IFN) may allow the HCV RNA within infected cells to continue to get packaged and secreted in virus particles. As a result, the earlier estimates of c based on the assumption that most virus production was shutoff may not be correct. If continuing viral production was reflected in the rate of viral load decay, then the underlying viral clearance must be more rapid.

The HCV NS5A protein has no known enzymatic function but appears to play roles in both viral RNA replication and in assembly/secretion of viral particles. Guedj et al. (2013) hypothesized that the NS5A inhibitor daclatasvir inhibited both functions. They constructed a model that incorporated intracellular HCV RNA replication, degradation, and assembly/secretion of viral RNA in virions and coupled that to the standard viral dynamic model (Figure 7). They used this model not only to fit VK data obtained with daclatasvir, but also to fit data obtained with IFN and telaprevir, and showed that all three data sets were consistent with $c=23$/day when the various drugs intracellular effects were more precisely modeled. Interestingly, this clearance rate corresponds to a $t_{1/2}=45$ min, which is also the best estimate of the HIV half-life in circulation. The model also predicted that both daclatasvir and telaprevir had an effect in blocking both viral assembly/secretion and viral replication, whereas IFN mainly affected replication. These predictions have subsequently been validated in vitro.

These analyses show that the basic viral dynamic model is a good starting point in modeling any viral infection but the details of how drugs act and how a virus replicates may be crucial in getting a more accurate and complete understanding of observed viral dynamics.

3. DRUG RESISTANCE

When HIV-infected or HCV-infected patients are treated with a single agent, such as the HIV protease inhibitor ritonavir or the HCV protease inhibitor telaprevir, drug-resistant variants are frequently selected, causing the viral load to rebound. In order to understand why this occurs, consider an antiviral drug against HCV in which a single nucleotide change in the drug target makes the drug less effective, that is generates some resistance. Due to the rapid replication of HCV, all possible single point mutations of any dominant viral variant are generated multiple times per day. To see

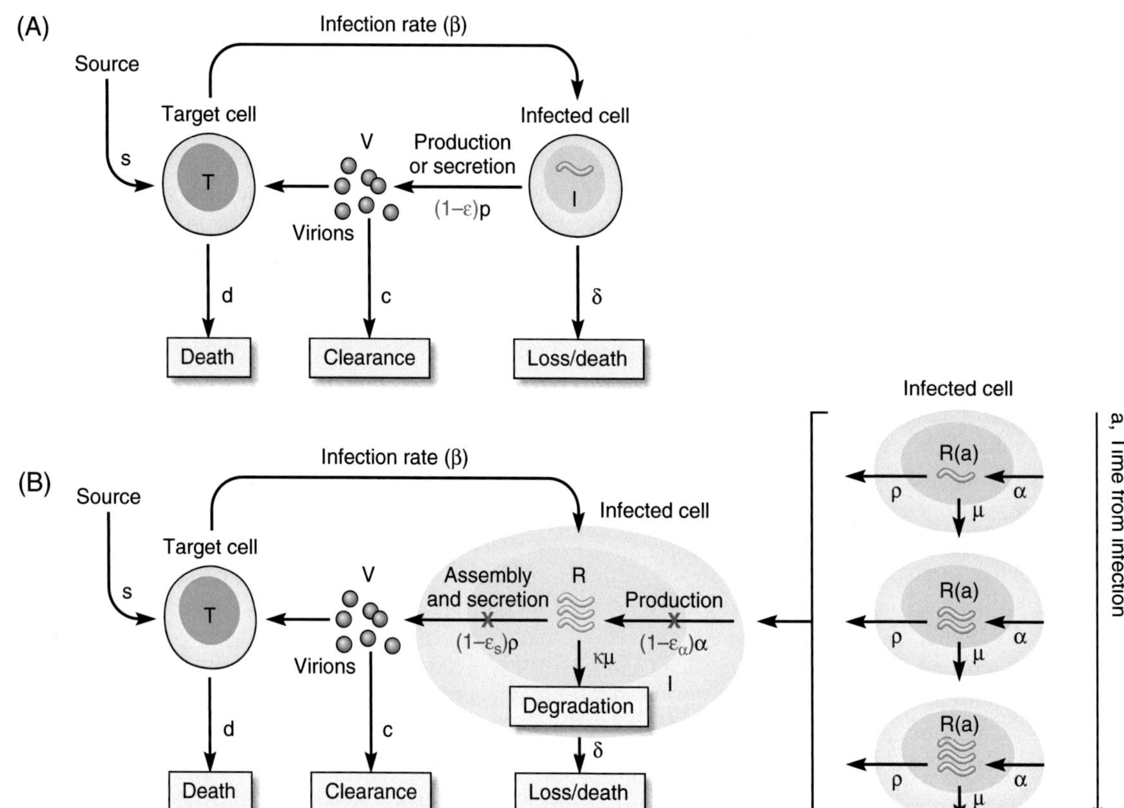

FIGURE 7 Comparison of basic model of viral infection (A) in which therapy is assumed to reduce virion production from infected cells by a factor of $(1-\varepsilon)$ with a multiscale model (B) designed to account for essential features of intracellular HCV RNA (vRNA) replication, R, that is production, degradation, and assembly/secretion with rate parameters α, μ, and ϱ, respectively. The vRNA level within an infected cell (colored oval) is assumed to increase with time since infection and reach a steady state. Treatment (parameters in red) may block vRNA production with effectiveness ε_α and/or virion assembly/secretion with effectiveness ε_s, and/or enhance the degradation rate of vRNA by a factor κ. *Reproduced with permission from Guedj J, Dahari H, Rong L, Sansone N, Nettles RE, Cotler SJ, Layden TJ, Uprichard SL, Perelson AS. Modeling shows that the NS5A inhibitor daclatasvir has two modes of action and yields a shorter estimate of the hepatitis C virus half-life. Proc Natl Acad Sci USA 2013;110:3991–3996.*

this, note that the HCV genome is approximately 10^4 bases long. Since there are 4 letters in the genetic code, each base can mutate to one of the three others. There are therefore only 30,000 possible single point mutations that can occur. With HCV, each virus contains a single RNA molecule, which was produced by the action of an error-prone RNA-dependent RNA polymerase, with an in vivo error rate estimated as 2.5×10^{-5} per base copied (Ribeiro et al., 2012). Thus, each time an HCV RNA is copied on average 0.25 bases are miscopied.

However, some HCV RNAs will have no bases miscopied, some will have one miscopied, some will have two miscopied, etc. The expected number with k bases miscopied can be calculated from the binomial distribution or its Poisson approximation (Table 1). The number of HCV RNAs produced per day was estimated by Neumann et al. (1998) using $c=6$/day as $\sim1.3\times10^{12}$. With the updated estimate of $c=23$/day, the daily production rate is now estimated as 5×10^{12}. In Table 1, we show the number of virions expected to be produced each day with 0, 1, 2, and 3 nucleotide changes and compare them to the number of all possible mutants with 1, 2, or 3 nucleotide changes. Given that there are only 30,000 possible single point mutations, one can compute from the table that each point mutation is made more than 10 million times per day. The number of possible double mutations, that is pairs of bases miscopied, is $10{,}000\times9{,}999\times3^2/2=4.5\times10^8$, since if one has n objects the number of distinct pairs is $n(n-1)/2$. The table shows that each possible double mutant is made hundreds of times per day. The table also shows that only a tiny fraction of all possible triple mutants are made each day. Consequently, if an HCV-infected individual is treated with a drug or set of drugs for which two mutations can generate highly resistant viruses fit enough to grow, drug resistance will probably be observed, whereas if three mutations are needed resistance is unlikely to be observed, assuming the treated individual is compliant with therapy. Thus, combinations of drugs with a "genetic barrier" of greater than two mutations need to be employed. A very similar argument was made about the need for combination therapy for HIV and influenza A virus infection, with the exception that for HIV mutations generally occur during reverse

TABLE 1 Probabilities and Rates of Generation of Various Hepatitis C Virus Mutants (See Text for Explanation)

Number of Nucleotide Changes	Probability	Number of Virions Generated per Day	Number of All Possible Mutants	Fraction of All Possible Mutants Created per Day
0	0.779	3.9×10^{12}		
1	0.195	9.8×10^{11}	3.0×10^{4}	1
2	0.024	1.2×10^{11}	4.5×10^{8}	1
3	0.002	1.0×10^{10}	4.5×10^{12}	2.3×10^{-3}

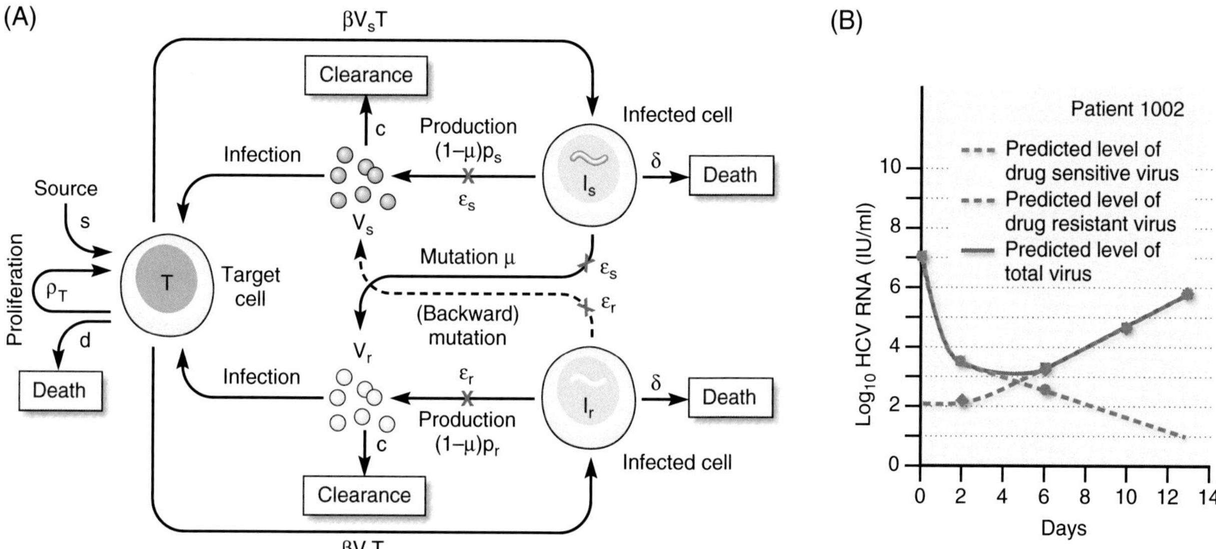

FIGURE 8 (A) Viral dynamic model with target cells (T), drug-sensitive virus (V_s), drug-resistant virus (V_r), cells infected with drug-sensitive virus (I_s), and cells infected with resistant virus (I_r). The parameters, s, ϱ_T, and d are the recruitment rate, maximum proliferation rate, and death rate of target cells, respectively; β is the infection rate of target cells by virus; δ is the death rate of infected cells; p_s and p_r are the viral production rates of the two strains; ε_s and ε_r are the drug efficacies of telaprevir in reducing viral production; μ is the mutation rate from the drug-sensitive to drug-resistant strain; and c is the viral clearance rate. The red crosses represent the effect of treatment in blocking viral production. (B) The best fit of the model to data collected in a patient. The green line is predicted level of drug-sensitive virus, the blue line the predicted level of drug-resistant virus and the circles are measured values of drug-sensitive and drug-resistant virus. *Figure reproduced with permission from AAAS. Rong L, Dahari H, Ribeiro RM, Perelson AS. Rapid emergence of hepatitis C virus protease inhibitor resistance. Sci Transl Med 2010;2:30–32.*

transcription rather than during production of each viral RNA (Perelson et al., 1997, 2012).

In deciding about appropriate drug combinations to be used in antiviral therapy, the fitness of mutants also needs to be considered. For example, the one known mutation that generates drug resistance to the HCV nucleotide polymerase inhibitor sofosbuvir, S282T, is extremely unfit and is generally not observed even in patients treated with sofosbuvir and a single other drug.

When quantitative information is known about the replicative fitness and fold-resistance of viral variants, models can be used to predict the possible outgrowth of resistant variants. Figure 8(A) shows such a model, which is a generalization of the basic viral dynamic model that includes both drug sensitive virus, V_s, and drug resistant virus, V_r, and cells infected by the two types of virus. Figure 8(B) shows how the model using best-fit parameters can predict the kinetics of outgrowth of resistant variants.

4. CONCLUDING REMARKS

Models are a simplification of reality that attempt to focus our attention on key aspects of a biological process and help us reason about them. They provide a dynamic quantitative analysis of infection, which adds a critical dimension to viral pathogenesis. A model can take the form of diagram, such as Figure 2, which can then be made more rigorous and converted into a set of equations. These equations can then be fit to data, allowing parameters to be estimated and the goodness of fit of the model to the data evaluated. The model equations can also be used to make predictions, such as, if an HIV protease inhibitor is given then the viral load will decrease at a rate given by Eqn (2). Model predictions need to be tested and while the predictions from Eqn (2), agreed with short-term data, that is what we now call first-phase declines, when data were collected for longer periods of time, the model needed to be modified multiple times to account for the second, third, and fourth phases of decline. Thus, like everything in biology, models evolve. As the statistician George Box has said, "Essentially, all models are wrong, but some are useful." In the case of HIV, models provided insight into why resistance was rapidly seen when patients were treated with single drugs and thus helped usher in the age of cART. In the case of HCV, models showed how one could evaluate the antiviral effectiveness of a drug that blocks viral replication with a very short-term clinical trial, making it relatively inexpensive to move many investigational compounds into human trials.

Models can also be used to test hypotheses. For example, one could test the hypothesis that type I IFN acts by establishing an antiviral state in cells and thus prevents their infection with HCV. As discussed in this chapter, this hypothesis does not explain the data, while the hypothesis that IFN partially blocks HCV production from already infected cells is compatible with the data. Models also allow one to compute useful quantities, such as the likelihood that HIV or HCV will generate drug-resistant variants over a certain period of time if new virus is produced at a certain rate and if its polymerase has a certain copying error rate.

Model development in virology is still in its early stages. As shown in Figure 7, multiscale models linking the intracellular life cycle of a virus to its systemic effects are under development. Coupling such models to the immune response to infection has not been discussed in this chapter, but is another area of great activity and one where many practical benefits may ultimately accrue (Conway and Perelson, 2015).

REFERENCES

Archin NM, Liberty AL, Kashuba AD, et al. Administration of vorinostat disrupts HIV-1 latency in patients on antiretroviral therapy. Nature 2012;487(7408):482–5.

Chomont N, El-Far M, Ancuta P, et al. HIV reservoir size and persistence are driven by T cell survival and homeostatic proliferation. Nat Med 2009;15(8):893–900.

Conway JM, Perelson AS. Post-treatment control of HIV infection. Proc Natl Acad Sci USA 2015;112(17).

Dahari H, Lo A, Ribeiro RM, Perelson AS. Modeling hepatitis C virus dynamics: liver regeneration and critical drug efficacy. J Theor Biol 2007a;247(2):371–81.

Dahari H, Ribeiro RM, Perelson AS. Triphasic decline of hepatitis C virus RNA during antiviral therapy. Hepatology 2007b;46(1):16–21.

Guedj J, Perelson AS. Second-phase hepatitis C virus RNA decline during telaprevir-based therapy increases with drug effectiveness: implications for treatment duration. Hepatology 2011;53(6):1801–8.

Guedj J, Dahari H, Rong L, et al. Modeling shows that the NS5A inhibitor daclatasvir has two modes of action and yields a shorter estimate of the hepatitis C virus half-life. Proc Natl Acad Sci USA 2013;110(10):3991–6.

Herrmann E, Zeuzem S, Sarrazin C, et al. Viral kinetics in patients with chronic hepatitis C treated with the serine protease inhibitor BILN 2061. Antiviral Ther 2006;11(3):371–6.

Hinrichsen H, Benhamou Y, Wedemeyer H, et al. Short-term antiviral efficacy of BILN 2061, a hepatitis C virus serine protease inhibitor, in hepatitis C genotype 1 patients. Gastroenterology 2004;127(5): 1347–55.

Ho DD, Neumann AU, Perelson AS, Chen W, Leonard JM, Markowitz M. Rapid turnover of plasma virions and CD4 lymphocytes in HIV-1 infection. Nature 1995;373(6510):123–6.

Kim H, Perelson AS. Viral and latent reservoir persistence in HIV-1-infected patients on therapy. PLoS Comput Biol 2006;2(10):e135.

Markowitz M, Louie M, Hurley A, et al. A novel antiviral intervention results in more accurate assessment of human immunodeficiency virus type 1 replication dynamics and T-cell decay in vivo. J Virol 2003;77(8):5037–8.

Neumann AU, Lam NP, Dahari H, et al. Hepatitis C viral dynamics in vivo and the antiviral efficacy of interferon-alpha therapy. Science 1998;282(5386):103–7.

Palmer S, Maldarelli F, Wiegand A, et al. Low-level viremia persists for at least 7 years in patients on suppressive antiretroviral therapy. Proc Natl Acad Sci USA 2008;105(10):3879–84.

Perelson AS, Neumann AU, Markowitz M, Leonard JM, Ho DD. HIV-1 dynamics in vivo: virion clearance rate, infected cell life-span, and viral generation time. Science 1996;271(5255):1582–6.

Perelson AS, Essunger P, Cao Y, et al. Decay characteristics of HIV-1-infected compartments during combination therapy. Nature 1997;387(6629):188–91.

Perelson AS, Essunger P, Ho DD. Dynamics of HIV-1 and CD4+ lymphocytes in vivo. AIDS 1997;11(Suppl. A):S17–24.

Perelson AS, Rong L, Hayden FG. Combination antiviral therapy for influenza infection: Predictions from modeling human infections. J Infect Dis 2012;205(11):1642–5.

Powers KA, Dixit NM, Ribeiro RM, Golia P, Talal AH, Perelson AS. Modeling viral and drug kinetics: hepatitis C virus treatment with pegylated interferon alfa-2b. Semin Liver Dis 2003;23(Suppl. 1):13–8.

Ramratnam B, Bonhoeffer S, Binley J, et al. Rapid production and clearance of HIV-1 and hepatitis C virus assessed by large volume plasma apheresis. Lancet 1999;354(9192):1782–5.

Reilly C, Wietgrefe S, Sedgewick G, Haase A. Determination of simian immunodeficiency virus production by infected activated and resting cells. AIDS 2007;21(2):163–8.

Ribeiro RM, Li H, Wang S, et al. Quantifying the diversification of hepatitis C virus (HCV) during primary infection: estimates of the in vivo mutation rate. PLoS Pathog 2012;8(8):e1002881.

Rong L, Perelson AS. Asymmetric division of activated latently infected cells may explain the decay kinetics of the HIV-1 latent reservoir and intermittent viral blips. Math Biosci 2009;217(1):77–87.

Shudo E, Ribeiro RM, Talal AH, Perelson AS. A hepatitis C viral kinetic model that allows for time-varying drug effectiveness. Antiviral Ther 2008;13(7):919–26.

Snoeck E, Chanu P, Lavielle M, et al. A comprehensive hepatitis C viral kinetic model explaining cure. Clin Pharmacol Ther 2010;87(6):706–13.

Stafford MA, Corey L, Cao Y, Daar ES, Ho DD, Perelson AS. Modeling plasma virus concentration during primary HIV infection. J Theor Biol 2000;203(3):285–301.

Talal AH, Ribeiro RM, Powers KA, et al. Pharmacodynamics of PEG-IFN alpha differentiate HIV/HCV coinfected sustained virological responders from nonresponders. Hepatology 2006;43(5):943–53.

FURTHER READING

Reviews, Chapters, and Books

Canini, L. and Perelson, A. S. Viral kinetic modeling: State of the art. J. Pharmacokinet. Pharmacodyn. 2014; 41: 431–43.

Chaterjee, A., Smith, P. F. and Perelson, A. S. Hepatitis C viral kinetics: The past, present and future. Clin. Liver Dis. 2013; 17, 13–26.

Nowak, M. A. and May, R. M. Virus Dynamics: Mathematical Principles of Immunology and Virology. Oxford University Press, Oxford, 2000.

Perelson, A. S. Modelling viral and immune system dynamics. Nature Rev. Immunol. 2002; 2: 28–36.

Perelson, A. S. and Nelson, P. Mathematical analysis of HIV-1 dynamics in vivo. SIAM Rev. 1999.

Perelson, A. S. and Ribeiro, R. M. (2013). Modeling the within-host dynamics of HIV infection. BMC Biology 2013; 11:96.

Selinger, C. and Katze, M. G. Mathematical models of viral latency. Curr. Opin. Virol. 2013; 3:402–7.

Part III

Emergence and Control of Viral Infections

The first two Parts of our book describe our knowledge from "classical" approaches to viral pathogenesis, and the introduction of systems and computational biology that has generated a dynamic and holistic approach to virus–host interactions. In Part III we ask: how does the transition from reductionism to systems biology impact practical issues in infectious disease?

The chapters in Part III are driven by the double imperative of emerging viral diseases and the burden of long-existing viral infections. Opportunities to treat, prevent, and even eradicate virus infections offer a challenge that distinguishes these illnesses from many non-communicable diseases. Part III highlights the role of viral pathogenesis in establishing a knowledge platform that supports the development of antiviral therapy and viral vaccines.

We open with a chapter on *emerging viral diseases* that describes the various ways in which a new viral disease can appear. In some instances, the discovery of a virus leads to the ability to identify the causal agent of an existing disease of unknown etiology; in other instances, changes in virus virulence or host susceptibility converts a trivial infection into one of epidemic importance; alternatively, a virus may "fade out" of a population only to cause an epidemic when it is re-introduced. Currently, the most important cause of emerging viral diseases is zoonotic infections that cross from animals into the human population. The ongoing AIDS epidemic and the 2014 Ebola pandemic exemplify this danger.

Central to the emergence of many new viruses, particularly RNA viruses, is a high frequency of mutation that reflects the absence of cellular RNA editing mechanisms. *Viral evolution* explains how to measure mutation rates and the rate of the virus-specific evolutionary "clock". The major determinants of mutation rates are multiplicity of infection, rate of replication, and transmission "bottlenecks". Viral evolution can be integrated with epidemiology, using phylodynamics, a new discipline that seeks to understand the impact of molecular epidemiology upon real-life viral outbreaks. These considerations set the stage for a discussion of the implications of viral evolution for emerging and ongoing viral diseases.

Viral epidemiology begins with an explanation of the parameters that are used to describe infections in a population, including incidence, prevalence, and transmitted versus common source outbreaks. Classical epidemiology characterizes outbreaks in terms of person, place, and time. Parallel to the paradigm shift in biomedical science, there is an ongoing evolution of methods in epidemiology, based on computer modeling, Internet data sources, and GPS-based geographical information. These developments provide a more real-time and nuanced view of outbreaks, as illustrated by a series of visual illustrations for diseases such as influenza and dengue.

Part III concludes with two chapters devoted to the control of viral diseases. *Viral vaccines* explains the principles of prevention, and how a vaccine needs to be tailored to the pathogenesis of each virus disease. Vaccines for acute infections usually do not produce "sterilizing immunity" but down-modulate recurrent infections, which become subclinical. Most established vaccines work mainly by inducing humoral immunity (circulating neutralizing antibodies) although some may act via mucosal immunity or effector T lymphocytes. Platforms for vaccines include live attenuated viruses, inactivated viruses, recombinant proteins, various vectors, replicons, and DNAs, many of which can be enhanced by adjuvants. There are over 15 established viral vaccines, perhaps the single most cost-effective prevention in all of medical science. The chapter concludes with the challenges of vaccines yet-to-be formulated, including HIV/AIDS, dengue, and Ebola hemorrhagic fever.

Antiviral therapy has made major strides in recent years, led by the development of a vast number of anti-HIV drugs (highly active antiviral therapy or HAART) that have converted a fatal disease into a chronic manageable infection. There are several major strategies for the development of antiviral therapy, including direct-acting antivirals that target viral enzymes, proteins, or nucleic acids; drugs that target cellular proteins or processes essential for viral replication; and passive antibodies that can neutralize circulating viruses. Recent developments in systems biology have opened the way for two new classes of potential drugs: (i) Repurposed approved drugs that have anti-cellular activity and shortcut the exorbitant cost of new drug development; and (ii) Drugs that modify immune activation and other host responses that can be detrimental and enhance viral illness or mortality. Antiviral therapy is an area of active research and development that continues to yield important advances, exemplified by newly introduced treatments for HCV, a persistent infection causing chronic liver disease.

Chapter 16

Emerging Viral Diseases

Why We Need to Worry about Bats, Camels, and Airplanes

James W. Le Duc[1], Neal Nathanson[2]

[1]*Galveston National Laboratory, University of Texas Medical Branch, Galveston, TX, USA;* [2]*Department of Microbiology, Perelman School of Medicine, University of Pennsylvania, Philadelphia, PA, USA*

Chapter Outline

One of the most dramatic aspects of virology is the emergence of new virus diseases, which often receives widespread attention from the scientific community and the lay public. Considering that the discipline of animal virology was established over 100 years ago, it may seem surprising that new virus diseases are still being discovered. How this happens is the subject of this chapter.

1. HOW DO NEW VIRAL DISEASES EMERGE?

There are many recent books and reviews (see Further Reading) that list the plethora of determinants that can lead to the emergence of infectious diseases (Table 1). In this chapter, we concentrate on those determinants that relate to viral pathogenesis and deal only briefly with the many societal and environmental factors that can be instrumental in disease emergence.

1.1 Discovery of the Etiology of an Existing Disease

In some instances, the "emergence" of a viral disease represents the first identification of the cause of a well-recognized disease. An example is La Crosse virus, a mosquito-transmitted bunyavirus that was first isolated from a fatal case of encephalitis in 1964. The isolation of the causal agent and the development of serological tests made it possible to distinguish La Crosse encephalitis from the rubric of "arbovirus encephalitis, etiology unknown." Since that time, about 100 cases have been reported annually, without any significant increase since the 1970s. It appears that the emergence of this "new" disease reflected only the newfound ability to identify this etiologic entity, rather than any true change in its occurrence.

Hantavirus pulmonary syndrome is another example of the "emergence" of an existing but previously unrecognized disease. In 1993, in the four corners area of the southwestern United States, there occurred a small outbreak of cases of acute pulmonary illness with a high mortality. Epidemiologic and laboratory investigation rapidly identified the causal agent, a previously unknown hantavirus, now named Sin Nombre virus (SNV). SNV is an indigenous virus of deer mice (*Peromyscus maniculatus*) that are persistently infected and excrete the virus. Apparently deer mice produce virus-infected excreta and, when they infest human dwellings, aerosolized fomites can result in occasional human infections. The 1993 outbreak is thought to reflect

Viral Pathogenesis. http://dx.doi.org/10.1016/B978-0-12-800964-2.00016-1

TABLE 1 Some of the Factors that Lead to Emergence or Reemergence of Viral Diseases

Factors Leading to Emergence	Determinant
Economic and social development	Population growth, density, distribution
	Environmental changes such as deforestation, dam building, global warming
	Increased global travel
	Increased international commerce
	Agribusiness, food processing, distribution
Poverty	Inadequate public health systems
	Open defecation
	Lack of safe water
Societal breakdown	Civil chaos
	War
Human factors	Sexual activity
	Substance abuse
Biological factors	Natural mutation
	Antimicrobial resistance
	Immunosuppression

a transient rise in deer mouse populations associated with an unusual crop of pine nuts, a major food source for these rodents. The recognition of SNV soon led to the discovery of other heretofore unrecognized hantaviruses in North, Central and South America, many of which also cause serious human disease.

1.2 Increase in Disease Caused by an Existing Virus

On occasion, a virus that is already widespread in a population can emerge as a cause of epidemic or endemic disease, due to an increase in the ratio of cases to infections. Such an increase can be caused by either an increase in host susceptibility or enhancement of the virulence of the virus. Although counterintuitive, there are some dramatic instances of such phenomena.

Increase in host susceptibility. Poliomyelitis first appeared as a cause of summer outbreaks of acute infantile paralysis in Sweden and the United States late in the nineteenth century (Figure 1). Isolated cases of infantile paralysis had been recorded in prior centuries, and an Eygptian tomb painting indicates that poliomyelitis probably occurred in early recorded history. Why then did poliomyelitis emerge abruptly as an epidemic disease? When personal hygiene and public health were primitive, poliovirus circulated as a readily transmitted enterovirus, and most infants were infected while they still carried maternal antibodies (up to 9–12 months of age). Under these circumstances, the virus produced immunizing infections of the enteric tract, but passively acquired circulating antibodies prevented invasion of

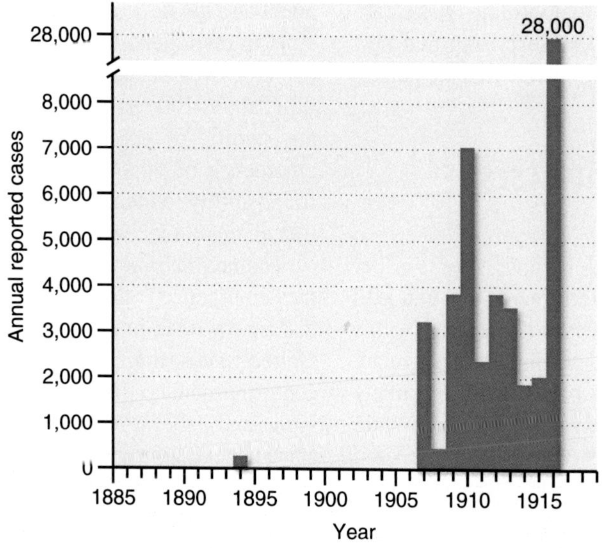

FIGURE 1 The appearance of epidemic poliomyelitis in the United States, 1885–1916. The graph is based on reported cases (mainly paralytic) during an era when reporting was estimated at about 50%. *Data from Lavinder CH, Freeman SW, Frost WH. Epidemiologic studies of poliomyelitis in New York City and the northeastern United States during the year 1916. Washington: United States Public Health Service (1918).*

the spinal cord. With the improvement of personal hygiene in the late-nineteenth century, infections were delayed until 1–3 years of age, after the waning of maternal antibodies. Infections now occurred in susceptible children, resulting in outbreaks of infantile paralysis. This reconstruction is supported by seroepidemiological studies conducted in North Africa in the 1950s, when epidemic poliomyelitis first emerged in this region.

Increase in viral virulence. Viruses may undergo sudden increases in virulence resulting in emergence of dramatic outbreaks. An outbreak of lethal avian influenza in Pennsylvania in 1983 is one documented example. In eastern Pennsylvania, avian influenza appeared in chicken farms early in 1983, but the virus was relatively innocuous and most infections were mild. However, in the fall of that year a fatal influenza pandemic spread rapidly through the same farms. When viruses from the spring and fall were compared, it appeared that both isolates had almost identical genomes. The fall virus had acquired a single point mutation in the viral hemagglutinin that facilitated the cleavage of the hemagglutinin. The virus could now replicate outside the respiratory tract, markedly increasing its virulence (discussed in Chapter 7, Patterns of infection). This point mutation led to the emergence of an overwhelming epizootic, which was only controlled by a widespread slaughter program involving millions of birds. Similar outbreaks of avian influenza have occurred subsequently in other countries.

1.3 Accumulation of Susceptible Hosts and Viral Reemergence

A virus that is endemic in a population may "fade out" and disappear, because the number of susceptibles has fallen below the critical level required for perpetuation in that population. If the population is somewhat isolated, the virus may remain absent for many years. During this interval, there will be an accumulation of birth cohorts of children who are susceptible. If the virus is then reintroduced, it can "reemerge" as an acute outbreak. In the years 1900–1950, Iceland had a population of about 200,000, which was too small to maintain measles virus, and measles periodically disappeared. When travelers to Iceland reintroduced the virus, measles reemerged in epidemic proportions.

1.4 Virus New to a Specific Population

On occasion, a virus can enter and spread in a region where it had never previously circulated, leading to the emergence of a disease new to that locale. A dramatic example is afforded by the emergence of West Nile virus (WNV) in the United States, beginning in 1999 (Figure 2). WNV, like most arboviruses, is usually confined to a finite geographic area, based on the range of its vertebrate reservoir hosts and permissive vectors. In an unusual event, WNV was imported into New York City, probably by the introduction of infected vector mosquitoes that were inadvertent passengers on a flight from the Middle East, where the virus is enzootic. This hypothesis was supported by the finding that the genomic sequence of the New York isolates was closely related to the sequence of contemporary isolates from Israel. Some American mosquito species were competent vectors for WNV, and certain avian species such as American crows were highly susceptible. As a result, West Nile encephalitis emerged as a significant disease new to the United States. Over a period of several years WNV spread across the continent, finally reaching the west coast and many areas in Latin America.

Chikungunya virus (CHIKV), another mosquito-borne arbovirus, has been endemic for many years in regions of Africa and Asia where it periodically caused outbreaks of a febrile illness associated with severe arthritis and arthralgia. In 2005–2006, CHIKV caused a large epidemic on the island of Reunion in the Indian Ocean, apparently associated with a mutation that allowed the virus to be more efficiently transmitted by vector mosquitoes. Chikungunya subsequently spread to India and elsewhere in Asia, followed by the Americas. As of November 2014, transmission of CHIKV had been documented in 40 countries or territories in the Caribbean, Central, South and North America, resulting in nearly one million suspected cases each year. It appears that CHIKV may become established in the New World, where virtually the entire human population currently lacks immunity.

2. ZOONOTIC INFECTIONS AS A SOURCE OF EMERGING VIRAL DISEASES

Zoonotic infections of animals that can be transmitted to humans are a major cause of emerging virus diseases of humans. These viruses are transmitted by direct contact, by virus-laden droplets or aerosols, or by insect vectors. All zoonotic viruses have one or more animal reservoir hosts, which play an important role in the epidemiological dynamics of human infections. Although many zoonotic viruses can be transmitted to humans on occasion, their relative ability to spread from human to human determines whether or not they emerge as significant new virus diseases of mankind (Table 2).

2.1 Dead-end Hosts

Most zoonotic viruses that are transmitted to humans cannot be spread directly from person to person, so humans are considered to be "dead-end hosts." One familiar example is rabies, which is enzootic in several animal hosts, such as dogs, skunks, foxes, raccoons, and bats. Humans are infected by bite of a rabid animal or by aerosol exposure

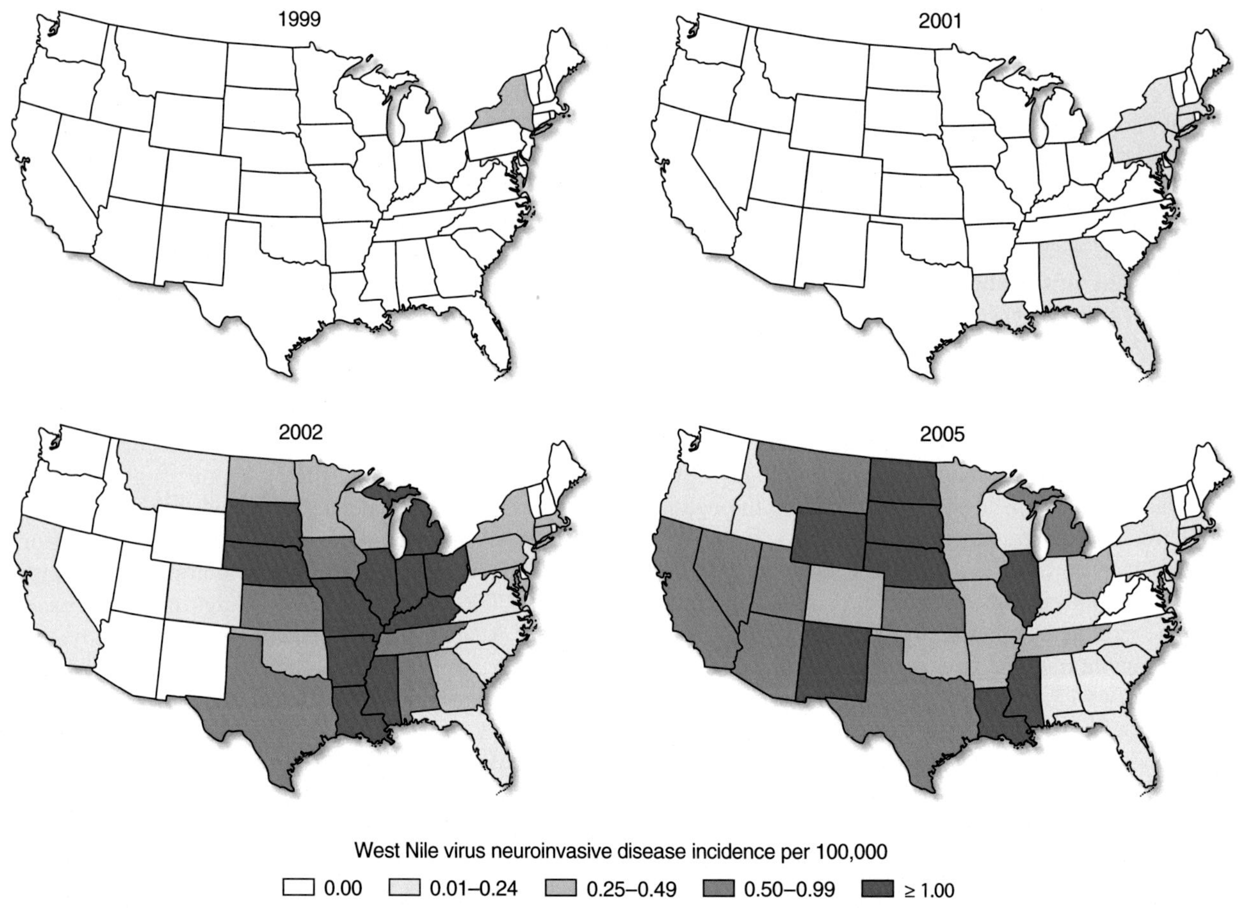

FIGURE 2 The emergence of West Nile virus in the United States, 1999–2005. West Nile virus neuroinvasive disease incidence reported to CDC ArboNET, by state, United States, 1999, 2001, 2002, and 2005. *Accessed via http://www.cdc.gov/westnile/statsMaps/finalMapsData/index.html, December 20, 2014.*

(in caves with roosting bats). Several zoonotic arenaviruses, such as Lassa, Machupo (Bolivian hemorrhagic fever), and Junin (Argentine hemorrhagic fever) viruses, are likely transmitted from the reservoir host (wild rodents) by inhalation of contaminated aerosols.

There are more than 500 viruses—belonging to several virus families—that are also classified as arboviruses (arthropod-borne viruses), based on a vertebrate–arthropod maintenance cycle in nature. Arboviruses replicate in both the vertebrate host and the arthropod vector (mosquitoes, ticks, sandflies, and others), and transmission occurs when the vector takes a blood meal. Typically, arboviruses have only a few vertebrate hosts and are transmitted by a narrow range of arthropods. Humans are not the reservoir vertebrate hosts of most arboviruses, but can be infected by many of these viruses, if they happen to be bitten by an infected vector. In most instances, arbovirus-infected humans are dead-end hosts for several reasons. Many arthropod vectors competent to transmit a zoonotic arbovirus prefer nonhuman hosts as a blood source, reducing the likelihood of transmission from human to vector. Also, infected humans are usually not sufficiently permissive to experience a high titer viremia, so they cannot serve as effective links in the transmission cycle. There are only a few exceptions: in urban settings, dengue, urban yellow fever, Oropouche, and Chikungunya viruses can be maintained by an arthropod vector–human cycle.

2.2 Limited Spread among Humans

As Table 2 shows, a few zoonotic viruses can be transmitted directly from human to human, at least for a few passages, and can emerge as the cause of outbreaks involving a few to several hundred cases. Since many viruses in this group cause a high mortality in humans, even a small outbreak constitutes a public health emergency. These viruses belong to many different virus families, and there is no obvious biological clue why they should be able to spread from human to human, in contrast to other closely related viruses. Typically, infections are mainly limited to caregivers or family members who have intimate contact with patients, often in a hospital setting. However, transmission is marginal, so that most outbreaks end after fewer than 5–10 serial transmissions, either spontaneously or due to infection-control practices.

TABLE 2 Zoonotic Virus Infections of Humans and the Extent of Their Human-to-Human Transmission

Extent of Human-to-Human Spread	Virus	Maintenance Cycle in Nature
Not contagious from human-to-human (humans are dead-end hosts; representative examples); the most frequent pattern	West Nile (flavivirus)	Mosquitoes, birds
	Yellow fever (flavivirus)	Mosquitoes, primates
	La Crosse encephalitis (bunyavirus)	Mosquitoes, wild rodents
	Rabies (rhabdovirus)	Raccoons, skunks, bats, dogs
Limited (<10) human-to-human transmissions; an uncommon pattern	Crimean Congo hemorrhagic fever (bunyavirus) 1–3 serial infections	Ticks, agricultural and wild animals
	Lassa, Machupo, Junin (arenavirus) 1–8 serial infections	Rodents
	Monkeypox (poxvirus) <6 serial infections	Rodents
	Ebola, Marburg (filovirus) 1–4 serial infections*	Bats?
	MERS (coronavirus)	Bats? camels?
	Nipah (paramyxovirus)	Bats, pigs
Unlimited human-to-human transmission (a new human virus); a rare pattern	HIV (lentivirus)	Monkeys, apes
	SARS (coronavirus)	Bats, other animals?
	Influenza (type A influenza virus)	Wild birds, domestic poultry, domestic pigs
	Ebola, Marburg (filovirus) >10 serial infections	Bats?

**Ebola in West Africa involved unlimited human-to-human transmission for over one year (see following).*

2.3 Crossing the Species Barrier

In the history of modern virology (the last 50 years) there are very few documented instances where zoonotic viruses have established themselves in the human population and emerged as new viral diseases of mankind (Table 2). Most viruses have evolved to optimize their ability to be perpetuated within one or a few host species, and this creates what is sometimes called the "species barrier." In most instances, a virus must undergo some adaptive mutations to become established in a new species.

SARS coronavirus. In November, 2002, an outbreak of severe acute respiratory disease (SARS) began in Guangdong Province, in southeast China near the Hong Kong border. In retrospect, the first cases were concentrated in food handlers, who then spread the virus to the general population in that region. Although not recognized immediately as a new disease, the outbreak continued to spread both locally and in other parts of China. In February, 2003, a physician who had been treating likely SARS patients, traveled to Hong Kong, where he transmitted SARS to a large number of contacts in a hotel. These persons, in turn, spread the infection to Singapore, Taiwan, Vietnam, and Canada, initiating a global pandemic that eventually involved almost 30 countries. From patient samples, several research groups isolated a novel coronavirus, which has been named the SARS coronavirus (SARS CoV). Clearly, this virus is new to the human population, and there is circumstantial evidence that it was contracted from exotic food animals that are raised for sale in restaurants in Guangdong Province. Recent studies suggest that horseshoe bats (genus *Rhinolophus*) may be the reservoir hosts and palm civets, consumed as food in China, may be the intermediary hosts for SARS CoV.

SARS CoV went through a large number of human passages (perhaps 25) before being contained by primary control measures, such as respiratory precautions, isolation, and quarantine. The virus has been eliminated from the human population, but the 2003 outbreak showed that it could be maintained by human-to-human transmission. From that perspective, it is potentially capable of becoming an indigenous virus of humans. Since many coronaviruses infect the respiratory system and are transmitted by the respiratory route, the SARS virus did not have to undergo any change in its pathogenesis. However, the virus did have to replicate efficiently in cells of the human respiratory tract and it is unknown whether this required some adaptive mutations from the virus that is enzootic in its reservoir hosts.

Middle Eastern respiratory syndrome. MERS is an acute respiratory disease of humans that was first recognized

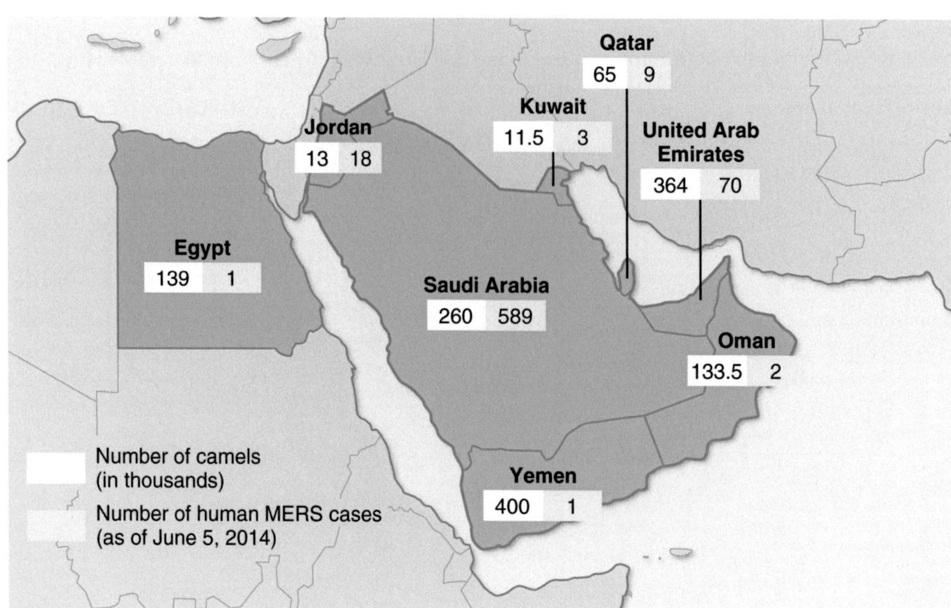

FIGURE 3 Distribution of reported cases of MERS, December, 2014. *Redrawn from Enserik, M. Mission to MERS. Science 2014, 346: 1218–1220.*

in 2012, when a novel coronavirus (MERS-CoV) was isolated from two fatal human cases in Saudi Arabia and Qatar. Since that time, more than 1400 clinical cases of MERS have been recognized, the great majority in Saudi Arabia but also in most of the other countries in the Arabian Peninsula and beyond (Figure 3). Hospitalized cases of MERS are characterized by an acute respiratory disease syndrome (ARDS) with a fatality rate over 25%. However, evolving evidence suggests that there may be many additional human infections with little or no associated illness.

What is the origin of this new disease? Based on fragmentary data now available, it appears that MERS-CoV is likely to be an enzootic virus of one or more species of bats. Camels are probably an intermediary host, and humans in close contact with camels can acquire infection. Also, MERS-CoV can spread from human to human, particularly to caretakers or others in close contact with acutely ill patients. However, there are many unknowns in this speculative reconstruction: How do camels become infected? Are camels the main source of human infections? What has caused this disease to emerge in 2012, or had it preexisted but was just recognized at that time? Is this an evolving outbreak, and if so, what is driving it?

Several relevant facts have been ascertained from basic studies. MERS-CoV replicates well in cell cultures obtained from many species, including bats and humans. The pantropic nature of this virus is explained in part by its cellular receptor, dipeptidyl peptidase 4 (DPPT4), a widely distributed cell surface molecule. This would facilitate the transmission of this zoonotic infection from bats to camels to humans. Of interest, MERS-CoV infects type II alveolar pneumocytes while SARS CoV uses a different cell receptor (angiotensin-converting enzyme 2, ACE2) and infects type I alveolar pneumocytes. However, both viruses appear to cause a similar ARDS.

Type A influenza virus. Genetic evidence strongly implicates avian and porcine type A influenza viruses as the source of some past pandemics of human influenza. It appears that new epidemic strains are often derived as reassortants between the hemagglutinin (and the neuraminidase in some cases) of avian influenza viruses with other genes of existing human influenza viruses. The new surface proteins provide a novel antigenic signature to which many humans are immunologically naïve. The human influenza virus genes contribute to the ability of the reassortant virus to replicate efficiently in human cells. It is thought that reassortment may take place in pigs that are dually infected with avian and human viruses.

Currently, there is concern that a new pandemic strain of Type A influenza could emerge as a derivative of highly virulent avian H5N1 or H7N9 influenza viruses now causing epidemics in domestic chickens in southeast Asia. There have been several hundred documented human infections with each of these avian viruses in recent years—mainly among poultry workers—with significant mortality. However, few if any of these infections have spread from human to human, perhaps because the infecting avian virus has not undergone reassortment with a human influenza virus. A critical determinant is that avian influenza viruses preferentially attach to α2-3 sialic acid receptors while human viruses attach to α2-6 sialic acid receptors.

The 1918 pandemic of influenza is presumed to be an example of a zoonotic influenza virus that crossed the species barrier and became established in humans where it caused an excess mortality estimated at 20–40 million persons. Recent viral molecular archaeology has recovered the sequences of the 1918 H1N1 influenza virus from the tissues of patients who died during the epidemic.

TABLE 3 Genetic Determinants of the Virulence of the 1918 Type A Influenza Virus (Spanish Strain) Based on Intranasal Infection of Mice with Reassortant Viruses. The M88 Isolate is a Human Type A Influenza Virus Which is Relatively Avirulent in Mice, Typical of Human Isolates. The Hemagglutinin (H) of the 1918 "Spanish" Virus Confers Virulence Upon the M88 Isolate and the Neuraminidase (N) Does Not Appear to Enhance the Effect of the Hemagglutinin. MLD 50: Mouse 50% Lethal Dose Determined by Intranasal Infection with Serial Virus Dilutions

	Origin of Viral Genes				
Virus Strain	H	N	Others	Log 10 MLD 50	Log 10 Virus Titer in Lung (Day 3 after Infection)
M88	M88	M88	M88	>6.2	2.9
M88/H^{sp}	Sp (1918)	M88	M88	4.4	5.1
M88/H^{sp}/N^{sp}	Sp (1918)	Sp (1918)	M88	5.2	4.7

After Kobasa D, Takada A, Shinya K, et al. Enhanced Virulence of Influenza a Viruses with the Haemagglutinin of the 1918 Pandemic Virus. *Nature*, 2004, 431: 703–707, with Permission

All of the genes of the reconstructed virus are avian in origin, but it is unknown whether the virus underwent mutations that enhanced its ability to be transmitted within the human population. The reconstructed hemagglutinin of the 1918 virus has been inserted into recombinant influenza viruses, and—in mice—markedly increases the virulence of primary human isolates of influenza virus (Table 3), but the full virulence phenotype appears to require many of the avian influenza genes. Several physiological factors play a role in disease enhancement, including increased replication in pulmonary tissues and an enhanced ability to stimulate macrophages to secrete pro-inflammatory cytokines which, in turn, cause severe pneumonitis.

Human immunodeficiency virus. HIV has emerged as the greatest pandemic in the recent history of medical science. Modern methods have made it possible to reconstruct its history in great detail (see Chapter 9, HIV/AIDS). A provisional reconstruction of the sequence of events is summarized in Figure 4. The emergence of HIV may be divided into two phases: What was the zoonotic source of HIV and when did it cross into humans? And when did HIV spread from the first human cases to become a global plague?

There have been several transmissions of simian immunodeficiency virus (SIV) to humans (Sharp and Hahn, 2011), and this account will focus on HIV-1 which appears to have been transmitted to humans in at least four separate instances, identified by individual HIV-1 lineages called groups (M, N, O, P). Of these, the most important was the M group of HIV-1, which has been responsible for the vast majority of human infections. Furthermore, HIV-1 is most closely related to SIVcpz, the SIV strain infecting two subpopulations of chimpanzees. Different segments of the SIVcpz genome, in turn, are closely related to genome segments of two SIVs of African monkeys, red-capped monkeys and *Cercopithecus* monkeys. It is hypothesized that chimpanzees, which regularly kill and eat monkeys, were infected during consumption of their prey; and that a recombination event produced SIVcpz, which was derived from parts of the genomes of the two acquired monkey viruses. It is speculated that a further transmission from chimpanzees may have occurred during the butchering of nonhuman primates, which occurs in rural Africa (Figure 5).

Amazingly, genomic similarities tentatively map the substrain of SIVcpz that is the ancestor of the M group of HIV-1, to chimpanzees in the southeastern corner of Cameroon, a small country in West Africa. Using a sequence analysis to compare diversity within current isolates, the common parent of the M group can be reconstructed. A molecular clock, derived from dated isolates, indicates that this virus was transmitted to humans during the period 1910–1930.

The period from 1930 to 1980 is a mystery (Pepin, 2011), but there are fragmentary data suggesting that the virus persisted as a rare and unrecognized infection in residents of jungle villages in West Africa during this time. It has also been proposed that the reuse of unsterilized needles—a frequent practice during the period of colonial rule—could have inadvertently helped to spread the virus. Starting about 1980, the virus began to spread more rapidly. It appears that accelerated spread began in the region centered on Kinshasa (previously Leopoldville) in the Democratic Republic of the Congo (previously the Belgian Congo, then Zaire) and Brazzaville, just across the Congo River in Congo. Transmission was exacerbated by the chaos in postcolonial Zaire.

During the period 1985–2004, HIV infection spread widely in Africa as shown in Figure 6. In the countries worst affected, the prevalence of infection among adults aged 15–49 years reached levels higher than 30%. The rapid spread was driven by many factors among which were: (1) a high frequency of concurrent sexual contacts in some segments of the population and the hidden nature of sexual networks; (2) the long asymptomatic incubation period during which infected individuals able to transmit the virus were sexually active; (3) the spread along commercial routes of travel within Africa; (4) the failure of health systems to

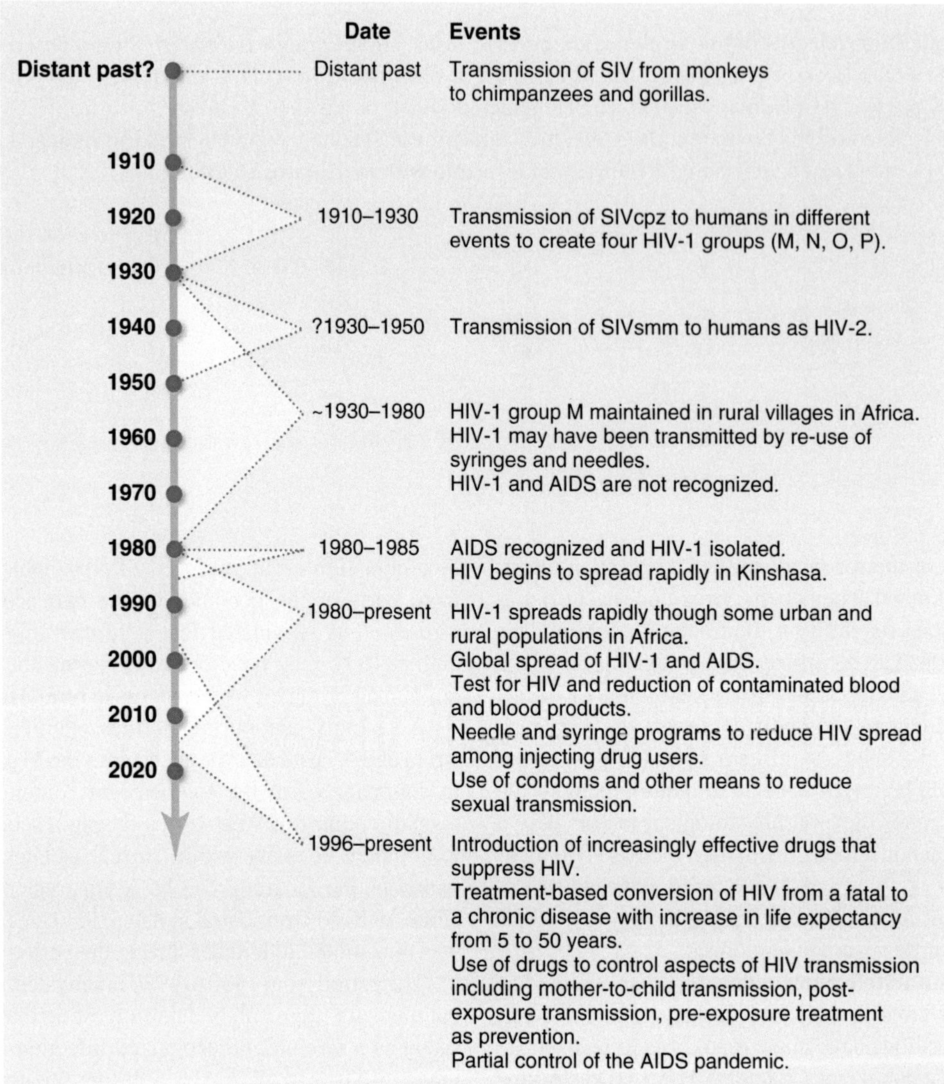

FIGURE 4 Speculative reconstruction of events following the transmission of SIVcpz to humans. *This reconstruction is based on data in Sharp and Hahn (2011), Pepin (2011).*

FIGURE 5 Bushmeat is part of the diet in rural Africa. *Photograph courtesy of Billy Karesh (2015).*

publicize the risks and the underutilization of condoms and other measures to reduce transmission; and (5) the slow introduction of antiviral treatment after it became available in the northern countries about 1996.

Concomitantly with the spread of HIV in Africa, the M group of HIV-1 evolved into nine different subtypes (A–D, F–H, J, K), based on sequence diversity. During the spread within Africa, there were population bottlenecks that resulted in the predominance of different M group subtypes in different regions. Subtype C is most frequent in southern Africa, and subtypes A and D are most frequent in eastern Africa. During the 1980s, HIV also spread globally, although prevalence rates did not reach the levels seen in some African countries. Subtype B is dominant in the western hemisphere and Europe, while subtype C is most frequent in India and some other Asian countries. This implies that each of these regional epidemics was initiated

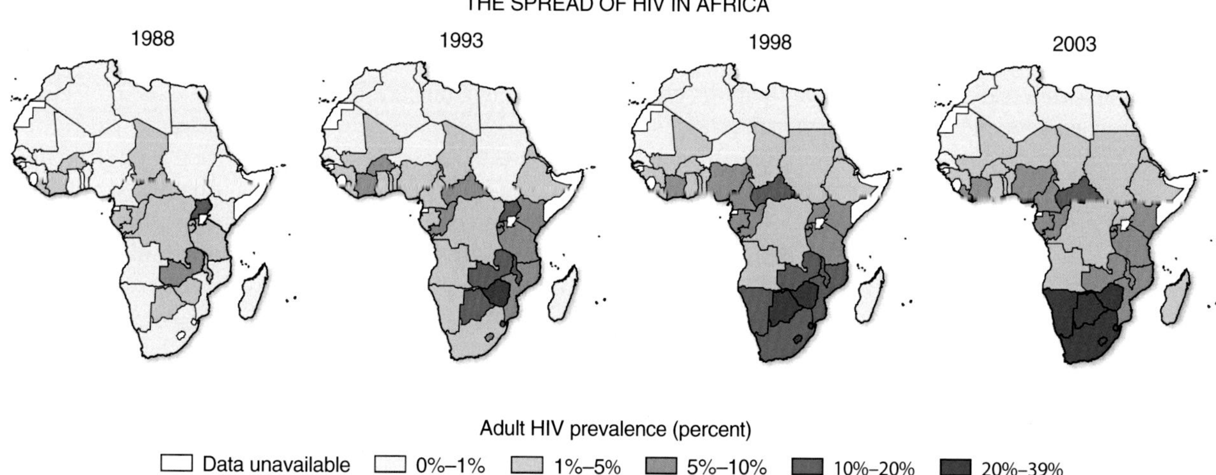

FIGURE 6 The spread of HIV in Africa, showing the prevalence of HIV in the adult population, 1988–2003. The map ends in 2003, but the prevalence has remained very similar during the period 2004–2015. (After UNAIDS 2004 report on global AIDS epidemic.)

by a small number of founder strains of HIV-1, improbable though that may seem.

Thirty years after its appearance as a global disease, almost 40 million persons have died, and there are more than 30 million people living with HIV/AIDS. Although the global incidence of HIV has fallen slightly since 2010, there are still more than two million new infections each year (2014).

Ebola hemorrhagic fever. The Ebola pandemic of 2014–15 is the most recent emerging virus disease that has riveted the attention of the world. Where did it come from? Why did it cause a pandemic? Why did the international health community fail to control it? Where will it end? These questions are discussed below.

Ebola is a filovirus indigenous to Africa that is maintained in one or more reservoir species of wild animals, among which bats are a likely host. The transmission cycle is not well documented but it is thought that humans get infected, either by direct contact with bats, or while slaughtering infected wild animals who may act as intermediate hosts. Human-to-human spread can then occur.

The pathogenesis of Ebola virus infection may be briefly summarized. Presumably Ebola enters the human host via the mucous membranes or cuts in the skin. The virus infects mononuclear cells including macrophages and dendritic cells and traffics to lymph nodes, whence it spreads to target organs including the liver, spleen, and adrenal glands. It causes a high titer viremia and a dysregulation of the innate immune system. Clinically, Ebola patients undergo severe vomiting and diarrhea with massive fluid losses and become very dehydrated, with a mortality that varies from 25% to 75%. In fatal cases, the infection of the liver leads to disseminated intravascular coagulopathy, a shock syndrome, and multiorgan failure, although the sequential details are not well understood. Infection is transmitted between humans by contact with bodily fluids of patients, but not by aerosols. Therefore, the virus is spread most frequently to caregivers or others who are in intimate contact with patients and their fomites. Rituals associated with funerals and burial practices often serve to transmit the virus.

Ebola virus was first isolated in 1976 in an outbreak near the Ebola River in the Democratic Republic of the Congo (then called Zaire) and almost concurrently in a second outbreak in southern Sudan. Viruses recovered from these two outbreaks were subsequently shown to be different and are now known as Ebola-Zaire and Ebola-Sudan. Since that time there have been more than 25 individual outbreaks of Ebola disease, mainly in central Africa. Past outbreaks have been controlled by use of protective garments by caregivers and quarantine of infected or potentially infected contacts. These controlled outbreaks have been limited to no more than ~5 serial human-to-human transmissions, and mainly ranged from about 25 to 300 cases.

In December, 2013, an Ebola-Zaire outbreak began in Guinea, West Africa. It appears that the initial case was in an infant who may have been infected by contact with bats. The outbreak then spread to two contiguous countries, Liberia and Sierra Leone. By the fall, 2014, the epidemic had become a catastrophe, and was raging out of control in several parts of these three countries. Although the data are incomplete, it is estimated that there have been at least 20,000 cases (with at least 50% mortality) through February, 2015 (Figure 7). The infection has spread to Nigeria, Senegal, Mali, the United States, and a few European countries, but these invasions have so far been controlled, with limited secondary cases. A global effort to control this pandemic was initiated, with participation from Doctors without Borders, the Red Cross, other non-government organizations, the World Health Organization, and the U.S. Centers for Disease Control and Prevention. As of March, 2015, it appeared that the epidemic was coming under control. Aggressive border screening, both on exit from the affected countries and on arrival at destinations, has limited spread by air travelers, but the porous land borders remain areas of concern.

From an epidemiological viewpoint, why did this outbreak of Ebola explode into a massive pandemic, in contrast to the many prior outbreaks that were limited to no more than a few hundred cases at most? The outbreak began in Guinea and was mainly confined to that country for about 6 months before it spread to neighboring Liberia and Sierra Leone (Figure 8). During this first 6 months, incidence in Guinea varied from a few to about 50 cases a week, and cases were concentrated in rural areas. From a public health viewpoint, this represented a missed opportunity to contain the outbreak. Contributing to this omission were a combination of factors: a weak health system fragmented by social disruption, a failure of local health authorities to recognize or respond to the outbreak, the failure of international health organizations such as WHO to take aggressive action, and local societal norms that brought family and friends into close contact with Ebola victims, during their illness and at their funerals (Washington Post, October 5, 2014; Cohen, 2014). Once Ebola infections spread from rural villages to urban centers, the outbreak exploded.

Beginning in the fall of 2014, it was recognized that the pandemic was a global threat. In response, a number of countries and international organizations provided resources to Africa, including building facilities, sending equipment, and recruiting personnel to work in the pandemic areas. A major effort was made to get the local population to temporarily change some of their normal social responses to

FIGURE 7 The journalistic face of the 2013–2015 Ebola epidemic. *Photograph courtesy of Tom Ksiazek, 2014.*

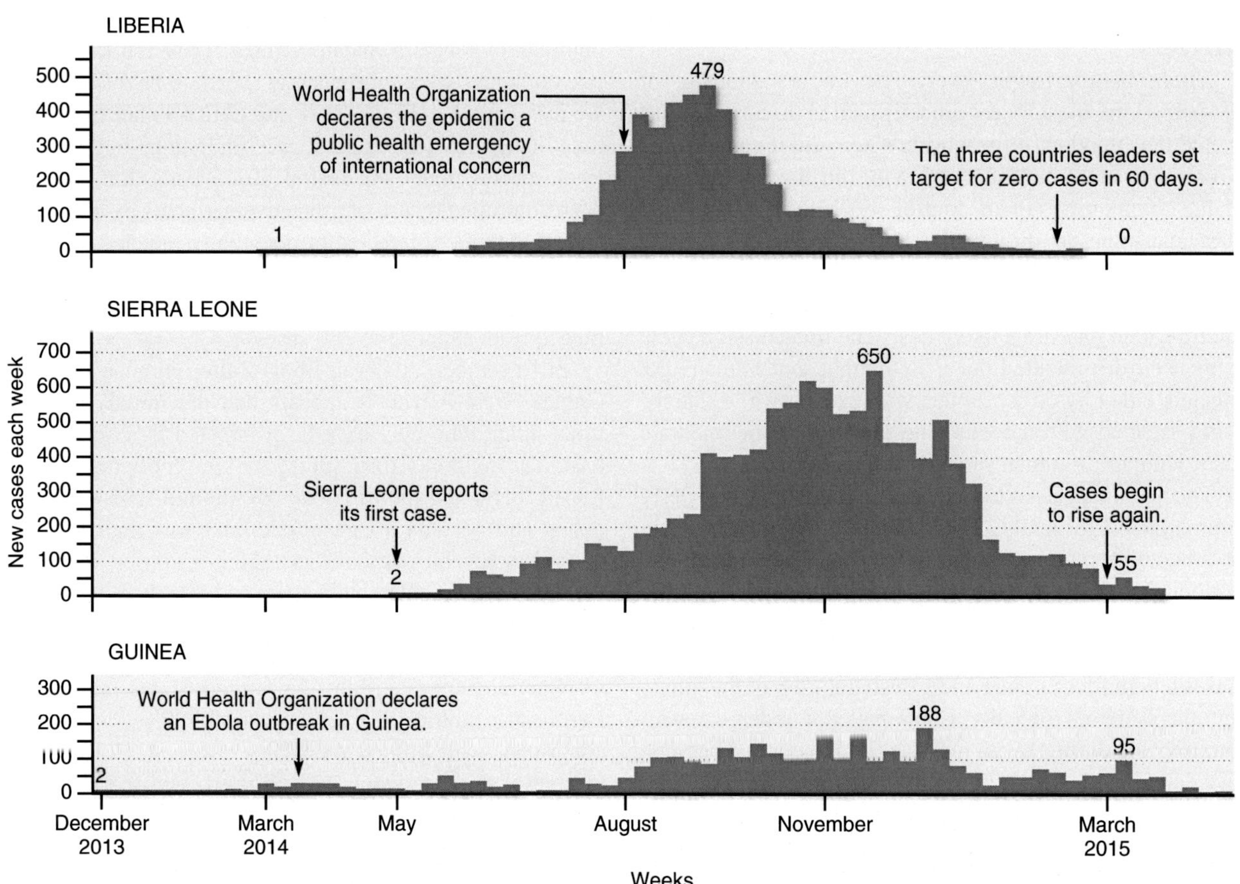

FIGURE 8 Ebola pandemic, by country, through April, 2015. After WHO: Ebola situation report, 29 April, 2015.

illness and death, to reduce the spread within the population. Combined with the efforts of the affected countries, these initiatives started to take hold about January, 2015. However, as of March, 2015, the epidemic was not yet terminated. In retrospect, the international community has acknowledged that it lacks a contingency plan to respond to global outbreaks wherever they may occur. Futhermore, because Ebola was a "neglected" disease, the tools to combat it had not been developed. The Ebola pandemic has spurred crash programs to develop a rapid diagnostic test, drugs and antibodies for treatment, and a vaccine for prevention (Product, 2015; Mire et al., 2015).

Canine parvovirus. CPV is another example of a disease that emerged due to the appearance of a virus new to its host species. In the late 1970s, a highly lethal pandemic disease appeared in the dog populations of the world. The etiologic agent was a parvovirus previously unknown in canines. The sequence of CPV is almost identical to that of feline panleukopenia virus (FPV), an established parvovirus of cats, which causes acutely fatal disease in kittens. CPV has a few point mutations that distinguish it from FPV, and these mutations permit CPV to bind to and infect canine cells, a property not possessed by FPV. It is presumed that these mutations permitted the emergence of a new virus disease of dogs.

2.4 The Species Barrier and Host Defenses

Many mammalian viruses have evolved with their hosts so that different members of a virus family are associated with each host species. Furthermore, under natural circumstances, each member of the virus family usually "respects" the species barrier and does not cross into other species, although it spreads readily between individual animals within its host species. Diverse mechanisms contribute to the species barrier, including host defenses and viral genes. For a zoonotic virus to establish itself in humans or another new species, it is probable that mutations are needed for full adaptation. This requirement is likely part of the explanation for the rarity of such events.

One of the best-studied examples is the transmission of SIVcpz, a virus of chimpanzees, to HIV, a human virus. Several recent studies have shown that APOBEC3 and tetherin are two very important gate keepers for transmission of lentiviruses between different primate species (reviewed in Sharp and Hahn, 2011). APOBEC3 proteins represent a powerful first line of defense, believed to be responsible for preventing the transmission of SIVs from monkeys to chimpanzees and from monkeys to humans (Etienne et al., 2013). However, APOBEC3 proteins can be counteracted by the HIV/SIV viral infectivity factor (Vif), albeit generally in a species-specific fashion. Thus, mutations in Vif were required for monkey SIVs to be able to infect chimpanzees and then humans (Letko et al., 2013). Tetherin is a second line of defense, and only HIVs that have evolved an effective tetherin antagonism have spread widely in humans (Kluge et al., 2014). Different lentiviruses use different viral proteins to antagonize tetherin: HIV-1 group M uses the viral protein U (Vpu); HIV-2 group A uses the envelope glycoprotein (Env), and HIV-1 group O uses the negative regulatory factor (Nef). In contrast, HIV-1 groups N and P, whose spread in the human population has remained very limited, are unable to counteract tetherin efficiently.

Turning to another virus, recent studies using a mouse-adapted strain of Ebola virus to infect a panel of inbred mice provided by the Collaborative Cross (see Chapter 10, Animal models) found striking variation in the response to Ebola virus infection, from complete resistance to severe hemorrhagic fever with 100% mortality. This underlines the role of host genetic background as a determinant of susceptibility. Consistent with this are comments of clinicians that there are unpredictable differences in the outcome of individual Ebola cases. These studies suggest a dynamic relationship between host and pathogen that may determine when a virus can cross the species barrier and create a new virus disease of humankind.

3. WHY VIRAL DISEASES ARE EMERGING AT AN INCREASING FREQUENCY

Although difficult to document in a rigorous manner, it does appear that new virus diseases of humans (and perhaps of other species) are emerging at an increased tempo. There are a number of reasons for this trend (Table 1).

3.1 Human Ecology

The human population is growing inexorably, and is becoming urbanized even faster. As a result, there are an increasing number of large-crowded cities, which provide an optimal setting for the rapid spread of any newly emergent infectious agent.

In the nineteenth century, it was noteworthy that someone could circumnavigate the globe in 80 days, but now it can be done in less than 80 h. However, the incubation period of viral infections (several days to several months) has stayed constant. Someone can be infected in one location and—within a single incubation period—arrive at any other site on earth. This enhances the opportunity for a new human virus to spread as a global infection before it has even been recognized, markedly increasing the opportunity for the emergence of a new disease. The same dynamics also apply to viral diseases of animals and plants, which has important economic and social consequences for humankind. The SARS pandemic of 2002–2003, described above, is an example of how rapidly and widely a new virus disease can emerge and spread globally. In this instance, it is extraordinary that the disease was brought under control—and eradicated from the human population—by the simple methods of isolation, quarantine, and respiratory precautions. Although conceptually simple, a heroic effort was required for success. Another example is the 2009 emergence of a novel H1N1 strain of influenza A virus that spread rapidly around the world before a vaccine could be produced.

Remote areas of the world are now being colonized at a high frequency, driven by population pressures and economic motives, such as the reclamation of land for agriculture or other uses, and the harvest of valuable trees and exotic animals. The construction of new dams, roads, and other alterations of the natural environment create new ecological niches. It is possible that this is the origin of urban yellow fever. Yellow fever virus is an arbovirus maintained in a monkey–mosquito cycle in the jungles of South America and Africa. Humans who entered jungle areas became infected. When they returned to villages or urban centers, an urban cycle was initiated, where *Aedes aegypti* mosquitoes—that are well adapted to the urban environment and preferentially feed on humans—maintained the virus.

In the last 25 years, agriculture has undergone a dramatic evolution with the development of "agribusiness." Food and food animals are now raised on an unprecedented scale and under very artificial conditions, where the proximity of many members of a single plant or animal species permits an infection to spread like wildfire. Furthermore, increasing numbers of plants, animals, and food products are rapidly transported over large distances; we enjoy fresh fruits and vegetables at any time, regardless of the season. International shipment of plants and animals can import viruses into new settings where they may lead to the emergence of unforeseen diseases. One example is the 2003 outbreak of monkeypox that caused about 80 human cases in the United States. This was traced to the importation from Africa of Gambian giant rats as exotic pets; several rodent species in west Africa appear to be reservoir hosts of this poxvirus. Monkeypox spread from these animals to pet prairie dogs and from prairie dogs to their owners. Because monkeypox causes infections in humans that resemble mild cases of smallpox, this outbreak was a major cause of public health concern.

3.2 Deliberate Introduction of a Virus New to a Specific Population

On occasion, a virus has been deliberately introduced into a susceptible population where it caused the emergence of a disease epidemic. For sport, rabbits were imported from Europe into Australia in the mid-nineteenth century. Because of the absence of any natural predator the rabbits multiplied to biblical numbers, and threatened natural grasslands and agricultural crops over extensive areas of southern Australia. To control this problem, myxomatosis virus was deliberately introduced in 1950 (Fenner, 1983). This poxvirus is transmitted mechanically by the bite of insects and is indigenous to wild rabbits in South America, in which it causes nonlethal skin tumors. However, myxomatosis virus causes an acutely lethal infection in European rabbits, and its introduction in Australia resulted in a pandemic in the rabbit population.

Following the introduction of myxomatosis virus in 1950, co-evolution of both virus and host were observed. The introduced strain was highly virulent and caused epizootics with very high mortality. However, with the passage of time field isolates exhibited reduced virulence, and there was a selection for rabbits that were genetically somewhat resistant to the virus. Strains of moderate virulence probably became dominant because strains of lowest virulence were less transmissible and strains of maximum virulence killed rabbits very quickly.

Concern has also been raised about disease emergence due to the deliberate introduction of viruses into either human or animal populations, as acts of bioterrorism.

3.3 Xenotransplantation

Because of the shortage of human organs for transplantation recipients, there is considerable research on the use of other species—particularly pigs—as organ donors. This has raised the question whether known or unknown latent or persistent viruses in donor organs might be transmitted to transplant recipients. Since transplant recipients are immunosuppressed to reduce graft rejection, they could be particularly susceptible to infection with viruses from the donor species. In the worst scenario, this could enable a foreign virus to cross the species barrier and become established as a new human virus that might spread from the graft recipient to other persons.

4. HOW ARE EMERGENT VIRUSES IDENTIFIED?

The impetus to identify a new pathogenic virus usually arises under one of two circumstances. First, a disease outbreak that cannot be attributed to a known pathogen may set off a race to identify a potentially new infectious agent. Identification of the causal agent will aid in the control of the disease and in prevention or preparedness for potential future epidemics. SARS coronavirus, West Nile virus, and Sin Nombre Virus are examples of emergent viruses that were identified in the wake of outbreaks, using both classical and modern methods. Alternatively, an important new virus may be discovered as a serendipitous by-product of research directed to a different goal, as was the case with hepatitis B virus (HBV). In this instance, a search for alloantigens uncovered a new serum protein that turned out to be the surface antigen (HBsAg) of HBV, leading to the discovery of the virus.

When a disease outbreak cannot be attributed to a known pathogen, and where classical virus isolation, propagation, and identification fail, molecular virology is required. Hepatitis C virus (HCV), Sin Nombre and other hantaviruses, certain rotaviruses, and Kaposi's sarcoma herpesvirus (HHV8), are examples of emergent viruses that were first discovered as a result of molecular technologies. Below, we briefly describe some methods of viral detection and identification. More detailed information and technical specifics can be found in several current texts.

4.1 Classic Methods of Virus Discovery

The first question that confronts the investigator faced with a disease of unknown etiology is whether or not it has an infectious etiology? Evidence that suggests an infectious etiology is an acute onset and short duration, clinical similarity to known infectious diseases, a grouping of similar illnesses in time and place, and a history of transmission between individuals presenting with the same clinical picture. For chronic illnesses, the infectious etiology may be much less apparent and a subject for debate.

Faced with a disease that appears to be infectious, the next question is whether it is caused by a virus. A classical example that predates modern virology is the etiology of yellow fever. In a set of experiments that would now be prohibited as unethical, the Yellow Fever Commission, working with the US soldiers and other volunteers in Cuba in 1900, found that the blood of a patient with acute disease could transmit the infection to another person by intravenous injection. Furthermore, it was shown that the infectious agent could pass through a bacteria-retaining filter and therefore could be considered a "filterable virus."

Virus isolation in cell culture and animals. The first step in identification of a putative virus is to establish a system in which the agent can be propagated. Before the days of cell culture, experimental animals were used for this purpose. Many viruses could be isolated by intracerebral injection of suckling mice, and some viruses that did not infect mice could be transmitted to other experimental animals. Human polioviruses—because of their cellular receptor requirements—were restricted to old world monkeys and great apes; the virus was first isolated in 1908 by intracerebral injection of monkeys and was maintained by monkey-to-monkey passage until 1949 when it was shown to replicate in primary cultures of human fibroblasts.

The modern era of virology (beginning about 1950) can be dated to the introduction of cultured cells as the standard method for the isolation, propagation, and quantification of viruses. There are now a vast range of cell culture lines that can be used for the isolation of viruses, and currently this is the first recourse in attempting to isolate a suspected novel virus. Some viruses will replicate in a wide variety of cells but others are more fastidious and it can be hard to predict which cells will support their replication.

It is also important to recognize that some viruses will replicate in cell culture without exhibiting a cytopathic effect. An important example is the identification of simian virus 40 (SV40). Poliovirus was usually grown in primary cell cultures obtained from the kidneys of rhesus monkeys, but SV40 had escaped detection because it replicated without causing a cytopathic effect. When poliovirus harvests were tested in similar cultures prepared from African green monkeys, a cytopathic effect (vacuolation) was observed, leading to the discovery of SV40 virus in 1960. Because inactivated poliovirus vaccine produced from 1955 to 1960 had been prepared from virus grown in rhesus monkey cultures, many lots were contaminated with this previously unknown virus, which inadvertently had been administered to humans. Since that time, viral stocks and cell cultures have been screened to exclude SV40 and other potential virus contaminants.

A number of methods are available to detect a noncytopathic virus that is growing in cell culture. These include visualization of the virus by electron microscopy, detection of viral antigens by immunological methods such as immunofluorescence or immunocytochemistry, the agglutination of erythrocytes of various animal species by virus bound on the cell surface (hemagglutination), the production of interferon or viral interference, and the detection of viral nucleic acids.

Detection of nonreplicating viruses. During the period from 1950 to 1980, there was a concerted effort to identify the causes of acute infections of infants and children. In seeking the etiology of diarrheal diseases of infants, it was hypothesized that—in addition to bacteria, which accounted for less than half of the cases—one or more viruses might be responsible for some cases of infantile diarrhea. Numerous unsuccessful attempts were made to grow viruses from stools of patients with acute diarrhea. It was conjectured that it might be possible to visualize a putative fastidious virus by electron microscopy of concentrated fecal specimens. When patients' convalescent serum was added to filtered and concentrated stool specimens, aggregates of 70 nm virions were observed in stools from some infants with acute gastroenteritis. The ability of convalescent but not acute illness serum to mediate virion aggregation provided a temporal association of the immune response with an acute diarrheal illness. Within 5 years, rotavirus was recognized as the most common cause of diarrhea in infants and young children worldwide, accounting for approximately one-third of cases of severe diarrhea requiring hospitalization.

Once an emergent virus has been identified, it is necessary to classify it, in order to determine whether it is a known virus, a new member of a recognized virus group, or represents a novel virus taxon. This information provides clues relevant to diagnosis, prognosis, therapy, and prevention.

In 1967, an outbreak of acute hemorrhagic fever occurred in laboratory workers in Marburg, Germany, who were harvesting kidneys from African green monkeys (*Chlorocebus aethiops*, formerly *Cercopithecus aethiops*). In addition, the disease spread to hospital contacts of the index cases, with a total of over 30 cases and 25% mortality. Clinical and epidemiological observations immediately suggested a transmissible agent, but attempts to culture bacteria were unsuccessful. However, the agent was readily passed to guinea pigs which died with an acute illness that resembled hemorrhagic fever. After considerable effort, the agent was adapted to tissue culture and shown to be an RNA virus. When concentrated tissue culture harvests were examined by electron microscopy, it was immediately recognized that this agent differed from known families of RNA viruses, since the virions consisted of very long cylindrical filaments about 70 nm in diameter. This was the discovery of Marburg virus, the first recognized member of the filoviruses, which now include Marburg and Ebola viruses.

4.2 The Henle–Koch Postulates

Isolation of a virus from patients suffering from an emergent disease provides an association, but not proof of a causal relationship. Formal demonstration that an isolated virus is the causal agent involves several criteria formulated over the past 100 years. These are often called the Henle–Koch postulates, after two nineteenth-century scientists who first attempted to enunciate the rules of evidence. Sidebar 1 summarizes these postulates, which have been modernized in view of current knowledge and experimental methods.

The classic version of the Henle–Koch postulates required that the causal agent be grown in culture. However, as discussed below, a number of viruses that cannot be grown in culture have been convincingly associated with a specific disease. Usually, this requires that many of the following criteria can be met: (1) Viral sequences can be found in the diseased tissue in many patients, and are absent in appropriate control subjects; (2) Comparison of acute and convalescent sera document induction of an immune response specific for the putative causal virus; (3) The disease occurs in persons who lack a preexisting immune response to the putative virus, but not in those who are immune; (4) The implicated virus or a homologous virus causes a similar disease in experimental animals; and (5) Epidemiological patterns of disease and infection are consistent with a causal relationship.

4.3 Methods for Detection of Viruses that Are Difficult to Grow in Cell Culture

Some very important human diseases—such as hepatitis B and hepatitis C—are caused by viruses that cannot readily be grown in cell culture. Experiences with these viruses have given credibility to the view that an infectious etiology can be inferred by clinical and epidemiological observations in the absence of a method for growing the causal agent. Also, they have stimulated researchers to devise novel techniques that bypass the requirement for replication in cell culture. Furthermore, the application of molecular biology, beginning about 1970, has led to an array of new methods—such as the polymerase chain reaction (PCR), deep sequencing, and genomic databases—that can be applied to the search for unknown viruses. Several case histories illustrate the inferences that lead to the hypothesis of a viral etiology, the strategy used to identify the putative causal agent, and the methods exploited by ingenious and tenacious researchers.

Sin Nombre virus. Hantavirus pulmonary syndrome was described above, as an example of an emerging virus disease. The disease was first reported in mid-May, 1993,

Sidebar 1 The Henle–Koch postulates (requirements to identify the causal agent of a specific disease)

The Henle–Koch postulates were formulated in 1840 by Henle, revised by Koch in 1890, and have undergone periodic updates to incorporate technical advances and the identification of fastidious agents with insidious disease pathogenesis.

Many of the following criteria should be met to establish a causal relationship between an infectious agent and a disease syndrome.

1. The putative causal agent should be isolated from patients with the disease; or the genome or other evidence of the causal agent should be found in patients' tissues or excreta; and less frequently from appropriate comparison subjects. Temporally, the disease should follow exposure to the putative agent; if incubation periods can be documented they may exhibit a log-normal distribution.
2. If an immune response to the putative agent can be measured, this response should correlate in time with the occurrence of the disease. Subjects with evidence of immunity may be less susceptible than naïve individuals.
3. Experimental reproduction of the disease should occur in higher incidence in animals or humans appropriately exposed to the putative cause than in those not so exposed. Alternatively, a similar infectious agent may cause an analogous disease in experimental animals.
4. Elimination or modification of the putative cause should decrease the incidence of the disease. If immunization or therapy is available, they should decrease or eliminate the disease.
5. The data should fit an internally consistent pattern that supports a causal association.

and tissues and blood samples from these cases were tested extensively, but no virus was initially isolated in cell culture. However, when sera from recovered cases were tested, they were found to cross-react with a battery of antigens from known hantaviruses, providing the first lead (in June, 1993). DNA primers were then designed, based on conserved hantavirus sequences, and these were used in a PCR applied to DNA transcribed from RNA isolated from tissues of fatal cases. Sequence of the resulting amplicon suggested that it was a fragment of a putative new hantavirus (July, 1993), yielding a presumptive identification of the emerging virus within 2 months after the report of the outbreak. An intense effort by three research teams led to the successful isolation of several strains of SNV by November, 1993. SNV is a fastidious virus that replicates in Vero E6 cells but not in many other cell lines.

Kaposi's sarcoma herpesvirus (HHV8). KS was described over 100 years ago as a relatively uncommon sarcoma of the skin in older men in eastern Europe and the Mediterranean region. In the 1980s, KS emerged at much higher frequency, as one of the diseases associated with AIDS. Furthermore, KS exhibited an enigmatic epidemiological pattern, since its incidence in gay men was more than 10-fold greater than in other AIDS patients, such as injecting drug users and blood recipients. These observations led to the hypothesis that KS was caused by a previously undetected infectious agent that was more prevalent among gay men than among other HIV risk groups. However, researchers were unable to isolate a virus from KS tissues.

Searching for footprints of such a putative agent, Chang and colleagues used the method of representational difference analysis to identify DNA sequences specific for KS tumor tissue. Several DNA fragments were identified, and found to be homologous with sequences in known human and primate herpesviruses. In turn, these sequences were used to design primers to obtain the complete genome of a previously undescribed herpesvirus, since named HHV8, human herpesvirus 8. To this date, HHV8 defies cultivation in tissue culture.

4.4 Computer Modeling of Emerging Infections

Is computer modeling a useful adjunct to the analysis or control of emerging viral diseases? The 2014–15 Ebola pandemic in West Africa offers an interesting case study. As the epidemic unfolded, several groups attempted to project how it would evolve (Meltzer et al., 2014; Butler, 2014). These projections had very wide confidence limits, and several of them had upper limits in the range of 100,000 or more cases. What the modelers could not foretell was that the infection did not spread across sub-Saharan Africa, and that several introductions (into countries such as Nigeria and Mali) were controlled by case isolation, contact tracing, and quarantine. However, modeling did contribute useful insights. Dobson (2014) suggested that rapid quarantine (within 5–7 days) of contacts of Ebola patients could be critical in epidemic control.

5. REPRISE

One of the most exciting current issues in virology is the emergence of new viral diseases of humans, animals, and plants. Even though the era of modern virology has been well established for more than 65 years, virus diseases continue to appear or reemerge. The Ebola pandemic of 2014–15 highlights the associated dangers and obstacles to control.

There are several explanations for emergence: (1) discovery of the cause of a recognized disease; (2) increase in disease due to changes in host susceptibility or in virus virulence; (3) reintroduction of a virus that has disappeared from a specific population; (4) crossing the barrier into a new species previously uninfected. Many zoonotic viruses that are maintained in a nonhuman species can infect humans, but most cause dead-end infections that are not transmitted between humans. A few zoonotic viruses can be transmitted between humans but most fade out after a few person-to-person transmissions. Rarely, as in the case of HIV, SARS coronavirus, and Ebola filovirus, a zoonotic virus becomes established in humans, causing a disease that is truly new to the human species.

There are many reasons for the apparent increase in the frequency of emergence of new virus diseases, most of which can be traced to human intervention in global ecosystems. Emergent viruses are identified using both classical methods of virology and newer genome-based technologies. Once a candidate virus has been identified, a causal relationship to a disease requires several lines of evidence that have been encoded in the Henle–Koch postulates, guidelines that are periodically updated as the science of virology evolves. Identifying, analyzing, and controlling emerging viruses involve many aspects of virological science. Virus–host interactions play a key role, to explain persistence in zoonotic reservoirs, transmission across the species barrier, and establishment in human hosts. Thus, the issues discussed in many other chapters contribute to our understanding of emerging viral diseases.

FURTHER READING

Reviews, Chapters, and Books

Eaton BT, Broder CC, Middleton D, Wang L-F. Hendra and Nipah viruses: different and dangerous. *Nature Reviews Microbiology* 2006, 4: 23-35.

Fenner F. Biological control as exemplified by smallpox eradication and myxomatosis. *Proceedings of the Royal Society of London, Part B Biological Sciences*, 1983, 218: 259-285.

Fredericks DN, Relman DA. Sequence-Based Identification of Microbial Pathogens: a Reconsideration of Koch's Postulates. *Clinical Microbiology Reviews* 1996, 9:18–33.

Howard CR, Fletcher NF. Emerging virus diseases: can we ever expect the unexpected? Emerging Microbes and Infection. 2012, 1, e46 doi. 10.1038/emi.2012.47.

Hayes EB, Komar N, Nasci RS, Montgomery SP, O'Leary DR, Campbell GL. Epidemiology and transmission dynamics of West Nile virus disease. *Emerging Infectious Diseases* 2005, 11: 1167-1173.

Lederberg J, Shope RE, Oaks Jr, SC, editors. *Emerging Infections*. National Academy Press, 1992, Washington, DC.

Parrish CR. The emergence and evolution of canine parvovirus - an example of recent host range mutation. *Semin Virol* 1994, 5:121-132.

Weiss RA, McMichael AJ. Social and environmental risk factors in the emergence of infectious diseases. *Nature Medicine Supplement* 2004, 10: S70-S76.

Wilson MR, Peters DJ. Diseases of the central nervous system caused by lymphocytic choriomeningitis virus and other arenaviruses. Chapter 33, in AC Tselis and J Booss, editors, Handbook of Clinical Neurology, Elsevier, New York, 2014.

Original Investigations

Abdel-Moneim AS. Middle East respiratory syndrome coronavirus (MERS-CoV) evidence and speculation. Archives of Virology 2014, 159: 1575-1584.

Di Giulio DB, Eckburg PB. Human monkeypox: an emerging zoonosis. *Lancet Infectious Diseases*, 2004, 4: 15-25.

Elliott LH, Ksiazek TG, Rollin PE, Spiropoulou CF, Morzunov S, Monroe M, Goldsmith CS, Humphrey CD, Zaki SR, Krebs JW, et al. Isolation of the causative agent of hantavirus pulmonary syndrome. *Am J Tropical Medicine and Hygiene* 1994, 51: 102-108.

Enders JF, Weller TH, Robbins FC. Cultivation of the Lansing strain of poliomyelitis virus in cultures of various human embryonic tissue. *Science* 1949, 109: 85-9.

Enserik, M. Mission to MERS. Science 2014, 346: 1218-1220.

Gao F, Bailes E, Robertson DL, Chen Y, Rodenburg CM, Michael SF, Cummins LB, Arthur LO, Peeters M, Shaw GM, Sharp PM, Hahn BM. Origin of HIV-1 in the chimpanzee Pan troglodytes. *Nature*, 1999;397:436-441.

Gire SK, Goba A, Andersen KG, Sealfon RSG, Park DJ et al. (2014) Genomic surveillance elucidates Ebola virus origin and transmission during the 2014 outbreak. Science 345(6202):1369-1372 doi: 10.1126/science.1259657.

Graham RL, Donaldson EF, Baric, RS. A decade after SARS: strategies for controlling emerging coronaviruses. Nature Reviews Microbiology 2013, 11: 836-848.

Josseran L, Paquet C, Zehgnoun A, Caillere N, Le Tertre A et al. (2006) Chikungunya disease outbreak, Reunion Island. Emerg Infect Dis 12 (12):1994-5.

Kobasa D, Takada A, Shinya K, Hatta M, Halfmann P, Theriault S, Suzuki H, Nishimura H, Mitamura K, Sugaya N, Usui T, Murata T, Maeda Y, Watanabe S, Suresh M, Suzuki T, Suzuki Y, Feldmann H, Kawaoka Y. Enhanced virulence of influenza A viruses with the haemagglutinin of the 1918 pandemic virus. *Nature*, 2004, 431: 703 707

Kupferschmidt K. MERS surges again, but pandemic jitters ease. Science 2015, 347: 1297-1297.

Lanciotti RS, Roehrig JT, Deubel V, et al. Origin of the West Nile virus responsible for an outbreak of encephalitis in the northeastern United States. *Science*, 1999, 286: 2333-2337.

Li KS, Guan Y, Wang J, Smith GJD, Xu KM, Duan L, Rahardjo AP, Puthavathana P, Buranathai C, Nguyen TD, Estoepangestie ATS, Chalsingh A, Auewarakui P, Long HT, Hang NTH, Webby RJ, Poon LLM, Chen H, Shortridge KF, Yuen KY Webster RG, Peiris JSM. Genesis of a highly pathogenic and potentially pandemic H5N1 influenza virus in eastern Asia. Nature, 2004, 430: 209-213.

Li W, Shi Z, Yu M, Ren W, Smith C, Epstein JH, Wang H, Crameri G, Hu Z, Zhang H, Zhang J, McEachern J, Field H, Daszak P, Eaton BT, Zhang S, Wang LF. Bats are natural reservoirs of SARS-like coronavirus. Science 2005, 310: 676-678.

Messina JP, Brady OJ, Pigott DM, et al. The many projected futures of dengue. Nature Reviews Microbiology 2015, 13: 230-240.

Moore P, Chang Y. Kaposi's sarcoma (KS), KS-associated herpesvirus, and the criteria for causality in the age of molecular biology. Am Journal of Epidemiology, 1998, 147: 217-221.

Nichol ST, Spiropoulou CF, Morzunov S, Rollin PE, Ksiazek TG, Feldmann H, Sanchez A, Childs J, Zaki S, Peters CJ. Genetic identification of a hantavirus associated with an outbreak of acute respiratory illness. *Science* 1993, 262: 914-917.

Rasmussen AL, Okumura A, Ferris MT, Green R, Feldmann F, et al. Host genetic diversity enables Ebola hemorrhagic fever pathogenesis and resistance. Science 2014 346(6212):987-991. Doi 10.1126/science.1259595.

Reed W. Recent researches concerning etiology, propagation and prevention of yellow fever, by the United States Army Commission. *Journal of Hygiene* 1902, 2:101-119.

Rouvinski A, Guardado-Calvo P, Barba-Spaeth G, et al. Recognition determinants of broadly neutralizing human antibodies against dengue viruses. Nature 2015, 520: 109-113.

Tsetsarkin KA, Vanlandingham DL, McGee CE, Higgs S. (2007) A single mutation in Chikungunya virus affects vector specificity and epidemic potential. PLoS Pathog 3(12):e201. Doi:10.1371/journal.ppat.0030201.

Tumpey TM, Basler CF, Aguilar PV, Zeng H, Solorzano A, Swayne DE, Cox NJ, Katz JM, Taubenberger JK, Palese P, Garcia-Sastre A. Characterization of the reconstructed 1918 Spanish influenza pandemic virus. *Science* 2005, 310: 77-80.

Webster RG, Kawaoka Y, Bean WJJ. Molecular changes in A/chicken/Pennsylvania/83(H5N2) influenza virus associated with acquisition of virulence. *Virology* 1986;149:165-173.

Ebola References

http://www.washingtonpost.com/sf/national/2014/10/04/how-ebola-sped-out-of-control/.

Butler D. Models overestimate Ebola cases. Nature 2014, 515: 18.

Cohen J. Breakdown of the year: Ebola. Science 2014, 346 1450-1451.

Dobson A. mathematical models for emerging disease. Science 2014, 346: 1294-1295.

Epstein JM, et al. Mobilizing Ebola survivors to curb the epidemic. Nature 2014 516: 323-325.

Feldmann H, Geisbert TW. Ebola hemorrhagic fever. Lancet 2011, 377: 849-862.

Gire SK, Goba A, Andersen KG, Sealfon RSG, Park DJ et al. Genomic surveillance elucidates Ebola virus origin and transmission during the 2014 outbreak. Science 2014, 345(6202):1369-1372 doi: 10.1126/science.1259657.

Kortepeter MG, Bausch DG, Bray M. Basic clinical and laboratory features of filoviral hemorrhagic fever. Journal of Infectious Diseases 2011, 204: S810-S816.

Marzi A, Feldmann H. Ebola virus vaccines: an overview of current approaches. Expert Review of Vaccines 2014, 13: 521-531.

Meltzer MI, CY, Santibanez S, Knust B, Petersen BW, Ervin ED, Nichol ST, Damon IK, Washington ML. Estimating the Future Number of Cases in the Ebola Epidemic — Liberia and Sierra Leone, 2014–2015. Morbidity and mortality weekly report. 2014; 63:1-14.

Mire CE, Matassov D, Geisbert JB, et al. Single-dose attenuated Vesiculovax vaccines protect primates against Ebola Makona virus. Nature 2015, doi:10.1038/nature14128, Published online, 08 April 2015.

Paessler S, Walker DH. Pathogenesis of the viral hemorrhagic fevers. Annual Review of Pathology 2013, 8: 411-440.

Product: ReEBOV™ Antigen rapid test kit. http://www.who.int/medicines/ebola-treatment/1st_antigen_RT_Ebola/en/, 2015.

Rasmussen AL, Okumura A, Ferris MT, Green R, Feldmann F, et al. Host genetic diversity enables Ebola hemorrhagic fever pathogenesis and resistance. *Science* 2014, 346(6212):987-991. Doi 10.1126/science.1259595.

Vogel G. A reassuring snapshot of Ebola. Science 2015, 347: 1407.

WHO Ebola Response Team. Ebola virus disease in West Africa—The first 9 months of the epidemic and forward projections. N Engl J Med 2014; 371:1481-1495.

WHO: EBOLA SITUATION REPORT, 29 April, 2015.

HIV References

Etienne, L., Hahn, B.H., Sharp, P.M., Matsen, F.A., and Emerman, M. Gene loss and adaptation to hominids underlie the ancient origin of HIV-1. *Cell Host Microbe* 2013, 14: 85-92.

Hahn BH, Shaw GM, De Cock KM, Sharp PM AIDS as a zoonosis: scientific and public health implications. *Science* 2000, 287: 607-614.

Kluge, S.F., Mack, K., Iyer, S.S., Heigele, A., Learn, G.H., Usmani, S.M., Sauter, D., Joas, S., Hotter, D., Pujol, F.M., Bibollet-Ruche, F., Plenderleith, L., Peeters, M., Sharp, P.M., Fackler, O.T., Hahn, B.H. and Kirchhoff, F. Nef proteins of epidemic HIV-1 group O strains antagonize human tetherin. *Cell Host Microbe*, 2014, 16: 1-12.

Letko, M., Silvestri, G., Hahn, B.H., Bibollet-Ruche, F., Gokcumen, O., Simon, V. and Ooms, M. Vif proteins from diverse primate lentiviral lineages use the same binding site in APOBEC3G. *J. Virol.*, 87:11861-11871, 2013. PMCID: PMC3807359.

Pepin J. The origins of AIDS. Cambridge University Press, Cambridge, UK, 2011.

Sharp, P.M. and Hahn, B.H. Origins of HIV and the AIDS pandemic. *Cold Spring Harb. Perspect. Biol.,* 1: a006841, 2011, 1-22.

Chapter 17

Viral Evolution

It Is All About Mutations

Adi Stern[1], Raul Andino[2]

[1]Department of Molecular Microbiology and Biotechnology, Tel-Aviv University, Tel-Aviv, Israel; [2]Department of Microbiology and Immunology, University of California, San Francisco, CA, USA

Chapter Outline

1. INTRODUCTION

The extremely high mutation rates of viruses are not matched by any other organism in the kingdom of life. The high mutation rates of viruses, coupled with short generation times and large population sizes, allow viruses to rapidly evolve and adapt to the host environment. This has important implications for the pathogenesis of viral infections. In the course of the chapter, we will address a number of questions about virus evolution: (1) How are mutation rates defined and what are substitution rates? (2) Why are mutation rates so high? How do they differ for RNA and DNA viruses? (3) How are these rates measured and what are the shortcomings of standard measures? (4) Why is multiplicity of infection so relevant to the accumulation of mutations? (5) What are phylodynamics? What is a molecular clock and how is it estimated? (6) What drives virus evolution? What role does the host response play in virus evolution? How does this impact pathogenesis? Do viruses evolve to a benign relationship with their hosts? This chapter sets forth the basic tenets that govern the evolution of viruses: mutation rates, population size, selection, and the MOI. We will explore how viruses evolve within a host, during transmission to novel susceptible hosts, and establish infections in new host species.

Virus mutations create genetic diversity, which is subject to the opposing actions of selection and random genetic drift, both of which are directly affected by the size of the virus population. When the population size is large, selection will be predominant and random drift less common. This means that deleterious alleles will be efficiently removed from the population, while adaptive alleles will have an opportunity to take over the population. However, when the population size is small, random effects may obscure the effects of selection. Under these conditions, slightly deleterious alleles may rise to an unexpectedly high frequency in the population, and adaptive alleles may be lost by chance.

High mutation rates create many viral variants. During an infection with human immunodeficiency virus (HIV), all genotypes that are one mutation away from the infecting genotype will be created every day. The rich cloud of mutants, often termed a "quasispecies," has the potential to encode viruses with elevated resistance to a drug, or the ability to evade neutralizing antibodies created by the host. As a corollary, this complicates efforts to design effective vaccines, as evolution can greatly increase the number of virus serotypes that circulate in human populations. Furthermore, the unique ability of viruses to change allows them to cross species barriers, resulting in zoonotic infections.

Virus evolution is further characterized by additional layers of complexity. One unique characteristic of viruses is their MOI, which is the ratio between the number of viruses and the infecting cells. MOI has several consequences for evolution that are discussed in a later section, and these are subject to the constantly changing size of the virus population. The typical view of viral evolution is that viruses create huge population sizes within the infected host. However, this huge population size is punctuated by frequent bottlenecks during host-to-host transmission, and population

Viral Pathogenesis. http://dx.doi.org/10.1016/B978-0-12-800964-2.00017-3

structure within an infected host, where different organs and tissues may support different independently replicating populations. These differences in population size will affect both the selection-drift balance mentioned above, and the MOI of different virus subpopulations. In the rest of this chapter, we discuss the different factors affecting the virus population, and how these factors intertwine to shape virus evolution.

2. MUTATION AND SUBSTITUTION RATES IN VIRUSES

How are mutation rates defined, and what are substitution rates? Why are mutation rates so high? How do they differ for RNA and DNA viruses?

Mutation rate is typically defined as the average number of errors created in genomes of viral progeny, per base, per replication cycle (mut/nuc/rep). Viruses possess mutation rates that are orders of magnitude higher than any other replicating entity (Table 1). These rates range from approximately 1.5×10^{-3} mut/nuc/rep in the RNA bacteriophage Qβ (Batschelet et al., 1976) to $\sim10^{-8}$ mut/nuc/rep in the DNA virus Herpes simplex (Drake and Hwang, 2005). These examples highlight the interesting difference between RNA viruses, which replicate with their own RNA-dependent RNA-polymerase (RdRp), and DNA viruses which replicate with either their own or the cellular DNA-dependent DNA polymerase. RdRps all lack the proofreading capabilities present in DNA polymerases, and thus RNA viruses have much higher mutation rates than DNA viruses. Strikingly, it has been found that both increasing and decreasing the mutation rate of a virus leads to reduced virulence of the virus population (Pfeiffer and Kirkegaard, 2005; Vignuzzi et al., 2006). This suggests there is a close link between the mutation rate of a virus, the diversity created in a virus population, and pathogenesis in an infected host.

While mutations create raw genetic diversity, it is the coupled action of mutation and selection that will determine which mutations will *persist* in the viral population. The rate at which mutations fix in a population is termed the *substitution rate*, or evolutionary rate, which is measured by comparing the genomes of different isolates of a virus collected at several different time points (Duffy et al., 2008). Once again, RNA viruses possess much higher substitution rates than DNA viruses, ranging from 0.01 substitutions per site per year (sub/site/yr) in the RNA poliovirus type 1 to 7×10^{-7} in the DNA virus monkeypox. As suggested by theory, in most viruses substitution rates correlate well with mutation rates (Table 1). This suggests that the short-term mutation rate is an important determinant of the rate of long-term molecular evolution. However, for the fastest mutators (mostly RNA viruses), there appears to be an upper limit to the rate of evolution. This is due to the exceptionally high load of deleterious variant viruses in these small RNA viruses, which slows down their rate of molecular evolution. This high load dictates a threshold beyond which populations may go extinct. Indeed, it has been shown that by artificially increasing the mutation rate of different RNA viruses, the population will collapse through a process termed lethal mutagenesis. This finding has led to the development of therapeutic drugs that induce lethal mutagenesis, which are used to treat a variety of viral infections such as Hepatitis C and West Nile virus (Beaucourt and Vignuzzi, 2014).

2.1 Measurement of Mutation Rates

How are mutation rates measured and what are the shortcomings of standard measures?

It is important to define mutation rate in a consistent and unbiased manner. Mutation rates refer to the rate of mutation per site per genome replication, or to the rate of mutation per site per round of viral replication (Duffy et al.,

TABLE 1 Some Examples of a Range of Mutation Rates and Evolutionary Rates Determined for Different Groups of Viruses

Group	Virus	Mutation Rates (Mutations per Nucleotide per Replication Cycle)	Evolutionary Rate (Substitutions per Nucleotide Site per Year)
Positive-stranded RNA	Poliovirus 1	2.2×10^{-5}–3×10^{-4}	1.17×10^{-2}
Negative-stranded RNA	Influenza A virus	7.1×10^{-6}–3.9×10^{-5}	9×10^{-4}–7.84×10^{-3}
Retrovirus	Human immunodeficiency virus 1	7.3×10^{-5}–1.0×10^{-4}	1.13×10^{-3}–1.08×10^{-2}
Single-stranded DNA	Bacteriophage phiX174	1×10^{-6}–1.3×10^{-6}	Unknown
Double-stranded DNA	Herpes simplex 1	5.9×10^{-8}	8.21×10^{-5}

Data taken from the VirMut website (http://www.uv.es/rasanve2/virmut.htm) (Sanjuan et al., 2010; Sanjuan, 2012).

2008; Sanjuan et al., 2010). Critically, these two measures can differ when viruses use *stamping machine* replication versus *geometric genome* replication (Figure 1). During stamping machine replication one single virus genome is used as the template for replication, leading to linear accumulation of mutations. With geometric genome replication, progeny strands can become templates for replication themselves, and thus there is an exponential (or geometric) increase in progeny genomes. This will lead to a completely different distribution of mutations in the progeny genomes.

Classically, the mutation rate of an organism is determined in one of the two ways: the Lurie–Delbruck fluctuation test or measurement of mutation accumulation. We describe both methods, their caveats, and present a novel sequencing technique, which has the potential to alleviate some of these problems.

Luria–Delbruck fluctuation assay. In this method, a number of parallel populations are grown in a nonselective environment. Next, a selective environment is used to measure a given phenotype caused by a single mutation (e.g., resistance to phage). In each clonal population, the frequency of mutants with the phenotype is measured. Since mutations arose spontaneously in the parallel clonal populations, it is possible to use the mutant frequency to backtrack the mutation rate. If the number of rounds of replication can be estimated and the number of initial input genomes is known, it is possible to obtain a mutation rate per site per round of replication. There are several caveats in this assay leading to a biased estimation of the mutation rate: Mutations at multiple sites may alter the mutant phenotype; or the number of rounds of replication may be incorrectly estimated when the mode (stamping machine versus geometric) of genome replication is unknown.

Mutation accumulation studies. Multiple lineages of one progenitor strain are propagated over many generations, often with severe bottlenecks between propagations. These bottlenecks should reduce the effectiveness of selection, and thus mutations are expected to accumulate at the unbiased basic mutation rate. Sequencing the input and output viruses identifies the number of accumulated mutations; with the size of the genome known, and the number of rounds of replication estimated, the mutations per site per round of replication can be calculated. The major caveat here is the underlying assumption: If selection does operate during this propagation scheme, even at a minor level, this will skew the calculated mutation rate, often in an unknown way.

Mutation composition in a virus population. With the advent of next-generation sequencing (NGS), it is possible to capture accurate information on rare mutations present in a population. Low MOI will tend to select for viruses that are the "fittest" under the specific growth conditions. Using a sequencing technique that reduces the high error rate of prevalent NGS techniques, Acevedo et al. (2014) could accurately record the frequency of lethal mutations, which are expected to be present in a population at a frequency equal to the basic unbiased mutation rate (Figure 2). Applying this method to Poliovirus 1 populations confirmed a mean mutation rate of 3.97×10^{-4}, consistent with previous measurements. However, these results yield a level of detail previously less appreciated: different pairs of bases are replaced at different rates. A further intriguing study has shown that measurements of viral mutation rates vary substantially when measured across different cell types. Thus, there are previously unappreciated layers of complexity in the ascertainment of viral mutation rates.

3. MULTIPLICITY OF INFECTION

Why is multiplicity of infection so relevant to the accumulation of mutations?

The MOI is the ratio between the number of viruses and the number of cells. When MOI is high, cells are coinfected with multiple viruses, and when MOI is one or lower, each cell is most likely infected by one virus only. High MOI leads to a myriad of complex and contrasting effects (Table 2). First, in recombining viruses (or those that undergo reassortment), high MOI will lead to increased levels of recombination.

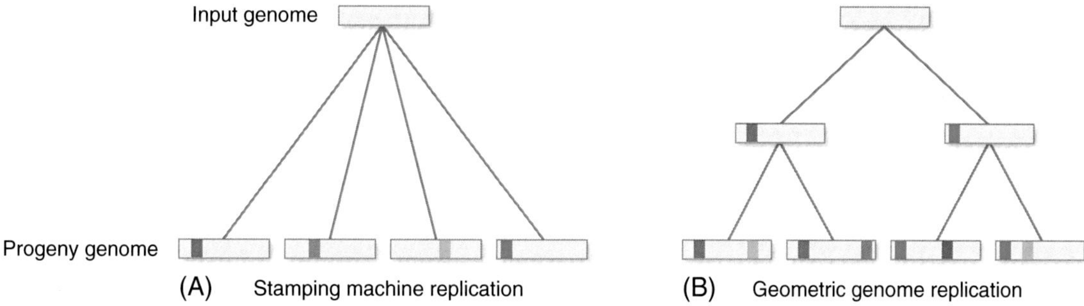

FIGURE 1 An illustration of (A) stamping machine replication, and (B) geometric genome replication. Yellow bars represent genomes, colored boxes represent mutations. In stamping machine replication, a single virus genome is used as the template for replication, leading to linear accumulation of mutations. With geometric genome replication, progeny strands can become templates for replication themselves, and thus there is an exponential increase in the number of mutations in the genomes of newly synthesized virions.

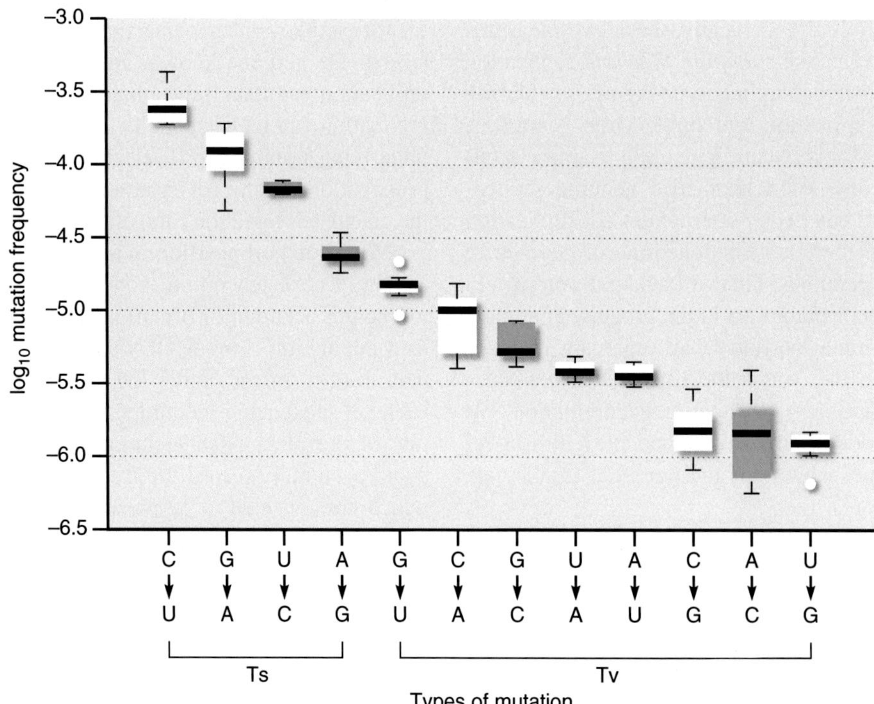

FIGURE 2 An improved method for the determination of virus mutation rates, as applied to poliovirus. This method is based on the premise that sequence data can be used to identify potential lethal mutations: in this case either stop codons within proteins or amino acid changes within catalytic sites of viral enzymes were enumerated. Based on population-genetics theory, the rate of lethal mutations was assumed to be equal to the basic unbiased mutation rate. Open boxes used both kinds of mutations and shaded boxes used only catalytic site mutations. Ts refers to a transition mutation; Tv refers to a transversion mutation (purine to pyrimidine or the reverse). *After Acevedo et al. (2014).*

TABLE 2 The Impact of Multiplicity of Infection (MOI) on Virus Genomic Evolution

Effect on Genome	High MOI	Low MOI
Recombination or reassortment	Frequent	Rare
Rescue of inferior genomes	Frequent	Rare
Rescue and maintenance of defective interfering genomes	Frequent	Rare
Selection for fittest genomes	Mixed	High
Bacteriophage infection	Lysogenic	Lytic

Frequent recombination may lead to more efficient selection, allowing the efficient removal of deleterious alleles and the incorporation of adaptive alleles. High rates of recombination/reassortment may also lead to the emergence of strains with a more virulent phenotype, as is thought to occur in cross-species transmission events of influenza virus. However, high MOI may also have contrary effects, whereby inferior genotypes are rescued, and maintained in the population, by products of superior genotypes. Complementation at high MOI also leads to propagation of defective particles. Thus under high MOI conditions, the beneficial effect of adaptive alleles may be masked (Stern et al., 2014).

High MOI also produces a higher gene copy number, that is multiple genomic copies of the same gene in one infected cell. In phages, copy number variation is highly influential: when copy number is one, phages will be lytic and kill the host cell, whereas if it exceeds one, phages become lysogenic and the bacterial host cell remains alive. Finally, competition for resources at high MOI may also have complex effects on viral replication. In fact, following complementation, viruses enter a "Prisoner's dilemma" regime where selfish genotypes evolve reducing the mean fitness of the viral population (had complementation been absent). Of these contrasting effects, the negative effects of complementation are more dominant that the positive effects of recombination, at least for bacteriophage Φ6 (Froissart et al., 2004) and for polioviruses (Stern et al., 2014). To summarize, it is evident that high MOI leads to complex and conflicting effects on genome selection.

There are a number of other implications of MOI. (1) The distribution of viral particles at different sites of an infection is unknown, and will also directly affect MOI and the efficiency of selection. (2) The MOI and ensuing diversity of the transmitting population affect the probability of establishing an infection in a novel host. Interestingly, the population with highest fitness in the original host does not necessarily fare well in new hosts, while low-frequency genotypes from the original host may be the ones that prevail in the new host. (3) It is likely that different types of viruses will be affected differently by MOI. For example, persistent viruses will likely replicate at an MOI that is lower than viruses that cause acute infections. Since MOI has such important effects on viral replication, it will require additional research to precisely determine how it affects selection in vivo, viral virulence, and the course of infection.

4. PHYLODYNAMICS: EVOLUTION IN A HOST POPULATION

What are phylodynamics? What is a molecular clock and how is it estimated?

There is an increasing interest, combined with new tools and methodology, to investigate transmission networks caused by viral epidemics. Such studies represent a collaboration between epidemiology and evolutionary biology, and the term phylodynamics has been coined to describe them. Phylodynamic methods are rooted in the powerful methodologies of phylogenetics, which emphasize the phylogenetic tree as key to investigating evolutionary processes. The ever-increasing availability of viral sequences has fueled this field and has made it possible to address a range of questions such as: "when did a virus emerge?," "what is the progenitor strain of a circulating epidemic?," and "what is the timing of the spread of a virus across countries and continents?" The phylodynamic approach has yielded remarkable insights into viral evolution.

Molecular clocks are based on previous observations that the number of nucleotide substitutions accumulates roughly linearly over time. This will be true when most nucleotide substitutions are neutral, and are driven directly by the mutation rate. Phylogeny provides a practical method to calibrate the molecular clock. When a viral phylogeny is reconstructed, it furnishes the distance between an ancestral sequence and an extant sequence in units of nucleotide substitutions. When the nucleotide difference (from the ancestral sequence) of each extant sequence is plotted against time, it is possible to infer the rate of nucleotide substitution and thus track back the date at which the ancestral sequence emerged. As an example, Kew et al. (1995) collected a set of isolates of poliovirus from a 10-year sequential infection chain in South America, and used this to determine the number of mutations per year (9×10^{-3} nucleotide substitutions per site per year for a 150 nucleotide window within the ~7500 nucleotide genome) (Figure 3). Most of these changes were synonymous mutations that would have little, if any, influence on the biological properties of the virus.

Based on the branching of the viral phylogeny, Holmes (2008) has suggested that a number of different patterns can explain viral dynamics across the globe. Examples include the following:

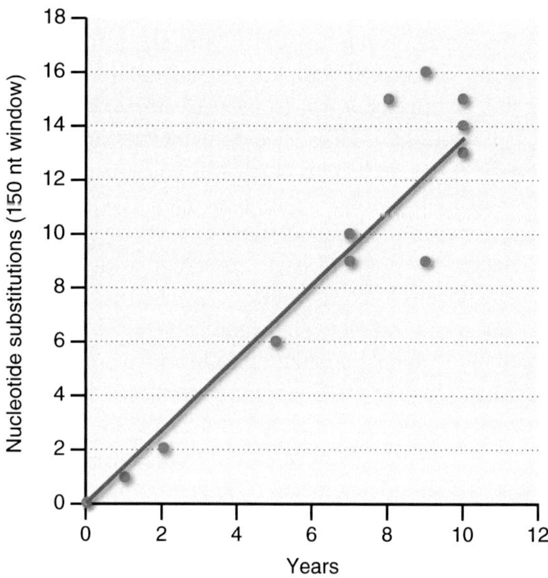

FIGURE 3 Deriving the molecular clock, the substitutions per nucleotide per year, in a host population. In this instance, a wild strain of poliovirus was introduced into a population and isolates were available over a 10-year time period. For each isolate, the number of nucleotide substitutions (from the original introduced strain) was plotted against the date of the isolate, to determine the substitution rate (substitutions per 150 nucleotides per year). *After Kew et al. (1995).*

- *Random mixing*: frequent viral traffic among different regions of the world cause lack of spatial structure in the tree, with no correlation between the geographic location of a sample and its location on the phylogeny
- *Population subdivision*: strong spatial subpopulations of viruses localized to certain regions of the world, suggesting that these populations do not mingle
- *Source–sink transmission*: one viral population acts as the source for all other viral populations in the world.

Different viruses follow distinctly different patterns. For influenza A, a source–sink model of viral population structure was found to best describe the global evolution of the virus. Accordingly, a persistent reservoir in south-east Asia continually seeds epidemics worldwide and drives viral diversity around the globe. On the other hand, hepatitis C virus displays a pattern of population subdivision, consisting of well-defined subtypes with distinct geographical locations.

The phylogenetic analysis yields some of the most striking examples of cross-species emergences. For HIV, as well as for other viruses where sufficient sequence data are available, it is now possible to trace back the evolution of a virus and how it acquired the ability to replicate efficiently in human cells (Sharp and Hahn, 2010). To date, four groups of HIV-1 have been identified: M, N, O, and P, with group M being responsible for the vast majority of infections worldwide. Phylodynamics showed that the progenitor of HIV-1 was an SIV strain from the chimpanzee strain *Pan t. troglodytes*, denoted SIVcpz-Ptt, as illustrated in Figure 4. Using this phylogeny, it was further possible to identify the consensus ancestor of HIV-1. Using sequences of dated isolates of HIV-1, a molecular clock was calibrated, and was used to calculate that the common ancestor of HIV-1 group M arose between 1910 and 1930.

Phylogenetic studies have provided insights about known pathogens such as SARS Coronavirus, the H5N1 avian influenza, monkeypox, and others, which have emerged in the human population following a cross-species transfer (see Chapter 16, Emerging viral diseases).

In a recent epidemic of Ebola virus, NGS compared the sequence of viruses obtained at different times during an outbreak in Mali (Hoenen et al., 2015). It appeared that the mutation rate was similar to that observed in prior outbreaks of Ebola virus. Importantly, there was no evidence that the 2013 epidemic strain of Ebola virus was evolving toward a strain with increasing transmissibility or virulence.

Multiple examples, together with several other lines of evidence (Parrish et al., 2008) highlight some important concepts in cross-species transfers: (1) The host range of a virus is usually well defined, and viruses only rarely gain the ability to spread efficiently to a new host. (2) The ability to spread to a new host species is compounded of many different factors, ranging from the probability of contact, demographic factors, the ability to bind a receptor, and the ability to overcome intracellular host restriction factors (Strebel, 2013). (3) Cross-species transmission events often involve crucial genetic adaptations on the part of the virus.

4.1 Drivers of Virus Evolution

What drives virus evolution? What role does the host response play in virus evolution? How does this impact pathogenesis?

As already noted, there are potentiators on viral evolution that have important implications for the control of viral

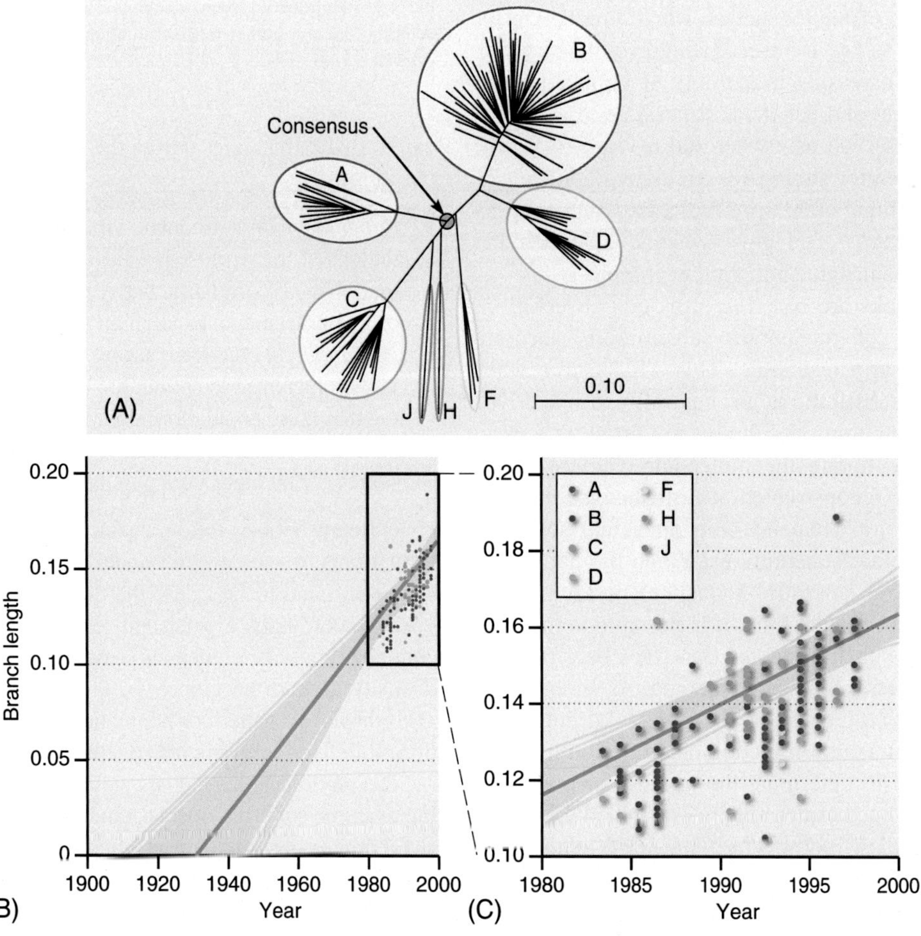

FIGURE 4 A phylogenetic tree of gp160 (the HIV surface glycoprotein) sequences from HIV-1 group M isolates. These data were used to identify an ancestral consensus sequence. Panel B: The nucleotide distances of individual virus isolates from the consensus sequence are plotted against their date of isolation. These data are used to project back to the date of the consensus sequence, giving a range from about 1915 to 1940. Panel C: A magnified view of the box in panel B. *Redrawn from Korber et al. (2000).*

diseases of humans. Two notable examples are escape from the host immune response and from antiviral drugs.

Immune escape mutants. A large armamentarium of effective viral vaccines have been developed over the last 75 years (see Chapter 19, Viral vaccines). These vaccines are based on "ancient" isolates (often at least 50 years old) of viruses which induce homologous neutralizing antibodies. Since these vaccines continue to provide protection against currently circulating viral strains, it must be concluded that the cognate viruses cannot evolve antibody-escape mutants that are sufficiently "fit" to survive and replace the original circulating strains. There are two notable exceptions to this general rule, influenza and HIV viruses. In both instances, antibody-escape mutants can survive to circulate in the population. For this reason, influenza virus continues to "drift" and new virus isolates must be used each year to produce protective vaccines. HIV has resisted the development of an effective vaccine mainly due to the difficulty of inducing "broadly neutralizing" antibodies that are effective against the population of escape mutants that have evolved over the last 30 years. In both instances, there is a structural explanation. Neutralizing antibodies to HIV and influenza viruses are not necessarily directed at the receptor-attachment site on the viral envelope, so neutralization escape mutants can still attach to cellular receptors, replicate, and persist (Figure 5).

Antiviral drugs. A vast experience with antiretroviral drugs for the treatment of HIV has shown that the virus swarm contains about 10^{-5} variants that will resist any single drug. Many such mutants will be produced each day in an HIV-infected patient (see Chapter 20, Antiviral therapy and Chapter 15, Mathematical modeling). For this reason, it has been found that effective control of HIV infections in humans usually requires the simultaneous use of at least three drugs, each of which requires a different set of mutations to escape drug control. The frequency of three escape mutations in a single genome is so low (~10^{-15}) that the drug combination will suppress HIV replication.

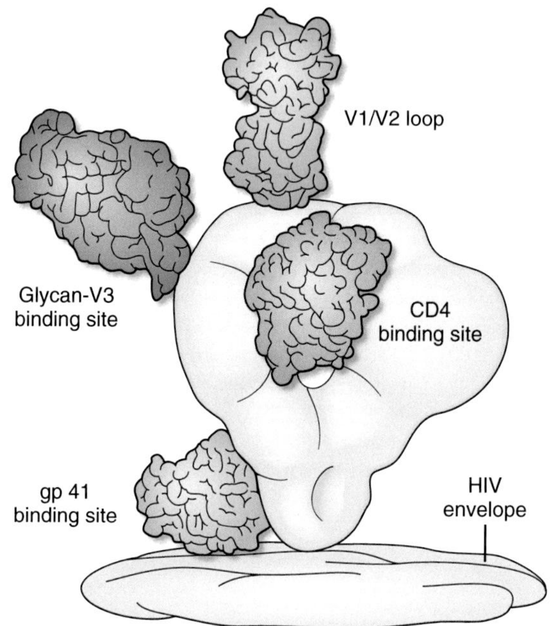

FIGURE 5 Neutralizing epitopes can be located distant from the cellular receptor-binding site on the virion, in which case antibody-escape variants can replicate and persist. This example is a cartoon of the principal neutralizing epitopes on the HIV envelope protein. Orange: CD4 binding site; purple: Glycan-V3 binding site; green: V1/V2 loop; gray: gp41 binding site. *Redrawn from Klein et al. (2013).*

Do viruses evolve to a benign relationship with their hosts?

How does virulence influence viral evolution? It has been proposed that when a virus and host undergo coevolution over a long period of time, a benign relationship will develop such that the virus will not cause disease. SIV is cited as an example. SIV strains are relatively benign in their natural simian hosts, where they have presumably been long established. However, when the SIVsmm strain crossed from its natural host—sooty mangabeys—into macaques, a novel host, it caused an AIDS syndrome.

However, a survey of mammalian viruses and their cognate hosts suggest that there is no necessary correlation between virulence and coevolution. Long-established virus infections range from inapparent to fatal. For instance, in the prevaccine era, poliovirus paralyzed only 1 person in every 150 infected (149 infections were inapparent), while rabies is 100% fatal in most of its animal and human hosts. Prior to smallpox eradication, there were two strains of variola virus. Variola major caused 30% mortality but variola minor only 1% mortality; yet each strain was maintained in the human population. It appears that viruses have used many strategies to perpetuate themselves in their host populations. Some strategies are benign, while others cause serious disease in their hosts.

5. SUMMARY AND CHALLENGES AHEAD

This chapter addresses some of the key features in viral evolution. (1) Viruses have mutation rates that are higher than any other member of the kingdom of life. This gives them the ability to evolve, even within the course of a single infection, and to evade multiple host defenses, thereby impacting pathogenesis. (2) There are several methods to estimate the mutation rates of viruses (mutations per nucleotide per genome replication), but each has its limitations. New improved methods utilize next-generation sequencing to take into account the large "swarm" of viral quasispecies. (3) The MOI has a critical effect on viral evolution. Low MOI will favor selection of the fittest viruses, while high MOI can have several opposing effects, and the net result is difficult to predict. (4) Phylodynamics, the collaboration between evolutionary biology and epidemiology, has generated data on the molecular clock that captures the rate of viral genome evolution (nucleotide substitutions per base per year) in host populations. (5) The drivers of viral evolution include evasion of host defenses, escape from antiviral drugs, and circumvention of vaccine-induced immunity.

What does the future hold for the study of virus evolution? For the first time, virus evolution can be informed by computational modeling based on experimental data. Evolutionary studies have now begun to compare related viruses that infect phylogenetically similar species, yielding a wealth of insights in viral evolution and host responses (Daugherty and Malik, 2012; Sawyer and Elde, 2012). A recent study has shown that evolution of viruses is dictated by strict protein constraints (Wylie and Shakhnovich, 2011), underlining the impact of protein structure on viral evolution. The ability of viruses to create antigenic variation is key to understanding epidemiologic dynamics, and the striking differences between viruses in their ability to escape immune responses reflects underlying structural and genomic determinants yet to be explained.

The increasing numbers of viral sequences has led to unprecedented observation of viruses as they evolve, from laboratory experiments, from evolution within a host, and from epidemiological sequences of patients around the globe. The integration of rich NGS data on evolving virus populations is opening the door to a better understanding of factors that facilitate adaptation and lead to disease. By establishing the rules that govern viral evolution, research is empowering the design of new strategies that control, treat, and possibly eradicate viral threats.

FURTHER READING

Acevedo, A., Brodsky, L., and Andino, R. (2014). Mutational and fitness landscapes of an RNA virus revealed through population sequencing. Nature 505, 686–690.

Batschelet, E., Domingo, E., and Weissmann, C. (1976). The proportion of revertant and mutant phage in a growing population, as a function of mutation and growth rate. Gene 1, 27–32.

Beaucourt, S., and Vignuzzi, M. (2014). Ribavirin: a drug active against many viruses with multiple effects on virus replication and propagation. Molecular basis of ribavirin resistance. Current opinion in virology 8C, 10–15.

Coffin, J.M. (1995). HIV population dynamics in vivo: implications for genetic variation, pathogenesis, and therapy. Science 267, 483–489.

Daugherty, M.D., and Malik, H.S. (2012). Rules of engagement: molecular insights from host-virus arms races. Annu Rev Genet 46, 677–700.

Drake, J.W., and Hwang, C.B. (2005). On the mutation rate of herpes simplex virus type 1. Genetics 170, 969–970.

Duffy, S., Shackelton, L.A., and Holmes, E.C. (2008). Rates of evolutionary change in viruses: patterns and determinants. Nat Rev Genet 9, 267–276.

Froissart, R., Wilke, C.O., Montville, R., Remold, S.K., Chao, L., and Turner, P.E. (2004). Co-infection weakens selection against epistatic mutations in RNA viruses. Genetics 168, 9–19.

Hoenen T.1, Safronetz D.1, Groseth A.1, Wollenberg K.R.2, Koita O.A.3, Diarra B.3, Fall I.S.4, Haidara F.C.5, Diallo F.5, Sanogo M.3, Sarro Y.S.3, Kone A.3, Togo A.C.3, Traore A.5, Kodio M.5, Dosseh A.6, Rosenke K.1, de Wit E.1, Feldmann F.7, Ebihara H.1, Munster V.J.1, Zoon K.C.8, Feldmann H.9, Sow S.10. Virology. Mutation rate and genotype variation of Ebola virus from Mali case sequences. Science. April 3, 2015;348(6230):117–119. doi: 10.1126/science.aaa5646. Epub 2015 Mar 26.

Holmes, E.C. (2008). Evolutionary history and phylogeography of human viruses. Annu Rev Microbiol 62, 307–328.

Kew O.M., Mulders M.N., Lipskaya G.Y., da Silva E.E., Pallansch M.A. Molecular epidemiology of poliovirus. Virology 1995, 6: 401–414.

Kirchhoff, F. (2010). Immune evasion and counteraction of restriction factors by HIV-1 and other primate lentiviruses. Cell host & microbe 8, 55–67.

Klein F., Mouquet H., Dosenovic P., et al. Antibodies in HIV-1 vaccine development and therapy. Science 2013 341: 1199–1204.

Korber B., Muldoon M., Theiler J., et al. Timing the ancestor of the HIV-1 pandemic strains. Science 2000, 288: 1789–1796.

Parrish, C.R., Holmes, E.C., Morens, D.M., Park, E.C., Burke, D.S., Calisher, C.H., Laughlin, C.A., Saif, L.J., and Daszak, P. (2008). Cross-species virus transmission and the emergence of new epidemic diseases. Microbiol Mol Biol Rev 72, 457–470.

Sanjuan, R. (2012). From molecular genetics to phylodynamics: evolutionary relevance of mutation rates across viruses. PLoS pathogens 8, e1002685.

Sanjuan, R., Nebot, M.R., Chirico, N., Mansky, L.M., and Belshaw, R. (2010). Viral mutation rates. Journal of virology 84, 9733–9748.

Sawyer, S.L., and Elde, N.C. (2012). A cross-species view on viruses. Current opinion in virology 2, 561–568.

Sharp, P.M., and Hahn, B.H. (2010). The evolution of HIV-1 and the origin of AIDS. Philosophical transactions of the Royal Society of London Series B, Biological sciences 365, 2487–2494.

Sharp, P.M. and Hahn, B.H. Origins of HIV and the AIDS pandemic. Cold Spring Harb. Perspect. Biol., 1: a006841, 2011, 1–22.

Stern, A., Bianco, S., Yeh, M.T., Wright, C., Butcher, K., Tang, C., Nielsen, R., and Andino, R. (2014). Costs and benefits of mutational robustness in RNA viruses. Cell reports 8, 1026–1036.

Strebel, K. (2013). HIV accessory proteins versus host restriction factors. Current opinion in virology 3, 692–699.

Wylie, C.S., and Shakhnovich, E.I. (2011). A biophysical protein folding model accounts for most mutational fitness effects in viruses. Proc Natl Acad Sci USA 108, 9916–9921.

Chapter 18

Viral Epidemiology

Tracking Viruses with Smartphones and Social Media

Kaitlin Rainwater-Lovett[1], Isabel Rodriguez-Barraquer[2], William J. Moss[3]

[1]Department of Pediatrics, Johns Hopkins School of Medicine, Baltimore, MD, USA; [2]Department of Epidemiology, Johns Hopkins Bloomberg School of Public Health, Baltimore, MD, USA; [3]Departments of Epidemiology, International Health and Molecular Microbiology and Immunology, Johns Hopkins Bloomberg School of Public Health, Baltimore, MD, USA

Chapter Outline

The science of epidemiology has been developed over the last 200 years, using traditional methods to describe the distribution of diseases by person, place, and time. However, in the last several decades, a new set of technologies has become available, based on the methods of computer sciences, systems biology, and the extraordinary powers of the Internet. Technological and analytical advances can enhance traditional epidemiological methods to study the emergence, epidemiology, and transmission dynamics of viruses and associated diseases. Social media are increasingly used to detect the emergence and geographic spread of viral disease outbreaks. Large-scale population movement can be estimated using satellite imagery and mobile phone use, and fine-scale population movement can be tracked using global positioning system (GPS) loggers, allowing estimation of transmission pathways and contact patterns at different spatial scales. Advances in genomic sequencing and bioinformatics permit more accurate determination of viral evolution and the construction of transmission networks, also at different spatial and temporal scales. Phylodynamics links evolutionary and epidemiological processes to better understand viral transmission patterns. More complex and realistic mathematical models of virus transmission within human and animal populations, including detailed agent-based models, are increasingly used to predict transmission patterns and the impact of control interventions such as vaccination and quarantine. In this chapter, we will briefly review traditional epidemiological methods and then describe the new technologies with some examples of their application.

1. HISTORY

Insight into the epidemiology of viral infections long preceded the recognition and characterization of viruses as communicable agents of disease in humans and animals, extending at least as far back as the treatise of Abu Becr (Rhazes) on measles and smallpox in the tenth century. Successful efforts to alter the epidemiology of viral infections can be traced to the practice of variolation, the deliberate inoculation of infectious material from persons with smallpox (see chapter on the history of viral pathogenesis). Documented use of variolation dates to the fifteenth century in China. Edward Jenner greatly improved the practice of variolation in 1796 using the less-virulent cowpox virus,

Viral Pathogenesis. http://dx.doi.org/10.1016/B978-0-12-800964-2.00018-5

establishing the field of vaccinology. An early example of rigorous epidemiological study prior to the discovery of viruses was the work of the Danish physician Peter Panum who investigated an outbreak of measles on the Faroe Islands in 1846. Through careful documentation of clinical cases and contact histories, Panum provided evidence of the contagious nature of measles, accurate measurement of the incubation period, and demonstration of the long-term protective immunity conferred by measles. The discovery of viruses as "filterable agents" in the late-nineteenth and early twentieth centuries greatly enhanced the study of viral epidemiology, allowing the characterization of infected individuals, risk factors for infection and disease, and transmission pathways.

1.1 Traditional Epidemiological Methods

Traditional epidemiological methods measure the distribution of viral infections, diseases, and associated risk factors in populations in terms of person, place, and time using standard measures of disease frequency, study designs, and approaches to causal inference. Populations are often defined in terms of target and study populations, and individuals within study populations in terms of exposure and outcome status. The purpose of much traditional epidemiological research is to quantify the strength of association between exposures and outcomes by comparing characteristics of groups of individuals. Exposures or risk factors include demographic, social, genetic, and environmental factors, and outcomes include infection or disease. In viral epidemiology, infection status is determined using diagnostic methods to detect viral proteins or nucleic acids, and serologic assays to measure immunologic markers of exposure to viral antigens. Infection status can be defined as acute, chronic, or latent.

Standard measures of disease frequency include incidence, the number of new cases per period of observation (e.g., 1000 person-years), and prevalence, the number of all cases in a defined population and time period. Prevalence is a function of both incidence and duration of infection and can increase despite declining incidence, as observed with the introduction of antiretroviral therapy for human immunodeficiency virus (HIV) infection in the United States. Although the number of new cases of HIV infection declined, the prevalence of HIV infection increased as treated individuals survived longer. Commonly used study designs include,

- cross-sectional studies in which individuals are sampled or surveyed for exposure and disease status within a narrow time frame,
- cohort studies in which exposed and unexposed individuals are observed over time for the onset of specified outcomes,
- case-control studies in which those with and without the outcome (infection or disease) are compared on exposure status, and
- clinical trials in which individuals are randomized to an exposure such as a vaccine or drug and observed for the onset of specified outcomes

Appropriate study design, rigorous adherence to study protocols, and statistical methods are used to address threats to causal inference (i.e., whether observed associations between exposure and outcome are causal), such as bias and confounding.

Much can be learned about the epidemiology of viral infections using such traditional methods and many examples could be cited to establish the importance of these approaches, including demonstration of the mode of transmission of viruses by mosquitoes (e.g., yellow fever and West Nile viruses), the causal relationship between maternal viral infection and fetal abnormalities (e.g., rubella virus and cytomegalovirus), and the role of viruses in the etiology of cancer (e.g., Epstein–Barr and human papilloma viruses).

1.2 Methods in Infectious Disease Epidemiology

The epidemiology of communicable infectious diseases is distinguishable from the epidemiology of noncommunicable diseases in that the former must account for "dependent happenings." This term was introduced by Ronald Ross to capture the fact that infectious agents are transmitted between individuals or from a common source. Traditional epidemiological and statistical methods often assume disease events in a population are independent of one another. In infectious disease epidemiology, individuals are defined in terms of susceptible, exposed, infectious, and recovered or immune. Key characteristics of viral infections that determine the frequency and timing of transmission, and thus the epidemiology, include the mode of transmission (e.g., respiratory, gastrointestinal, sexual, bloodborne, and vector-borne), whether infection is transient or persistent, and whether immunity is short or long lasting. Temporal changes in the transmission dynamics of viral infection can be displayed with epidemic curves, by plotting the number or incidence of new infections over time to demonstrate outbreaks, seasonality, and the response to interventions.

Key metrics in infectious disease epidemiology that capture the dependent nature of communicable diseases include: (1) the latent period, the average time from infection to the onset of infectiousness; (2) the infectious period, the average duration of infectiousness; (3) the generation time, the average period between infection in one individual and transmission to another; and (4) the basic reproductive number (R_0), the average number of new infections initiated by a single infectious individual in a completely susceptible population over the course of that individual's infectious period.

If R_0 is larger than one, the number of infected individuals and hence the size of the outbreak will increase. If R_0 is smaller than one, each infectious individual infects on average less than one other individual and the number of infected individuals will decrease and the outbreak ceases. The reproductive number (R) is a function not only of characteristics of the viral pathogen (e.g., mode of transmission), but also the social contact network within which it is transmitted and changes over time in response to a decreasing number of susceptible individuals and control interventions. An important concept related to the interdependence of transmission events is herd immunity, the protection of susceptible individuals against infection in populations with a high proportion of immune individuals because of the low probability of an infectious individual coming in contact with a susceptible individual.

The concepts and methods of infectious disease epidemiology provide the tools to understand changes in temporal and spatial patterns of viral infections and the impact of interventions. Traditional epidemiological methods provide powerful analytical approaches to measure associations between exposures (risk factors) and outcomes (infection or disease). Recent technological advances enhance these methods and permit novel approaches to investigate the emergence, epidemiology, and transmission dynamics of viruses and associated diseases.

2. SURVEILLANCE

Expanded access to the Internet and social media has revolutionized outbreak detection and viral disease surveillance by providing novel sources of data in real time (Chunara, 2012). Traditional epidemiologic surveillance systems rely on standardized case definitions, with individual cases typically classified as suspected, probable, or confirmed based on the level of evidence. Confirmed cases require laboratory evidence of viral infection. Surveillance systems are either active or passive. Active surveillance involves the purposeful search for cases within populations whereas passive surveillance relies on routine reporting of cases, typically by health care workers, health care facilities, and laboratories. Data acquired through active surveillance are often of higher quality because of better adherence to standardized case definitions and completeness of case ascertainment but are more expensive and resource intensive. However, both active and passive surveillance are prone to delays in data reporting.

The major advantage of using the Internet and social media to monitor disease activity is that the signal can be detected without the lag associated with traditional surveillance systems. Influenza is the most common viral infection for which the Internet and social media have been used for disease surveillance because of its high incidence, wide geographic distribution, discrete seasonality, short symptomatic period, and relatively specific set of signs and symptoms. However, the Internet and social media have several limitations compared to traditional active and passive surveillance systems and complement rather than replace these methods. These limitations include lack of specificity in the "diagnosis," and waxing and waning interest and attention in social media independent of disease frequency.

2.1 Novel Data Streams

In 2008, the Internet company Google developed a web-based tool called Google Flu Trends, for early detection of influenza outbreaks. Google Flu Trends is based on the fact that millions of people use the Google search engine each day to obtain health-related information (Ginsberg, 2009). Logs of user key words for pathogens, diseases, symptoms, and treatments, as well as information on user location contained in computer Internet Protocol (IP) addresses, allow temporal and spatial analyses of trends in search terms (Figure 1). Early results suggested that Google Flu Trends

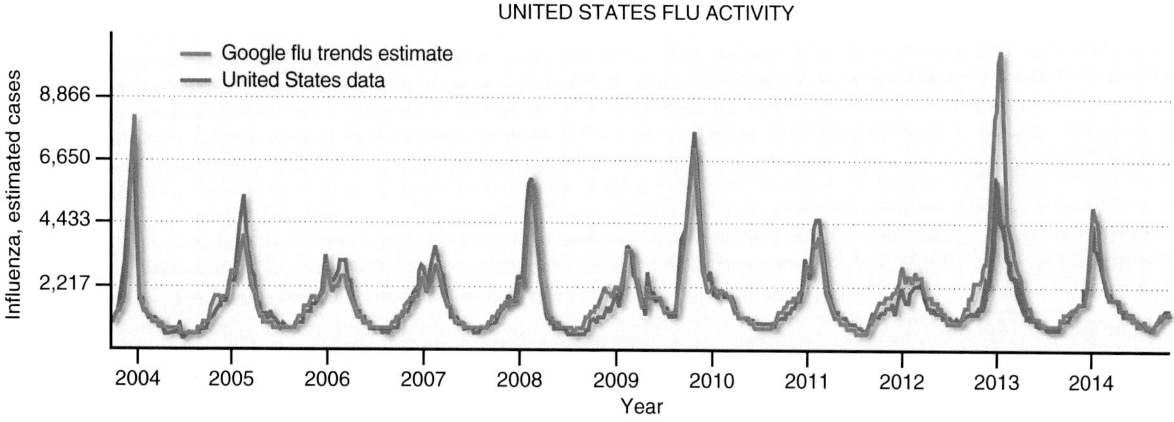

FIGURE 1 Number of cases of influenza in the United States from 2004 to 2013 estimated by Google Flu Trends and the Centers for Disease Control and Prevention. The number of cases of influenza estimated by Google Flu Trends closely approximated the number of cases estimated by the Centers for Disease Control and Prevention. *Redrawn from http://www.google.org/flutrends/about/how.html.*

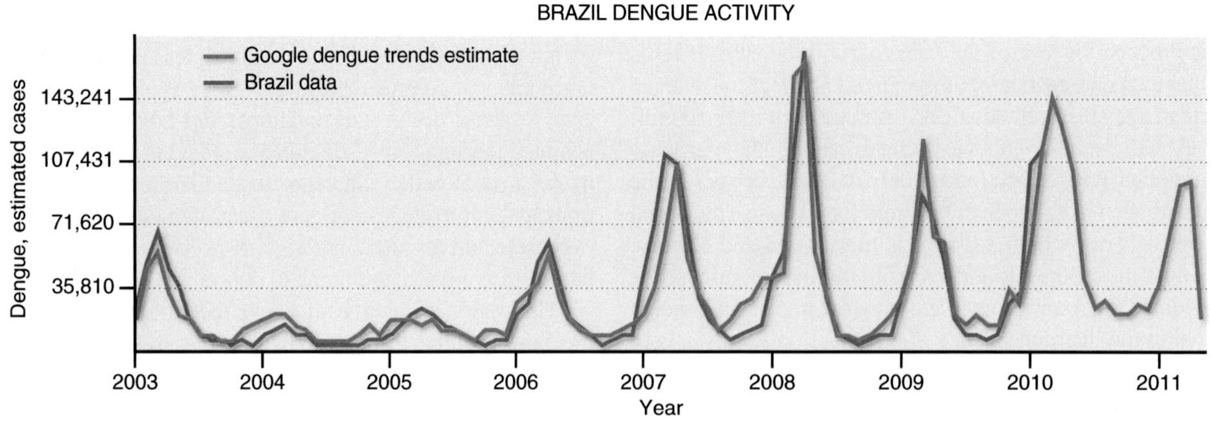

FIGURE 2 Number of cases of dengue in Brazil from 2003 to 2011 estimated by Google Dengue Trends and the Ministry of Health, Brazil. The number of cases of dengue estimated by Google Dengue Trends closely approximated the number of cases estimated by the Ministry of Health of Brazil. *Redrawn from http://www.google.org/denguetrends/about/how.html.*

detected regional outbreaks of influenza 7–10 days before conventional surveillance by the Centers for Disease Control and Prevention (Carneiro, 2009). However, accurate prediction was not as reliable as initially thought, and Google estimates did not closely match measured activity during the 2012–2013 influenza season. Google now reevaluates estimates using data from traditional surveillance systems (specifically those of the Centers for Disease Control and Prevention) to refine model and parameter estimates. These refinements more accurately capture the start of the influenza season, the time of peak influenza virus transmission, and the severity of the influenza season.

A similar approach, called Google Dengue Trends, is used to track dengue virus infections by aggregating historical logs of anonymous online Google search queries associated with dengue, using the methods developed for Google Flu Trends. Early observations suggest Google queries are correlated with national-level dengue surveillance data, and this novel data source may have the potential to provide information faster than traditional surveillance systems (Figure 2).

Other Internet sources are being explored to enhance viral surveillance. Wikipedia is a free, online encyclopedia written collaboratively by users and is one of the most commonly used Internet resources since it was started in 2001. As with Google searches, the use of disease-specific queries to Wikipedia are expected to correlate with disease activity. The number of times specific influenza-related Wikipedia sites were accessed provided accurate estimates of influenza-like illnesses in the United States 2 weeks earlier than standard surveillance systems and performed better than Google Flu Trends (McIver, 2014).

Similarly, social media data are being evaluated for surveillance purposes. Twitter is a free social networking service that enables users to exchange text-based messages of up to 140 characters known as tweets. As with Google Flu Trends, the number of tweets related to influenza activity is correlated with the number of symptomatic individuals. Several published studies reported correlations between Twitter activity and reported influenza-like illnesses (Chew, 2010; Signorini, 2011; Figure 3).

Limitations to using social media, such as Twitter, to monitor disease activity are illustrated by the Ebola virus outbreak in West Africa in early 2014. Despite the fact that Ebola had not yet occurred in the United States, posts to Twitter on Ebola rose dramatically, likely in response to intense media coverage and fear. Clearly, such tweets could not be interpreted to indicate Ebola disease activity in the United States. Studies reporting misleading associations, or the lack of correlation between social media and disease activity, are rarely published, providing a cautionary note.

While initial efforts using data from the Internet for viral disease surveillance offer promising results, concerns have been raised regarding the utility and robustness of these approaches (Lazer, 2014). Integration into existing surveillance frameworks will be necessary to maximize the utility of these data streams.

2.2 Communicating Surveillance Data

The Internet allows rapid processing and communication of health-related information, including the aggregation and display of surveillance data for viral infections. Traditional surveillance networks can be linked through the Internet to allow rapid integration and dissemination of information. Information on viral disease outbreaks available through Internet postings of health care agencies such as the World Health Organization (WHO) and Centers for Disease Control and Prevention (CDC), as well as press reports and blogs, can provide data that are more current than traditional surveillance systems. Information from these online sources can be made available to a large, global audience. Several of the most

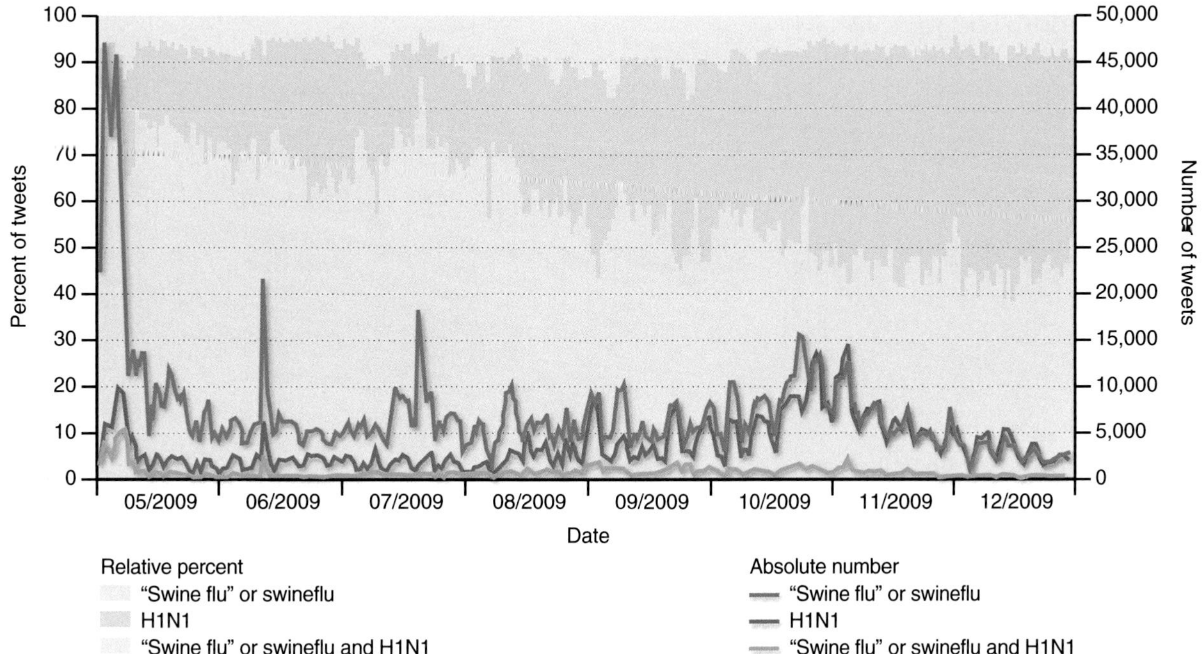

FIGURE 3 Absolute number and relative percent of tweets containing H1N1, swine flu, or both from May to December 2009. The lines show the absolute number of tweets and the shading shows the relative percent of tweets. The blue lines and shading indicate tweets specific for swine flu, the red lines and shading indicate tweets specific for H1N1 and the green lines and shading indicate tweets containing both terms. The data highlight monthly variation in tweets and the increase in the relative percent of tweets specific for H1N1 over the study period. *Redrawn from Chew C, Eysenbach G. Pandemics in the age of Twitter: content analysis of Tweets during the 2009 H1N1 outbreak.* PLoS One *2010; 5: e14118.*

Sidebar 1 Sources of information on outbreaks of viral diseases

ProMED, the Program for Monitoring Emerging Diseases, is an Internet-based reporting system established in 1994 that compiles information on outbreaks of infectious diseases affecting humans, animals, and food plants. ProMED relies on official announcements, media reports, and local observers, including the network of subscribers. A team of experts screen, review, and investigate reports before posting and often provide commentary. Reports are distributed by email to direct subscribers and posted on the ProMED-mail Web site. ProMED-mail currently reaches over 60,000 subscribers in at least 185 countries.

Started in 2006 by epidemiologists and software developers at Boston Children's Hospital, HealthMap monitors disease outbreaks and provides real-time surveillance of emerging public health threats, including viral infections (Figure 4). HealthMap organizes and displays data on disease outbreaks and surveillance using an automated process. Data sources include online news aggregators, eyewitness reports, expert-curated discussions and validated official reports.

Internet Resources

Google flu trends	http://www.google.org/flutrends
Google dengue trends	http://www.google.org/denguetrends
HealthMap	http://healthmap.org
ProMED	http://www.promedmail.org

commonly used surveillance sites report animal as well as human diseases (see Sidebar 1 and Figure 4).

3. VIRAL INFECTIONS IN SPACE AND TIME

Mapping spatial patterns of disease and relationships with environmental variables preceded the development of modern epidemiology. The classic example is John Snow's hand-drawn map of London cholera cases of 1854. However, routine mapping of health data only became commonplace in the 1990s after desktop geographic information systems became widely available. Combined with satellite imagery and remotely sensed environmental and ecological data, spatial mapping of viral infections is a powerful tool for surveillance and epidemiological research.

Spatial epidemiology is typically used to identify and monitor areas of differential risk. An early example was a large outbreak of St. Louis encephalitis virus infection in Houston, Texas in 1964. Spatial analysis showed that the outbreak was concentrated in the city center, with lower incidence at the outskirts. Further investigation revealed that the city center was associated with the lowest economic strata, unscreened windows, lack of air-conditioning and pools of standing water, factors facilitating virus transmission.

Investigation into the spatiotemporal dynamics of viral diseases at smaller spatial scales has become

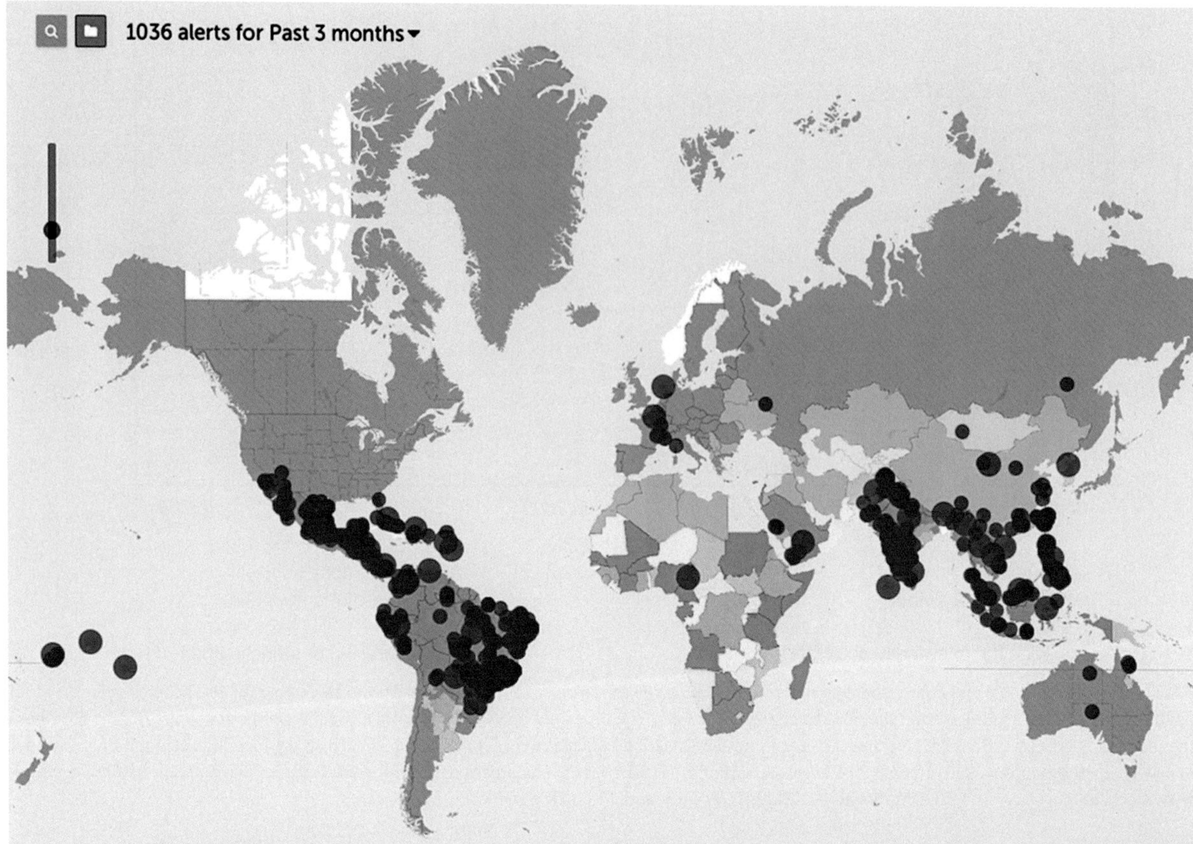

FIGURE 4 HealthMap display showing reports of global dengue cases from official sources and media reports. The large red circles indicate reports at the country level and the small red circles indicate reports at the local level. The risk of dengue at a country level is indicated by the color: blue absent or unlikely; yellow uncertain; orange likely; purple known. From *http://healthmap.org/dengue/en/*.

possible with increasing availability of global positioning systems devices and geocoding algorithms. Such studies have revealed spatial heterogeneity in the local transmission of some directly (e.g., HIV and influenza) and indirectly (e.g., dengue and chikungunya) transmitted viruses. For example, clustering analyses of the residential locations of people with dengue in Bangkok over a 5-year period showed evidence of localized transmission at distances less than 1 km (Salje, 2012; Figure 5). Analyses of data from a large population-based cohort of HIV-infected persons in Rakai District, Uganda revealed strong within-household clustering of prevalent and incident HIV cases as well clustering of prevalent cases up to 500 m (Grabowski, 2014).

Beyond descriptive applications, mapping spatiotemporal patterns of viral infections can provide fundamental insights into transmission dynamics at different spatial scales. Traveling waves from large cities to small towns were shown to drive the spatiotemporal dynamics of measles in England and Wales (Xia, 2004). The incidence of dengue hemorrhagic fever across Thailand manifested as a traveling wave emanating from Bangkok and moving radially at a speed of 148 km/month (Cummings, 2004).

3.1 Remotely Sensed Data

Insight into the spatial epidemiology of viral infections and associations with environmental risk factors can be greatly enhanced when information on the spatial location of cases is combined with remotely sensed environmental data (Rodgers, 2003). The spatial coordinates of cases can be overlaid on satellite imagery to demonstrate relationships with environmental features—such as bodies of water—and formally analyzed using spatial statistical techniques. Satellite sensors that detect reflected visible or infrared radiation provide additional information on temperature, rainfall, humidity, and vegetation among other variables, which are particularly important for the transmission dynamics of vector-borne viral infections. Satellite data for epidemiologic analyses are provided by a number of sources such as: (1) earth-observing satellites with high spatial resolution (1–4 m) but low repeat frequencies such as Ikonos and Landsat satellites; (2) oceanographic and atmospheric satellites such as MODIS and ASTER with lower spatial resolution (0.25–1 km) that provide images of the Earth surface twice a day; and (3) geostationary weather satellites such as GEOS with large spatial resolution (1–8 km).

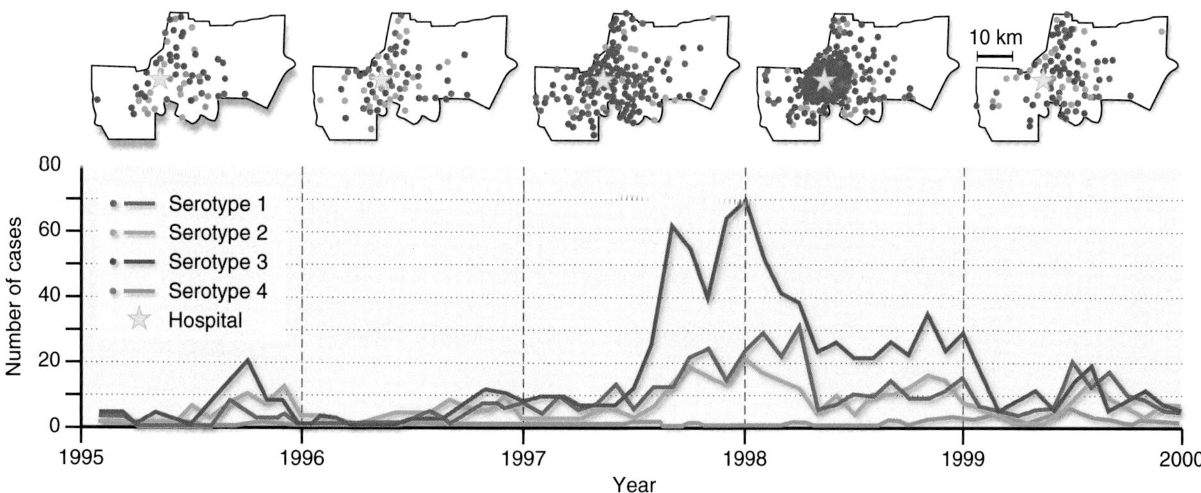

FIGURE 5 Spatial and temporal distribution of clinical cases of dengue disease by month at Queen Sirikit Hospital between 1995 and 2000 showing spatial heterogeneity in local transmission. The border in each map represents the Bangkok provincial boundary. *Redrawn from Salje, H, Lessler J, Endy TP, Curriero FC, Gibbons RV, Nisalak A, Nimmannitya S, Kalayanarooj S, Jarman RG, Thomas SJ, Burke DS, Cummings DA. Revealing the microscale spatial signature of dengue transmission and immunity in an urban population.* Proc Natl Acad Sci U S A *2012; 109: 9535–8.*

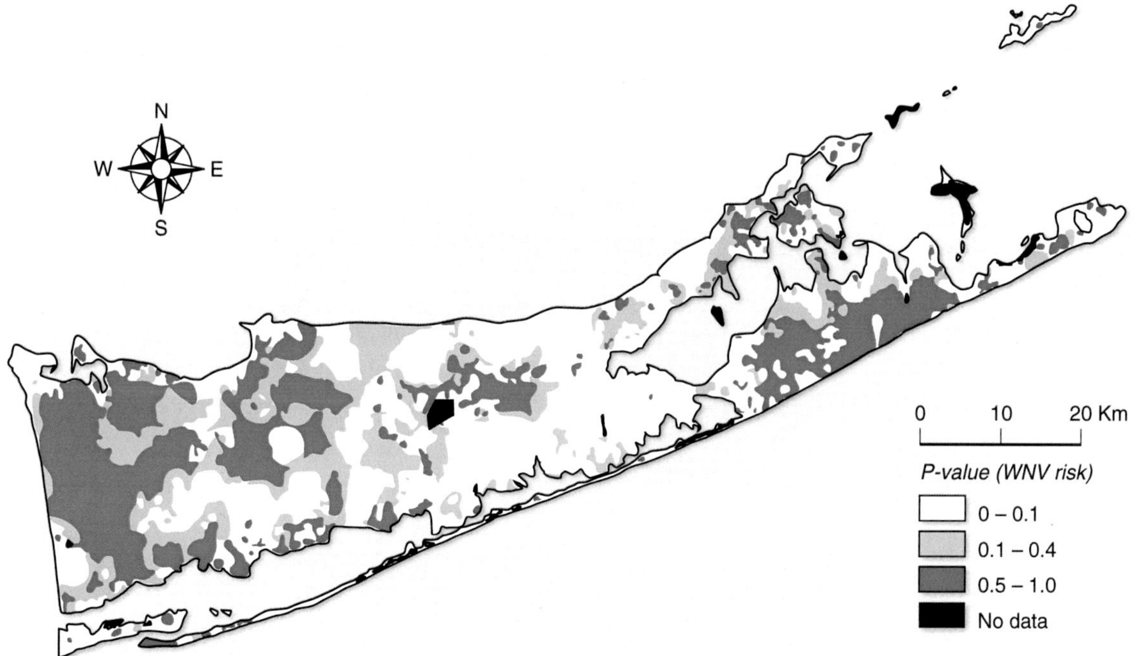

FIGURE 6 Risk map for West Nile virus in Suffolk County, New York. The risk of West Nile virus is displayed as a probability ranging from low (white), medium (blue) to high (red) based on data on vector habitat, landscape, virus activity, and socioeconomic variables derived from publicly available data sets. *Redrawn from Rochlin I, Turbow D, Gomez F, Ninivaggi DV, Campbell SR. Predictive mapping of human risk for West Nile virus (WNV) based on environmental and socioeconomic factors.* PLoS One *2011; 6: e23280.*

The statistical relationships between cases and environmental risk factors can be used to construct risk maps. Risk maps display the similarity of environmental features in unsampled locations to environmental features in locations where the disease is measured to be present or absent. Spatial analysis of the initial cases of West Nile virus infection in New York City in 1999 identified a significant spatial cluster (Brownstein, 2002). Using models incorporating measures of vegetation cover from satellite imagery, the risk of West Nile Virus could be estimated throughout the city. A more recent risk map for West Nile virus in Suffolk County, New York, was generated with data on vector habitat, landscape, virus activity, and socioeconomic variables derived from publicly available data sets (Rochlin, 2011; Figure 6).

4. POPULATION MOVEMENT

Population movement plays a crucial role in the spread of viral infections. In the past, quantifying the contribution of movement to viral transmission dynamics at different spatial scales was challenging, due to limited data. As an early example, the impact of restrictions of animal movement on transmission of foot-and-mouth disease in 2001 was estimated, using detailed contact-tracing data from farms in the United Kingdom (Shirley, 2005). However, such detailed data are rarely available for patterns of human movement. Studies have attempted to model the impact of long-range human movement on the spread of viral diseases using measures such as distance between cities, commuting rates, and data on air travel. This approach has been used to explain regional and interregional spread of influenza viruses. Data on air traffic volume, distance between areas, and population sizes have been invoked to describe and predict local and regional spread of chikungunya virus in the Americas (Tatem, 2012).

4.1 Movement Loggers

New technologies have greatly enhanced the capacity to study the impact of human movement on transmission dynamics of infectious diseases. Data from mobile phones and GPS loggers can be used to characterize individual movement patterns and the time spent in different locations (Figure 7). Individual movement patterns can be overlaid on risk maps to quantify movement to and from areas of high (sources) and low risk (sinks) as well as to estimate potential contact patterns. GPS data loggers generated 2.3 million GPS data points to track the fine-scale mobility patterns of 582 residents from two neighborhoods in Iquitos, Peru, to better understand the epidemiology of viral infections (Vazquez-Prokopec, 2013). Most movement occurred within 1 km of an individual's home. However, potential contacts between individuals were irregular and temporally unstructured, with fewer than half of the tracked participants having a regular, predictable routine. The investigators explored the potential impact of these temporally unstructured daily routines and contact patterns on the simulated spread of influenza virus. The projected outbreak size was 20% larger as a consequence of these unstructured contact patterns, in comparison to scenarios modeling temporally structured contacts.

5. TRANSMISSION NETWORKS

In addition to identifying individual and environmental characteristics associated with temporal and spatial patterns of viral infections, transmission networks are critical drivers of the dynamics of viral infections. Analysis of transmission networks defines the host contact structure within which directly transmitted viral infections spread. Network theory and analysis are complex subjects with a long history in mathematics and sociology, but have recently been adapted by infectious disease epidemiologists. The epidemiologic study of social networks is facilitated by unique study designs, including snowball sampling or respondent-driven sampling, in which study participants are asked to recruit additional participants among their social contacts. Differing sexual contact patterns serve as an example of the importance of contact networks to the understanding of viral epidemiology. Concurrent sexual partnerships amplify the spread of HIV compared with serial monogamy. This could partially explain the dramatic differences in the prevalence of HIV in different countries. Social networks were shown to affect transmission of the 2009 H1N1 influenza virus, and were responsible for cyclical

FIGURE 7 Movement density map of residents of Choma District, Zambia who carried a GPS logger for 1 month, overlaid on a satellite image of the area. The yellow shading shows the movement paths and the height in red indicates the time spent at location, with the taller peaks indicating more time spent in that location. By overlaying the movement density map on the satellite image as shown, investigators can explore relationships between movement, time and duration of stay, and environmental risk factors for disease. Image courtesy of Kelly Searle.

patterns of transmission between schools, communities, and households. Technological advances in quantifying contact patterns, with wearable sensors and the use of viral genetic signatures, have greatly enhanced the ability to understand complex transmission networks.

5.1 Contact Patterns

Self-reported contact histories and contact tracing are the traditional epidemiological methods to define transmission networks. Contact tracing has a long history in public health, particularly in the control of sexually transmitted diseases and tuberculosis, and is critical to the control of outbreaks of viral infections such as the Middle East respiratory syndrome coronavirus (MERS-CoV) and Ebola virus. To better understand the nature of human contact patterns, sensor nodes or motes have been used to characterize the frequency and duration of contacts between individuals in settings such as schools and health-care facilities. These technologies offer opportunities to validate and complement data collected using questionnaires and contact diaries.

As an example, investigators used wireless sensor network technology to obtain data on social contacts within 3 m for 788 high school students in the United States, enabling construction of the social network within which a respiratory pathogen could be transmitted (Salathe, 2010). The data revealed a high-density network with typical small-world properties, in which a small number of steps link any two individuals. Computer simulations of the spread of an influenza-like virus on the weighted contact graph were in good agreement with absentee data collected during the influenza season. Analysis of targeted immunization strategies suggested that contact network data can be employed to design targeted vaccination strategies that are significantly more effective than random vaccination.

6. MOLECULAR EPIDEMIOLOGY

Advances in nucleic acid sequencing and bioinformatics have led to major advances in viral epidemiology. Population (Sanger) sequencing has been the standard method for DNA sequencing but is increasingly replaced by deep sequencing in which variants within a viral swarm are distinguished. Sequencing allows for the detection of single nucleotide polymorphisms (SNPs) and nucleotide insertions or deletions ("indels"), analysis of synonymous and nonsynonymous mutations, and phylogenetic analysis (see chapter on virus evolution).

Sequencing techniques can be applied to both viral and host genomes. SNPs may be associated with changes in viral pathogenesis, virulence, or drug resistance. Molecular techniques applied to pathogens also have been fundamental to the study of the animal origins of many viral infections including HIV and MERS. Phylogeographic approaches were used to trace the origins of the HIV pandemic to spillover events in central Africa (Sharp, 2010). More recently, sequence data were used to track the animal reservoirs of MERS-CoV associated with the 2014 outbreaks (Haagmans, 2014), and to compare the Ebola virus strain circulating in the 2014 West Africa outbreak to strains from prior outbreaks (Gire, 2014)

Epidemiologic studies that probe host genomes can be either candidate gene studies or genome-wide association studies. The goal of these studies is to link specific changes with an increased risk of infection or disease. As an example, a small subset of individuals who failed to acquire HIV infection despite exposure, prompted studies to determine how these individuals differed from those who acquired infection. A 32-base-pair deletion in the human CCR5 gene, now referred to as CCR5-delta 32, accounted for the resistance of these subjects. Individuals who are CCR5-delta 32 homozygotes are protected against HIV infection by CCR5-tropic HIV strains, while heterozygotes have decreased disease severity.

7. PHYLODYNAMICS

Infectious disease epidemiologists are increasingly linking evolutionary, immunologic, and epidemiological processes, a field referred to as phylodynamics (Grenfell, 2004; Voltz, 2013). Because of the high mutation rates of viral pathogens, particularly RNA viruses, evolutionary and epidemiological processes take place on a similar timescale (see chapter on virus evolution). According to this framework, phylodynamic processes that determine the degree of viral diversity are a function of host immune selective pressures and epidemiological patterns of transmission (Figure 8).

Intrahost phylodynamic processes begin with molecular characteristics of the virus as well as the host's permissiveness and response to infection. For example, a single amino acid substitution in Epstein–Barr virus was shown to disrupt antigen presentation by specific human leukocyte antigen polymorphisms (Liu, 2014). This resulted in decreased T-cell receptor recognition and successful viral immune escape. The virus must also induce an "optimal" host immune response to maximize transmission to new hosts. If the virus induces a strong, proinflammatory immune response not balanced by the appropriate anti-inflammatory responses, the host may succumb to the overabundance of inflammation and cannot propagate viral transmission. Alternatively, a virus that fails to stimulate an immune response may also replicate uninhibited, overwhelming, and killing the host prior to transmission. Selective pressures maximize replication while sustaining transmission between hosts.

Interhost dynamics are affected by several factors including evolutionary pressures, timescales of infection, viral latent periods, and host population structures. Typically, only a small number of virions are transmitted

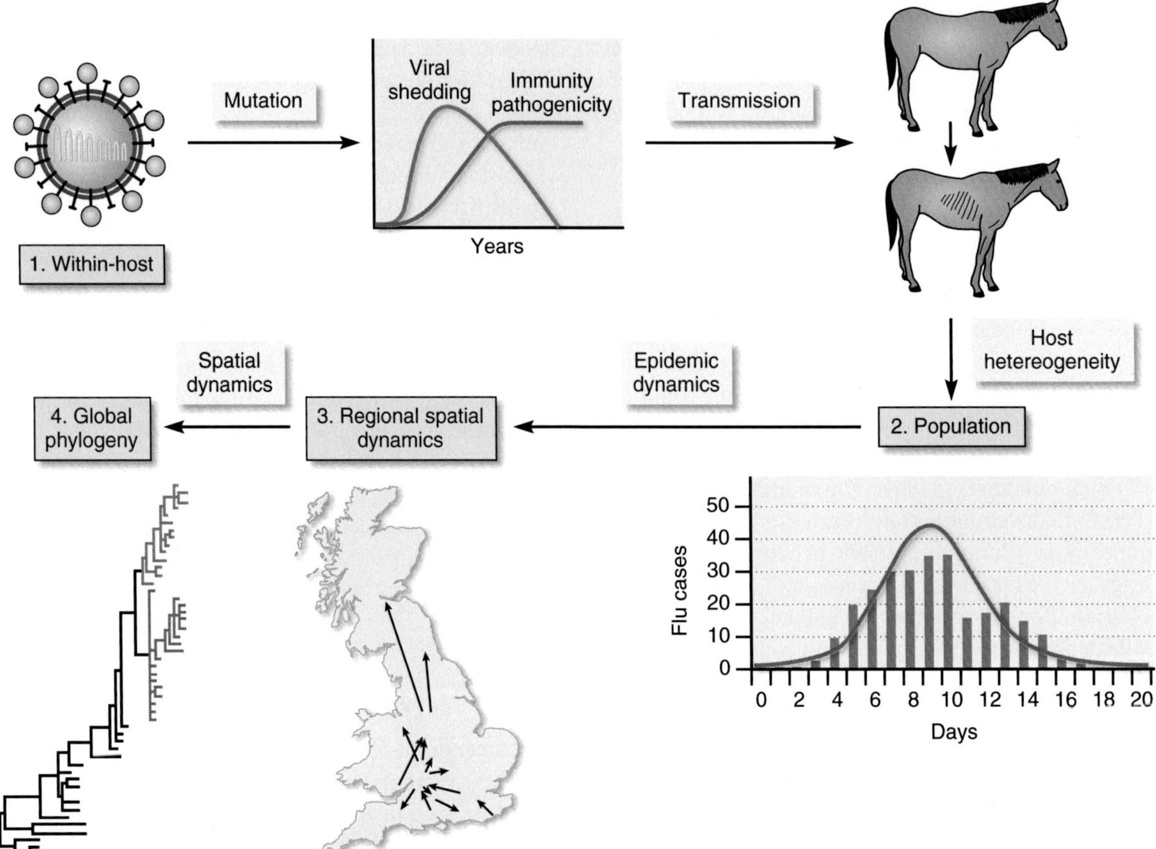

FIGURE 8 Phylodynamics links evolutionary, immunologic, and epidemiologic processes to explain viral diversity, as shown here for equine influenza virus. For viral evolution, these processes take place on a similar timescale. Within host mutations (1) result from an interplay between optimization of viral shedding, immunologic selective pressures and host pathogenicity. Transmission bottlenecks and host heterogeneity (2) further determine the population genetic structure of the virus, which in turn influences and is determined by the epidemic dynamics. Larger scale spatial dynamics at local, regional, and global levels (3), determines the global viral phylogenetic structure. *Revised and redrawn from Grenfell BT, Pybus OG, Gog JR, Wood JL, Daly JM, Mumford JA, Holmes EC. Unifying the epidemiological and evolutionary dynamics of pathogens.* Science *2004; 303:327–32.*

between hosts, creating a genetic bottleneck that limits viral diversity. A virus that mutates to cause highly pathogenic disease but is not transmitted cannot propagate its pathogenicity. Cross-immunity between viral strains also precludes the replication of particular viral lineages. Influenza vaccine strains require annual changes due to new circulating influenza strains that have escaped immune pressures through high mutation rates and gene re-assortment. The strong selection pressure of cross-immunity is reflected in the short branch lengths in a phylogenetic tree of influenza viruses isolated from infected individuals. Thus, the selection of influenza strains for future vaccines is partly determined by cross-immunity to prior circulating strains, because influenza viral strains that circulated in the past may elicit immune protection against currently circulating strains.

At the population level, phylodynamic methods have been used to estimate R_0 for HIV and hepatitis C virus, for which reporting and surveillance data are often incomplete (Volz, 2013). Phylodynamic and phylogeographic models also have been useful in reconstructing the spatial spread of viruses to reveal hidden patterns of transmission. For example, epidemiological and molecular studies of influenza virus transmission were compared at different spatial scales to highlight the similarities and differences between these data sources (Viboud, 2013). The findings were broadly consistent with large-scale studies of interregional or inter-hemispheric spread in temperate regions with multiple viral introductions resulting in epidemics followed by interepidemic periods driven by seasonal bottlenecks. However, at smaller spatial scales—such as a country or community—epidemiological studies revealed spatially structured diffusion patterns that were not identified in molecular studies. Phylogenetic analyses of gag and env genes were used to assess the spatial dynamics of HIV transmission in rural Rakai District, Uganda, using data from a cohort of 14,594 individuals residing in 46 communities (Grabowski, 2014). Of the 95 phylogenetic clusters identified, almost half comprised two individuals sharing a household. Among the remaining clusters, almost three-quarters involved individuals living in different communities, suggesting transmission chains frequently extend beyond local communities in rural Uganda.

The timescale of infection is also important for viral diversity and transmission dynamics. Some viruses are capable of initiating an acute infection that is cleared within days, while other viral infections are chronic and persist for a lifetime. The duration of infection impacts how quickly a virus must be transmitted and has implications for the infectious period and the potential to be transmitted to new hosts. Viruses with long latent periods create interhost phylogenetic trees with longer branch lengths. The long duration between infection and transmission permits accumulation of viral changes through many rounds of viral replication before transmission to the next host. Examples include hepatitis B virus, hepatitis C virus, and human immunodeficiency virus.

8. MATHEMATICAL MODELS OF TRANSMISSION DYNAMICS

Availability of computational resources allows widespread use and development of classic approaches to the mathematical modeling of viral transmission dynamics, such as compartmental, metapopulation and network models, to address epidemiologic questions (see chapter on mathematical methods). These models have been used extensively in the study of viral dynamics and to explore the potential impact of control interventions. New sources of high-resolution spatial, temporal, and genetic data create opportunities for models that integrate these data with traditional epidemiological data. Such analyses improve estimates of key transmission parameters and understanding of the mechanisms driving virus spread.

8.1 Agent-Based Models

Agent-based models (also known as individual-based models) can now be run using desktop computers, and offer advantages over more traditional mathematical models. Because each unit in a population is modeled explicitly in space and time and assigned specific attributes, agent-based models can reproduce the heterogeneity and complexity observed in the real world. More traditional compartmental, differential equation models often require simplifying assumptions that limit applicability. Agent-based models have been used to study the spread of viruses in populations as well as the evolution of viruses within and across populations. While agent-based models are intuitive and easy to formulate, these models are often difficult to construct due to the large number of parameters necessary to describe the behavior and interaction between individual units. Commercial frameworks that offer large computational power and intuitive user interphases have also become increasingly available. The Global Epidemic and Mobility Model (GLEAM) on the gleamviz platform (www.gleamviz.com), for example, contains extensive data on populations and human mobility, and allows stochastic simulation of the global spread of infectious diseases using user-defined transmission models.

9. SYNTHESIS

Our understanding of the epidemiology of viral infections is being revolutionized by the integration of traditional epidemiological information with novel sources of data.

- Data streams from the Internet are promising sources to enhance traditional surveillance but have yet to be fully validated.
- Molecular data on viral genomic sequences provide unprecedented opportunities to characterize viral transmission pathways.
- Phylodynamic and phylogeographic models have been used to estimate R_0, and characterize the spatial spread of viruses.
- Network analysis reveals hidden patterns of transmission between population subgroups that are not easy to capture with traditional epidemiological methods.
- Novel analytical and computational resources are playing a key role in integrating information from multiple large data banks.

These more comprehensive methods improve our ability to estimate the impact of infection control measures. The combination of traditional and evolving methodologies is closing the gap between epidemiological studies and viral pathogenesis. These developments have laid the foundation for exciting future research that will complement other approaches to the pathogenesis of viral diseases.

With these evolving technologies in mind, it is timely to ask: Is the world able to control viral diseases more effectively? It is a mixed score card. On the one hand, smallpox has been eradicated and we are on the verge of elimination of wild polioviruses. Furthermore, deaths of children under the age of 5 years (which are mainly due to viral and other infectious diseases) have decreased by almost 50% in the last few decades. On the other hand, the AIDS pandemic continues to rage in low-income countries, with only a slight reduction in the annual incidence of new infections. The United States has not done any better in reducing HIV incidence which has been unchanged for at least 20 years. The 2014–15 Ebola pandemic in West Africa reflects the limited capacity for dealing with new and emerging viral diseases on a global basis. In conclusion, epidemiological science continues to advance with evolving new technologies, but their application to public health remains a future challenge and opportunity.

FURTHER READING

Review, Articles, and Books

Chunara R, Freifeld CC, Brownstein JS. New technologies for reporting real-time emergent infections. Parasitology 2012;139:1843–51.

Carneiro HA, Mylonakis E. Google trends: a web-based tool for real-time surveillance of disease outbreaks. Clin Infect Dis 2009;49:1557–64.

Grenfell BT, Pybus OG, Gog JR, Wood JL, Daly JM, Mumford JA, Holmes EC. Unifying the epidemiological and evolutionary dynamics of pathogens. Science 2004;303:327–32.

Rogers DJ, Randolph SE. Studying the global distribution of infectious diseases using GIS and RS. Nat Rev Microbiol 2003;1:231–7.

Tufte ER. The Visual Display of Quantitative Information. 2nd edition. 2001. Graphics Press, Cheshire, Connecticut.

Original Research Reports

Brownstein JS, Rosen H, Purdy D, Miller JR, Merlino M, Mostashari F, Fish D. Spatial analysis of West Nile virus: rapid risk assessment of an introduced vector-borne zoonosis. Vector Borne Zoonotic Dis 2002;2:157–64.

Chew C, Eysenbach G. Pandemics in the age of Twitter: content analysis of Tweets during the 2009 H1N1 outbreak. PLoS One 2010;5:e14118.

Cummings D, Irizarry RA, Huang NE, Endy TP. Travelling waves in the occurrence of dengue haemorrhagic fever in Thailand. Nature 2004; 427:344–7.

Ginsberg J, Mohebbi MH, Patel RS, Brammer L, Smolinski MS, Brilliant L. Detecting influenza epidemics using search engine query data. Nature 2009;457:1012–4.

Gire SK, Goba A, Andersen KG, Sealfon RS, Park DJ, Kanneh L, Jalloh S, Momoh M, Fullah M, Dudas G, Wohl S, Moses LM, Yozwiak NL, Winnicki S, Matranga CB, Malboeuf CM, Qu J, Gladden AD, Schaffner SF, Yang X, Jiang PP, Nekoui M, Colubri A, Coomber MR, Fonnie M, Moigboi A, Gbakie M, Kamara FK, Tucker V, Konuwa E, Saffa S, Sellu J, Jalloh AA, Kovoma A, Koninga J, Mustapha I, Kargbo K, Foday M, Yillah M, Kanneh F, Robert W, Massally JL, Chapman SB, Bochicchio J, Murphy C, Nusbaum C, Young S, Birren BW, Grant DS, Scheiffelin JS, Lander ES, Happi C, Gevao SM, Gnirke A, Rambaut A, Garry RF, Khan SH, Sabeti PC. Genomic surveillance elucidates Ebola virus origin and transmission during the 2014 outbreak. Science 2014;345:1369–72.

Grabowski MK, Lessler J, Redd AD, Kagaayi J, Laeyendecker O, Ndyanabo A, Nelson MI, Cummings DA, Bwanika JB, Mueller AC, Reynolds SJ, Munshaw S, Ray SC, Lutalo T, Manucci J, Tobian AA, Chang LW, Beyrer C, Jennings JM, Nalugoda F, Serwadda D, Wawer MJ, Quinn TC, Gray RH; Rakai Health Sciences Program. The role of viral introductions in sustaining community-based HIV epidemics in rural Uganda: evidence from spatial clustering, phylogenetics, and egocentric transmission models. PLoS Medicine 2014;11:e1001610.

Haagmans BL, Al Dhahiry SH, Reusken CB, Raj VS, Galiano M, Myers R, Godeke GJ, Jonges M, Farag E, Diab A, Ghobashy H, Alhajri F, Al-Thani M, Al-Marri SA, Al Romaihi HE, Al Khal A, Bermingham A, Osterhaus AD, AlHajri MM, Koopmans MP. Middle East respiratory syndrome coronavirus in dromedary camels: an outbreak investigation. Lancet Infect Dis 2014;14:140–5.

Lazer, DM, Kennedy R, King G, Vespignani A. The parable of Google Flu: traps in big data analysis. Science 2014;343:1203–5.

McIver DJ, Brownstein JS. Wikipedia usage estimates prevalence of influenza-like illness in the United States in near real-time. PLoS Comput Biol 2014;10:e1003581.

Liu YC, Chen Z, Neller MA, Miles JJ, Purcell AW, McCluskey J, Burrows SR, Rossjohn J, Gras S. A molecular basis for the interplay between T cells, viral mutants, and human leukocyte antigen micropolymorphism. J Biol Chem 2014;289:16688–98.

Paz-Soldan VA, Stoddard ST, Vazquez-Prokopec G, Morrison AC, Elder JP, Kitron U, Kochel TJ, Scott TW. Assessing and maximizing the acceptability of global positioning system device use for studying the role of human movement in dengue virus transmission in Iquitos, Peru. Am J Trop Med Hyg 2010;82:723–30.

Rochlin I, Turbow D, Gomez F, Ninivaggi DV, Campbell SR. Predictive mapping of human risk for West Nile virus (WNV) based on environmental and socioeconomic factors. PLoS ONE 2011;6:e23280.

Salathé M, Kazandjieva M, Lee JW, Levis P, Feldman MW, Jones JH. A high-resolution human contact network for infectious disease transmission. Proc Natl Acad Sci U S A 2010;107:22020–5.

Salje, H, Lessler J, Endy TP, Curriero FC, Gibbons RV, Nisalak A, Nimmannitya S, Kalayanarooj S, Jarman RG, Thomas SJ, Burke DS, Cummings DA. Revealing the microscale spatial signature of dengue transmission and immunity in an urban population. Proc Natl Acad Sci U S A 2012;109:9535–8.

Sharp PM, Hahn BH. The evolution of HIV-1 and the origin of AIDS. Philos Trans R Soc Lond B Biol Sci 2010;365:2487–94.

Shirley MD1, Rushton SP. Where diseases and networks collide: lessons to be learnt from a study of the 2001 foot-and-mouth disease epidemic. Epidemiol Infect 2005;133:1023–32.

Signorini A, Segre AM, Polgreen PM. The use of Twitter to track levels of disease activity and public concern in the U.S. during the influenza A H1N1 pandemic. PLoS One 2011;6:e19467.

Tatem AJ, Huang Z, Das A, Qi Q, Roth J, Qiu Y. Air travel and vector-borne disease movement. Parasitology. 2012;139:1816–30.

Vazquez-Prokopec GM, Stoddard ST, Paz-Soldan V, Morrison AC, Elder JP, Kochel TJ, Scott TW, Kitron U. Usefulness of commercially available GPS data-loggers for tracking human movement and exposure to dengue virus. Int J Health Geogr 2009;8:68.

Vazquez-Prokopec GM, Bisanzio D, Stoddard ST, Paz-Soldan V, Morrison AC, Elder JP, Ramirez-Paredes J, Halsey ES, Kochel TJ, Scott TW, Kitron U. Using GPS technology to quantify human mobility, dynamic contacts and infectious disease dynamics in a resource-poor urban environment. PLoS One 2013;8:e58802.

Viboud C, Nelson MI, Tan Y, Holmes EC. Contrasting the epidemiological and evolutionary dynamics of influenza spatial transmission. Philos Trans R Soc Lond B Biol Sci 2013;368:20120199.

Volz EM, Koelle K, Bedford T. Viral phylodynamics. PLoS Comput Biol 2013;9:e1002947.

Xia Y, Bjørnstad ON, Grenfell BT. Measles metapopulation dynamics: a gravity model for epidemiological coupling and dynamics. Am Nat 2004;164:267–81.

Chapter 19

Viral Vaccines

Fighting Viruses with Vaccines

Juliet Morrison[1], **Stanley Plotkin**[2]

[1]Department of Microbiology, University of Washington, Seattle, WA, USA; [2]University of Pennsylvania, Philadelphia, PA, USA

Chapter Outline

1. INTRODUCTION

Vaccines are biological preparations that stimulate protective immune responses against pathogens. They have had a profound impact on human health: decreasing illness, extending life spans, and improving quality of life. Edward Jenner's introduction of vaccinia immunization to ameliorate smallpox is an eighteenth-century landmark in public health and one of the earliest demonstrations of the basic principles of immunology. However, immunization may have been practiced as early as 1000 CE in Asia and Africa; there is evidence that the Chinese inoculated people against smallpox by variolation, a procedure in which a small amount of material from smallpox pustules was inoculated into the skin or nostrils of naïve individuals. Vaccinia also provided a precedent for the use of live attenuated viruses to induce effective long-lasting protection, an example that even today inspires vaccinologists. During the last half of the twentieth century, a large number of safe and effective viral vaccines were developed for use in humans and animals. Vaccine development has been largely an empirical science, but systems biology approaches are helping to unravel mechanisms of protection and to predict vaccine efficacy.

This chapter is based on the premise that vaccine-induced protection can best be understood in the context of viral pathogenesis, which identifies potential steps in the infectious process where immunity might intervene to prevent disease. Importantly, the pathogenesis of a specific viral disease helps to determine the immunobiological requirements for a vaccine to protect against that particular infection. We first describe the major vaccine modalities, with their strengths and limitations, followed by an analysis of the mechanisms of vaccine-induced protection as exemplified by a few of the best-studied vaccines. Finally, we explore the utility of systems approaches to vaccine characterization and development.

Viral Pathogenesis. http://dx.doi.org/10.1016/B978-0-12-800964-2.00019-7

2. VACCINE MODALITIES

There are certain immunological principles that govern the induction of protective responses by any vaccine modality (Sidebar 1). Delivery of an immunogen to professional antigen-presenting cells (APCs) is the most effective way to initiate immune induction, which can be modulated to emphasize either cellular or humoral responses. There is a physiological limit to the expansion of naïve T lymphocytes during the primary response but, once rested, committed memory lymphocytes can be restimulated to undergo further expansion (often called an anamnestic or booster response). Adjuvants can bring professional APCs into contact with antigens through their proinflammatory action or exploit cytokines to increase proliferation of antigen-responsive lymphocytes. Newer vaccine modalities attempt to exploit these immunological principles to both enhance and focus the immune response to maximize protective efficacy.

Most vaccines licensed today protect through the induction of functional antibodies. The reason for this is that the diseases for which we have vaccines are largely those in which the agent replicates on the mucosa where antibodies can prevent implantation, or in which the agent disseminates from the mucosa through the bloodstream. Antibodies in the blood can neutralize those viruses that spread via a cell-free viremia and prevent invasion of organs. However, if vaccination does not entirely prevent infection and spread, cellular immune responses may then kill infected cells and thus reduce viral replication. Most effective viral vaccines in use are directed against acute infections, and they do not give 100% protection. Vaccinated individuals—when exposed to a wild virus—may undergo a modest infection that is subclinical and is evidenced only by an anamnestic jump in antibody titer.

Vaccine modalities fall into three broad categories: attenuated live viruses, nonreplicating inactivated viruses or purified antigens, and vectors with limited replicative capacity. Each of these modalities has its advantages and disadvantages, and it is unpredictable which one will produce the most successful vaccine for a given viral disease (Table 1). For instance, in the case of poliomyelitis, there are two vaccines: both the inactivated poliovirus vaccine (IPV) and the oral poliovirus vaccine (OPV) have their complementary advantages.

The earliest vaccines were attenuated viruses derived using chemicals or oxygen to weaken them, often leading to the development of attenuated strains that were relatively safe. Later vaccines were derived by serial passage of viruses in animals or cell cultures to select for attenuated mutants. In some instances, molecular sequencing and virus cloning have been used to produce improved versions. With the beginnings of experimental virology, technology was developed that led to the earliest nonreplicating viral vaccines, formulated by chemical or physical inactivation of virulent viruses. Further advances permitted the production of recombinant viral proteins that could be used as immunogens. Most recently, a variety of vector systems have been introduced to express viral proteins, and these are currently under active development as potential vaccine modalities.

2.1 Attenuated Viruses

Attenuated viruses produce infections that are milder than the illnesses produced by the virulent wild-type counterparts from which they are derived. Attenuated variants may differ in several ways from wild-type isolates. They are often host range mutants so that their replicative capacity—relative to their wild-type counterparts—is high in selected cell culture systems but much lower in vivo. Also, attenuated vaccine viruses are selected for differential tropism in vivo compared to their virulent parents. For instance, the cold-adapted viruses that constitute the live attenuated influenza vaccine (LAIV) will replicate quite well at 33 °C but poorly at 37 °C. In vivo, the cold-adapted virus replicates in the upper respiratory tract (nasal epithelium) but very little in the lower respiratory tract (alveolar epithelium), whereas the virulent

Sidebar 1 Principles of immune induction relevant for vaccine efficacy

- Immune induction is more efficient if an immunogen is presented by professional APCs, such as macrophages and dendritic cells.
- There is a relationship between the amount of antigen presented and the number of naïve lymphocytes that are induced to respond. The number of T lymphocytes induced during the active response determines the number of antigen-specific memory T lymphocytes that are generated.
- Following immune induction, about 10–15 cell divisions occur in antigen-responsive T lymphocytes at which time there is no further proliferation. After a "rest" of weeks to months, antigen-committed T cells may then be induced to proliferate again to produce an anamnestic immune response.
- For many viral infections, immunoglobulin and cellular effector systems can both participate in protective immunity, but their relative role varies for different viruses.
- Immune induction can be manipulated to favor either T_H1 (cellular) or T_H2 (antibody) responses, by formulation of immunogen, route of immunization, and the use of adjuvants.
- Adjuvants can enhance the immune response in a variety of ways, mediated by their induction of proinflammatory cytokines.
- Presentation of antigen to the mucosa-associated lymphoid system can induce local immunity, which may provide an effective barrier to viruses that invade via mucosal tissues.

virus replicates well in both sites. Attenuated OPV exhibits a different pattern of tropism than does wild-type poliovirus, since it replicates well in the gastrointestinal tract but poorly in the central nervous system (CNS). In contrast, wild-type virus replicates robustly in both sites.

Attenuated viruses used as vaccines depend for their efficacy on replication of the agent, which generates antibody and cellular immunity, as well as innate immune responses. In the case of measles, mumps, rubella, varicella, OPV, smallpox, and yellow fever vaccines it is mainly serum IgG antibody that prevents disease. Mucosal immune responses, particularly IgA but including transuded IgG, play the major role in protection afforded by rotavirus and LAIV. OPV prevents disease through elicitation of IgG serum antibodies, but also protects against intestinal infection through local induction of an IgA response. CD4+ T helper cells are critical to the B cell response, but cellular immune responses also contribute to protection from clinical measles, varicella, and smallpox if the wild virus infects.

Two current live attenuated vaccines are genetic reassortants: influenza and one of the rotavirus vaccines. These vaccines are made possible by the segmentation of the viral genomes. In the case of influenza, both live and inactivated, the RNA segments coding for hemagglutin and neuraminidase (of currently circulating strains) are reassorted with RNA segments coding for the six other viral proteins that are obtained from attenuated strains. Thus, the reassortant is attenuated but induces antibody responses against the two viral surface proteins. In the case of the rotavirus, there are also two surface proteins on the virus: VP4, the protease-cleaved protein (P) and VP7, the glycoprotein (G). The pentavalent rotavirus vaccine contains 10 double-stranded RNA fragments from a bovine rotavirus that is attenuated for humans and a single segment coding for both of the common P and G proteins, in order to induce antibodies that will protect by homotypic or heterotypic neutralization.

The search for an acceptable attenuated vaccine strain requires identification of variants that fall in a putative window of robust immunogenicity with minimal disease potential. In spite of diligent efforts to achieve complete safety, some attenuated vaccine viruses retain residual pathogenicity. For instance, OPV causes an occasional case of paralytic poliomyelitis, at a frequency of about 2 cases per 1,000,000 primary immunizations. In addition, some attenuated vaccine viruses may revert in virulence during passage in the primary vaccine recipient, which can be a problem if the virus is excreted. Thus, OPV often increases in virulence upon a single human passage due to revertant mutations. Type 3 OPV strains isolated from vaccine-associated poliomyelitis cases contain uracil to cytosine reversions at nucleotide 472 that restore neurovirulence. In the period 2000–present, more than 10 small outbreaks of poliomyelitis have been traced to reverted strains of vaccine virus that spread from person to person.

Another problem with live virus vaccines is that they may be inadvertently contaminated with adventitious

TABLE 1 Vaccine Modalities

	Live Attenuated Viruses	Inactivated or Subunit Viruses and Recombinant Proteins
Safety advantages	None	Avoids dangers of attenuated viruses
Safety disadvantages	Residual pathogenicity	Potential residual infectious pathogenic virus
	Reversion to increased pathogenicity	Safety tests difficult and expensive
	Unrecognized adventitious agents	Induction of unbalanced immune response
	Possible persistence	
Efficacy advantages	Local immunity at portal of entry	No viral interference
	Cellular and humoral immunity induction	
	Long-lasting immune response	Avoids limitations of attenuated viruses
	Herd immunity	
	Less expensive to manufacture	
Efficacy disadvantages	Interference between serotypes	No induction of local immunity
		Poor induction of cellular immunity
	Interference by adventitious viruses	May not mimic native epitopes for humoral immunity
	Loss of infectivity on storage	Short duration immunity (some products)
	Cold chain required to maintain infectivity	More expensive to manufacture

agents. For instance, yellow fever vaccine produced a massive epidemic of hepatitis B in the 1940s that was traced to a batch of vaccine that contained human serum obtained from an asymptomatic individual who was later shown to be a carrier of hepatitis B virus (HBV). Another contaminant of yellow fever vaccine was avian leukosis virus, acquired from the eggs used to prepare chick embryo cultures in which the vaccine virus was grown; this problem has now been eliminated by using leukosis-free eggs.

The extent of replication determines the level of the immune response. Increasing attenuation unfortunately is correlated with lower responses, presumably because of reduced antigen presentation and lower induction of innate immune responses. Thus, attenuation must establish a balance between safety and immunogenicity. For some candidate vaccines, this has been sought through making the agent replication-defective: in other words, permitting one cycle of replication to stimulate the immune response, but preventing the production of live virus. However, no replication-defective vaccine is yet licensed.

2.2 Inactivated Viruses

A number of chemical and physical methods can be used to inactivate viruses without destroying the integrity of the virus particle or much of its antigenicity. For instance, IPV is manufactured by treating the virus with dilute formalin (formaldehyde gas dissolved in water) at 37 °C for several weeks. The chemical treatment denatures the outer capsid protein sufficiently to prevent viral attachment and entry, while retaining epitopes that induce neutralizing antibodies. Beta propriolactone is another chemical that acts in a manner similar to formalin, and has been used to prepare inactivated rabies virus vaccines. An alternative is a so-called "split product" vaccine, produced by treatment of the virion with mild detergent or ethyl ether that dissociates the particle to yield a suspension of proteins and nucleic acids that are noninfectious but retain antigenicity. This method has been used to produce influenza virus vaccines.

Inactivated virus vaccines are often formulated from pathogenic virus strains, and their safety is contingent upon total inactivation. On occasion, failures in inactivation have caused cases of disease, such as occurred during the "Cutter incident" that confounded the introduction of IPV. A related problem is that "over-inactivation," done to insure safety, can compromise the immunogenicity of inactivated vaccines.

On rare occasions, inactivated vaccines can induce an "imbalanced" immune response that leads to untoward effects. For instance, inactivated measles virus elicited an immune response that resulted in enhanced disease. When children immunized in this manner were exposed to natural measles, they were not protected but developed "atypical" measles with unusual symptoms. Similarly, early trials of an inactivated vaccine against respiratory syncytial virus, an important respiratory virus of children, resulted in enhanced disease rather than protection.

2.3 Recombinant Proteins

A modern alternative to inactivated viruses is the preparation of a recombinant viral protein for use as an immunogen. Since the efficacy of the vaccine is based upon antibodies that target one or two of the viral proteins, there is no need to use the complete virion as an immunogen. However, recombinant proteins must retain their "native" conformation so that they elicit protective antibodies. For instance, the human papillomavirus (HPV) vaccine is based on in vitro synthesis of the major capsid protein, L1. Purified L1 proteins assemble into virus-like particles, which elicit antibodies that prevent attachment of the virus to the basement membrane of the mucosal epithelium. Another example is the recombinant hepatitis B vaccine that consists of the surface antigen of the virus produced in yeast or insect cells. However, industrial-scale production, purification, and stabilization of recombinant proteins are a daunting challenge, and such products are often expensive to manufacture.

2.4 Vectors

In the last few years, there has been a burst of research activity dedicated to novel modes of antigen presentation, sometimes called vectors or "platforms." These new approaches include recombinant viruses, replicons, and purified DNA.

2.4.1 Recombinant Viruses

Genetic engineering has allowed the development of vector-based strategies for immunization, in which the coding sequence for a protective protein is inserted into a nonpathogenic virus that expresses the protein of interest. Although many virus genomes can be manipulated to express foreign antigens, the largest viruses, such as poxviruses and herpesviruses, are most suitable for this purpose. Poxviruses have been used more frequently than other viruses, and vaccinia virus is the basis for some licensed animal vaccines, such as a rabies virus vaccine that has been deployed for the successful immunization of wildlife. A recombinant poxvirus was used in the HIV vaccine trial conducted in Thailand that provided the first evidence for modest efficacy in humans (see later section).

There are several considerations in selecting a replicating virus for use as a vaccine platform, including safety, immunogenicity, and prior immunity of the target population. Current safety standards make it much more acceptable to use a virus that has already had widespread use in the human populations, such as vaccinia virus or 17D, the attenuated vaccine strain of yellow fever virus. Even here, there are safety problems, since vaccinia causes serious complications albeit at low frequency. Thus, certain attenuated strains of vaccinia virus, such as MVA (modified virus Ankara) or NYVAC are preferred to standard vaccinia virus.

The immunogenicity of a recombinant virus depends in part on the cells that it targets. Some viruses infect macrophages and dendritic cells, and this maximizes their ability to deliver proteins to professional APCs, thereby enhancing the immunogenicity of the recombinant proteins that they encode. Since many recombinant constructs are based on human viruses, vaccinees may have been previously infected with the wild-type counterpart, and this preexisting immunity can reduce the replication of the recombinant virus and compromise its immunogenicity. For instance, recombinant vaccinia viruses are somewhat less immunogenic in persons who were previously vaccinated than in vaccinia-naïve subjects. Recombinant adenoviruses have proven to be highly immunogenic vectors, but are less effective in subjects already immune to the serotype used in the vaccine construct.

2.4.2 Replicons

Replicons are virus-like particles that will enter a target cell, undergo limited transcription and translation to synthesize encoded proteins, but will not produce infectious progeny. Replicons consist of a virus genome that has been engineered to insert a new protein and to delete some of the genes of the parent virus. Such genomic constructs often lack the genes for their envelope spike, and are transfected into packaging cell lines that provide a viral envelope in trans. This permits the assembly of a virus-like particle with the cellular specificity associated with the envelope. Replicons cannot spread beyond the cells that they initially "infect," and are a lower risk platform than recombinant viruses. They can exploit the attributes of many wild-type viruses that would be unacceptable for use as an infectious recombinant virus.

The efficacy of replicons depends upon their ability to reach a sufficient number of target cells, to produce enough novel immunogen, and to deliver the immunogen to professional APCs. In addition, it may be difficult to produce certain replicons on the industrial scale needed for vaccine deployment. Finally, replicons must pass safety tests to ensure that they will not recombine with cellular sequences to reconstitute the potentially pathogenic viruses from which they are derived. Only future investigation will determine whether replicons are a practical platform for vaccine formulation.

2.4.3 DNA Vaccines

It was first discovered in the early 1990s that a DNA plasmid, encoding a protein, could be used as an immunogen by simple injection of the "naked" DNA. This novel technology is currently under active investigation. DNA vaccine plasmids usually use a promoter such as the cytomegalovirus (CMV) promoter, which is highly active in most eukaryotic cells, driving a genetic insert expressing the gene of interest, followed by a transcriptional terminator and a polyadenylation sequence. Modifications of the protein sequence, such as addition of a signal sequence or a transmembrane domain, can be used to influence how the protein is processed in APCs.

DNA constructs are usually administered intramuscularly using a hypodermic needle or into the epidermis using a gene gun, which bombards the skin with gold beads coated with DNA. To be immunogenic, the DNA-encoded protein must be presented by professional APCs. Proteins produced in epithelial cells would be taken up by APCs via the exogenous pathway, while proteins produced in APCs could enter the endogenous pathway. Gene gun injections induce responses with less DNA than is required for soluble DNA, but tend to induce T_H2 responses biased toward antibody. DNA immunogens may be enhanced by the use of adjuvants. For instance, unmethylated CpG motifs in plasmid DNA provide a T_H1-biased adjuvant effect through toll-like receptors (TLRs). Also, DNA can be adjuvanted with plasmids encoding cytokines such as IL-2. DNA-based immunogens have shown modest immunogenicity, but have been more effective when used to prime an immune response followed by boosting with another vaccine modality, a type of vaccination called heterologous prime/boost.

As a vaccine, DNA possesses several advantages. First, it represents a well-defined and stable immunogen that can be precisely characterized and controlled, and produced on a large scale at relatively low cost. It appears to be biologically safe, assuming that it is adequately purified, and it avoids some of the dangers intrinsic in attenuated viruses, inactivated viruses, and certain vectors. Also, DNA immunogenicity is not inhibited by preexisting immunity, a problem with some viral vectors such as recombinant adenoviruses.

2.5 Adjuvants

Adjuvants, sometimes called "the immunologist's dirty little secret," have long been known to enhance the immunogenicity of antigens, particularly foreign proteins. The classic adjuvant is Freund's complete adjuvant (FCA), an oil–water emulsion containing inactivated tubercle bacillus, and the selected foreign protein. However, FCA caused granulomas at the site of injection and is not acceptable for use in humans. Aluminum oxides (alum) are much less irritating and are used in some human vaccines. Recent understanding of the innate immune system (see Chapter 4) has illuminated the mechanisms by which adjuvants appear to operate. Most of them bind to one or more of the TLRs, thereby activating dendritic cells and increasing the production of proinflammatory cytokines, as well as drawing macrophages to the site of antigen deposition. This amplifies the amount of the antigen that is bound by professional APCs and increases the number of antigen-specific T cells that respond to the antigen. Thus, adjuvants enhance innate immune responses that go on to initiate adaptive immune responses.

3. MECHANISMS OF PROTECTION BY ESTABLISHED VACCINES

A large number of viral vaccines have been developed, licensed, and are in use for the prevention of disease in humans (Table 2). These successful established products provide examples of the mechanisms of vaccine-conferred protection (Sidebar 1).

3.1 Poliovirus

The pathogenesis of poliovirus is understood at an organ level, although many of the specific cellular details have never been elucidated. When the virus is ingested, it invades via the tonsils and the lymphoid tissue of the small intestine, spreads to regional lymph nodes, and is transmitted through efferent lymphatics into the blood, where it circulates as a cell-free plasma viremia. Blood-borne virus invades the CNS either directly across the blood–brain barrier or indirectly by invading peripheral nerves or peripheral ganglia followed by neuronal spread to the CNS. Early studies demonstrated that after injecting a virulent wild-type virus into macaques, viremia is observed for about 1 week, followed by the appearance of neutralizing antibody, simultaneous with the disappearance of infectious virus.

These considerations, along with observations from the gamma globulin trial, led to the formulation by Jonas Salk of an inactivated preparation of poliovirus (IPV) as a candidate immunogen. The 1954 field trial of IPV provided an opportunity to test the hypothesis that neutralizing antibody could account for protection. There was a good correlation between the proportion of vaccinees who responded at a titer of 1:4 or greater and the estimated efficacy of the vaccine (~65%). This correlation suggests that a minimal level of neutralizing antibody can account for protection, not by preventing infection, but by preventing invasion of the CNS.

When attenuated strains of poliovirus (developed by Albert Sabin) were licensed as an OPV in the early 1960s, it became possible to compare IPV and OPV. IPV conferred minimal protection against enteric infection but OPV reduced fecal excretion significantly. It is likely that OPV generates local immunity by inducing antibody production by B cells in the gut-associated lymphoid tissue, although there is little direct evidence for this speculation.

Can the efficacy of poliovirus vaccines be attributed entirely to neutralizing antibody? It was noted above that OPV causes rare cases of poliomyelitis in vaccine recipients (about two per million vaccine recipients). Uniformly, these children have been diagnosed as hypo- or agammaglobulinemic. Strikingly, children with inherited T-cell defects (such as DiGeorge syndrome) do not seem to be at risk of vaccine-associated poliomyelitis. The absence of any data on the development of cellular immune responses to poliovirus vaccines precludes definitive conclusions, but there is little suggestion that CD8-mediated mechanisms play a role in protective immunity against poliovirus.

3.2 Rotavirus

Rotaviruses are an important cause of infant diarrhea and death, particularly in developing countries. These viruses have double-stranded 11-segmented RNA genomes, and genetic reassortants are readily obtained from mixed infections. The pathogenesis of rotavirus disease is not completely understood, but at least two mechanisms have been identified. The virus infects and kills epithelial cells at the tips of intestinal villi, and an internal protein, NSP4, acts as an enterotoxin. Rotaviruses have triple-layered virions, with two outer proteins, VP4 and VP7, both of which are targets for neutralizing antibody. These proteins also determine serotype; the most common VP7 serotypes are G1–G4 and G9 (G, glycoprotein) and the most common VP4 serotypes are P1 and P2 (P, protease sensitive). Vaccine trials (see below) suggest that there is some degree of immunological cross-protection between the different serotypes. Neutralizing antibody appears to be the most important determinant of protection against re-infection, while both T and B cells are important in recovery from primary infection.

Three live rotavirus vaccines have been developed, Rotashield (Wyeth), RotaTeq (Merck), and Rotarix (GlaxoSmithKline). Rotashield and RotaTeq are reassortant viruses, based on animal rotaviruses with VP4 and VP7 genes derived from human rotaviruses. Rotashield has a simian rotavirus and RotaTeq, a bovine rotavirus backbone. By contrast, Rotarix is a single human rotavirus (serotype G1 P1) that was attenuated by passage in cell culture. The ability of these three vaccine viruses to replicate in the human enteric tract varies considerably, and the dose used for immunizing human infants is highest for Rotateq and lowest for Rotarix. These vaccines are administered in two or three oral doses, beginning at age of 2 months. The vaccines elicit intestinal IgA and vaccine "takes" are usually determined by detection of virus-specific serum IgA. In large-scale trials, all three vaccines have been >80% efficacious at preventing severe rotavirus diarrheal disease in young infants. In developed countries, the vaccines are more than 90% effective, but in the tropics efficacy is much lower for reasons that may have to do with the microbiome.

Rotashield was the first of these vaccines to be licensed, but was withdrawn in 1999 (9 months after it became

TABLE 2 Commonly Used Viral Vaccines

Date of US Approval	Virus and Disease	Vaccine Modality and Route of Administration	Use in the United States
Before 1900	Variola	Attenuated	Only in the event of exposure
	Smallpox	Intradermal	
~1939	Yellow fever	Attenuated	Only in the event of exposure
		Subcutaneous	
1955	Polio	Inactivated	Yes
	Poliomyelitis	Intramuscular	All infants
1963	Polio	Attenuated	Yes
	Poliomyelitis	Oral	Special circumstances
1963	Measles	Attenuated	Yes
		Subcutaneous	All infants
1967	Mumps	Attenuated	Yes
		Subcutaneous	All infants
1969	Rubella	Attenuated	Yes
	German measles	Subcutaneous	All infants
1971	Influenza	Inactivated	Yes
		Intramuscular	High risk only
1980	Rabies	Inactivated	Yes
		Intramuscular	High risk only
1981	Hepatitis B	Inactivated	No
		Intramuscular	No longer made
1986	Hepatitis B	Recombinant HBs protein	Yes
		Intramuscular	All infants
1995	Varicella	Attenuated	Yes
	Chickenpox	Subcutaneous	All infants
~1996	Hepatitis A	Inactivated virus	Yes
		Intramuscular	High risk only
2006	Rotavirus	Attenuated	Yes
	Infant diarrhea	Oral	Infants
2006	Varicella	Attenuated	Yes
	Shingles	Subcutaneous	
2006	Human papillomavirus	Recombinant L1 and L2 proteins	Yes
		Intramuscular	

available) because it caused intussusception (a telescoping of the small intestine causing gangrene and peritonitis, requiring surgical intervention). The excess of intussusception cases (about 1 case per 10,000 vaccinees) occurred mainly during the first 2 weeks after the first dose of vaccine. Although the etiology of intussusception is not known, it has been speculated that the vaccine virus causes transient inflammation and swelling of Peyer's patches (lymphoid follicles in the intestinal wall) and that peristalsis leads to mechanical internalization of an intestinal segment. The other two rotavirus vaccines have rarely been associated with intussusception. RotaTeq was licensed in the United States in 2006 and Rotarix in 2009.

Rotavirus vaccines raise provocative questions associated with mucosal immunity (see Chapter 5). In contrast to most ingested foreign proteins, why are the viral proteins immunogenic? This paradox is not completely understood, but it appears that there are several factors that favor immune induction. Rotaviral infection of the intestinal tract is an invasive process, in contrast to the passive presence of a foreign protein in the intestinal lumen. Rotavirions are taken up by activated dendritic cells in the intestinal epithelium, and invading viral RNA will bind to TLRs 3, 7, and 8, activating dendritic cells and facilitating immune induction. Antiviral IgA can be identified in the intestinal secretions of immunized infants and likely neutralizes ingested rotaviruses. However, rotaviruses also produce a transient viremia, and protection against severe disease may be partly due to circulating antiviral IgG. It is unclear whether cellular immune responses play a role in vaccine-induced protection against rotaviral disease.

3.3 Rabies Virus

Rabies virus presents a special challenge for immunization because of its unusual pathogenesis, and it is one of the few infections where postexposure vaccination is frequently used. Rabies virus is often acquired through the bite of a rabid animal. Following injection into muscle or other peripheral site, the virus replicates locally, crosses the neuromuscular junction, and travels by the neural route to the CNS where it produces a fatal encephalomyelitis. Importantly, rabies virus never produces viremia.

One peculiar aspect of rabies pathogenesis is the variability in the incubation period. The virus may transit to the CNS within a few days or may be sequestered in an extraneural site for weeks to months before it invades the nervous system. This variability in the length of the rabies incubation period is determined by a variety of parameters, particularly the strain of virus. Thus, a neuro-adapted rabies virus, CVS (challenge virus standard), produces rabies with a high frequency and a short incubation period, whereas a freshly isolated wild-type strain (a so-called "street" virus) usually produces a lower frequency of infections and a much longer incubation period.

The long incubation period following exposure to street rabies virus provides the opportunity for postexposure prophylaxis. In the United States, preexposure vaccination is limited to veterinarians or others who are at occupational risk. Because the general population is not routinely immunized, postexposure prophylaxis is the major mode of rabies prevention. The protective mechanisms of pre- and postexposure prophylaxis are somewhat different and are considered separately.

3.3.1 Preexposure Prophylaxis

It appears that neutralizing antibody plays an important role in preexposure prophylaxis. Passive administration of antibody protects animals against subsequent challenge with rabies virus, the degree of protection being correlated with the titer of antibody, the timing of administration, and the strain and dose of rabies virus used for infection. Vaccinia recombinant viruses or DNA constructs that express only the rabies virus envelope glycoprotein provide excellent protection, which is proportional to the titer of neutralizing antibody. It is likely that antibody acts at several different levels, at the site of virus injection, at the neuromuscular junction, and even within the CNS. Specific depletion of antibody responses, by treatment with anti-μ antiserum, potentiates intracerebral infection with an attenuated nonlethal rabies virus, implying that antibody can even reduce *trans*-synaptic transmission within the CNS.

3.3.2 Postexposure Immunization

Active immunization, begun just after infection with street rabies virus, reduces overall mortality, and passive antibody synergizes this protective effect, reducing mortality even further. Passive antibody, given shortly after infection with street rabies virus, does not reduce overall mortality but does prolong the incubation period. Therefore, this synergistic effect is likely due to the ability of antibody to delay virus spread, thereby providing the host an advantage in the "race" between the virus and induction of an active immune response. A person exposed to rabies virus (and who has never been vaccinated) typically receives a dose of rabies immune globulin and four doses of rabies vaccine (made from inactivated rabies virus).

Rabies immunization also illustrates a much-discussed but probably rare phenomenon, immune-mediated disease enhancement by use of a vaccine. In the mouse model of postinfection vaccination, the number of long-incubation period cases is markedly reduced, but there is an absolute increase in short-incubation period cases following vaccination. The excess of short-incubation period cases implies immune enhancement, although the mechanism awaits elucidation.

3.4 Hepatitis B Virus

The pathogenesis of HBV is characterized by a number of unusual features. The timing of events suggests that HBV is not cytopathic and that the acute hepatitis is caused by the cellular immune response (see Chapter 6). The course of acute infection in adults is marked by replication in the liver, rising levels of circulating hepatitis B surface antigen (HBsAg), the viral envelope protein, together with infectious virions (10^6 per ml of plasma). The resolution of infection is accompanied by acute hepatitis that ranges from subclinical to severe or even fatal. Concomitant with the resolution of infection, there is an immune response that leads to waning of liver infection and circulating HBsAg, and the appearance of anti-HBsAg antibodies.

Experimental evidence for immune-mediated viral clearance comes from a transgenic mouse model, in which mice express one or several HBV proteins in the liver. When these animals are adoptively immunized with HBsAg-specific T lymphocytes, the viral protein is cleared from hepatocytes, but treatment with anti-HBsAg antibody has no effect. CD8-initiated viral clearance is mediated by cytokines (IFNγ and TNF) secreted by effector cells that inhibit HBsAg expression, rather than by cytolysis, explaining how it occurs in the absence of overwhelming hepatitis. Thus infection of adult humans with HBV is an example of an immune response which both produces disease and clears the infection.

An alternative course of infection is seen frequently in infants infected during birth, who become persistent virus carriers, with high levels of virus in the liver and blood. Such persistent infections are not accompanied by acute hepatitis, strengthening the view that the virus infection alone does not cause hepatitis. However, neonatal infection carries a high risk of cirrhosis and hepatocellular carcinoma, which only develop decades later. It is likely that persistent infection represents a state of HBsAg immune tolerance due to "exhaustion" or "deletion" of HBsAg-reactive CD4+ or CD8+ T cells.

The HBV vaccine consists of a recombinant form of the HBsAg that induces "neutralizing" antibodies. Presumably these antibodies protect adults who are exposed to HBV, either through contaminated blood or blood products or (rarely) via sexual contact.

When infants born of mothers who are HBV carriers are immunized with the recombinant HBsAg vaccine at birth, a remarkable result is seen. A high proportion (~90%) of these infantile infections is "converted" from persistent to short duration, but without acute hepatitis. This is surprising, because the immune response to the vaccine only appears 1–3 months after birth (i.e., 1–3 months after infection). The sequence of events includes a transient HBs antigenemia. Since HBV can only replicate in hepatocytes, this implies that HBV infection is established in the liver and is subsequently cleared. Again, it is likely that a host cellular immune response, elicited by either the vaccine or by the active infection, plays a role in vaccine-induced protection. The synergistic cooperation of humoral and cellular immunity may therefore explain the efficacy of the HBV vaccine. Immune memory is also important in this case because half of vaccinees lose antibodies with time but are nevertheless protected by an anamnestic response.

3.5 Human Papillomavirus

HPV has evolved to replicate in a very specialized niche, that is the epithelium of skin and mucous membranes. There are over 100 HPV serotypes and a few of them (particularly types 16 and 18) are a significant cause of cervical cancer (see Chapter 8). Combining all serotypes worldwide, it is estimated that HPV causes at least 200,000 cervical cancer deaths annually.

The natural history of HPV is rather unusual. HPV is transmitted through sexual contact that deposits virions on the mucosal surface, and it invades through minute breaks in mucosal epithelia. Initially, the virions attach to the basal membrane that underlies the epithelial cell layers, and undergo an essential conformational change. The altered virions can now infect the basal stem cells that generate the overlying epithelia. The virus begins replication in these cells, and is carried within these differentiating cells toward the epithelial surface where mature virions are synthesized and released on the mucosal surface. Natural HPV infections persist for varying periods of time and cervical infections are often cleared in 1–2 years. In those infections with oncogenic types of HPV that do persist, cervical cancer develops in a series of steps progressing from initial infection, to persistent infection, to hyperplasia, to cervical intraepithelial neoplasia, to cervical cancer and metastatic spread. The whole process takes many years, but the early phases can be detected within 1–2 years of infection in some individuals.

HPV vaccines have been formulated to prevent or ameliorate infection with HPV, and are not directed against the oncogenic proteins (E6 and E7) of the virus. Instead vaccines are focused on the L1 protein, a major component of the outer capsid. When L1 is expressed as a recombinant protein, the monomers self-assemble into virus-like particles, and these particles induce serum neutralizing IgG when administered as a parenteral immunogen. Neutralizing antibodies and protection are mainly type specific, so that vaccines are formulated as multivalent products.

There are two L1 vaccines, Gardasil by Merck (licensed in 2006) and Cervarix by GlaxoSmithKline (licensed in 2009). The vaccines induce circulating neutralizing antibodies in a high proportion of vaccinees, and they also prevent the earliest oncogenic changes. These vaccines have shown a high degree of efficacy, and that on the surface represent a paradox that "contradicts" vaccinology dogma; that is, that protection against a mucosal infection requires a mucosal—not a parenteral—vaccine. However, it is now known that serum antibodies leak into the small injuries that allow HPV to reach the basement membrane. In addition to serum neutralizing IgG, circulating anti-HPV IgG antibodies also appear in the female genital tract at low levels, a process known as transudation. Genital tract antibodies reduce the frequency of infection, and if HPV infection is not prevented, these antibodies may reduce lateral spread of infection in the epithelium. These considerations provide at least a partial explanation for how a vaccine that induces serum IgG can provide effective protection against a mucosal infection acquired by sexual contact.

4. VACCINES MUCH NEEDED AND YET TO COME

Why do we not have certain needed vaccines? Despite the enormous success of vaccination since the time of Edward Jenner and Louis Pasteur, developing vaccines against some of today's pathogens is inhibited by several different problems. In some instances, the scientific challenges still elude solution. In other instances, the international community has not made a sufficient investment because the infection is considered a relatively rare "orphan" disease, or because it mainly impacts populations in low-income countries who do not represent an attractive market. We discuss a few prominent examples below.

4.1 HIV Vaccine: Why Do We Not Have One?

HIV was isolated and identified in 1983–84 as the cause of AIDS. Since that time, there has been a vast investment in the development of an effective vaccine, yet modern biomedical science is still being outwitted by 10,000 nucleotides. There are a number of reasons for the failure to develop an effective HIV vaccine, which illustrates some of the potential challenges in virus vaccinology:

- Most natural infections with wild viruses induce long-lasting, often life-long protection against a "second attack." They do not always produce "sterilizing" immunity against reinfection; but reinfections are reduced in magnitude and length, so that they are subclinical. In contrast, primary infections with HIV do not appear to prevent second infections or even ameliorate their magnitude. This is a poor augury for vaccine formulation.
- Viral diseases, even the most dreaded, cause less than 100% mortality (rabies is an exception), suggesting that there is a close balance between virus and host, a balance that could be tilted in favor of the host. HIV is a recent crossover from the chimpanzee, and that host has had a chance to evolve protection against it. In humans, HIV infections, if untreated, are 100% fatal. Another poor augury.
- For HIV, even a minimal inoculum involving a single infectious virion leads to a lethal infection. A protective vaccine, therefore, should provide "sterilizing" immunity. As noted above, this is a standard that few, if any, established effective viral vaccines meet.
- Although HIV infection does induce serum-neutralizing antibodies in the infected patient, these antibodies are "narrow"; that is, they will neutralize only the infecting virus strain, and few other HIV isolates. Furthermore, during the course of a single infection, neutralizing escape mutants are selected, so that the virus can continue to replicate in the face of an active immune response. These escape mutants are "fit," so that they can be transmitted to other uninfected individuals in the population. As a result, during the course of the AIDS pandemic, a very large number of antigenically distinct viruses have been generated. Among human patients, very few have raised antibodies capable of neutralizing this wide variety of mutants. This stands in contrast to most other human viruses, which are not capable of continuously generating new viable escape mutants. For instance, a single strain of measles vaccine virus, used for more than 50 years, will still induce antibodies that can neutralize current measles isolates from anywhere in the world.

Put together, these considerations constitute a set of daunting scientific challenges. One line of research seeks the "holy grail," that is, the development of an immunogen that can induce broadly neutralizing antibodies. Another effort uses gene therapy to endow recipient B cells with the ability to express rare antibody genes that will generate broadly neutralizing monoclonal antibodies. Researchers recently demonstrated that modification of HIV envelope-derived immunogens leads to preferential activation of B cells that produce broadly neutralizing antibodies over those that produce narrowly neutralizing antibodies.

Other research is more empirical, using trial and error to generate protective vaccine formulations. There have been six Phase III (efficacy) trials of candidate HIV vaccines (Table 3). Only one trial (the "Thai trial," RV144) has shown any inkling of success, about 30% protection versus placebo controls. A recombinant canarypox virus expressing the envelope protein of HIV was used to prime the immune system, followed by use of the envelope protein itself as a boost. Interestingly, protection appears to correlate with antibodies against two variable loops (V1 and V2) on the surface protein, rather than with serum-neutralizing activity. It is speculated that the anti-loop antibodies may have acted through antibody-dependent cellular cytotoxicity. However, in the Thai trial, only subjects who self-classified as "low or medium risk" (but not "high risk") showed evidence of protection, and protection appeared to wane after about 1 year. A repeat of this immunization regimen (with some modification) is under way, which will indicate if this empirical approach offers a pathway to success. There has also been a great deal of interest in a vaccine candidate that uses CMV as a vector because this strategy has been used

TABLE 3 Phase III Efficacy Trials of HIV Vaccines

Name of Trial	Vaccinees Risk of HIV Infection	Vaccine Construct	Efficacy	References
VAX003	High risk	AIDSVAXgp120	None	The rgp120 HIV Vaccine Study Group (2005)
VAX004	Injecting drug users High risk	AIDSVAXgp120	None	Pitisuttithum (2006)
Step	High risk	Ad5-gag-pol-nef	None Increased risk in some populations (see Section 5.3)	Buchbinder (2008)
Phambili	High risk	Ad5-gag-pol-nef	None	Gray (2011)
HVTN505	Mainly MSM High risk	DNA-Ad5-env-gag-pol	None	Hammer (2013)
RV144	General community Mixed risk	ALVAC-AIDS-VAX	31%	Rerks-Ngarm (2009)

to successfully abort SIV infection in nonhuman primates. A human version of the vaccine is currently being assessed in a phase I clinical trial in humans.

4.2 Dengue Virus

Dengue virus infections are transmitted mainly by *Aedes aegypti*, a peridomestic mosquito that is also the vector of urban yellow fever. Dengue fever is pandemic in many of the tropical parts of the world, with more than 50 million cases each year. It is an acute febrile infection with severe pain in muscles and joints (sometimes called "breakbone fever"). The majority of patients recover spontaneously, but a small proportion (less than 5%) develops hemorrhagic fever and shock syndrome (DHF/DSS), which has a fatality rate as high as 25%. Applied to the high incidence, this could result in as many as one million deaths annually.

Dengue virus is a flavivirus that occurs in four distinct serotypes (1–4). Infection with a specific serotype confers long-term immunity against that serotype but not against other serotypes. Immune protection appears to be conferred by circulating neutralizing antibodies, but the role of cellular immunity is unclear. The pathogenesis of dengue hemorrhagic fever is poorly understood, but it appears to be immune-mediated in part, since most severe cases occur in persons who are immune to at least one serotype.

The challenge for a safe and effective dengue vaccine is to induce protective antibodies against all four serotypes. A vaccine that induces antibodies against some but not all serotypes might not only fail to protect against the "missing" serotypes, but might enhance the risk of dengue DHF/DSS. A chimeric vaccine has been developed based on the live, attenuated 17D yellow fever virus. The premembrane and envelope genes from 17D have been deleted and replaced by those of each of the four dengue viruses, creating a quadrivalent replicating vaccine. It was thought that antibodies against the envelope of each virus would provide a highly effective vaccine, but when phase II and III studies were performed, efficacy against type 2 virus, and to a lesser extent type 1, was considerably less than efficacy against types 3 and 4. The reasons for these differences have not been fully elucidated, but it appears that the conformations of the types 1 and 2 envelopes in the chimera are significantly different from those in the native virus and thus the induced antibodies do not always neutralize the viruses injected by mosquitoes.

4.3 Ebola Virus

Ebola hemorrhagic fever is caused by a filovirus and is a prime example of an emerging viral disease (see Chapter 16). Although the mortality rate is above 50%, Ebola has historically been considered an "orphan" disease. There have been more than 25 outbreaks of Ebola disease in Africa since the 1970s. Prior to 2014, outbreaks had been relatively small (the largest no more than several hundred cases) and all had been controlled by quarantine. Several laboratories had been working on candidate immunogens as vaccine candidates, but there was little incentive for a full-blown vaccine development program, in either the public or private sector. All that changed with the 2014 epidemic in West Africa, which by the end of that year had caused over 10,000 cases with more than 5000 deaths. Furthermore, the importation of a few cases into high-income countries lifted Ebola to a global health problem.

Two Ebola vaccine candidates were moved into human trials in 2014. Each candidate is based on a recombinant vector (either adenovirus or vesicular stomatitis virus, VSV) that expresses the Ebola virus glycoprotein. When tested in nonhuman primates, both of these candidates provide 100% protection against a potentially lethal challenge with wild Ebola virus. As of August 2015, interim results from a phase 3 trial indicate that the VSV-based vaccine is highly efficacious. Absent untoward events, it is likely that a vaccine will become available by 2016.

5. SYSTEMS APPROACHES TO VACCINOLOGY

In addition to the examples described above, more general impediments to vaccine development include our limited understanding of the following:

- how vaccines induce a specific, potent, broad, and long-lived immune response;
- which pathogen-specific antigens are needed to confer protective immunity;
- the differences between naïve and immunized hosts in responses to infection; and
- ways in which to maximize vaccine efficacy in heterogeneous populations.

Systems vaccinology aims to address these challenges by characterizing the complexity of the host response to vaccination and by facilitating predictions of vaccine efficacy (Figure 1). Although systems approaches have only been applied to vaccinology in the past decade, they have already improved our understanding of how some vaccines provide protection. The majority of this work has been done with yellow fever and influenza vaccines, but the responses to HIV vaccines are also being explored, as are efforts to develop pan-vaccination signatures.

5.1 Yellow Fever Virus

The 17D vaccine is one of the most efficacious vaccines ever created and is considered a gold standard for vaccine development. Although this live-attenuated vaccine has been in use for over 50 years, the reason for its effectiveness was fully understood only recently. Transcriptional profiling of whole blood and peripheral blood mononuclear cells (PBMCs) from human vaccinees revealed that 17D activates multiple aspects of innate and adaptive immunity. The key to its success lies in its ability to activate dendritic cells through multiple TLRs leading to a mixed Th1/Th2 T-cell response. An early transcriptional signature that includes expression of complement gene C1qB and eukaryotic translation initiation factor 2 (EIF2AK2) is predictive of a strong CD8+ T-cell response. Another signature that includes expression of TNFRS17, a B-cell growth factor, is predictive of the strength of the humoral response. These data suggest that vaccines that can elicit these early signatures may activate protective immune responses.

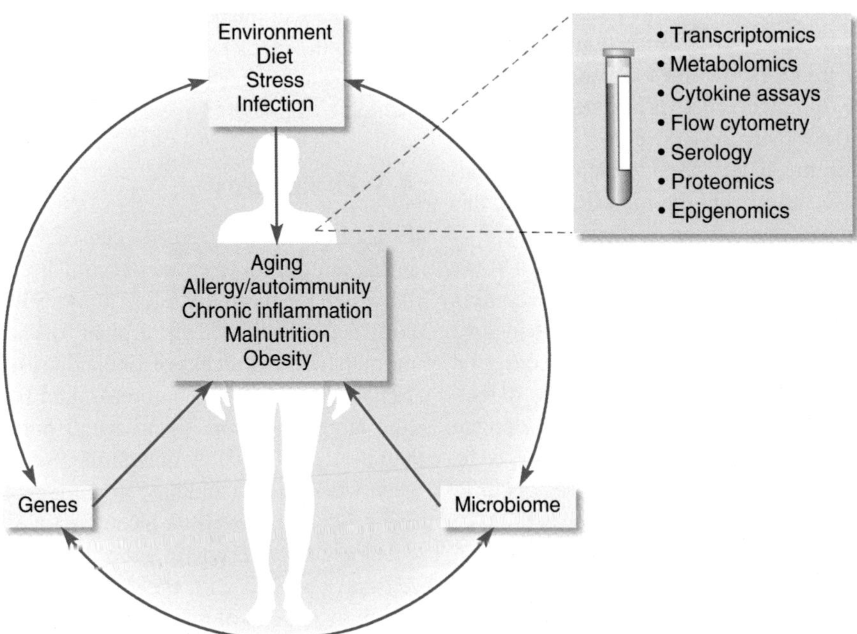

FIGURE 1 Systems vaccinology. Genes, the environment, and the microbiome are interdependent determinants of human physiology. Variations in each of these factors impact aging, immunity, inflammation, and nutritional status. Systems vaccinology seeks to understand the complexity and diversity of host determinants and immune responses to enable the rational design of vaccines. *Adapted from Pulendran (2014).*

5.2 Influenza Virus

There are two types of seasonal influenza vaccines, live attenuated and inactivated. LAIV contains replication-competent viruses of low pathogenicity that are administered intranasally. Inactivated influenza vaccine (IIV) is administered intramuscularly and consists of viruses that have been chemically denatured. LAIV and IIV come in trivalent or tetravalent formulations.

LAIV and IIV protect against influenza virus by different mechanisms and elicit vastly different transcriptional profiles in the blood of vaccinees. CamkIV and E2F2 expression are negatively associated with the magnitude of the humoral response to IIV and STAT1 expression is positively associated. Type-I IFN responses dominate with LAIV, whereas genes enriched in antigen secreting cells are important for IIV-induced immunity. A predictive signature has not been identified for LAIV, because the correlates of protection are less defined. However, serum antibody and IgA mucosal antibody both correlate with protection by LAIV. The US Food and Drug Administration defines seroconversion as an HAI titer of 1:40 or an at least fourfold increase in antibody titer after vaccination, but these numbers are rarely reached after LAIV vaccination. Since LAIV does not induce high antibody titers, it would be deemed inferior to IIV if it were judged solely on this basis. Thus, other correlates of protection need to be identified.

Systems approaches have also been used to compare the host responses of naive and vaccinated macaques to a wild-type influenza virus challenge. Protective vaccines do not necessarily induce sterilizing immunity, but they do alter the course of infection with a wild-type virus such that the infection is often subclinical. When naïve macaques and macaques immunized with a live influenza vaccine (attenuated through the truncation of the viral nonstructural 1 protein) were challenged with virulent influenza virus, the responses in vaccinated and naïve animals were drastically different. The lungs of the vaccinated animals had lower virus levels, less pathology, and lower expression of innate immune response and cytokine genes.

5.3 Human Immunodeficiency Virus

The failure of the MRKAd5/HIV vaccine (and possible enhancement of HIV infection) has been examined through a systems lens. This vaccine is comprised of a replication-incompetent adenovirus-serotype-5 vector expressing HIV gag, pol, and nef. In 2007, a clinical trial for MRKAd5/HIV efficacy was halted prematurely when data indicated that MRKAd5/HIV vaccination increased HIV-1 acquisition rates in vaccine recipients with high levels of antibodies against the Ad5 vector. Transcriptional profiling revealed that PBMCs isolated from Ad5-seropositive patients display an attenuated innate immune signature to MRKAd5/HIV compared to that of Ad5-seronegative patients. Down-regulation of RANTES and up-regulation of IFNλ2 are associated with induction of strong CD8+ T-cell responses in Ad5-seronegative patients, but these are muted in Ad5-seropositive patients. However, blood cell transcriptional profiling can only explain some differences, and additional work is needed to understand the possible enhancement of infection in Ad5-positive vaccinees.

5.4 Developing Pan-Vaccination Signatures

It is clear that different vaccines elicit different responses in the blood, but are there common signatures that could be predictive across vaccine types? To attempt to answer this question, one study has used publicly available human blood transcriptomic data from multiple vaccine trials. These data were used to generate gene co-expression networks and to form different gene expression modules. Correlating antibody titers with changes within a module increases prediction sensitivity because large changes in the expression of individual genes are not necessary for efficacy. Integrative network modeling of PBMC responses to LAIV, IIV, 17D, and two meningococcal vaccines (MCV4 and MPSV4) identified early transcriptional signatures that determine the magnitude of the antibody responses to these vaccines. MPSV4 and MCV4 elicit similar protection as measured by serum bactericidal activity even though they elicit different amounts of IgG. This is a common theme in vaccinology; antibody levels are not necessarily predictive of vaccine efficacy. There are numerous similarities and differences in the host response to LAIV, IIV, 17D, MCV4, and MPSV4, but there is currently no single gene signature that predicts responses to multiple vaccines (Figure 2).

5.5 Population Heterogeneity

One of the biggest challenges vaccine developers face is ensuring that vaccines will be effective in heterogeneous populations. Sex, age, ethnicity, and microbiota have a large impact on the host response to vaccination (Figure 3). Males and females respond differently to the yellow fever vaccine with more women than men reporting adverse events (AE). When the responses of male and female vaccinees were compared, it was found that 10-fold more genes are differentially expressed in the blood of female vaccinees. These genes are enriched for innate and adaptive immunity functions, suggesting that the increased incidence of AE in females is due to a more robust inflammatory response. Women also consistently have more AE in response to influenza vaccinations, and produce more robust antibody responses to IIV. In one study on sex-dependent differences in vaccination, researchers identified a cluster of lipid metabolism genes that are likely modulated by testosterone and whose expression

FIGURE 2 Transcriptional profiling of whole blood reveals distinct mechanisms of antibody response. Li et al. compiled over 30,000 human blood transcriptomes from over 500 studies to extract modules that contained genes that were co-expressed. These modules were then used to correlate the transcriptomic programs and antibody responses elicited by different vaccines. Each vaccine data set is shown as one of six segments on the circular plot. In each segment, the inner circular bands show an ordered list of all blood transcriptional modules, layered by histograms of modules significantly correlated to the antibody response, red for positive correlation and blue for negative correlation. Modules that are common between vaccines are linked by a color curve in the center. *From Li et al. (2014).*

correlate with the higher antibody-neutralizing response to IIV observed in females. Testosterone may act by decreasing expression of transcription factors such as FOS, JUNB, and JUND that, in turn, repress the expression of lipid metabolism genes that encode immunosuppressive activities. In fact, women develop antibody responses that are equal to those of men when given only half the standard vaccine dose, suggesting that vaccine regimens may need to be tempered in women (or boosted in men) to achieve equal efficacy or reduced AE.

Other factors, such as ethnicity and country of origin, also affect vaccination outcomes. This is likely due to a combination of environmental and genetic factors. A study of responses to 17D vaccination in subjects from Switzerland and Uganda found that 17D-induced B- and T-cell responses were significantly lower in Ugandan vaccinees. The Ugandan volunteers had higher frequencies of differentiated T- and B-cell subsets, proinflammatory monocytes, and exhausted and activated NK cells. This suggests that Ugandan patients had an activated immune microenvironment, and this is supported by the fact that 17D replicated to lower levels in this cohort. These finding suggest that 17D vaccine regimens might need to be boosted in African populations to achieve efficient immunity.

The effects of the host microbiome on vaccination outcome are also being explored. It was recently shown that

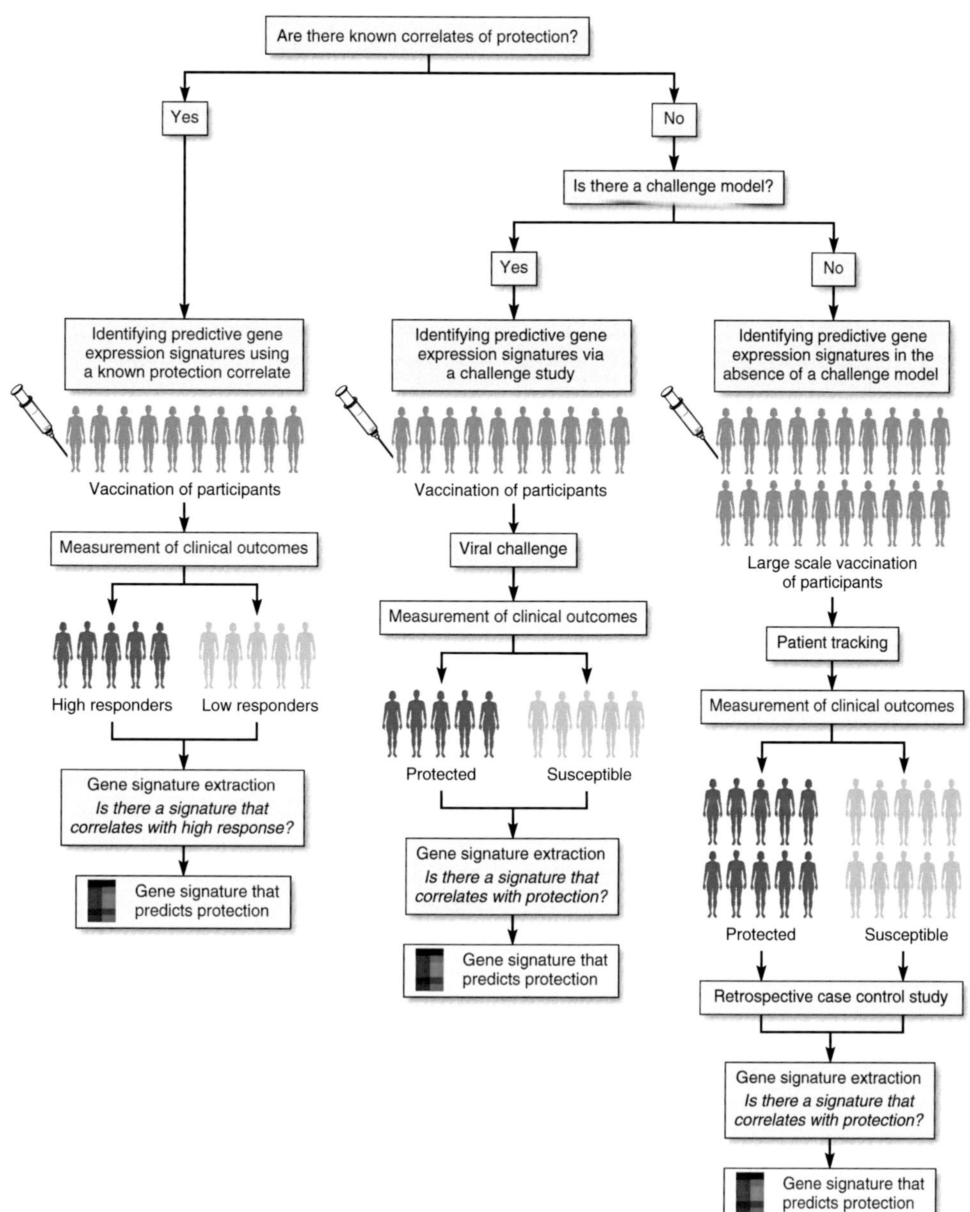

FIGURE 3 Systems approaches can be used to identify gene expression signatures predictive of vaccine protection. The ease of gene signature identification depends on whether correlates of protection from a disease are known and whether there is a human challenge model. When no established correlate of protection and no human challenge model exist, systems approaches can be applied to retrospective studies to identify novel correlates. *Adapted from Pulendran (2014).*

antibiotic-treated mice have a defect in the production of virus-specific humoral and cellular responses to influenza A virus infection. Neomycin-sensitive bacteria may be required for optimal lung immunity because they effectively "prime" the expression of proinflammatory cytokines, which in turn promote robust immunity to influenza. TLR5, a sensor of bacterial flagellin, also appears to have a role in vaccine immunity. TLR5 expression on day 3 after IIV vaccination correlates with the magnitude of the antibody response in humans, and TLR5$^{-/-}$ mice have weaker antibody responses to IIV vaccination than wild-type mice. Apparently, intestinal microbiota stimulate TLR5 leading to an enhancement of IIV immunity. Similar results were found for rotavirus vaccination where antibiotic treatment before vaccination results in a more durable rotavirus antibody response in mice.

5.6 The Future of Systems Vaccinology

Systems vaccinology studies generally rely upon taking measurements hours or days after vaccination, but it is possible that the quality of a vaccine response can be predicted using prevaccination information. When PBMC transcriptomic data, serum titers, cell subpopulation frequencies, and B-cell responses are assessed before and after vaccination in patients vaccinated with IIV, the frequency of various cell populations on day 1 can be used to predict the response.

Recent studies using a BALB/C mouse model of influenza infection suggest that peptide microarrays may also hold promise for predicting vaccine performance. Sera from mice vaccinated with live or killed influenza virus were screened by peptide microarray. The mice were then challenged with wild influenza virus to determine whether they were protected. Immunosignatures derived from peptide microarrays were more predictive of vaccine efficacy than ELISA, and could be used to identify the protective epitopes within a vaccine. This approach may potentially identify pathogen-specific antigens that are needed to confer protective immunity.

6. VACCINES AND PUBLIC HEALTH

There are about 20 safe and effective viral vaccines available for use throughout the world. This armamentarium represents one of the most cost-effective tools in public health and preventive medicine. Nevertheless, viral vaccines are underutilized in many parts of the world, mainly due to the absence of health systems for well-child care. For instance, it is estimated that worldwide more than 100,000 children die each year from measles, a totally preventable disease with a safe and inexpensive vaccine.

In developed countries, there are small groups of individuals who refuse to have their children immunized. These vaccine "refuseniks" are motivated by several different imperatives. Some base their attitudes upon religious beliefs and others upon the view that vaccines are a risk factor for diseases such as autism. Although there is strong scientific evidence against most purported vaccine-associated disease risks, the antivaccine movement remains robust.

7. REPRISE

The mechanisms whereby immunization protects against viral disease depend upon the pathogenesis of the specific infection. In some instances, preformed neutralizing antibody intercepts invading virus at the portal of entry and partially or (rarely) totally inactivates the viral inoculum. In other instances, circulating antibodies neutralize virus entering the blood, preventing dissemination to or within key target organs or tissues. In some instances, CD4+ or CD8+ T cells, B cells, and perhaps other lymphoreticular elements cooperate to provide vaccine-induced protection that is more effective than that mediated by any single component of the immune response. In most vaccine-protected individuals, exposure to wild-type virus usually initiates a mild infection that is rapidly cleared through CD8+ effector lymphocytes and antibody. However, there are many variants in the mechanisms by which effective vaccines protect, and some vaccines violate the general "rules" of vaccinology.

There are a number of modalities that have been used to formulate vaccines. Most human vaccines now in use are based on attenuated live virus strains, inactivated viruses, or viral proteins. However, other platforms can be used to present antigens, including recombinant viruses, replicons, and naked DNA. Undoubtedly, some of these will be used for future vaccines. Multiple parameters determine optimal vaccine modalities, including immunogenicity, safety, route of administration, public acceptability, and ease and cost of production. Experience suggests that different modalities will be best suited for different vaccines.

In the past, vaccine development has depended upon an empirical strategy involving repetitious trial and error; a cumbersome and inefficient process. Systems biology offers a new and important addition to the evaluation of candidate vaccines. Recent studies have developed omics profiles for successful existing vaccines, and these offer guidance for vaccines now under development. Systems approaches can also identify early innate responses, which are critical for an effective adaptive immune response. These new approaches and technologies therefore provide a potentially more efficient approach to vaccine development, which is badly needed for vaccines yet-to-be formulated, such as those against HIV or dengue. Such efforts are imperative, since vaccines are arguably the most effective approach to controlling the viral diseases of mankind and animals.

FURTHER READING

General

Offit P. Deadly choices: how the anti-vaccine movement threatens us all. Basic Books, New York, USA, 2011.

Plotkin SA, Orenstein WA, Offit P, editors. *Vaccines*, 6th edition, Saunders, Philadelphia, 2013.

Plotkin SA. Complex correlates of protection after vaccination. Clinical Infectious Diseases 2013, 56 (10): 1458–1465.

Sawyer WA, Meyer KF, Eaton MD, Bauer JH, Putnam P, Schwentker FF. Jaundice in army personnel in the western region of the United States and its relation to vaccination against yellow fever. American Journal of Hygiene 1944; 39:337–387.

Influenza

Kirchenbaum GA, Ross TM. Eliciting broadly protective antibody responses against influenza. Current Opinion in Immunology 2014, 28: 71–76.

Huber V. Influenza vaccines: from whole virus preparations to recombinant protein technology. Expert Reviews Vaccines 2014, 13: 31–42.

Poliomyelitis

Francis TJ Jr, Napier JA, Voight R, Hemphill FM, Wenner HA, Korns RF, Boisen M, Tolchinsky E, Diamond EL. Evaluation of the 1954 field trial of poliomyelitis vaccine. School of Public Health, University of Michigan, 1957, Ann Arbor, Michigan.

Henry JL, Jaikara ES, Davies JR, Tomlinson AJH, Mason PJ, Barnes JM, Beale AJA. A study of poliovaccination in infancy: excretion following challenge with live virus by children given killed or living poliovaccine. Journal of Hygiene 1966, 64: 105–120.

Offit P. the Cutter incident. Yale U Press, New Haven, 2005.

Sabin AB. Properties and behavior of orally administered attenuated poliovirus vaccine. Journal of the American Medical Association 1957, 164: 1216–1223.

Rabies

Baer GM, Cleary WF. A model in mice for the pathogenesis and treatment of rabies. Journal of Infectious Diseases 1972, 125: 520–532.

Jackson AC, editor. Rabies, 3rd edition, Academic Press, Oxford, UK, 2013.

Wiktor TJ. Cell-mediated immunity and postexposure protection from rabies by inactivated vaccines of tissue culture origin. Developmental Biological Standardization 1978, 40: 255–265.

Hepatitis B (HBV)

Beasley RP, Hwang LY, Stevens CE, Lin C-C, Hsih F-J, Wang K-Y, Sun T-S, Szmuness W. Efficacy of hepatitis B immune globulin for prevention of perinatal transmission of the hepatitis B virus carrier state: final report of a randomized double-blind placebo-controlled trial. Hepatology 1983, 3: 135–171.

Xu ZY, Liu CB, Francis DP, Purcell RK, Gun Z-L, Duan S-C, Chen R-J, Margois HS, Hugan C-H, Maynard JE. Prevention of perinatal acquisition of hepatitis B carriage using vaccine: preliminary report of a randomized double-blind placebo-controlled and comparative trial. Pediatrics 1985, 76: 713–718.

Rotavirus

Leshem E, Lopman B, Glass R, Gentsch J, Bányai K, Parashar U, Patel M. Distribution of rotavirus strains and strain-specific effectiveness of the rotavirus vaccine after its introduction: a systematic review and meta-analysis. Lancet Infectious Diseases 2014, 14: 847–856.

Human Papillomavirus (HPV)

Franco EL, Harper DM. Vaccination against human papillomavirus infection: a new paradigm in cervical cancer control. *Vaccine* 2005, 23: 2388–2394.

Harper DM, Franco EL, Wheeler C, Ferris DG, Jenkins D, Schuind A, Zahaf T, Innis B, Naud P, De Carvalho NS, Roteli-Martins DM, Teixeira J, Blatter MM, Kron AP, Quint W, Dubin G. Efficacy of a bivalent L1 virus-like particle vaccine in prevention of infection with human papillomavirus types 16 and 18 in young women: a randomized controlled trial. Lancet 2004, 364: 1757–1765.

Tyler M, Tumban E, Chackerian B. Second-generation prophylactic HPV vaccines: successes and challenges. Expert reviews Vaccines 2014, 13: 247–255.

Human Immunodeficiency Virus (HIV)

Barouch DM, Deeks SG. Immunologic strategies for HIV-1 remission and eradication. Science 2014, 345: 169–176.

Klein F, Mouquet H, Dosenovic P, Scheid JF, Scharf L, Nussenzweig MC. Antibodies in HIV-1 Vaccine Development and Therapy. Science 2013, 341: 1199–1204.

McGuire AT, Dreyer AM, Carbonetti S, et al. Antigen modification regulates competition of broad and narrow neutralizing HIV antibodies. Science 2014, 346: 1380–1383.

Richman DD, Wrin T, Little SJ, Petropoulos CJ. Rapid evolution of the neturalizing antibody response to HIV type 1 infection. Proceedings of the National Academy of Sciences, 2003, 100: 4144–4149.

West, Jr AP, Scharf L,1 Scheid JF, Klein F, Bjorkman PJ, Nussenzweig MC. Structural Insights on the Role of Antibodies in HIV-1 Vaccine and Therapy. Cell 2014, 156: 633–648.

HIV Vaccine Trials

Buchbinder SP, Mehrotra DV, Duerr A, et al. Efficacy assessment of a cell-mediated immunity HIV-1 vaccine (the Step Study): a double-blind, randomised, placebo-controlled, test-of-concept trial. Lancet 2008, 372: 1881–1893.

Gray GE, Allen M, Moodie Z, et al. Safety and efficacy of the HVTN 503/Phambili Study of aclade-B-based HIV-1 vaccine in South Africa: a double-blind, randomised, placebo-controlled test-of-concept phase 2b study. Lancet Infectious Diseases 2011, 11: 507–515.

Hammer SM, Sobieszczyk ME, Janes H, et al. Efficacy Trial of a DNA/rAd5 HIV-1 Preventive Vaccine. New England Journal of Medicine 2013, 369: 183–192.

Pitisuttithum P, Gilbert P, Gurwith M, et al. Randomized, Double-Blind, Placebo-Controlled Efficacy Trial of a Bivalent Recombinant Glycoprotein 120 HIV-1 Vaccine among Injection Drug Users in Bangkok, Thailand. Journal of Infectious Diseases 2006, 194: 1661–1671.

Rerks-Ngarm S, Pitisuttithum P, Nitayaphan S, et al. Vaccination with ALVAC and AIDSVAX to Prevent HIV-1 Infection in Thailand. New England Journal of Medicine 2008, 361: 2209–2220.

The rgp120 HIV Vaccine Study Group. Placebo-Controlled Phase 3 Trial of a Recombinant Glycoprotein 120 Vaccine to Prevent HIV-1 Infection. Journal of Infectious Diseases 2005, 191: 654.

Ebola

Feldmann H, Geisbert TW. Ebola hemorrhagic fever. Lancet 2011, 377: 849–862.

Gatherer D. The 2014 Ebola virus disease outbreak in West Africa. J General Virology 2014, 95: 1619–1624.

Marzi A, Feldmann H. Ebola virus vaccines: an overview of current approaches. Expert Reviews Vaccines 2014, 13: 521–31.

Dengue

Bhatt S, Gething PW, Brady OJ, Messina JP, Farlow AW, Moyes GL, et al. The global distribution and burden of dengue. Nature 2013 496: 504–507.

Simmons CP, Farrar JJ, Nguyen VV, Wills B. Dengue. NEJM 2012, 366: 1423–1432.

Halstead SB. Identifying protective dengue vaccines: guide to mastering an empirical process. Vaccine 2013, 31: 4501–4507.

Systems Vaccinology

Huber VC. Influenza vaccines: from whole virus preparations to recombinant protein technology. Expert Reviews Vaccines 2014, 13: 31–42.

Li S, Nakay HI, Kazmin DA, et al. Systems biological approach to measure and understand vaccine immunity in humans. Seminars in Immunology 2013, 25: 209–18.

Li S, Rouphael N, Duraisingham S, et al. Molecular signatures of antibody responses derived from a systems biology study of five hman vaccines. Nature Immunology 2014, 15: 195–204.

Pulendran B, Oh JZ, Nakaya HI, et al. Immunity to viruses: learing from successful human vaccines. Immunological Reviews 2013, 255: 243–255.

Pulendran B. Systems vaccinology: probing humanity's diverse immune systems with vaccines. Proceedings of the National Academy of Sciences 2014, 111: 12300–12306.

Chapter 20

Antiviral Therapy

Douglas D. Richman[1,2], **Neal Nathanson**[3]

[1]VA San Diego Healthcare System, San Diego, CA, USA; [2]University of California, San Diego, La Jolla, CA, USA; [3]Department of Microbiology, Perelman School of Medicine, University of Pennsylvania, Philadelphia, PA, USA

Chapter Outline

Antiviral therapy is one of the most exciting aspects of virology, since it has successfully employed basic science to generate very effective treatments for serious viral infections. Table 1 lists selected examples of those human viral diseases for which there are established antiviral drugs. Therapy for human immunodeficiency virus (HIV) infection has demonstrated the potential impact antivirals can have on a lethal, chronic infection with lifesaving therapy administered to more than 12 million individuals by 2015. This dramatic advance is about to be recapitulated for the treatment of hepatitis C virus (HCV) infection. The development of new antiviral drugs is very much a work in progress, with active drug discovery programs for filoviruses, coronaviruses, dengue, and others.

The conceptual approach to drug development is in flux. In the past, the primary focus has been upon virus targets, and this continues to be a very productive strategy. It is now being complemented by a wider set of approaches, so that present strategies include: compounds that target generic viral targets such as RNA or DNA synthesis and could be active against a range of different viruses and compounds that are directed against host cellular activities necessary for virus replication, which might target one or a spectrum of viruses (Figure 1). Furthermore, established methodologies for drug discovery are now supplemented with the use of large databases and evolving methods in computational biology. Finally, there is an increased emphasis on the repurposing of drugs already approved for human use, driven by the inordinate time and cost of drug development. This chapter explores all of these issues.

We begin by discussing the mechanisms by which antiviral agents act, illustrated by selected examples. The presentation attempts to highlight the importance of viral pathogenesis for designing different therapeutic strategies for individual viral diseases. We continue with a brief discussion of pharmacodynamics and toxicity, critical hurdles that a safe and effective drug must pass. This section closes with a discussion of the new horizons in drug development. The pathway to drug development with all its challenges is next described, followed by an overview of those virus infections for which the most effective therapy is available. We conclude with a section on the future of antiviral therapy.

1. PRINCIPLES OF ANTIVIRAL THERAPY

1.1 Virus Targets

Viral proteins. Current understanding of the molecular replication of individual viruses provides a detailed elucidation of the role of individual viral proteins. It is possible to map functional domains within viral proteins and to image their structures. These data can be used for "rational" drug

Viral Pathogenesis. http://dx.doi.org/10.1016/B978-0-12-800964-2.00020-3

TABLE 1 Viral Diseases for Which There Are Established Antiviral Drugs: Some Examples

Virus Family	Specific Virus and (Disease)	Example of Drug	Mechanism of Action
Orthomyxovirus	Influenza virus (influenza)	Amantadine	Binds and blocks the H+ ion channel formed by the viral M2 proteins, prevents RNA uncoating; type A viruses only
		Oseltamivir	Binds the enzymatic site on the viral neuraminidase, prevents cleavage of terminal sialic acid residues, and release of virions from infected cells; all influenza type A and B viruses
Retrovirus	HIV (AIDS)	Zidovudine (AZT)	Reverse transcriptase inhibitor; nucleoside analogue; prevents synthesis of DNA transcripts
		Nevirapine	Reverse transcriptase inhibitor; nonnucleoside analogue; prevents synthesis of DNA transcripts
		Atazanavir	Protease inhibitor; blocks processing of viral proteins
		Maraviroc	Entry inhibitor; binds host cell CCR5 to inhibit binding of R5-tropic HIV to this coreceptor
		Raltegravir	Integrase strand transfer inhibitor; blocks integration of linear dsDNA reverse transcript
Hepadnavirus	Hepatitis B virus (chronic hepatitis)	Tenofovir, emtricitabine	HBV DNA polymerase inhibitor as well as HIV reverse transcriptase inhibitor; nucleotide analogue; prevents synthesis of viral DNA
Hepacivirus	Hepatitis C virus (chronic hepatitis)	Sobosfuvir	Nucleoside analogue inhibitor of viral RNA polymerase (NS5)
		Simeprevir	Protease NS3 inhibitor—blocks processing of viral polypeptide
		Ledipasvir	Viral NS5A inhibitor—targets viral protein essential for replication but whose function is incompletely characterized
Herpesvirus	Herpes simplex (encephalitis)	Acyclovir	Viral DNA polymerase inhibitor; guanine derivative; prevents synthesis of DNA transcripts
	Cytomegalovirus (retinitis)	Ganciclovir, valganciclovir	Viral DNA polymerase inhibitor; acyclovir derivative; prevents synthesis of DNA transcripts
Poxvirus Adenovirus Polyoma virus	Variola (smallpox) Adenovirus viremia BK virus in renal transplant patients	Brincidofovir	Viral polymerase inhibitor; Cytosine derivative; prevents synthesis of DNA transcripts

design, either to synthesize small molecules that will bind to active sites on viral proteins, or to develop high-throughput screening procedures to test a very large battery of small molecules for those that block a specific activity.

HIV serves as a useful example, since there has been an exhaustive effort to develop antiviral drugs exploiting many of the viral proteins. Most of the anti-HIV drugs target one of the viral enzymes, either the reverse transcriptase, the protease, or the integrase (Table 2). In particular, there are many drugs that block reverse transcription, an enzymatic activity not expressed in normal cells. There are two classes of reverse transcriptase inhibitors: nucleoside and nonnucleoside (NRTIs and NNRTIs, respectively). NRTIs are compounds that are incorporated in to the nascent DNA chain and block its elongation. NNRTIs bind directly to the enzyme itself, inhibit its function, and may lead to its degradation. The other major enzymatic drug target is the viral protease that cuts gag and pol viral polypeptides to produce mature proteins. Protease inhibitors usually bind to the catalytic site on the protease molecule.

In addition, there are drugs that inhibit other steps in the HIV replication cycle. The initial step in HIV cellular entry is binding of the viral gp120 to the cellular coreceptor (see Chapter 9, HIV/AIDS). HIV gp41 then undergoes a conformational change that exposes its N terminal fusion domain, which inserts into the plasma membrane of the host cell. Close to the N terminus of gp41 is a heptad repeat (HR1) that forms a three-helix bundle. HR1 associates with another three-helix bundle (HR2) at the C terminus of gp41, which forces the molecule into a hairpin

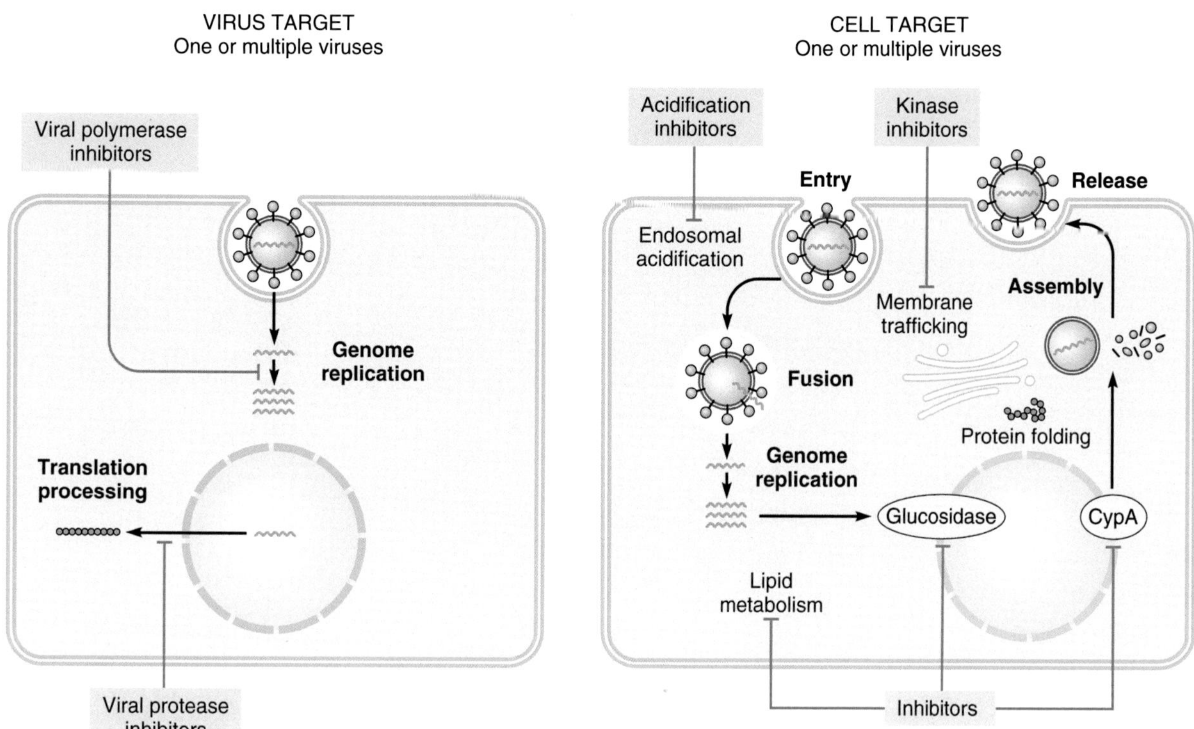

FIGURE 1 The various scientific strategies for development of antiviral drugs. Left panel: direct-acting antivirals that target a specific viral protein and are aimed at a single virus target or a target for multiple viruses. This cartoon shows inhibitors of viral polymerases or proteases but other viral proteins may also be targeted. Right panel: drugs that target cellular processes that are essential for replication of one of several viruses. The cartoon shows several classes of inhibitors but there are many other cellular functions that could be targets. *Reconceived after Bekerman and Einav (2015). CypA: cyclophilin A.*

TABLE 2 HIV Drugs Approved As of Early 2015

Year Approved	Generic Name	Manufacturer
NRTIs (nucleoside reverse transcriptase inhibitors)		
1987	Zidovudine (AZT)	GSK
1991	Didanosine (ddI)	BMS
1992	Zalcitabine (ddC)[a]	Roche
1994	Stavudine (d4T)	BMS
1995	Lamivudine (3TC)	GSK
1998	Abacavir	GSK
2001	Tenofovir	Gilead
2003	Emtricitabine (FTC)	Gilead
NNRTIs (nonnucleoside reverse transcriptase inhibitors)		
1996	Nevirapine	Boehringer Ingelheim
1997	Delavirdine	Pfizer
1998	Efavirenz	BMS, Merck
2002	Etravirine	Tibotec, J&J
2011	Rilpivirine	Tibotec, J&J

Continued

TABLE 2 HIV Drugs Approved As of Early 2015—cont'd

Year Approved	Generic Name	Manufacturer
PIs (protease inhibitors)		
1995	Saquinavir	Roche
1996	Ritonavir	Abbott
1996	Indinavir	Merck
1997	Nelfinavir	Pfizer
1999	Amprenavir[a]	GSK
2000	Lopinavir and ritonavir	Abbott
2003	Atazanavir	BMS
2003	Fosamprenavir	GSK
2005	Timpranavir	Boehringer Ingelheim
2006	Darunavir	Tibotec, J&J
Entry inhibitor		
2003	Enfuvirtide	Roche
CCR5 antagonist		
2007	Maraviroc	Pfizer
Integrase strand transfer inhibitors		
2007	Raltegravir	Merck
2012	Elvitegravir (as combination, see below)	Gilead
2013	Dolutegravir	GSK
Nucleoside reverse transcriptase inhibitor combinations		
1997	Zidovudine and lamivudine	GSK
2000	Abacavir, zidovudine, and Lamivudine	GSK
2004	Abacavir and lamivudine	GSK
2004	Tenofovir and emtricitabine	Gilead
Multiclass combinations		
2006	Efavirenz, emtricitabine, and tenofovir	BMS and Gilead
2011	Rilpivirine, emtricitabine, and tenofovir	Gilead
2012	Elvitegravir, eobicistat, emtricitabine, and tenofovir	Gilead

Manufacturer: GSK, GlaxoSmithKline; BMS, Bristol-Myers Squibb; J&J, Johnson & Johnson.
[a]*Drug has been withdrawn.*
Adapted from: http://www.fda.gov/ForPatients/Illness/HIVAIDS/Treatment/ucm118915.htm.

configuration. A synthetic oligopeptide analogue of HR2, enfuvirtide (originally called T20), can bind to HR1 and prevent this hairpin formation, thereby blocking HIV-1 cellular entry. Enfuvirtide has been shown to be active in HIV-1-infected patients who have "failed" other anti-HIV drug therapy (Lazzarin et al., 2003).

Viral mutagens. The survival of viruses depends in part on their ability to evolve in response to antiviral pressures, such as host immune responses. From this perspective, the rapid mutational rate of RNA viruses, in particular, facilitates the selection of fitness mutants among an ever-present swarm of genetic variants. In theory, the polymerases of RNA viruses have evolved to an optimal balance of processivity and fidelity (mutational rate), which permits the generation of large numbers of progeny with many genetic variants. These variants facilitate rapid adaptation to selective pressures, such as immune responses and drug treatment.

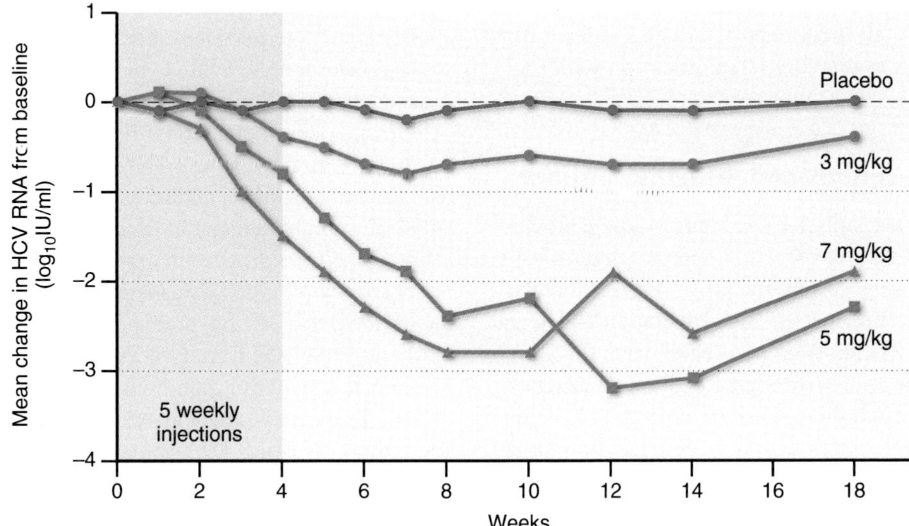

FIGURE 2 MicroRNA (miRNA) treatment has the potential as a treatment modality for selected human viral infections. This example is a trial in which microRNA-122 was targeted for the treatment of HCV. This miRNA is expressed by liver cells and the replication of HCV is dependent upon a functional interaction of its genome with miRNA-122. Shown are the mean changes in HCV RNA levels from baseline for patients receiving 3, 5, or 7 mg of miravirsen (an antisense oligonucleotide that sequesters miRNA-122) per kilogram of body weight, as compared with placebo. Miravirsen was administered in 5 weekly subcutaneous injections during the first 29 days of the study (shading). The dashed line indicates no change from baseline. *Redrawn from data in Janssen et al. (2013).*

A counterintuitive approach to antiviral drugs is the use of mutagens that can increase viral mutational rate, so that fit variants are overwhelmed by less fit or nonsense mutants leading to a lethal accumulation of errors or "error catastrophe" (discussed in Chapter 17, Virus Evolution). At least one antiviral drug, ribavirin, a nucleoside analogue, has been proposed to act through this mechanism. It has been shown to both increase mutation rate and decrease the production of infectious particles in several viruses including poliovirus and Hantaan virus.

RNA interference (RNAi). In the mid-1990s, RNA interference was discovered serendipitously when an attempt to overexpress specific plant genes, using viral vectors, instead resulted in the knockout or silencing of those genes. RNAi is described in more detail in Chapter 3, Concepts of viral pathogenesis. Its potential as an antiviral therapeutic has been shown in a proof-of-concept clinical trial for hepatitis C (Figure 2).

1.2 Cellular Targets

Cellular targets have a theoretical advantage over viral targets since they will not undergo escape mutations. On the other hand, targeting cellular molecules can interfere with vital host functions. This could be a problem for long-term treatment, but might not be a critical impediment for acute infections. Cellular targets have not been a major focus in the past but are now receiving increasing attention (Table 3). Techniques for identifying potential cellular targets are covered in Chapter 11, Systems Virology and Chapter 12, The Virus–Host Interactome.

CCR5 (chemokine receptor 5) is a coreceptor for HIV (see Chapter 9, HIV/AIDS). Some individuals are homozygous for a mutation in CCR5 (the "delta 32" mutation) that prevents the expression of this host gene. These persons do not show any ill effect from this mutation and are very resistant to HIV infection. These observations suggest that a small molecule that blocks the receptor domain on CCR5 might safely be used to treat HIV infection, without interfering with any essential cellular functions (Fatkenheuer et al., 2005). Maraviroc is such an inhibitor and has been proven effective for HIV treatment.

Cyclosporin A (CsA) is an approved drug that binds to—and inactivates—a cellular molecule, cyclophilin A. Cyclophilins are a family of *cis*–trans isomerases that convert prolines from the *trans* to the *cis* form. As a family, cyclophilins play an essential role in many cellular processes that require protein folding and trafficking, and this activity is essential for the replication of HCV. Interestingly, CsA, which is approved as an immunosuppressive drug, was discovered serendipitously to have strong activity against HCV (Lin and Gallay, 2013). Of note, mice with cyclophilin A knockouts are generally healthy, so this molecule does not seem to be essential for life (at least in a shoebox).

Another focus of cellular targets has been the attempt to develop a pan-virus strategy for antiviral therapeutics. One example was directed at phosphatidylserine, an anionic phospholipid that is located on the inner leaflet of the plasma membrane but exposed in virus-infected cells. An antibody directed against anionic phospholipids interfered with the replication of arenaviruses (Soares et al., 2008). Antibody treatment of infected guinea pigs spared them from a potentially lethal infection with Pichinde, an arenavirus.

1.3 Viral Pathogenesis and Antiviral Strategy

The pathogenesis, transmission, and epidemiological characteristics of individual viruses are important determinants of the potential efficacy of antiviral drugs. Viruses that have a very short incubation period and generation time, and spread very rapidly, tend to be poor candidates for antiviral treatment because it is difficult to complete diagnosis and initiate therapy in a timely fashion. Influenza is a good example of a serious illness with a short incubation period (18–72 h). Neuraminidase inhibitors are quite effective anti-influenza drugs but need to be given prior to infection or very soon after symptoms appear. This drawback may be overcome under certain circumstances; in the presence of a pandemic wave of influenza that is spreading across a community, antiviral drugs could be widely administered as a short-term prophylactic, thereby anticipating potential infection.

Persistent viral infections that cause significant chronic illness are attractive targets for antiviral treatment. Their slow course permits an accurate diagnosis and evaluation prior to initiating therapy. Furthermore, there are a few persistent infections, such as hepatitis B and hepatitis C, which carry a long-term risk of liver failure or hepatocellular cancer. With an estimated >150 million cases of hepatitis C and >350 million cases of hepatitis B globally, these diseases constitute significant opportunities for treatment. Effective therapeutic intervention, particularly long-lasting viral suppression or even viral clearance (a "cure"), would significantly reduce the disease burden.

In many persistent infections, there is a dynamic balance between the persistent virus and host defenses. Thus, some persons infected with hepatitis B virus (HBV) are able to clear the infection even after years of persistence. This pattern has two implications. First, it suggests that antiviral therapy might tip the balance in favor of the host and lead to viral clearance, and second, it suggests that antiviral antibodies might be used in synergy with antiviral drugs to improve the therapeutic outcome.

Therapeutic antibody. Neutralizing antibodies are a major mediator of the preexposure protection conferred by many established viral vaccines (see Chapter 19, Viral Vaccines). In addition, antibodies induced during primary infection play a role in clearance and recovery from certain acute viral diseases. Therefore, it is plausible that passive antibody administered during acute viral infection might be therapeutic.

Primary infection with West Nile virus (WNV), a flavivirus, is one example. Following transmission by mosquito bite, WNV initiates a plasma viremia followed by invasion of the central nervous system, resulting in potentially fatal encephalitis. Experiments in immunologically deficient mice have shown that both antibody and cellular immunity play a role in the outcome of infection. Surprisingly, antibody also plays a role in the clearance of virus from the central nervous system, even when infection has been well established in neurons. When administered to mice undergoing acute WNV encephalitis, passive antibody can markedly improve survival, at least under experimental conditions. Recently, it has been found that neutralizing antibodies are very effective when used for treatment of acute infection with Ebola virus in an animal model (see below).

Interferons (IFNs) and interferon inducers. Type 1 interferons (IFN-α and -β) are an important component of

TABLE 3 Antiviral Drugs Directed against Cellular Targets

Drug	Virus	Mechanism of Action	Clinical Status	References
Maraviroc	HIV	Blocks CCR5 coreceptor	Approved	Fatkenheuer et al. (2005)
Cyclosporin A and related compounds	HCV	Binds and inactivates cyclophilin A, a *cis*–trans isomerase	Phase III trials	Lin and Gallay (2013)
Deoxynojirimycin, Castanospermine	Enveloped RNA viruses	Binds and blocks ER-resident glucosidases essential for virus maturation	Approved for other indications Phase II trials	Chang et al. (2013)
Erlotinib Sunitinib	HCV	Binds and blocks AP2M1, a subunit of adaptor protein complex, required for maturation of HCV virions	Research	Neveu et al. (2012)
Nilotinib	Ebola	Blocks c-Abl/1 tyrosine kinase, required to phosphorylate Ebola virus proteins	Research	Garcia et al. (2012)

the innate immune response (discussed in Chapter 4, Innate Immunity). IFNs induce a complex pleiotropic response that inhibits viral replication in several different ways, in addition to activating antigen-specific adaptive immune responses. Use of exogenous IFN as an antiviral therapy has been tried for many viral infections but—to date—has been used mainly for the treatment of two persistent infections, HCV and HBV (discussed below). However, IFN therapy has substantial unwanted side effects in humans, which limit its practical utilization.

Cytokine storm. Several serious viral infections of humans produce disease via an excessive or imbalanced host response, leading to intense dysregulation of proinflammatory cytokines and chemokines. This cytokine storm can cause a range of serious disease manifestations. These include acute lung injury in association with respiratory viruses such as highly pathogenic influenza viruses (avian H5N1 and 1918 H1N1) and SARS, or a shock syndrome in fatal cases of Ebola virus infection. Recent studies of the host–virus interactome (see Chapter 12, The Virus–Host Interactome) have made it possible to identify some of the components of this pathological host response. Such insights may be useful in formulating new therapies to ameliorate an excessive deleterious innate response.

1.4 Drug-Resistance Mutations

Drug-resistant viral mutants constitute a major problem in antiviral therapy. The frequency of resistant mutants varies widely (see Chapter 17, Virus Evolution) and is determined by a number of factors (Sidebar 1).

1. RNA viruses have a mutation rate estimated at 10^{-4} (1 mutation in 10,000 base replications) that is much higher than the rate for DNA viruses (10^{-8}); the difference reflects poor fidelity, as well as the absence of cellular proofreading mechanisms for RNA polymerases.
2. The replication rate of the virus during a specific infection will vary widely and influence the rate at which mutant virions are produced. For instance, it has been estimated that during an HIV-1 infection, 10^8–10^{11} virions are produced daily; this would yield 10^4–10^7 virions with single point mutations (or an average 1–1000 mutants for each of the 10,000 bases) each day. At the other end of the scale, human papillomavirus (HPV, a DNA virus) replicates very slowly in vivo, so that very few mutant virions would be synthesized daily. These differences are reflected in the observation that individual primary HIV isolates consist of a "swarm" of viruses the sequences of which—after several years of infection vary from 5% to 10%, for different genes. In contrast, primary isolates of DNA viruses show much less variation.
3. Different classes of drugs target diverse viral functions that vary in their importance for viral replication, and individual drugs vary in the degree to which they can block their targeted function. Furthermore, resistant mutants vary in their ability to replicate in the presence of the drug and also—absent drug—in their replicative capacity or fitness. These nuances are reflected in the observation that different HIV-1 NRTIs—which are directed against the same viral function—select for different escape mutations.
4. The in vivo selective pressure of a specific drug will depend upon both its intrinsic ability to block an essential virus function and its pharmacodynamics, which will determine its actual concentration at sites of viral replication. As the selective pressure increases, the relative advantages of mutants increase, but the rate of replication of wild-type virus decreases. As replication diminishes, the probability that resistant viruses will emerge diminishes. The selection of escape mutants is maximized when the drug concentration is high enough to select for resistant mutants, but not so high that it substantially inhibits virus replication (see Chapter 17, Viral Evolution).

One important implication of the foregoing considerations is the potential advantage of multidrug therapy. If a virus has to replicate in the presence of three diverse drugs each of which select for different resistance mutations at a frequency of 10^{-4}, then triple mutants (assuming no interaction between various mutations) would occur at 10^{-12}, which might be a very rare phenomenon. In the case of HIV-1, there has been a comparison of multiple drug therapy as new compounds have been introduced (Tang and Shafer, 2012). There is a dramatic stepwise increase in efficacy with each additional drug (Figure 3). For HIV-1, triple drug therapy is

Sidebar 1 Determinants of antiviral drug resistance

- Variation in viral mutation rate, which is about 10,000-fold greater for RNA than DNA viruses.
- Variation in in vivo viral replication, which determines the rate at which mutants are generated.
- Variation in the structural mechanism of drug-mediated viral inactivation, which determines the frequency and fitness of resistant mutants.
- Variation in drug-mediated selective pressure in vivo, which determines the relative replication rates of wild-type and mutant viruses.
- Concurrent use of several drugs that act upon different viral functions will markedly reduce the frequency of resistant virions, since these must possess multiple mutations.

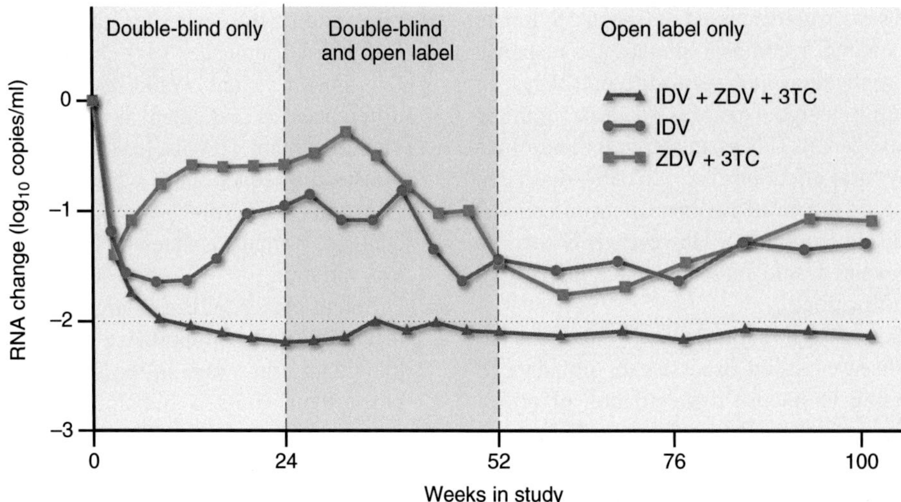

FIGURE 3 Virus mutants may escape a single or even two drugs, but are much less likely to escape from three or more drugs administered simultaneously. HIV-infected treatment-naïve patients were started in regimens of two nucleosides (azidothymidine (ZDV) and lamivudine (3TC)), the protease inhibitors indinavir (IDV), or the combination of all three drugs. The decrease in the level of blood HIV RNA from baseline is shown. The rises in plasma HIV RNA after several months in the nucleoside and protease only arms was associated with the emergence of resistance mutations in the genes of the respective target enzymes. *Drawn from data in Gulick et al. (1997).*

usually required to suppress viral replication and minimize escape mutations.

1.5 Pharmacodynamics

A critical aspect of drug efficacy is its behavior in vivo. Many compounds that appear active in cell culture systems fail when tested in animals. The pharmacodynamics of a drug depends on at least the following parameters: (1) Is the compound soluble? (2) Can it be absorbed if given by mouth or does it require injection or even intravenous administration? (3) Is the drug active in its administered formulation or does it require biochemical processing in the liver to be activated? (4) How fast is the compound released into the blood? Does it circulate as a free molecule in plasma or does it bind to albumin or other plasma proteins? (5) Where does the drug act, in blood or in specific target tissues? How fast does it enter target tissues? (6) What is the half-life of the compound in blood and in tissues? (7) Is the drug inactivated in the liver or excreted in the urine or intestinal tract? (8) Does the drug achieve therapeutic levels in blood or target tissues? What is the dosage regimen required to maintain therapeutic levels? Thus, a complex set of experiments must be conducted to determine whether a candidate compound meets pharmacodynamic criteria that make it practical for use.

Related to the pharmacodynamics and often studied in concert is toxicity. Usually, a standard battery of tests is conducted to detect unwanted side effects of a candidate drug. Toxic effects may be unpredictable from the mode of action of an antiviral compound. Both pharmacodynamic and toxicity studies are required by the Food and Drug Administration as part of the application for an IND (Investigational New Drug) which must be obtained prior to Phase I trials in humans (see below).

1.6 New Horizons for Drug Development

A new era for drug development is dawning, based on advances in computational biology and large public databases. Also, there is a broadened conceptual view, to include the host–pathogen interactome, cellular targets, and drug repurposing (Casadevall and Pirofski, 2015). Major incentives are the inefficiencies in established development pathways, the many years required, the inordinate cost, and the low yield, discussed below. Table 4 lists some recent publications where diverse applications of computational biology are proposed for drug discovery. Several examples will illustrate the approaches being taken.

One such computational approach was used by Josset et al. (2014) when studying the transcriptome response to virulent influenza A viruses, H7N9, H5N1, and seasonal H3N2 viruses, in a human lung epithelial cell system. To identify potential antiviral compounds, a data-based approach was used which relies on the assumption that an effective drug would have the inverse effect on cellular transcriptional response to that of the targeted virus. Using a publically available database, Connectivity Map, which contains thousands of gene expression profiles from over a thousand compounds, several drugs were identified as potential antivirals against the H7N9 virus. These included cellular kinase inhibitors as well as some FDA-approved drugs, such as troglitazone and minocycline. A similar approach has been taken for a number of other viruses,

TABLE 4 Computational and Big Data Strategies for Drug Discovery: Some Recent Publications

Category	Message	References
General	Big data: hype versus utility	Hu and Bajorath (2014)
General	Integration diverse data on human diseases: network-based models	Berg (2014)
General	Mathematical modeling: network-based multiscale strategy	Wangand Diesbock (2014)
General	Broad spectrum antiviral drugs	Beker et al. (2015)
General	Using systems biology to find therapeutic targets	Dopazo (2013)
Platform	Open innovative drug discovery platform for data mining	Alvim-Gaston et al. (2014)
Platform	Activity-based protein profiling to identify antiviral targets	Blais et al (2013)
Platform	Using host–pathogen interactome to identify antiviral targets	Brown et al. (2011)
Cellular antiviral	Genomics screen to identify antiviral proteins	Brass et al. (2009)
Cellular antiviral	Used siRNA and other omics to identify cellular antiviral factors	Munk et al. (2011)
Cellular antiviral	Genomic screen to identify cellular factors critical for virus replication	Schwegman et al. (2008)
Repurposing	Repurposing: computational methods	Jin and Wong (2014)
Repurposing	Computational data mining of publicly available databases	Law et al. (2013b)
Repurposing	Computational methods for data mining FDA-approved drugs	Ekins et al. (2011)
Repurposing	Used tissue culture of MERS coronavirus to screen FDA-approved drugs	Dyall et al. (2014)
Metabolomics	Inventory of drugs that alter metabolism	Fillet and Frederich (2014)
Toxicity	Used gene expression profiling to detect side effects	Verbist et al. (2015)
Structural	Example of computational designed inhibitor of a viral anti-host cell protein	Procko et al. (2014)
Structural	CANDO: inventory of protein–protein interactions	Minie et al. (2014)

such as dengue, HIV-1, and hepatitis C (Munk et al., 2011; Brass et al., 2009).

The CANDO (computational analysis of novel drug opportunities) platform focuses on potential interactions between small molecules and proteins of interest (Minie et al., 2014). Input comes from a variety of sources, including structural homologies, curated databases, and other information in the scientific literature. The goal is to identify small molecules that might interact with specific proteins and thereby interfere with their function. This could be applied to blocking viral functions or host proteins critical to viral replication.

A set of cell lines were used by Verbist et al. (2015) to test the effect of over 700 drug candidates on gene transcripts using a microarray readout. The focus was on a predetermined set of genes, some of which—if upregulated—could be toxic, while others—if upregulated—could be therapeutic. The data were used to make some go/no-go decisions to aid in selection of compounds for further investigation.

Broad-spectrum antiviral drugs. Some viruses share steps in their replication strategies so it would be possible to design direct-acting antiviral compounds that would inhibit families of viruses in contrast to a single target. Several compounds that inhibit RNA polymerases are being investigated for their potential to treat a number of viruses (Warren et al., 2014; Furuta et al., 2013). Brincidofovir, a nucleotide analogue that can block the action of DNA polymerases, has been proposed as a candidate treatment for several dsDNA viruses (Florescu et al., 2014).

Antiviral cellular targets. As noted above, there are examples of cellular targets that—if compromised—will inhibit individual or classes of viruses (Table 3). How does a system's approach exploit this potential opportunity for drug discovery? First, it offers a systematic approach to identifying cellular enzymes, proteins, or processes that are essential for the replication of a virus. Second, it provides a broad-based approach—using siRNA, knockouts, or other methods—to determining which potential cellular targets may be "expendable." Third, it will enable a search for potential inhibitors of these cellular targets. Fourth, it can be used for screens to select anticellular compounds for their ability to inhibit virus replication in vitro or in vivo.

Although these methods have not yet produced approved therapeutics, the various examples cited in Tables 3 and 4 demonstrate the potential for these newer approaches to drug discovery.

2. EXAMPLES OF ANTIVIRAL THERAPY

How are new drugs developed? We begin with a short discussion of the process of drug discovery. To give a sense of the scope and diversity of approved drugs, and the variables that influence their efficacy, this section presents some examples of approved antiviral therapy. More detailed information is available in clinical texts and reviews (see Further Reading).

2.1 Challenges of Drug Development and Utility of Drug Repurposing

New antiviral drugs follow a well-worn developmental pathway that is usually required before drugs are approved for use in human subjects. Overall this is a very slow, cumbersome, and expensive process, which can take at least 10 years and cost more than $1 billion. Figure 4 shows a typical pathway for drug development in the United States. What the figure does not show is the cost of all the candidate compounds that never make it to market, because they fail to pass one of the successive hurdles shown in Figure 4. It is guesstimated that only 1 in 20 compounds makes it from the beginning to the end of this process.

The drug development process can be divided into three phases: basic research, preclinical studies, and clinical trials. Basic research involves identification of a druggable target molecule or step in viral replication. Preclinical studies require an informed or high-throughput search for compounds that will inactivate the target molecule, and evidence that it will work in an animal model of the viral disease under study. In animal models, there are a complex set of pharmacological parameters that must meet practical standards, including the dose and route of administration, the pharmacodynamics and frequency of administration, the concentration in blood or key target organs, and possible requirement for activation in the liver. Care must be taken to search for possible toxic effects at therapeutic drug levels.

Once past these steps, the compound becomes eligible for testing in humans, assuming that an IND approval can be obtained from the USA Food and Drug Administration. Typically, there are three sequential phases of clinical trials: Phase I focuses on safety; Phase II on some parameter that can serve as a surrogate for efficacy; and Phase III-controlled trials in human populations to determine efficacy and safety.

There are several consequences of this slow and expensive process. In most countries, the early basic research that identifies potential druggable targets takes place at not-for-profit research institutions and is funded by both government and private organizations. Further drug development is usually conducted by for-profit pharmaceutical or biotechnology companies. Considering the cost, there is an understandable reluctance to develop drugs unless there is a remunerative potential market. AIDS and hepatitis C are good examples of diseases for which there is an attractive market, while Ebola is an example of an orphan disease for

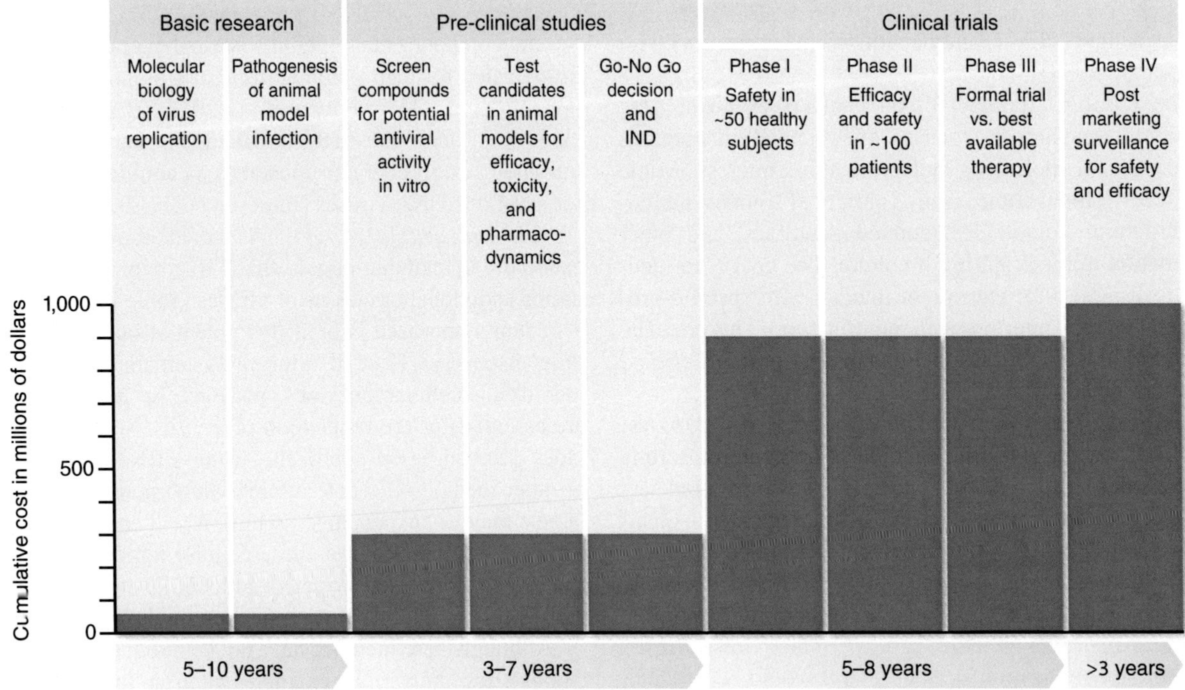

FIGURE 4 The pathway to development of a new antiviral drug is slow and expensive. The figure shows typical parameters for drug development. *Based on data in Tufts (2014).*

which there was a very limited potential market—at least until the pandemic of 2013–2015.

Drug repurposing. The impediments to new drug development have activated the field of drug repurposing. There are large numbers of drugs, directed against cellular pathways, which have been approved by the U.S. Food and Drug Administration for human use. It is possible that some of them could also have antiviral activity, either by blocking a cellular pathway critical for virus replication or by ameliorating the innate response to a virus which—in some instances—contributes to disease pathogenesis (Ekins et al., 2011). If a previously approved drug can be identified as an effective antiviral, this "shortcut" in the drug development pathway could be quicker and less expensive (Dyall et al., 2014).

There are a number of strategies for drug repurposing, some of which are noted in Table 4. One example is cyclosporin A, described above, as a compound that inhibits the maturation of HCV. A non-immunosuppressive derivative of cyclosporin A, that maintained activity against HCV was included in a trial of patients with HCV-induced inflammatory liver disease. It induced an antiviral response due to its anti-cyclophilin A activity.

2.2 Influenza Virus

Influenza is one of the most prevalent viral diseases, affecting an estimated 10–20% of the population annually, with 3–5 million cases of severe respiratory illness and 0.5–1 million deaths. In the United States alone, there is an annual excess mortality of about 35,000 attributed to influenza. As explained above, the pathogenesis of influenza—its very short incubation period of 18–72 h and its acute course—makes it a difficult target for antiviral therapy.

Inhibitors of viral entry (M2 inhibitors). As one step in its cellular entry pathway, influenza virus is endocytosed into an acidic vacuole; an H+ ion channel formed by the viral M2 proteins then facilitates acidification of the interior of the virion, which in turn permits dissociation of the matrix protein from the ribonucleoprotein viral core that enters the cytosol and initiates replication. Amantadine and rimantadine, related drugs, bind in the ion channel of influenza A viruses and prevent the final step in viral entry. These drugs have been proven effective for both prophylaxis and therapy; however, drug-resistant mutants are frequently isolated from patients after a few days of therapy. Moreover, all circulating influenza viruses over the past decade have maintained resistance to this class of drugs.

Inhibitors of virus release (neuraminidase inhibitors). As influenza virus buds from a host cell, the viral hemagglutinin binds to receptors on the cell surface, which contain *N*-acetylneuraminic (sialic acid) residues. Release of virions is accomplished by action of the neuraminidase on the viral surface, which cleaves terminal cellular sialic acid residues and frees virions to spread to adjacent uninfected cells. The neuraminidase inhibitors, oseltamivir and zanamivir, which are sialic acid analogues, bind to the catalytic site on the neuraminidase, thereby inhibiting viral release and spread. The active domain of the neuraminidase is highly conserved in order to maintain this enzymatic function; however, escape mutants emerge after monotherapy in patients with prolonged virus shedding (most often children and immunosuppressed individuals). Some influenza virus strains carrying resistance mutations to this class of drugs have circulated successfully.

Not surprisingly, influenza drugs have limited efficacy in clinical application. If administered within 48 h of the onset of symptoms, each of these drugs reduced the duration of symptoms by about 1 day and reduced the time to return to normal activity by about 1 day also.

2.3 Human Immunodeficiency Virus

There has been a greater research investment in antiviral drugs for HIV than for any other virus, because the large number of infected persons, the persistent nature of infection, and the 100% fatality rate among untreated patients created an ethical imperative and offered a very lucrative market for effective therapies (Gunthard et al., 2014). As a result, there is a panoply of FDA-approved products (Table 2). The action of these drugs has been described in the foregoing section on antiviral compounds. The ability to effectively treat this inevitably fatal infection is one of the great triumphs of scientific medicine. It is now guesstimated that a subject age 20 who is recently infected with HIV and is on optimal HAART (highly active antiretroviral therapy) has a life expectancy of 50 years (contrasted with 5 years absent treatment). However, present regimens that control HIV replication do not provide a "cure" since they fail to eradicate latent HIV genomes (see Chapter 9, HIV/AIDS).

HIV drug resistance. The loss of susceptibility of isolates from subjects treated with the first antiretroviral as monotherapy, AZT, was associated with the cumulative acquisition of mutations in the target gene of this NRTI, the reverse transcriptase. As explained above, a combination of three different drugs is needed (Figure 3). Protease or integrase inhibitors are usually included in combination with two NRTIs in antiretroviral regimens for initial therapy. The development of fixed dose combinations of three drugs in a single capsule is an important practical advance, which has enhanced patient compliance and become a standard regimen. At present, over 10 million people receive a single pill once daily to treat HIV (Table 2).

For HIV prevention, another important application of antiretroviral treatment is preexposure prophylaxis (PreP). In several trials, PreP appears to offer a potentially powerful adjunct to treatment of infected patients, particularly in a high exposure context (Cohen, 2015b).

Persistence of HIV genomes. When HIV-infected patients initiate potent combination therapy, the titers of viral RNA in plasma drop in a biphasic manner (see Figure 7 in Chapter 9, HIV/AIDS). There is an early and rapid decline lasting about 1 week and a slower decline lasting several weeks, which then plateaus often below the limits of detection. However, if treatment is stopped, viremia inevitably reappears, usually in a few weeks, indicating that viral eradication has not been achieved. Even if continuing low-level HIV replication is completely suppressed, persistence of HIV, in the form of latent proviral genomes in resting CD4+ T cells, precludes eradication with present regimens. Substantial investigative efforts have been initiated to identify approaches to eradicate the latent reservoir, which would require activating latent HIV genomes and killing the infected cells (Cohen, 2015a). However, this is at best a long-term initiative.

2.4 Hepatitis C Virus

HCV is an important cause of human disease, since it is estimated that there are >150 million persistently infected humans worldwide, many of whom develop chronic hepatitis, liver failure, or hepatocellular cancer. HCV is a hepacivirus, a positive-stranded RNA virus, in the flavivirus family. HCV has presented a difficult experimental challenge because there are limited cell culture systems (recently improved) and the only animal model is the chimpanzee.

The natural history of HCV is clouded by the insidious, asymptomatic nature of infection. About 25% of patients eliminate the virus within 3–24 months after infection. Of persistently infected patients, about 50% experience little chemical or clinical evidence of disease, while 50% have chronic hepatitis that can progress to end-stage cirrhosis or hepatocellular cancer over a period of 5–40 years. These data indicate that in many patients there is a delicate balance between HCV and host defenses, supporting the notion that antiviral therapy could lead to viral clearance.

Prior to 2014, treatment with HCV focused on a combination of ribavirin and pegylated (polyethylene glycol-conjugated) interferon. IFN-α2 produces a virological response that is sustained for many months after ending a course of treatment, with a substantial number of "cures" (about 75% of patients infected with genotype 1 HCV). However, this treatment strategy caused serious toxic side effects associated with long-term administration of interferon.

The recent discovery and development of drugs against HCV has produced a revolution in therapy (De Clercq, 2014; Rice and Saeed, 2014). These compounds are called direct-acting agents (DAA) to differentiate them from interferons. Interferon-free regimens of drugs that target HCV polymerase, protease, and the NS5A and NS5B proteins can result in viral clearance without relapse. This is effectively a cure in 90–100% of those treated (Petta and Craxi, 2015). These recent successes are energizing the development of other antiviral drugs.

2.5 Hepatitis B Virus

As mentioned above, the course of HBV infection is variable and a high proportion of patients spontaneously clear the virus (see Chapter 7, Patterns of Infection). Persistent HBV infections can be divided into several categories. Patients with lower virus titers (<10^5 HBV DNA copies per ml serum) and normal liver function tests (serum alanine aminotransferase) are not usually treated since they are at relatively low risk of end-stage liver disease (cirrhosis or hepatocellular carcinoma). Treatments, with an IFN and either a nucleoside or a nucleotide reverse transcriptase inhibitor, are used for patients with more severe disease. IFN-α administered parenterally for months to years induces a therapeutic response in about one-third of patients. However, present regimens do not clear the virus and replication resumes after interruption of drug therapy. In view of the toxic effects of prolonged interferon treatment and the low response rate of present therapies, there was a need for improved treatments. The development of potent nucleoside (entecavir) and nucleotide (tenofovir) HBV polymerase inhibitors has provided more tolerable and effective long-term treatments.

A major obstacle to a cure of HBV is the existence in infected hepatocytes of covalently closed circular (ccc) DNA, which maintains the viral genome as a nuclear episome. One approach is the use of the CRISPr/Cas9 system to introduce DNA nucleases specific for cccDNA (Kennedy et al., 2015; Dong et al., 2015; Ahmed et al., 2015), although the delivery of this system in vivo is a challenge. A number of drugs directed against HBV enzymes are in use or under development; they suppress HBV replication and reduce the level of cccDNA episomes (Shi et al., 2015; Tavis et al., 2015; Liu et al., 2015). Finally, there are newly developed systems for screening anti-HBV compounds (Suresh et al., 2015; Ishida et al., 2015). The estimated >350 million HBV infections worldwide provide a major incentive for research and development, and the current investment in the development of HBV drugs holds promise for the introduction of curative regimens.

2.6 Herpesviruses

There are several human herpesviruses that cause considerable morbidity (and rare mortality), including herpes simplex virus (HSV, cold sores and genital lesions), varicella zoster virus (VZV, chicken pox and herpes zoster), and cytomegalovirus (CMV, retinitis and other complications in immunosuppressed subjects). Herpesvirus infections offer special challenges for treatment because they cause persistent lifelong latent infections with intermittent activation of active replication and accompanying disease. During the latent phase viral DNA is maintained in the nucleus, making it unlikely that effective therapy can eradicate the virus. Therefore, treatment is directed at the symptomatic phase of infection.

Herpesviruses are large DNA viruses that encode their own DNA polymerases, which are required for transcription of their genomes. Acyclovir, the first highly effective compound for the treatment of HSV and VZV, is a nucleoside analogue DNA chain terminator. Acyclovir undergoes in vivo activation by addition of three phosphates to its side chain. The first phosphate is added by a virus-encoded thymidine kinase so that the drug is converted to its active triphosphate only in HSV- and VZV-infected cells. The second and third phosphates are added by cellular kinases. Acyclovir triphosphate is incorporated into nascent DNA chains but—absent a ribose—the DNA polymerase cannot add further nucleotides and the chain is terminated. This dual viral target (thymidine kinase and DNA polymerase) of acyclovir confers great specificity (and hence safety) to this class of drugs.

Several other closely related compounds with improved clinical activity have also been developed. The lipid-conjugated nucleotide, brincidofovir, is a DNA polymerase inhibitor active against HSV, VZV, and CMV. There are number of compounds directed against the viral polymerase or other viral proteins that are now under test as potential antiherpesvirus agents. Among the nonnucleoside compounds are helicase–primase inhibitors, that target other aspects of viral DNA synthesis.

Absent exogenous therapy, the host's virus-specific immune response plays an important role in the control of the active phase of herpesvirus infection and represents an important complementary arm of therapy. In fact, reimmunization of VZV-immune subjects reduces the incidence of herpes zoster in the elderly.

2.7 Ebola Virus

Ebola is a zoonotic filovirus that has caused multiple small outbreaks in Africa since 1976, and one large pandemic in West Africa that began in 2013 but finally waned in May, 2015 (see Chapter 16, Emerging Viral Diseases). From the viewpoint of treatment or prevention, it is an "orphan" or neglected disease. In response to the ongoing pandemic, a crash program to develop Ebola treatments has been initiated, and a few compounds are in early clinical trials (Table 5).

Recent experimental data have shown that a cocktail of neutralizing monoclonal antibodies (ZMapp) against Ebola virus can have a dramatic therapeutic effect in a nonhuman primate model that closely simulates Ebola disease in humans (Qui et al., 2014; Murin et al., 2014). Monkeys were challenged with a 100% lethal dose of Ebola virus that kills animals in 4–8 days. Antibody treatment was only initiated 3–5 days after infection and—amazingly—it protected all animals, which were reported to recover to normal health. A crash program to scale up the production of these antibodies has been initiated but it is not known—at this writing—when these will be available for use in humans. ZMapp or a similar antibody cocktail is a very promising therapy and appears to be the most efficacious of the interventions so far developed. A wide variety of compounds with different modes of action are under investigation as potential drugs to treat Ebola. One proposed strategy to develop algorithms to screen FDA-approved drugs for their anti-Ebola activity (Ekins and Coffee, 2015; Vejlkovic et al., 2015; Litterman et al., 2015). This field is moving rapidly and the data in Table 5 will be outdated by the time of publication.

An important advance for the treatment of Ebola infection is a rapid point-of-care diagnostic test kit that does not rely on electric power, refrigeration, or complex equipment and highly trained technicians. Such a test was approved in 2015 by the World Health Organization (First antigen rapid test for Ebola, 2015). The ability to determine whether patients presenting with fevers of unknown origin in an Ebola-endemic area need anti-Ebola treatment is a critical adjunct to effective therapeutic interventions.

It is also relevant to note that there are several promising Ebola vaccine candidates in human trials in West Africa in 2015 (see Chapter 19, Viral Vaccines). If approved for human use, they will also serve as a critical tool for control of Ebola outbreaks (Rampling et al., 2015; Agnandji et al., 2015; Lipsitch et al., 2015; Marzi et al., 2015).

On a practical note, one of the ironies in the development of Ebola treatments is that the pandemic was waning—May, 2015—just as new compounds were readied for clinical trials (Kuehn, 2015; Fleck, 2015; Kupferschmidt, 2015). Hopefully, it will be possible to test the safety of those interventions that appear most promising, so that they could be used on a compassionate basis when the next outbreak occurs.

3. THE FUTURE OF DRUG DEVELOPMENT

There is an exciting future for development of new antiviral drugs, which has been energized by the recent success with HCV. What are the considerations that will guide this field in the near future? Both scientific and public health issues are in play.

In the scientific arena, as set forth in Figure 1, there are a number of pathways for drug development: direct-acting antiviral compounds that are virus-specific; direct-acting compounds that are somewhat generic; cellular targets that are virus-specific; and broad cellular targets. The pathogenesis of different infections will have an important role in the selection and efficacy of such putative drugs. There is great

TABLE 5 Drugs in Ebola Virus Clinical Trials. This Table, Which is Updated on a Continuous Basis, Summarizes the Data on Drugs That Are Either Being Tested or Considered for Testing in Patients with Ebola Virus Disease. This Truncated Version is Limited to Those Drugs That Look Promising Based on High Efficacy in Nonhuman Primates and Low Toxicity in Nonhuman Primates or in Phase I Human Trials in April, 2015. A Full Version of This Table is Posted on the WHO Website, http //who.int/medicines/ebola-treatment/2015-0116_TablesofEbolaDrugs.pdf, accessed April, 2015

Drug/Company	Drug Type	Ebola Preclinical Data	Known Safety Issues	Availability And Logistical Considerations	Comments
Category A: drugs already under evaluation in formal clinical trials in West Africa					
Favipiravir (Fuji/ Toyama Japan)	Small molecule antiviral with activity against many RNA viruses. Functions through inhibiting viral RNA-dependent RNA polymerase. Approved in Japan for treating novel/pandemic influenza.	In vitro inhibition IC50 64 μM; higher than that needed for influenza. Mice: protected at 300 mg/kg. Nonhuman primate (NHP): antiviral effect seen; 2 log reduction in viremia. Model limitation due to frequent need to anesthe-tize NHP to administer drug orally.	Clinical use in healthy volunteers up to 3.6 g on first day followed by 800 mg twice daily (BID). No safety issues identified. Increased drug exposure in setting of hepatic dysfunction	200 mg tablets; dosing at 6 g/first day requires 30 tablets—potentially difficult to swallow. 1.6 million tablets available free (10,000 treatment courses). Thermostable.	4 patients received drug under compassionate use. No conclusions possible from these patients, but no obvious safety concerns identified. Clinical efficacy trial began in guinea in December 2014. Target 6 g dosing (day 1) followed by 2.4 g per day (day 2–10). Preliminary data presented in early February by investigators do not permit a firm conclusion regarding efficacy and more data are required.
Category B: drugs that have been prioritized for testing in human efficacy trials but for which such trials are not yet underway					
Zmapp (MappBio, USA)	Cocktail of three monoclonal antibodies produced in tobacco plants.	NHP: 100% survival when administered 5 days after virus challenge.	No formal safety studies in humans yet. Phase I safety study initiated in January 2015.	Supply reported to be 15 treatment courses every 6 weeks.	8 patients treated on compassion-ate grounds to date. No conclu-sion regarding safety or efficacy possible. Some adverse reactions noted—possibly due to immune complex formation with virus. Phase I safety/PK study started in January 2015. Efficacy study due to start in early 2015.
AVI-7537 (Sarepta, USA)	Antisense polymorpholino oligonucleotide. Inhibits Ebola virus replication by binding to RNA in sequence-specific manner to VP24 gene. Specific to this strain of Ebola.	NHP: 100% survival for Marburg virus (using Marburg sequence) and 50–60% survival for Ebola using Ebola sequence.	Phase I safety study completed. Tolerability demonstrated.	Limited no. of doses available.	No clinical trials planned at this time.

potential for drugs to treat persistent infections with serious long-term consequences, while drugs may be less practical for acute infections that resolve before a virus-specific diagnosis can be made. The molecular nature of the virus–host infection is important, since it can determine whether drugs that control an infection will also lead to a permanent cure. A comparison of HIV (control no cure) and HCV (control and cure) illustrates that point.

What will guide the search for effective antiviral compounds? The recent successes for HCV and HIV suggest that direct-acting antiviral molecules still remain the gold standard that will inform many initiatives. One Achilles heel of such compounds is viral escape, but extensive experience with HIV has shown that the use of triple combinations can overcome this problem. Recent experience with HCV has provided several lessons: understanding the virus replication cycle at a molecular level paves the way for identification of candidate compounds; these insights can be used to construct high-throughput screening assays; and this strategy can lead to products that are clinically effective and safe (Scheel and Rice, 2013; Kim et al., 2013; Rice and Saeed, 2014).

The burgeoning field of systems biology offers significant new strategies for identifying antiviral compounds, particularly in the intertwined fields of cellular targets and drug repurposing (Table 3). Drugs that target host functions always carry the risk of unwanted side effects, particularly for long-term drug treatment. However, toxicity may be less of an issue for drugs used to treat acute infections such as Ebola hemorrhagic fever. The utility of a drug directed against the CCR5 molecule (for HIV treatment) demonstrates that this is a viable strategy.

The era of personalized (or precision) medicine is in its infancy (see Gary Gilliland commentary in the Chapter 22, What Lies Ahead?). In fact, personalization of treatment regimens is already a standard practice for certain persistent viral infections such as hepatitis B and C (Lok, 2015). Future trials of candidate drugs will more and more include genetic data on the participants to determine if drug efficacy or toxicity is localized to subpopulations (Insel et al., 2015; Ramamoothry et al., 2015; Johannessen et al., 2015). This may "rescue" certain candidate compounds for use in preselected patients.

Given these considerations, which viral diseases may be targets for drug development in the near future? A list of persistent infections would include: HBV; HPV; and some of the herpesviruses. In each instance, there is a large global population of infected persons who would benefit from a cure. In addition, cure or control of large numbers of virus carriers should have an impact on prevalence in the population. Among acute infections, those that carry a high mortality would be candidates, particularly viruses with epidemic potential, such as Ebola and other hemorrhagic fevers (Table 4). Dengue and influenza type A viruses are also high on this list, and experience with existing influenza antiviral compounds indicates that viral escape is common, mandating the need for multiple drug therapy.

Using this information, pharmaceutical companies have the research tools that enable them to develop or select compounds that have significant antiviral activity. However, drug development has become a daunting challenge, due both to scientific hurdles and financial requirements. In vivo bioavailability, pharmacodynamics, and potential toxicity are additional major obstacles that must be overcome during the development of safe and effective antiviral chemotherapeutics. Development of a new drug from discovery to licensure often takes as long as 10 years and costs more than $1 billion dollars. Repurposing of drugs already approved for human use could help to address these obstacles.

Clearly, there is an exciting future for antiviral drug development (De Clercq, 2013). This field demonstrates the practical yield of research in basic virology, viral pathogenesis, and viral epidemiology. However, support for "orphan" drugs, particularly those for the great neglected diseases, remains a major ethical and strategic challenge for the high-income countries, both for government and private sectors alike (Karan and Pogge, 2015). To date, we have failed to implement an effective strategy for the development and distribution of therapeutics for the neglected infectious diseases that continue to plague low-income countries.

FURTHER READING

Reviews, Chapters, and Books

http://www.hcvguidelines.org/ (regularly updated expert guidelines in this rapidly changing field).

Ahmed M, Wang F, Levin A, et al. Targeting the Achilles heel of the hepatitis B virus: a review of current tratments against covalently closed cirulcar DNA. Drug discovery today 2015, doi: 10.1016/j.drudis.2015.01.008.

Bekerman E, Einav S. Combating emerging viral threats. Science 2015, 348: 282–283.

Brown JR, Magid-Slav M, Sanseau P, Rajpal DK. Computational biology approaches for selecting host-pathogen drug targets. Drug Discovery Today 2011, 16: 229–236.

Casadevall A, Pirofski L-A. Ditch the term pathogen. Nature 2014, 516: 165–166.

De Clercq E. Antivirals: past, present and future. Biochemical Pharmacology 2013, 85: 727–744.

De Clercq E. Current race in the development of DAAs (direct-acting antivirals) against HCV. Biochemical Pharmacology 2014, 89: 441–452.

Günthard HF, Aberg JA, Eron JJ, Hoy JF, Telenti A, Benson CA, Burger DM, Cahn P, Gallant JE, Glesby MJ, Reiss P, Saag MS, Thomas DL, Jacobsen DM, Volberding PA. Antiretroviral Treatment of Adult HIV-1 Infection: 2014 Recommendations of the International Antiviral Society–USA Panel. JAMA. 2014, 312: 410–25.

Insel PA, Amara SG, Blaschke TF. Introduction to the theme "precision medicine and prediction in pharmacology". Annual review of Pharmacology and toxicology 2015, 55: 11–14.

Johannessen CM, Clemons PA, Wagner BK. Integrating phenotypic small-molecule profiling and human genetics: the next pahse in drug discovery. Trends in Genetics 2015, 31: 16–23.

Karan A, Pogge T. Ebola and the need for restructuring pharmaceutical incentives. journal of global health 2015, 5: 1–4.

Law GL, Korth MJ, Benecke AG, et al. Systems virology: host-directed approaches to viral pathogenesis and drug targeting. Nature reviews microbiology 2013a, 11: 455–466.

Law GL, Tisoncik-Go J, Korth MJ, et al. Drug repurposing: a better approach for infectious disease drug discovery? Current Opinion in Immunology 2013b, 25: 588–592.

Menendez Arias L, Richman D, et al. Antivirals and Resistance. *Current Opinion in Virology* October 2014, 8: 1–116.

Ramamoothry A, Pacanowski MA, Zhang I. Racial/ethnic differences in drug disposition and response: review of recently approved drugs. Clinical pharmacology and therapeutics 2015, 97: 263–273.

Scheel TKH, Rice CM. Understanding the hepatitis C virus life cycle paves the way for highly effective therapies. Nature Medicine 2013, 19: 837–849.

Schwegman A, Brombacher F. Host-directed drug targeting of factors hijacked by pathogens. Science Signaling 2008, 1: 1–7 (re8).

Tang MW, Shafer RW. HIV-1 antiretroviral resistance. Drugs 2012, 72: e1–e25.

Wilson RC, Doudna JA. Molecular Mechanisms of RNA Interference. Annual Review of Biophysics 2013, 42: 217–239.

Original Contributions

Agnandji ST, Huttner A, Zinser ME, et al. Phase 1 trials of rVSV Ebola vaccine in Africa and Europe – preliminary report. NEJM 2015, doi: 10.1056/NEJMoa502924.

Alvim-Gaston M, Grese T, Mahoui A, et al. Open innovation drug discovery (OIDD): a potential path to novel therapeutic chemical space. Current topics in medicinal chemistry 2014, 14: 294–303.

Berg E. Systems biology in drug discovery and development. Drug discovery today 19: 113–125.

Blais DR, Nasher N, McKay CS, et al. Activity-based protein profiling of host-virus interactions. Cell 2012, 30: 89–97.

Brass AL, Huang I-C, Beita Y, et al. The IFITM proteins mediate cellular resistance to influenza A H1N1virus, West Nile virus, and dengue virus. Cell 2009, 139: 1243–1254.

Chang J Block TM, Guo J-T. Antiviral therapies targeting host ER alpha-glucosidases: current status and future directions. Antiviral research 2013, 99: 251–260.

Cohen J. Drug flushes out hidden AIDS virus. Science 2015a, 347: 1056.

Cohen J. Doubts dispelled about HIV prevention. Science 2015b, 347: 1055–1056.

Dong C, Qu L, Want H, et al. Targeting hepatitis B virus cccDNA by CRISPR/Cas9 nuclease efficiently inhibits viral replication. Antiviral research 2015, 118: 110–117.

Dopoazo J. Genomics and transcriptomics in drug discovery. Drug discovery today 2014, 19: 126–132.

Dyall J, Coleman CM, Hart BJ, et al. Repurposing of clinically developed drugs for treatment of Middle East respiratory syndrome coronavirus infection. Antimicrobial agents and chemotherapy 2014, 58: 4885–4893.

Ekins S, Coffee M. FDA approved drugs as potential Ebola treatments. F1000Research 2015, 4: 48.

Fatkenheuer G, Poznia AL, Johnson MA, et al. Efficacy of short-term monotherapy with maraviroc, a new CCR5 antagonist, in patients infected with HIV. Nature medicine 2005, 11: 1170–1172.

Fillet M, Frederich M. Ethe emergence of metabolomics as a keh disciploine the the drug discovery process. Drug discovery today 2015, doi: org/10.1016/j.ddtec.2015.01.006.

First Antigen Rapid Test for Ebola through Emergency Assessment and Eligible for Procurement. Accessed at http://www.who.int/medicines/ebola-treatment/1st_antigen_RT_Ebola/en/, April, 2015.

Fleck F. Tough challenges for testing Ebola therapeutics. Bulletin of the WHO 2015 93: 70–71.

Florescu DF, Keck MA. Development of CMX001 (Brincidofovir) for the treatment of serious conditions caused by dsDNA viruses. Expert review of anti-infective therapy 2014, 12: 1171–1178.

Furuta Y, Gowen BB, Takahashi K, et al. Favipiravir (T-705), a novel viral RNA polymerase inhibitor. Antiviral research 2013, 100: 446–454.

Garcia M, Cooper A, Shi W, et al. Productive replication of Ebola virus is regulated by the c-Abl1 tyrosine kinase. Science translational medicine 2012, 4: 123ra24.

Guidotti LG, Ando K, Hobbs MV, Ishikawa T, Rundel L, Schreiber RD, Chisari FV. Cytotoxic T lymphocytes inhibit hepatitis B virus gene expression by a noncytolytic mechanism in transgenic mice. Proceedings of the National Academy of Sciences, 1994, 91: 3764–3768.

Gulick RM, Mellors JW, Havlir D. Eron JJ, Gonzalez C, McMahon D, Richman DD, Valentine FT, Jonas L, Meibohm A, Emini EA, Chodakewitz JA. Treatment with indinavir, zidovudine, and lamivudine in adults with human immunodeficiency virus infection and prior antiretroviral therapy. New England Journal of Medicine 1997, 337: 734–739.

Hu Y, Bajorath J. Learning from "big data": compounds and targets. Drug Discovery today 2014, 19: 357–360.

Ishida Y, Yamasaki C, Yanagi A, et al. Novel robust in vitro hepatitis B virus infection model using fresh human hepatocytes isolated from humanized mice. American journal of pathology 2015, 185: 1275–1285.

Janssen H, Reesink HW, Lawitz EJ, Zeuzem S, Rodriguez-Torres M, Patel K, van der Meer AJ, Patick AK, Chen A, Zhou Y, Persson R, King BD, Kauppinen S, Levin AA, Hodges MR. Treatment of HCV Infection by Targeting MicroRNA. New England J Medicine 2013, 368: 1685–1694.

Jin G, Wong STC. Toward better drug repositioning: prioritizing and integrating exisint metods n efficient pipelines. Drug discovery today 2014, 19: 637–644.

Josset L, Zeng H, Kelly SM, et al. Transcriptomic characterization of the novel avian-origin influenza A (H7N9) virus: specific host response and responses intermediate between avian (H5N1 and H7N7) and human (H3N2) viruses and implication for treatment options. mbio 2014, 5: 1-12 e01102–13.

Kennedy EM, Bassit LC, Mueller H, et al. Suppression of hepatitis B virus DNA accumulation in chronically infected cells using a bacterial CRISPR/Cas RNA-guided DNA nuclease. Virology 2015, 476: 196–205.

Kim H-Y, Li X, Jones CT, et al. Development of a multiplex phenotypic cell-based high throughput screening assay to identify novel hepatitis C antivirals. Antiviral Research 2013, 99: 6–11.

Kuehn BM. As Ebola epidemic begins to slow, trials of drugs and vaccines speed up. JAMA 2015, 313: 1000–1002.

Kupferschmidt K. As Ebola wanes, trials jockey for patients. Science 2015, 348: 20.

Lazzarin A, Clotet B, Cooper D, Reynes J, Arasteh K, Nelson M, Katlama C, Stellbrink H-J, Delfraissy J-F, Lange J, Huson L, DeMasi R, Wat C, Delehanty J, Drobnes C, Salgo M. Efficacy of enfuvirtide in patients infected with drug-resistant HIV-1 in Europe and Australia. *New England J Medicine* 2003, 348: 2186–2185.

Lin K, Gallay P. Curing a viral infection by targeting the host: the example of cyclophilin inhibitors. Antiviral research 2013, 99: 68–77.

Lipsitch M, Eyal N, Halloran ME, et al. Ebola and beyond. Science 2015, 348: 46–48.

Litterman N, Lipinski C, Ekins S. Small molecules with antiviral activity against Ebola virus. F1000Research 2015, 4: 38.

Liu Y, Sheng J, Fokine A, et al. Structure and inhibition of EV-D68, a virus that causes respiratory illness in children. Science 2015, 347: 71–74.

Liu N, Zhao F, Jia H, et al. Non-nucleoside anti-HBV agents: adnvances in structural optimization and mechanism of action investigations. Medchemcomm 2015, 6: 521–535.

Lok AS. Personalized treatment of hepatitis B. Clinical and molecular hepatology 2015, 21: 1–6.

Marzi A, Halfmann P, Hill-Batorski L, et al. An Ebola whole-virus vaccine is protective in nonhuman primates. Science 2015, 348: 439–442.

Minie M, Chopra G, Sethi G, et al. CANDO and the infinite drug discovery frontier. Drug Discovery Today 2014, 19: 1353–1363.

Neveu G, Barouch-Bentov R, Ziv-Av A, et al. Identification and targeting of an interaction between a tyrosine motif within hepatitis C virus core protein and AP2M1 essential for viral assembly. PLOS pathogens 2012, 8: e1002845.

Murin CD, Fusco ML, Bornholdt ZA, et al. Structures of protective antibodies reveal sites of vulnerability on Ebola virus. PNAS 2014, 111: 17182–17187.

Petta S, Craxi A. Current and future HCV therapy: do we still need other anti-HCV drugs? Liver International 2015, 35: S1: 4–10.

Procko E, Berguig GY, Shen BW, et al. A computationally designed inhibitor of an Epstein-Barr viral Bcl-2 protein induces apoptosis in infected cells. Cell 2014, 157: 1644–1656.

Qui X, Wong C, Audit J, et al. Reversion of advanced Ebola disease in nonhuman primates with ZMapp. Nature 2014; 514: 47–53.

Rampling T, Ewer K, Bowyer G, et al. A monovalent chimpanzee adenovirus Ebola vaccine – preliminary report. NEJM 2015, doi: 10.1056/NEJMoa1411627.

Rice CM, Saeed M. Treatment triumphs. Nature 2014, 510: 43–44.

Shi M, Sun WL, Hua YY, et al. Effects of entecavir on hepatitis B virus covalently closed circular DNA in hepatitis B e Antigen-positive patients with hepatitis B. PLOSone 2015, doi: 10.1371/journal.pone.0117741.

Soares MM, King SW, Thorpe PE. Targeting inside-out phosphatidylserine as a therapeutic strategy for viral diseases. Nature Medicine 2008, 14: 1357–1362.

Suresh V, Krishnakumar KA, Asha VV. A new fluorescent based screening system for high throughput screening of drugs targing HBV-core and HBsAg interaction. Biomedicine and pharmacology 2015, 70: 305–316.

TablesofEbolaDrugs.pdf. http://who.int/medicines/ebola-treatment/2015-0116, accessed April, 2015.

Tavis JE, Lomonosova E. The hepatitis B virus ribonuclease H as a drug target. Antiviral research 2015, 118: 132–138.

Tufts Center for the Study of Drug Development (2014), *Cost to Develop and Win Marketing Approval for a New Drug Is $2.6 Billion*, http://csdd.tufts.edu/news/complete_story/pr_tufts_csdd_2014_cost_study.

Veljkovic V, Loiseau PM, Figadere B, et al. Virtual screen fo repurposing approved and experimental drugs for candidate inhibitors of Ebola virus infection. F1000Research 2015, 4: 34.

Verbist B, Klambauer G, Vervoort L, et al. Using transcriptomics to guide lead optimization in drug discovery projects: lesson learned from the QSTAR project. Drug discovery today 2015, doi: org/10.1016/j.drudis.2014.12.014.

Wang Z, Deisboeck TS. Mathematical modeling in cancer drug discovery. Drug discovery today 2014, 19: 145–150.

Warren TK, Wells J, Panchal RG, et al. Protection against figovirus disesaee by a nove broad spectrum nucleoside analogue BCX4430. Nature 2014, 508: 402–405.

Part IV

Prizes and Predictions for Viral Pathogenesis

We have seen how the transition from reductionism to systems biology has transformed the field of virology. In this last Part, we take a look at *past* breakthroughs in viral pathogenesis and then conclude with a glimpse of the *future* of the field.

We thought that the reader would be interested to learn about some of the important discoveries that have pioneered advances in viral pathogenesis over the last century. To this end we selected eight Nobel Prizes that are relevant to viral pathogenesis. The chapter, *Breakthrough: Nobel prizes*, lets each laureate speak for themselves: first, to describe their breakthrough experiments, and second, to tell something of their life in science, why each scientist entered the field and how they came to do their Prize-winning research. Also, we reflect on how each breakthrough advanced our field. Hopefully, these tales will encourage some of our readers to think out of the box and embark on research that no one else has considered or dared to undertake.

For the final chapter, *what lies ahead?*, we recruited 14 scientists and asked each one to give us—within their particular area of expertise—their vision of the future. Contributors were given the freedom to use the opportunity with no requirements other than the limit of 1000 words, which most respected. We hope our readers will find the resulting mélange of perspectives interesting and—perhaps in some instances—take inspiration from these projections.

This concludes our tour of viral pathogenesis, both the established and the new. We hope that—for some of our readers at least—we have achieved our goals: to set out established knowledge in viral pathogenesis, to explain the paradigm shift from reductionism to a systems approach, and to provide the background that will enable the next generation of scientists to do groundbreaking research in our field.

Chapter 21

Breakthrough

Nobel Prize Discoveries in Viral Pathogenesis

Neal Nathanson
Department of Microbiology, Perelman School of Medicine, University of Pennsylvania, Philadelphia, PA, USA

Chapter Outline

1. INTRODUCTION

Our knowledge of viral pathogenesis is founded upon experimental data generated by many investigators over at least the last century. Furthermore, ongoing research constantly revises and expands our understanding. To give readers a sense of the experimental basis for viral pathogenesis, we dedicate this chapter to examples of important breakthroughs in this field. Since there were many Nobel prizes for research relevant to viral pathogenesis, we decided to use some of these as illustrations (Table 1). In each instance, we describe the original experimental data, explain its relevance to viral pathogenesis, and provide a short biosketch of one of the Nobel awardees. The examples we have selected range over many aspects of pathogenesis, including virology, immunology, host responses, and vaccines. Much of this information is based on the lectures and other information generated as part of the Nobel ceremonies and set forth on the official site of the Nobel Foundation (Nobelprize.org). The original documents available online make fascinating reading, and the video lectures are a great introduction to the subjects. We hope that these individual stories will inspire some of our readers to consider the opportunities for future research in viral pathogenesis.

Chapter 21.1

Yellow Fever Vaccine

1. THE PRIZE CITATION

The Nobel Prize in Physiology or Medicine 1951 was awarded to Max Theiler *for his discoveries concerning yellow fever and how to combat it.*

2. THE ORIGINAL OBSERVATIONS

Some important discoveries about yellow fever preceded Max Theiler's work and should be mentioned to place his contributions in context. During the Spanish–American War of 1898–1901, yellow fever became a serious problem for military troops in Cuba and other Caribbean islands. A yellow fever commission was organized by the U.S. Department of the Army. Their investigations demonstrated—using volunteer soldiers—that this infection was transmitted by the bite of a mosquito, *Aedes aegypti*, and caused by a filterable virus (Reed, 1902). Research then languished until the problem was taken up by the Department of Medicine and Public

Viral Pathogenesis. http://dx.doi.org/10.1016/B978-0-12-800964-2.00021-5

TABLE 1 Nobel Prizes Relevant to Viral Pathogenesis, 1950–2013

Nobel Laureates	Topic	Date of Prize	Illustrated
	Viral Vaccines		
Theiler	Yellow fever vaccine	1951	X
	Enabling Advances		
Enders, Weller, Robbins	Polio in tissue culture	1954	X
Jacob, Monod, Lwoff	Genetics of enzymes and viruses	1965	
Delbruck, Hershey, Luria	Genetics of viruses	1969	
Arber, Nathans, Smith	Restriction enzymes	1978	
Fire, Mello	RNA interference	2006	
Capecchi, Evans, Smithies	Genetic modification of mice	2007	
	Unconventional Infectious Agents		
Prusiner	Prions	1997	X
	Viral Oncogenesis		
Rous	Tumor viruses	1966	X
Baltimore, Dulbecco, Temin	Reverse transcriptase of retroviruses	1975	
Blumberg, Gajdusek	Hepatitis B virus Transmissible spongiform encephalopathies	1976	
Bishop, Varmus	Viral oncogenes	1989	X
Hartwell, Hunt, Nurse	Cell cycle	2001	
Zur Hausen, Barré-Sinoussi, Montagnier	Human papillomavirus causes cervical cancer Discovery of human immunodeficiency virus	2008	
	Viral Immunology		
Burnet, Medawar	Immunological tolerance	1960	
Edelman, Porter	Structure of antibodies	1972	
Benacerraf, Dausset, Snell	Major histocompatibility complex	1980	
Jerne, Köhler, Milstein	Monoclonal antibodies	1984	
Tonegawa	Genetic diversity of antibodies	1987	
Doherty, Zinkernagel	MHC restriction of virus-specific T cells	1996	X
Beutler	Innate immunity	2011	X
Steinman	Dendritic cells	2011	X

Health of the Rockefeller Institute, which initiated studies in Africa in the 1920s. Those studies showed that the infection could be transmitted to rhesus monkeys, by injection of blood obtained from acutely ill patients; the monkeys developed yellow fever with fatal hepatic necrosis (Stokes et al., 1928). The identification of a laboratory animal host suddenly opened the field to experimental investigation.

Enter Max Theiler. Theiler was recruited in 1930 to be a part of the team assembled by the Rockefeller Foundation to make an attack on yellow fever, with the eventual goal of controlling this devastating disease. While at Harvard, Theiler had found that yellow fever could be transmitted to laboratory mice (Theiler, 1930). However, in contrast to monkeys, mice could only be infected by intracerebral injection and not by parenteral routes. Mice did not develop hepatitis or disease of other visceral organs but died of acute encephalomyelitis. Furthermore, the virus became "adapted" to mice on serial intracerebral passage, as evidence by shortening of the incubation period to death. After a number of passages, the virus was deemed to be "fixed" since the incubation period could not be further reduced in length.

When virus obtained at different mouse passage levels was tested by injection back into rhesus monkeys, it gradually lost the ability to cause acute hepatitis. This key observation triggered the idea that yellow fever virus might be sufficiently "attenuated" by some type of serial passage to make it a candidate for use as a live virus vaccine. However, it was early realized that this alteration in the properties of the virus was a "double-edged" sword. The mouse-adapted virus—while losing its ability to cause hepatitis in monkeys—had acquired "neurotropism"; that is, it would cause acute fatal encephalitis when injected in monkeys by the intracerebral route. When the mouse-passaged virus was injected by the subcutaneous route in monkeys, it usually caused a mild infection with the development of immunity against subsequent parenteral challenge with "wild-type" yellow fever virus. However, a few monkeys would develop lethal encephalitis, even after subcutaneous injection. Hence, the development of an attenuated strain of yellow fever virus for use as a human vaccine presented a complicated—perhaps insuperable—challenge (Theiler, 1951a).

During the 1930s, extensive efforts were made to modify the use of a mouse-adapted virus for immunization of humans, but—although used widely in Africa—this virus was considered (by the Rockefeller Foundation) to be excessively neurovirulent. Theiler therefore initiated a very extensive series of studies in the search for a laboratory-adapted strain that was safe for humans (Theiler and Smith, 1937a,b). Among many passage lines of virus, one line maintained in chick embryo cultures was found to have retained modest infectivity for humans while showing reduced ability to cause either hepatitis or encephalitis in monkeys (Figure 1). This virus, the 17D strain, was adopted as a vaccine and has been used to successfully immunize millions of humans. Theiler concluded that the 17D strain had undergone an unexplained mutation since parallel passage lines of the same virus failed to show reduced neurovirulence. However, the 17D virus has its limitations, since it is liable to develop neurovirulent revertants during serial passage in tissue culture (Fox et al., 1942). The fascinating history of yellow fever research and control in 1950 is well described in a monograph published by the Rockefeller Foundation (Strode, 1951).

3. SIGNIFICANCE FOR VIRAL PATHOGENESIS

The story of the discovery of the 17D strain of yellow fever virus is more than a tale about vaccine development (Norrby, 2007). It illustrates many aspects of virus variation and pathogenicity, a key variable in viral pathogenesis. What are the lessons learned?

- Laboratory passage can alter the biological properties of a virus so that it no longer can be used as a surrogate for fresh isolates from field cases of infection or disease (often called "field" or "wild" virus strains). Interesting, this lesson has been repeatedly ignored or forgotten. Even as recently as the 1980s, it was not recognized that when HIV isolates were passaged into cell culture (using human T lymphocytes), the virus was altered in significant ways. The T-cell adapted virus was relatively easy to neutralize in vitro. This led to the premature conclusion that it would be straight forward to induce a protective response against HIV in human subjects.

TABLE VIII

A Comparison of the Pathogenicity of Cultivated Yellow Fever for Rhesus Monkeys and Hedgehogs

Tissues used in culture medium	Results in monkeys		Results in hedgehogs by subcutaneous inoculation
	By extraneural inoculation	By intracerebral inoculation	
Whole mouse embryo—virus 17 E	Monkeys survive, but show a considerable amount of virus in the circulating blood.	Death from encephalitis	Survive
Mouse and guinea pig testicular tissue—virus 17 AT	Monkeys survive and show only traces of virus in the circulating blood	" " "	Not tested
Chick embryo tissue with head and spinal cord removed—virus 17 D	Monkeys survive and show only traces of virus in the circulating blood	Non-fatal encephalitis	Survive
Unmodified Asibi virus	About 95 per cent of monkeys die of yellow fever showing typical visceral lesions; virus present in the circulating blood in high concentration	Death usually from generalized infection with typical visceral lesions. Death from encephalitis only when immune serum is given intraperitoneally at the time of intracerebral injection of virus	Death from yellow fever in 3 to 7 days with typical visceral lesions

FIGURE 1 Theiler and associates conducted a series of passages of yellow fever virus in various tissue culture systems, hoping that one of these lines would be attenuated for primates and could be used as a vaccine. They succeeded with a culture line called 17D, but not with other culture lines, as shown in this table from their original 1937 publication. *Table 8 in Theiler and Smith (1937a), with permission.*

- Passage by intracerebral injection often selects for "neuro-adapted" strains of virus that have lost their "pantropic" properties. Such neurotropic strains are mainly limited to replication in the central and peripheral nervous system or cultures of neural cells. This phenomenon has been seen with a number of quite different virus groups, including rabies virus, poliovirus, influenza virus, bunyavirus, as well as yellow fever flavivirus. However, the molecular basis of this phenomenon has yet to be elicited.
- Passage in cell culture (as contrasted with passage in animals) also can select for virus strains that depart from field isolates. The properties of such passage strains differ from neuro-adapted viruses, and may exhibit reductions in pathogenicity for animals or humans. In some instances (such as oral poliovirus (OPV) vaccine), it has been possible to map the multiple point mutations responsible for these changes in biological properties (Minor, 1992). However, when such attenuated viruses are further passaged in cell culture or used as live virus vaccines in humans, they are likely to revert toward a wild-type virulent phenotype (Fox et al., 1942).
- Fast forward to the present. The 17D vaccine virus continues to yield new lessons using the current tools of molecular and systems biology (Hahn et al., 1987; Beck et al., 2014; Tangy and Desprès, 2014). The 17D and parent Asibi strains differ by less than 1% in their RNA and amino acid sequences. However, deep sequencing reveals that the wild-type Asibi strain is a quasispecies containing a wide variety of slightly different genetic species. By contrast, 17D strain exhibits very restricted genetic variation. Interestingly, the mutational differences seen in the 17D strain are not found among the minority sequences of the wild virus; this indicates that 17D was derived by de novo mutations during long-term passage in tissue culture rather than selection of a preexisting minority population in the wild quasispecies. This significant finding also suggests that the 17D strain is less likely to revert to a wild-type phenotype, in contrast to another attenuated vaccine virus, OPV, which reverts with a very high frequency (Minor, 1980).
- Further questions remain for future research, including: (1) Will an infectious DNA clone of wild yellow fever virus confer the fully virulent phenotype? (McGee et al., 2008a,b) or does it require a quasispecies, as suggested by Andino and his colleagues (Vignuzzi et al., 2006)? (2) Which viral genes are responsible for the attenuated phenotype of the 17D strain (McGee et al., 2008a,b)? (3) What can "omics" analysis of the responses of humans infected with 17D—and nonhuman primates infected with 17D and Asibi strains—tell us about the mechanisms that elicit such different pathological outcomes (Pulendran, 2009; Law et al., 2013; Campi-Azevedo et al., 2012; Li et al., 2014)?
- In summary, the story of yellow fever vaccine recapitulates the transition from basics to systems biology, a central theme of this book.

4. BIOSKETCH: MAX THEILER (1899–1972)

Max Theiler was born in South Africa into an academic family (Theiler, 1951b). His father, Sir Arnold Theiler, was educated in Switzerland but shortly thereafter emigrated to South Africa where he spent his career. Sir Arnold was a distinguished veterinary scientist who described Theiler's disease of horses and was the founding director of the first veterinary school in South Africa. Max Theiler had three siblings, all of whom were health professionals with an interest in infectious diseases. Theiler began his medical education in South Africa but completed it in London, with dual degrees at St Thomas' Hospital and the London School of Tropical Medicine. Shortly thereafter, at the age of 23, he took a position in the Department of Tropical Diseases at Harvard Medical School where he served for 7 years. During this time, he initiated his research on yellow fever and found that yellow fever could be transmitted to laboratory mice (Theiler, 1930). Theiler soon discovered that the virus could also be transmitted to laboratory mice by intracerebral injection, and this initiated his lifelong studies of this agent. In 1931, Theiler joined the yellow fever program at the Division of Medicine and Public Health at the Rockefeller Institute, where he spent the rest of his career, until his death in 1972.

REFERENCES

Beck A, Tesh RB, Wood TG, et al. Comparison of the live attenuated yellow fever vaccine 17D-2014 strain to its virulent parental strain Asibi by deep sequencing. J Infect Dis 2014;209:334–44.

Campi-Azevedo AC, Araujo-Porto LP, Luiza-Silva M, et al. 17DD and 17D-213/77 yellow fever substrains trigger a balanced cytokine profile in primary vaccinated children. PLoS One 2012;7(1):e49828.

Fox JP, Lennette EH, Manso C, Souza Aguiar JR. Encephalitis in man following vaccination with 17D yellow fever virus. Am J Hyg 1942;36:117–42.

Hahn GS, Dalrymple JM, Strauss JH, et al. Comparison of the virulent Asibi strain of yellow fever virus with the 17D vaccine strain derived from it. Proc Natl Acad Sci USA 1987;84:2019–23.

Law GL, Korth MJ, Benecke AG, et al. Systems virology: host-directed approaches to viral pathogenesis and drug targeting. Nat Rev Microbiol 2013;11:455–66.

Li S, Rouphael N, Duraisingham S, et al. Molecular signatures of antibody responses derived from a systems biology study of five human vaccines. Nat Immunol 2014;15:195–204.

McGee CE, Lewis MG, St Claire M, et al. Recombinant chimeric virus with wild-type dengue 4 virus premembrane and envelope and virulent yellow fever Asibi backbone sequences is dramatically attenuated in nonhuman primates. J Infect Dis 2008a;197:693–7.

McGee CE, Tsetsarkin K, Vanlandingham DL, et al. Substitution of wild-type yellow fever Asibi sequences for 17D vaccine sequences in ChimeriVax-Dengue 4 does not enhance infection of *Aedes aegypti* mosquitoes. J Infect Dis 2008b;197:686–92.

Minor PD. The molecular biology of poliovaccines. Journal of General Virology 1992;73:3065–77.
Norrby E. Yellow fever and Max Theiler: the only Nobel Prize for a virus vaccine. J Exp Med 2007;204:2779–84.
Pulendran B. Learning immunology from the yellow fever vaccine: innate immunity to systems vaccinology. Nat Rev Immunol 2009;9:741–7.
Reed W. Recent researches concerning etiology, propagation and prevention of yellow fever by the United States Army Commission. J Hyg 1902;2:101–19.
Stokes A, Bauer JH, Hudson NP. Experimental transmission of yellow fever to laboratory animals. Am J Trop Med 1928;8:103–64.
Strode GK, editor. Yellow fever. New York: McGraw-Hill; 1951.
Tangy F, Desprès P. Yellow fever vaccine attenuation revealed: loss of diversity. J Infect Dis 2014;209:318–20.
Theiler M. Susceptibility of white mice to virus of yellow fever. Science 1930;71:367.
Theiler M, Smith HH. Effect of prolonged cultivation in vitro upon pathogenicity of yellow fever virus. J Exp Med 1937a;65:767–86.
Theiler M, Smith HH. Use of yellow fever virus modified by in vitro cultivation for human immunization. J Exp Med 1937b;65:787–800.
Theiler M. Nobel prize lecture. http://www.nobelprize.org/nobel_prizes/medicine/laureates/1951/theiler-lecture.html; 1951a.
Theiler M. Max Theiler biographical. http://www.nobelprize.org/nobel_prizes/medicine/laureates/1951/theiler-bio.html; 1951b.
Vignuzzi M, Stone JK, Arnold JJ, et al. Quasispecies diversity determines pathogenicity through cooperative interactions in a viral population. Nature 2006;439:344–8.

Chapter 21.2

Cultivation of Polioviruses

1. THE PRIZE CITATION

The Nobel Prize in Physiology or Medicine 1954 was awarded jointly to John Franklin Enders, Thomas Huckle Weller, and Frederick Chapman Robbins *for their discovery of the ability of poliomyelitis viruses to grow in cultures of various types of tissue.*

2. THE ORIGINAL OBSERVATIONS

The original report that a strain of poliovirus had been cultivated in suspension cultures of nonneural human embryonic tissues appeared in Science magazine in 1949 (Enders et al., 1949). The experiments were stimulated by the prior failed attempts of Sabin and Olitsky (1936) to cultivate the MV strain (a neuro-adapted strain) of poliovirus in nonneural tissue cultures. Also, important were reports that poliovirus could be transmitted to experimental animals from fecal samples obtained from paralyzed patients, suggesting that the virus could multiply in nonneural tissues.

The reported experiments were quite simple. Suspensions of human fetal tissues from skin, muscle, and connective tissue of arms and legs were maintained in a medium consisting of balanced salt solution and ox serum filtrate. The cultures were maintained for several months by replacing the medium every 4–7 days, which permitted an estimate of the cumulative replication of the original inoculum. When infectivity was still found after a cumulative calculated dilution of 10^{-15} or so, the investigators concluded that the virus was multiplying (Figure 2). Critical to the success of these experiments was the use of the Lansing strain of type 2 poliovirus, which had been adapted to mice by intracerebral inoculation. Apparently, this laboratory-adapted strain of poliovirus was no longer an obligatory neurotrope. Furthermore, mice could be used to assay infectivity at each exchange of culture medium. To confirm the identity of the infectivity obtained at the end of the experiment, the investigators cited several lines of evidence: the virus still produced paralysis in mice, was neutralized by antiserum against the Lansing virus, and caused polio-like paralysis when injected into rhesus monkeys. One final important result was the observation that the virus produced a cytopathic effect in the tissue culture fragments.

TABLE 1

MULTIPLICATION OF LANSING POLIOMYELITIS VIRUS IN TISSUES OBTAINED FROM THE EXTREMITIES OF HUMAN EMBRYOS

Culture set	No. of nutrient fluid changes prior to subculture		Day of incubation subculture done		Mouse LD_{50} of pooled fluids used to inoculate subcultures		Calculated dilution of original inoculum at time of subculture	
	Exp. 1	Exp. 2	Exp. 1	Exp. 2	Exp. 1	Exp. 2	Exp. 1	Exp. 2
Original*	3	4	20th	20th	$10^{-2.0}$	$10^{-1.57}$	10^{-5}	$10^{-6.2}$
1st subculture	2	2	19th	12th	$10^{-1.57}$	$10^{-1.71}$	$10^{-8.8}$	$10^{-10.0}$
2nd subculture	2	4	12th	20th	$10^{-0.63}$	$10^{-1.34}$	$10^{-12.6}$	$10^{-16.2}$
3rd subculture	3	..	16th	....	$10^{-0.16}$	.....	$10^{-17.7}$	.....

* The LD_{50} of the suspension of mouse brain used as the inoculum in the first experiment was 10^{-2}; that of the suspension employed in the second experiment was $10^{-1.01}$.

FIGURE 2 Enders and colleagues showed that the Lansing strain of poliovirus could replicate in cultures of human embryonic tissues. They selected the Lansing strain because it also would infect and kill mice when injected intracerebrally (in contrast to most poliovirus isolates that will only replicate in primates). This attribute provided an independent method to test the infectious titer of virus grown in human tissues. *Table 1 in Enders et al. (1949), with permission.*

As described by John Enders in his Nobel lecture, these observations were quickly confirmed by a number of other investigators, and led to rapid improvements in the methods for cell culture, the detection of cytopathic effects, and the evidence that cultured virus could be used to induce protective immunity (Enders et al., 1954). Finally, it was also found that tissue culture passage could reduce the neurovirulence of poliovirus without affecting its ability to replicate in cell cultures, an observation that stimulated the successful development of attenuated strains of poliovirus to use as a vaccine (oral poliovirus vaccine, or Sabin vaccine).

3. SIGNIFICANCE FOR VIRAL PATHOGENESIS

At first glance, the discovery that polioviruses could be propagated in cultures of human cells might seem mundane by comparison with advances in molecular biology and genomics. However, viewed in a historical context, the observations take on much more significance. Poliovirus was first "isolated" by Landsteiner and Popper in Vienna in 1908 (Landsteiner and Popper, 1909). These workers reported that they could produce clinical paralysis in monkeys if they injected them with a homogenate of the spinal cord removed from a patient recently dead of acute paralytic poliomyelitis. These observations met the Henle–Koch criteria for the isolation of the causal agent of an infectious disease, and were accepted as demonstrating that a filterable virus was the cause of this disease. But there is more to the story.

In the days prior to tissue culture, the standard method to propagate poliovirus was by the harvest of spinal cord from a monkey with experimental paralytic poliomyelitis, and the injection of that material by the intracerebral route into another monkey. What these early workers could not recognize was that this CNS passage protocol selected for a virus variant that was exclusively neuronotropic, that is, could only replicate in cells of the peripheral and central nervous system. Experiments—in monkeys with this neuronotropic virus—designed to simulate the pathogenesis of human poliomyelitis were doomed to failure, and they misled the field for many years. Using the neuronotropic virus, it was not possible to infect by virus feeding, which discredited the prior (and correct) hypothesis that polio was an enteric infection transmitted by the oral route. The only noninvasive route that led to infection with the neuro-adapted virus was intranasal insufflation, leading to the hypothesis that polio was transmitted by primary infection of the first-order sensory neurons in the nasal mucosa. This view of poliomyelitis was summarized in 1931 by Simon Flexner, Director of the Rockefeller Institute, in an influential review in Science magazine.

The neurotropic hypothesis led to a human trial of a nasal astringent spray, which deliberately cauterized the nasal mucosa (and caused many cases of anosmia) to prevent acquisition of infection. The total failure of this trial (Tisdall et al., 1937) led to a reexamination of pathological lesions in patients who died during an acute attack of poliomyelitis. There was no pathological evidence of infection of the olfactory bulbs, in contrast to Flexner's monkey model, and this led to a re-thinking of the pathogenesis.

Enter the Enders observations. Using the newly reported methods for growing poliovirus in cell cultures prepared from a variety of human tissues, it was now possible to make primary virus isolates from humans. These experiments showed that virus could be grown from both oral secretions and fecal samples of acutely infected patients, strongly affirming the enteric nature of the infection. Furthermore, these primary isolates would readily infect chimpanzees (and selected species of monkeys) after feeding. The resulting revised view of the pathogenesis of poliomyelitis was summarized in a 1955 review—again in Science magazine—by David Bodian, a leading polio researcher at the Johns Hopkins University. Critical to this reconstruction was the role that viremia played in the spread of systemic infection to the central nervous system. In turn, that insight provided the scientific basis for the potential efficacy of an immunogen that would induce neutralizing antibodies in the circulation (Nathanson, 2005). Luckily, it turned out that chemical inactivation of poliovirus grown in cell culture could create a safe and effective immunogen, leading to the development of inactivated poliovirus vaccine (IPV, or Salk vaccine) first deployed in 1955 (Francis et al., 1957).

4. BIOSKETCH: JOHN F. ENDERS (1897–1982)

John Enders was born into an affluent family in Hartford, Connecticut (Enders, 1954). After graduating from Yale University at the age of 23, he spent 7 years trying various careers before he enrolled in a PhD program in bacteriology at Harvard Medical School. He worked on a variety of bacterial projects and only began studies of mammalian viruses in 1938. Due to the interruptions occasioned by the Second World War, he did not become fully engaged in virology until 1946 when he was asked to start a research laboratory in the Children's Hospital next to Harvard Medical School. He and his colleagues developed or adopted cell culture systems to grow several human viruses, including mumps, measles, vaccinia, and varicella zoster. Beginning in 1948, application of these methods to poliovirus yielded immediate positive results, and led to intensive studies (1948–1952) on how best to cultivate poliovirus, with immediate exploration of potential application to vaccine development. Enders continued a productive research career with many human viruses until he retired in 1977 at the age of 80. He was an outstanding research mentor, and a number of prominent virologists trained in his laboratory.

REFERENCES

Bodian D. Emerging concept of poliomyelitis infection. Science 1955;122:105–8.

Enders JF, Weller TH, Robbins FC. Cultivation of the Lansing strain of poliomyelitis virus in cultures of various human embryonic tissues. Science 1949;109:85–7.

Enders JF, Robbins FC, Weller TH. Nobel prize lecture. The cultivation of the poliomyelitis viruses in tissue culture. http://www.nobelprize.org/nobel_prizes/medicine/laureates/1954/enders-lecture.html; 1954.

Enders JF, John F. Enders biographical. htttp://www.nobelprize.org/nobel_prizes/medicine/laureates/1954/enders-bio.html; 1954.

Flexner S. Poliomyelitis (infantile paralysis). Science 1931;74:251–4.

Francis Jr TJ, Napier JA, Voight R, Hemphill FM, Wenner HA, Korns RF, et al. Evaluation of the 1954 field trial of poliomyelitis vaccine. Ann Arbor, Michigan: School of Public Health, University of Michigan; 1957.

Landsteiner K, Popper E. Ubertragung der Poliomyelitis acuta auf Affen. Z Immunitatstorsch Orig 1909;2:377–90.

Nathanson N. David Bodian's contribution to the development of poliovirus vaccine. Am J Epidemiol 2005;161:207–12.

Sabin AB, Olitsky PK. Proc Soc Exp Biol Medi 1936;34:357–61.

Tisdall FF, et al. Zinc sulphate spray in prophylaxis of poliomyelitis: observation of group of 4,713 children, 3010 years, during epidemic in Toronto, Canada. Can J Public Health 1937;28:523–31.

Chapter 21.3

Discovery of Prions

1. THE PRIZE CITATION

The Nobel Prize in Physiology or Medicine 1997 was awarded to Stanley B. Prusiner *for his discovery of Prions—a new biological principle of infection.*

2. BACKGROUND COMMENT

Because the transmissible spongiform encephalopathies are such a unique set of diseases, an introductory note may be helpful. The spongiform encephalopathies are a set of fatal progressive degenerative neurological diseases, which are grouped together because they share similar pathological features: noninflammatory spongiform alterations in the nervous system, often accompanied by deposits of amyloid-like material. The first of these diseases to be recognized was scrapie, a fatal neurological disease of sheep, described at least 200 years ago by sheep herders in England. It was subsequently noted that Creutzfeldt–Jakob disease, a rare sporadic human syndrome, shared a similar pathology. Kuru, a comparable affliction, was subsequently observed as an epidemic in a small stone-age linguistic group in the eastern highlands of New Guinea. These diseases also had the unique characteristic that they appeared to be both inheritable and transmissible, and—in their transmissible form—exhibited some unusual physical and chemical properties.

3. THE ORIGINAL OBSERVATIONS

The biochemical and molecular nature of prions were unraveled through a sequence of experimental breakthroughs, many of which preceded the characterization of the prion protein (PrP). The first step was the recognition that scrapie, a degenerative neurological disease of sheep, could be transmitted by injection of a brain homogenate from an affected sheep to a normal sheep (Greig, 1950). The sheep model was not very tractable for experimental purposes, so the first advance was the transmission of scrapie to mice by Chandler (1963). Although mice could be used to measure prion infectivity, this was a relatively cumbersome system, taking up to 24 months and 50–100 mice for each titration. Subsequently, it was found that the infectious agent could be transmitted to hamsters, with a 2-month readout using only 4 animals per titration (Kimberlin and Walker, 1977; Prusiner et al, 1982). The hamster system provided a tractable bioassay, which was essential to support efforts to concentrate and characterize the infectious agent, which had resisted innumerable efforts at purification.

Enter Stanley Prusiner. Beginning in 1972, Prusiner embarked on a program to purify scrapie infectivity. Classical biochemical techniques, such as gradient centrifugation, were frustrated since infectivity was associated with lipid membranes and was spread across the gradient. However, it was found that mild proteolysis could eliminate 98% of contaminating proteins and the hamster assay permitted relatively rapid titration of the residual infectivity, which appeared to be associated with a single protein. Prusiner's key breakthrough was biochemical, that is he spearheaded a collaborative effort that succeeded in sequencing the NH2-terminal of a concentrated protease-digested oligopeptide obtained from a gradient fraction with maximum infectivity. However, the protease treatment produced "ragged" ends to the NH_2-terminal of the oligopeptide. Alignment of the different amino acid sequences revealed a consensus order, from which it was possible to deduce the cognate DNA sequence of a small section of the putative implicated protein. Nucleotide probes complementary to this sequence were used to identify and clone the gene that encoded a single protein (Figure 3), which Prusiner named the protease-resistant or prion protein (PrP).

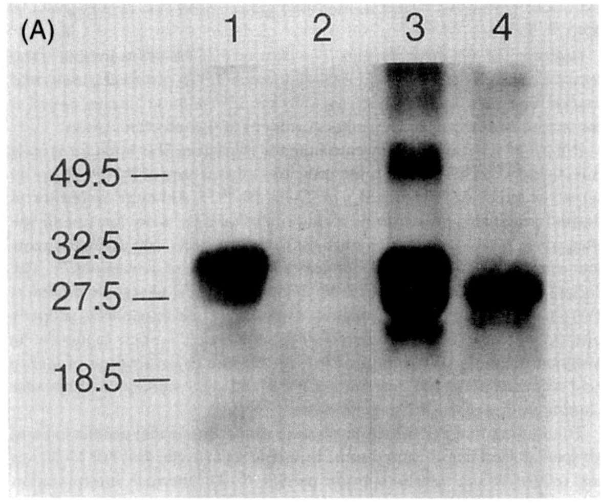

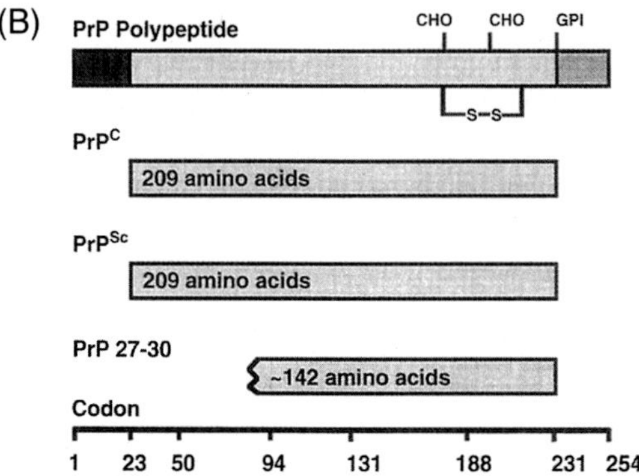

FIGURE 3 Prion protein isoforms. (**A**) Western immunoblot of brain homogenates from uninfected (lanes 1 and 2) and prion-infected (lanes 3 and 4) Syrian hamsters. Samples in lanes 2 and 4 were digested with 50 μg/ml of proteinase K for 30 min at 37°C. PrP^c in lanes 2 and was completely hydrolyzed under these conditions, whereas approximately 67 amino acids were digested from the NH_2-terminus of PrP^{Sc} to generate PrP 27–30. After polyacrylamide gel electrophoresis (PAGE) and electrotransfer, the blot was developed with anti-PrP R073 polyclonal rabbit antiserum. Molecular size markers are in kilodaltons (kD). (**B**) Bar diagram of SHaPrP which consists of 254 amino acids. After processing of the NH_2- and COOH- termini, both PrP^c and PrP^{Sc} consist of 209 residues. After limited proteolysis, the NH_2-terminus of PrP^{Sc} is truncated to form PrP 27–30, which is composed of approximately 142 amino acids. *From Prusiner (1997), Nobel lecture, with permission.*

Interestingly, PrP turned out to be a "normal" protein of unknown function, which was found in similar concentration in the tissues of normal and diseased animals. This was consistent with the hypothesis that there were at least two different conformations of PrP, PrP^C (normal cellular isoform) and PrP^{Sc} (scrapie isoform). An essential part of the prion hypothesis was that a posttranslational modification of the normal protein—designated PrP^{Sc}—was responsible both for the disease and for its transmissibility (Prusiner, 1997a).

It now remained to show that the implicated PrP met two stringent criteria: that it was both necessary and sufficient to transmit a spongiform encephalopathy. Work by many investigators since 1982 has generated a large body of data to meet these criteria (Prusiner, 1997). A few of these experiments may be mentioned. First, knockout animals that did not express PrP resisted transmission of a PrP^{Sc} inoculum that produced disease in normal animals; furthermore, mice or hamsters engineered to over-express PrP exhibited enhanced susceptibility to transmission (Bueler et al., 1993; Prusiner et al., 1993). More recently, laboratory-synthesized PrP protein that was free of extraneous nucleic acid was shown to initiate transmissible encephalopathy in mice (Zhang et al., 2013).

4. SIGNIFICANCE FOR VIRAL PATHOGENESIS

Prusiner's work was ground-breaking for several different reasons. As he points out in his Nobel lecture (Prusiner, 1997), the "protein-only" hypothesis had been put forth decades before he entered the field. However, it was only one among a wide variety of candidate explanations for the enigmatic properties of the scrapie agent. Furthermore, this hypothesis directly contradicted the "central dogma" of molecular biology, that is, DNA encodes RNA encodes proteins. The idea that post-translational alterations in protein could encode unique transmissible information was revolutionary in its day, before current developments in epigenetics. Therefore, the protein-only hypothesis was regarded as wild speculation supported merely by circumstantial evidence (such as the resistance of scrapie to inactivation by radiation that would damage nucleic acids). The identification of PrP, grounded in studies that involved classical biochemical and genetic methods, gave tangible support to the protein-only hypothesis.

In addition to advancing and supporting a revolutionary hypothesis in the life sciences, the identification of PrP and its different conformers enabled new and extended studies of the pathogenesis of the transmissible spongiform encephalopathies (Aguzzi, et al., 2013). At a cellular level, it was now possible to identify PrP as a major (perhaps the only essential) component of the protein aggregates that were one characteristic of the cellular pathology of encephalopathy. Furthermore, antibodies against PrP^C could be used to track the sequential spread of the pathological conformers of PrP, and demonstrate that there was an extraneural phase of infection that often preceded neural involvement. In turn, these studies showed that—in addition to transmission by direct injection of the pathological protein—PrP^C could be transmitted by extraneural routes such as ingestion. As a corollary, it was discovered that infection could be spread via the blood. It has gradually become clear that the transmissible "phase" of the spongiform encephalopathies shows an amazing similarity to the pathogenesis of conventional viruses.

Furthermore, deciphering the DNA sequence of PrP has informed the pathogenesis of certain familial varieties of spongiform encephalopathy in humans. In affected families, specific inheritable point mutations in PrP markedly enhance the spontaneous conversion of PrP^{C} into PrP^{Sc}. Also, certain mutations in PrP turned out to be risk factors for "spontaneous" or acquired cases of Creutzfeldt–Jakob disease. All of these insights followed Prusiner's identification of the prion protein.

More recently, the prion hypothesis has led to a major extension in views about the pathogenesis of other degenerative neurological diseases of humans (Aguzzi and Falsig, 2012; Guo and Lee, 2014; Prusiner, 2012). It appears that, in addition to PrP, selected other normal proteins can also undergo conformational changes, which are autocatalyic and lead to accumulations of "amyloid-like" material in the central nervous system. Among diseases in this category are Alzheimer's, Parkinson's, and amyotrophic lateral sclerosis. Together, these diseases constitute an increasing burden in the aging human population. If further research supports this view of the pathogenesis of many neurodegenerative diseases, the prion hypothesis will take on an even greater significance.

5. BIOSKETCH: STANLEY PRUSINER (1942–PRESENT)

Stanley Prusiner was raised in the Midwest in a middle-class Jewish family that clearly valued education and intellectual pursuits. The first indication of his future interest in biomedical research came during his college years at the University of Pennsylvania, where he was able to engage in a research project in the laboratory of Sidney Wolfson, a Penn professor. This honed his appetite for experimental research, and led to his matriculation as a medical student, also at the University of Pennsylvania. Once again, he found an opportunity to engage in research in the laboratory of Britton Chance, a distinguished investigator. In retrospect, Prusiner was already embarked on the course that he would follow for the rest of his career. Following graduation, he spent 3 years as a "yellow beret," a public health service officer at the National Institutes of Health. During these years in the laboratory of Earl Stadtman, a distinguished biochemist, Prusiner acquired a strong grounding in many aspects of experimental investigation. Interestingly, he then elected to complete his clinical training in neurology, at the University of California in San Francisco, a leading academic medical center where research was a core activity.

A patient with Creutzfeldt–Jakob disease, a fatal spongiform encephalopathy, triggered Prusiner's imagination, particularly when he learned that this was one of a family of "spongiform encephalopathies" that included scrapie of sheep, which had been transmitted to laboratory mice. Enchanted by the mystery and challenge of these agents, which exhibited many enigmatic biological properties, he set out to apply his newly acquired biochemical skills to purifying the transmissible agent. Little did he dream where his voyage of discovery would take him.

In 1974, Prusiner began his studies of the scrapie agent at UCSF. During the next 8 years he persevered, but had to struggle to obtain funding for his research and support from his academic institution. In particular, the challenge to characterize the scrapie agent, as described above, was a daunting one that many earlier researchers had failed to overcome. However, based on some experimental breakthroughs described above, he published a seminal review in Science in 1982 (Prusiner, 1982) that described evidence for the "protein-only" hypothesis. As described vividly in his Nobel biosketch (Prusiner, 1997), this article created a firestorm from many directions. At the core was disbelief that posttranslational modifications in a protein could carry transmissible information, information which was not encoded in the host genome. Prusiner was challenged to prove a negative, that is, that there was no nucleic acid associated with the scrapie agent.

The 10 years, 1982–1992, saw a series of important scientific advances, with the identification of the prion protein (PrP), the cloning of its gene (*nPrP*), the development of knockout and transgenic mice, and the discovery of mutations associated with selected familial types of spongiform encephalopathies. Many investigators—as described in Prusiner's Nobel lecture and biosketch—made critical contributions. In the aggregate, these led to increasing acceptance of the protein-only hypothesis.

In the last 25 years, Stanley Prusiner has not "rested on his oars" but has chosen to continue his active engagement in research, through a very large and productive laboratory. The prion discovery has spawned a worldwide constantly expanding set of engaged researchers. As he has described (Prusiner, 2012), these efforts have recently led to the concept that many degenerative human neurological diseases are caused by aberrations of other cellular proteins, by mechanisms similar to that which explains scrapie.

REFERENCES

Aguzzi A, Falsig J. Prion propagation, toxicity and degradation. Nat Neurosci 2012;15:936–9.

Aguzzi A, Nuvolone M, Zhu C. The immunobiology of prion diseases. Nat Rev Immunol 2013;13:888–902.

Bueler H, Aguzzi A, Sailer A, Greiner R-A, Autenried P, Aguet M, et al. Mice devoid of PrP are resistant to scrapie. Cell 1993;73:1339–47.

Chandler RL. Experimental scrapie in the mouse. Res Vet Sci 1963;4:276–85.

Greig JR. Scrapie in sheep. J Comp Pathol 1950;60:263–6.

Guo JL, Lee VML. Cell-to-cell transmission of pathogenic proteins in neurodegenerative diseases. Nat Med 2014;20:130–8.

Kimberlin R, Walker C. Characteristics of a short incubation model of scrapie in the golden hamster. J Gen Virol 1977;34:295–304.

Prusiner SB. Nobel lecture. Prions. http://www.nobelprize.org/nobel_prizes/medicine/laureates/1997/prusiner-lecture.html; 1997a.

Prusiner SB. Nobel bio. http://www.nobelprize.org/nobel_prizes/medicine/laureates/1997/prusiner-bio.html; 1997b.

Prusiner SB. Novel proteinaceous infectious particles cause scrapie. Science 1982;216:136–44.

Prusiner SB. A unifying role for prions in neurodegenerative diseases. Science 2012;336:1511–3.

Prusiner SB, Cochran SP, Groth DF, Downey DE, Bowman KA, Martinez HM. Measurement of the scrapie agent using an incubation time interval assay. Annals of Neurology 1982;11:353–8.

Prusiner SB, Groth D, Serban A, Koehler R, Foster D, Torchia M, et al. Ablation of the prion protein (PrP) gene in mice prevents scrapie and facilitates production of anti-PrP antibodies. Proc Natl Acad Sa USA 1993;90:10608–12.

Zhang Z, Zhang Y, Wang F, Wang X, Xu Y, Yang H, et al. De novo generation of infectious prions with bacterially expressed recombinant prion protein. Federation of American Societies for Experimental Biology (FASEB) Journal 2013;27:4768–75.

Chapter 21.4

Tumor-Inducing Viruses

1. THE PRIZE CITATION

The Nobel Prize in Physiology or Medicine 1966 was divided equally between Peyton Rous *for his discovery of tumor-inducing viruses* and Charles Brenton Huggins *for his discoveries concerning hormonal treatment of prostatic cancer.*

2. THE ORIGINAL OBSERVATIONS

In retrospect, Rous' original observations seem very simple but highly significant (Rous, 1910, 1911, 1966a). A hen from an inbred stock of Plymouth Rock chickens was noted to have a tumor on the breast. When examined pathologically, it consisted of spindle cells and was classified as a sarcoma. Tissue fragments from this tumor were injected into a few hens from the same stock and these hens soon developed tumors at the site of injection. Tumors were not transmissible to other breeds of chickens but eventually became transmissible to Plymouth Rock chickens from other sources. In a subsequent set of experiments, tumor tissue was thoroughly ground, centrifuged repeatedly, and the supernatant fluid passed through a bacteria-retaining Berkefeld filter (Figure 4). The filtrate when injected initiated sarcomas at the site of injection, although more slowly than did transplants of whole tumor tissue. As before, transmission was most readily accomplished in chickens from the original stock. Chicken breeds other than Plymouth Rocks were not susceptible to the filtrate although there was no exhaustive attempt to test many breeds.

Rous conducted a number of subsequent follow-up studies (Rous, 1913; Rous and Murphy, 1914a,b) on the filterable agent that became known as Rous sarcoma virus. The salient findings may be summarized: (1) A number of different chicken tumors were identified, three of which differed strikingly in their pathological appearance. These tumors "bred true" in that they maintained their distinctive histological appearance through many animal-to-animal passages. (2) A filterable agent could be obtained from each tumor of different appearance, and would also "breed true," initiating a consistent pathological picture. Rous concluded that a family of similar agents was causing all of these neoplasms. Furthermore, his studies of immunity reinforced the distinction between the agents causing these three distinct tumor phenotypes. (3) Tumor tissue could be inactivated (unable to cause transplantable tumors) by drying in a vacuum or by limited exposure to ultraviolet radiation but these treatments did not inactivate the filterable agent. These observations reinforced the identity of an infectious agent in contrast to viable tumor cells. (4) Some chickens exhibited "spontaneous" resistance to these tumors; of these, some chickens resisted transplanted tumor tissue but not the filterable agent, while other chickens showed the reverse pattern. Also, chickens resistant to one pathological tumor type might be susceptible to another. Again, these findings reinforced the distinctions between tumor cells and the filterable agent, and between the different filterable agents themselves. (The explanation for some of these specific observations remains obscure, but the overall conclusions appear justified.) (5) Some pathological tumor types grew progressively till they killed their host, but others spontaneously regressed. Furthermore, it was possible to encourage regression by using very small inocula. (6) Rechallenge experiments—in chickens whose tumors had regressed—were used in attempts to demonstrate acquired immunity against the tumor cells. Immunity appeared to be inducible for some tumor types but not for others. Furthermore, "cross-challenge" experiments indicated that—when present—acquired immunity was tumor-type specific. (7) Immunity against the filterable agents (in contrast to the tumors) was also demonstrated, in cross-challenge experiments. Groups of chickens, each bearing a different pathotype of tumor, were resistant to the agent associated with that tumor but susceptible to the agent associated with another tumor pathotype.

3. SIGNIFICANCE FOR VIRAL PATHOGENESIS

Rous' papers of 1910–1914 provide very clear evidence of the existence of a group of agents, almost certainly viruses, which could induce malignant neoplasms in chickens. However, the significance of these experiments was not

"Experiment iv.-In this experiment the material was never allowed to cool. About fifteen grams of tumor from chicken 140 (7th generation, B) was ground in a warm mortar with warm sand; mixed with 200 c.c. of heated ringer's solution; shaken for thirty minutes within a thermostat at 39° C; centrifugalized; and the fluid passed through a filter similar to that used in experiment iii. Both before and after the experiment, this filter was tested and found to hold back bacillus prodigiosis. The filtration of the fluid was done at 38.5° C., and its injection followed immediately. In four of ten fowls inoculated into the muscle of each breast with 0.2 to 0.5 c.c. of the filtrate, there developed a sarcoma at one of the points of inoculation; and though the growths required several weeks to appear, their subsequent enlargement was of average rapidity. Pieces removed at operation showed the characteristic structure, and transplantation into other chickens proved successful. Three of the hosts have died, and in two profuse metastases were found."

FIGURE 4 Rous showed that homogenates of a sarcoma of chickens were transmissible to other chickens of the same breed. In the experiment reported, he took the discovery one step further, and showed that the tumor-producing agent could be passed through a filter that retained bacteria, thereby concluding that the sarcoma was caused by a "filterable virus." *Report of an experiment from Rous (1911), with permission.*

recognized for many years, as evidenced by the extraordinary gap of more than 50 years between his publications and his Nobel Prize. Why so?

Rous and others have provided several explanations for this phenomenon. First, virology was in its infancy in 1911, and there were no tools to identify and classify individual agents, let alone to visualize these submicroscopic entities, which appeared to violate the biologists' definition of life itself. Second, it was asserted that since chickens were not mammals, their tumors could not be relevant to human cancer. (In that era, the universality of many basic life processes was not understood.) At this early period, the nature of cancer was poorly understood and the idea that an infectious agent could cause seemed inconceivable. Lastly, and perhaps most important, observations that establish important new scientific paradigms are often met with disbelief and denial, followed by grudging acceptance, before they become part of the canon of accepted knowledge.

Rous comments in his Nobel lecture that he abandoned his work on chicken tumors because he was so discouraged by the cold reception his observations received within the scientific community. However, he never lost his interest in tumor viruses, and returned to this field in collaboration with his colleague Richard Shope, who discovered the Shope papilloma virus of rabbits.

In historical perspective, what is the significance of Rous's observations and the avian tumor viruses? In addition to establishing the existence of oncogenic viruses, a reading of his original publications shows that he recognized several important attributes of these agents. There appeared to be several agents each causing a specific tumor phenotype. There existed both "spontaneous" and acquired resistance to these tumors, and there was a distinction between immunity to the tumor cells and immunity to the causal agents.

These observations provided a robust foundation for investigators such as Harry Rubin and Howard Temin, who re-opened the field 30 years later. They used then-current methods to isolate and characterize the causal viruses, to introduce a quantitative method to assay the oncogenic activity of Rous sarcoma virus in cell culture, and establish the existence of defective and helper viruses. This system led Howard Temin to postulate the existence of a cellular copy of the oncogenic information, and the discovery—with David Baltimore—of reverse transcription. Corollary to this leap of insight were concepts such as the integration of viral genes in host DNA, and the existence of the whole family of retroviruses. Michael Bishop and Harold Varmus, using the Rous sarcoma model, then established the relationship of viral oncogenes to host-encoded growth factors. Many of these outstanding advances (Coffin et al., 1997) were themselves worthy of Nobel prizes.

In retrospect, Rous' pioneering work initiated the field of RNA tumor viruses, and provided a model system that eventually led to much of our present knowledge of the molecular pathogenesis of cancer.

4. BIOSKETCH: PEYTON ROUS (1879–1970)

Peyton Rous was born in Baltimore into an educated family (Rous, 1966a). However, his father died young and the family was poor. Nevertheless, Rous entered the Johns Hopkins University where he received a bachelor's and a medical degree. In medical school, he acquired a local infection with tuberculosis which caused a year out during which he had an exhilarating and maturing experience on his uncle's cattle ranch in Texas. Following graduation from medical school, Rous interned at Hopkins; during this experience he concluded that he was not suited for the practice of medicine. Seeking an opportunity in medical research, he took a position in the Department of Pathology at the University of Michigan. Through the good auspices of Professor Alfred Warthin, the head of department, he spent some months in Germany and then received a "beginners" grant from the nascent Rockefeller Institute. His success investigating lymphocytes led to a junior position at the Institute. Simon Flexner, Director of the Rockefeller Institute, then asked Rous to take over the laboratory for cancer research, which led to his investigations of the spontaneous tumors of chickens. He had immediate success demonstrating that these tumors were transmitted by a filterable virus, and his work in this field, from 1909 to 1915, is a classic.

Rous then attempted to duplicate his work with tumors of mice, and his failure to do so led him to turn to other areas of biomedical research, particularly those dealing with blood transfusion. During World War I, he developed a method for storing whole blood for use on the battlefield. 20 years later, in 1934, he returned to tumor viruses as a collaborator with Richard Shope, who discovered the Shope papilloma virus. Interestingly, this virus turned out to be a prototype of papillomaviruses, the human representatives of which cause cervical carcinoma (work that led to a Nobel Prize for Harold zur Hausen, in 2008). Rous spent his career at the Rockefeller Institute and continued to work in the laboratory until his death in 1970.

REFERENCES

Coffin JM, Hughes SH, Harold E, Varmus HE, editors. Retroviruses. Cold Spring Harbor, New York: Cold Spring Harbor Laboratory Press; 1997.

Rous P. A transmissible avian neoplasm. (Sarcoma of the common fowl.). J Exp Med 1910;12:696–705.

Rous P. A sarcoma of the fowl transmissible by an agent separable from the tumor cells. J Exp Med 1911;13:397–411.

Rous P. Resistance to a tumor-producing agent distinct from resistance to the implanted tumor cells. J Exp Med 1913;18:416–27.

Rous P. Peyton Rous biographical. http://www.nobelprize.org/nobel_prizes/medicine/laureates/1966/rous-bio.html; 1966a.

Rous P. Nobel prize lecture. http://www.nobelprize.org/nobel_prizes/medicine/laureates/1966/rous-lecture.html; 1966b.

Rous P, Murphy JB. On the causation by filterable agents of three distinct chicken tumors. J Exp Med 1914a;19:52–69.

Rous P, Murphy JB. On immunity to transplantable chicken tumors. Journal of Experimental Medicine 1914b;20:419–42.

Chapter 21.5

Cellular Origin of Retroviral Oncogenes

1. THE PRIZE CITATION

The Nobel Prize in Physiology or Medicine 1989 was awarded jointly to J. Michael Bishop and Harold E. Varmus *for their discovery of the cellular origin of retroviral oncogenes.*

2. THE ORIGINAL OBSERVATIONS

As described by Harold Varmus in his Nobel lecture (Varmus, 1989a), a small group of investigators were working intensely in the 1960s and 1970s to unravel the secrets of Rous sarcoma virus, following their intuition that this obscure chicken tumor virus was a "Rosetta stone" that would yield important insights about the molecular basis of cancer. These workers shared hypotheses, data, and reagents that facilitated specific advances, and the following brief account will not cite all of the individual contributions or researchers. Furthermore, Bishop and Varmus worked as a closely knit team, but for the sake of simplicity this summary focuses on Varmus.

To appreciate the key experiment, some background information is important. In 1969, one critical hypothesis was that RSV genome included two functional components, a "virogene" that was necessary for the replication of the virus, and an "oncogene" that transformed normal chicken cells in culture and was responsible for tumor induction (Huebner and Todaro, 1969). Evidence for this hypothesis included the discovery of temperature-sensitive mutants that would replicate at the restrictive temperature but would only transform at the permissive temperature (Martin, 1970). The other key hypothesis was Howard Temin's proposal that genetic information for RSV existed as a "provirus"; a DNA copy of the RNA genome of RSV, which was maintained in infected cells and served as the template for production of new virus particles (and also accounted for the stable transformation phenotype). The discovery of reverse transcriptase independently in 1969 by Temin and Baltimore, in chicken and mouse retroviruses, respectively, confirmed Temin's provirus hypothesis (Temin and Mizutani, 1970; Baltimore, 1970). Like many seminal discoveries, it led to a series of new questions; in this instance, the scientific community was challenged to elucidate the detailed mechanism of reverse transcription.

Enter Bishop and Varmus. Armed with Varmus' prior experience using molecular hybridization in Ira Pastan's laboratory, they set out to characterize the steps involved in reverse transcription. Another goal was to identify and characterize the postulated "oncogene." Radiolabelled DNA probes that were transcribed from viral RNA using reverse transcriptase detected RSV genetic information in normal chicken cells, even though they were not producing infectious virus nor exhibiting the transformed phenotype. This made it difficult to follow the steps of infection in normal chicken cells and raised questions about the identity of the RSV sequences that they carried.

Varmus and Bishop used two new experimental approaches. (1) A key discovery that supported the evolving RSV hypotheses was the isolation of transformation-defective (td) mutants of RSV that would replicate but not transform cultured cells. (Vogt, 1971) (These mutants should be distinguished from Steve Martin's temperature-sensitive mutant described above.) The Vogt mutants were shown to have an RNA genome that was 15% shorter than transforming clones of RSV; they were presumed to include the "virogene" but lack the postulated "oncogene." These mutants provided a key reagent for hybridization studies, and made it possible to construct two probes, one specific for the virogene and one specific for the oncogene. To obtain a molecular probe specific for the oncogene, they prepared a probe from transcripts of the whole RSV genome, and then used the RNA from the td mutants to "subtract" those fragments that hybridized to the virogene, yielding a probe that was specific for the oncogene. (2) They began to use cells from other avian species (quail, duck, emu) that were negative for RSV sequences, but could be infected with RSV. This system was used to follow the molecular steps in reverse transcription using probes specific for virogenes and oncogenes.

All such experiments need to be anchored by various negative controls (Figure 5). In this instance, the oncogene probe was shown to be specific because it would hybridize with RSV RNA from a transforming isolate but not with RNA from the transformation-defective mutant virus RNA. Both probes hybridized well to normal chicken DNA, confirming prior results (and indicating that normal uninfected chickens carry genomic information for both virogene and oncogene, although not exhibiting a transformed phenotype or producing infectious RSV). So far so good. When the probes were tested against DNA from other avian species, the td probe was negative, indicating that these species did not carry ancestral fragments of RSV in their genomes. However, and unexpectedly, the oncogene probe hybridized robustly to DNA from all of these avian species (Stehelin et al., 1970).

What to make of this unexpected result, a control that was predicted to be negative but was positive instead? Here is where Bishop and Varmus introduced a game-changing hypothesis. But let Varmus speak for them both (Varmus, 1989a).

"From these findings, we drew conclusions that seem even bolder in retrospect, knowing they are correct, than they did at the time (Stehelin, Varmus, Bishop, 1976). We said that the RSV transforming gene is indeed represented in normal cellular DNA, but not in the form proposed by the virogene-oncogene hypothesis. Instead, we argued, the cellular homolog is a normal cellular gene, which was introduced into a retroviral genome in slightly altered form during the genesis of RSV. Far from being a noxious element lying in wait for a carcinogenic signal, the progenitor of the viral oncogene appeared to have a function valued by organisms, as implied by its conservation during evolution. Since the viral src gene allows RSV to induce tumors, we speculated that its cellular homolog normally influenced those processes gone awry in tumorigenesis, control of cell growth or development."

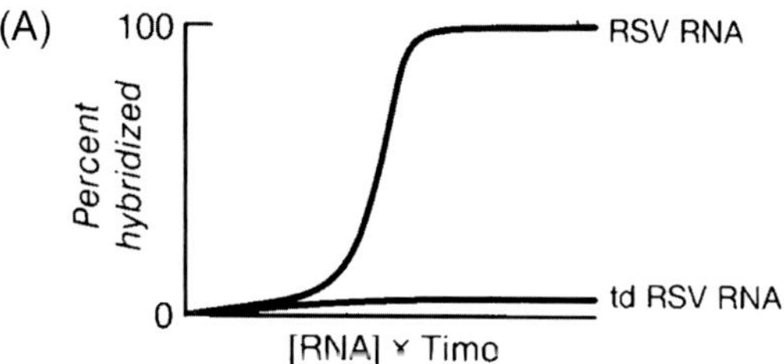

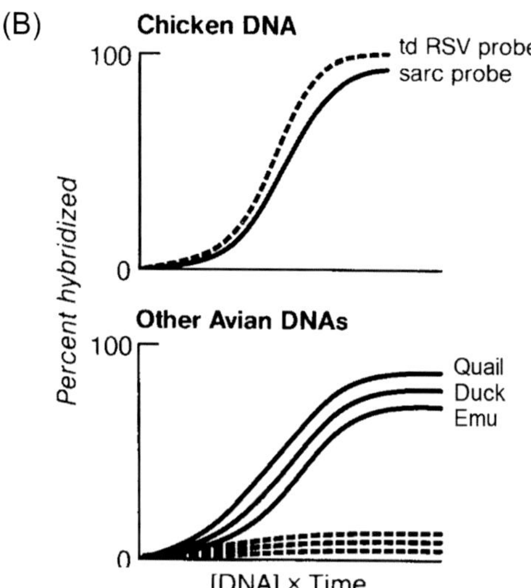

FIGURE 5 Summary of experiment with the oncogene (sarc) probe. This figure shows the results of hybridization experiments done with two radiolabeled DNA probes, one for a transformation-defective (but replication competent) mutant of Rous sarcoma virus (RSV) (td RSV or "virogene" probe) and one for the transforming or "oncogene" (sarc probe). Panel A. The sarc probe is specific for the putative "oncogene" since it will hybridize to RSV RNA but not to RNA from a td mutant of RSV. Panel B, top. Both probes hybridize to normal chicken DNA, indicating that normal chickens carry genomic sequences for both virogene and oncogene of RSV (although these cells do not produce infectious RSV or show a transformed phenotype). Panel B, bottom. The virogene probe (td) does not hybridize to DNA from several avian species, indicating that they do not carry copies of RSV in their genome. However, unexpectedly, the oncogene (sarc) probe does hybridize. The investigators postulated that this was because viral oncogenes are perturbed copies of normal cellular genes that are being detected by the sarc probe. *From Varmus (1989a), with permission.*

3. SIGNIFICANCE FOR VIRAL PATHOGENESIS

The hypothesis that viral oncogenes were perturbed versions of normal cellular genes involved in growth and development opened a new vista on the mechanisms of carcinogenesis (Bishop, 1989). The identification of the sarc protein as the product of the RSV oncogene (v-src) and its cellular counterpart (the product of c-src) confirmed the hypothesis (Brugge and Erikson, 1977; Collett et al., 1978), which was quickly accepted by the scientific community. The 1976 paper by Stehelin, Varmus, and Bishop led to an explosion of work that eventually uncovered a large array of oncogenes that fell into a number of distinct functional categories, depending upon their mechanism of action (Bishop, 1989). The impact of their discovery was reflected in the launch of new journals dedicated solely to oncogenes. Furthermore, it stimulated an intensive investigation of the normal cell cycle, since many oncogenes acted on specific steps in the cycle, resulting in the removal of normal "brakes" on cell proliferation.

In summary, the discovery of cellular "proto-oncogenes" initiated our understanding of the molecular basis of carcinogenesis, a field that continues to yield important insights as well as providing a rich lode to mine for potential therapies.

4. BIOSKETCH: HAROLD E. VARMUS (1939–PRESENT)

Harold Varmus was born into a middle-class Jewish family on the south shore of Long Island, in New York State (Varmus, 1989b). His father was a physician in general practice and his mother a social worker; both had been trained in elite institutions. After a comfortable high school education, Varmus was educated at Amherst College in Massachusetts. Following an abortive trial as a graduate student in English, he went to medical school at Columbia University in New York City; during medical school he was stimulated by lectures in basic medical sciences. After a residency at Columbia-Presbyterian Hospital, facing a draft obligation, he became a "yellow beret" and joined Ira Pastan's laboratory at the U.S. National Institutes of Health (NIH) as a clinical fellow. Here he had his "first serious exposure to laboratory science" working on bacterial gene regulation for 2 years. This led him to seek further research training and he made the critical decision to take a postdoctoral fellowship in tumor virology with Mike Bishop at the University of California in San Francisco (UCSF). Here he collaborated with Bishop on the seminal studies that led to the recognition that viral oncogenes were related to cognate host genetic elements.

Varmus was a very productive investigator at UCSF, working on various problems in molecular and tumor virology from 1969 to 1993, when he was recruited to be the Director of the NIH. He has continued to play a leadership role in science administration, first at NIH (1993–1999), then as President of the Sloan Kettering Cancer Institute (2000–2010), and more recently as Director of the National Cancer Institute, again at NIH (2010–2015). He has leveraged his Nobel Prize to make notable contributions as a leader of the scientific establishment in the United States.

REFERENCES

Baltimore D. RNA-dependent DNA polymerase in virions of RNA tumour viruses. Nature 1970;226:1209–11.

Bishop JM. Nobel Prize lecture. http://www.nobelprize.org/nobel_prizes/medicine/laureates/1989/bishop-lecture.html; 1989.

Brugge JS, Erikson RL. Identification of a transformation-specific antigen induced by an avian sarcoma virus. Nature 1977;269:346–8.

Collett MS, Brugge JS, Erikson RL. Characterization of a normal avian cell protein related to the avian sarcoma virus transforming gene product. Cell 1978;15:1363–9.

Huebner RJ, Todaro GJ. Oncogenes of RNA tumor viruses as determinants of cancer. Proceedings of the National Academy of Sciences (USA) 1969;64:1087–94.

Martin GS. Rous sarcoma virus: a function required for the maintenance of the transformed state. Nature 1970;227:1021–3.

Stehelin D, Varmus HE, Bishop JM, Vogt PK. DNA related to the transforming gene(s) of avian sarcoma viruses is present in normal avian DNA. Nature 1970;260:170–3.

Temin HM, Mizutani S. RNA-dependent DNA polymerase in virions of Rous sarcoma virus. Nature 1970;226:1211–3.

Varmus HE. Nobel prize lecture. http://www.nobelprize.org/nobel_prizes/medicine/laureates/1989/varmus-lecture.html; 1989a.

Varmus HE. Harold varmus biography. http://www.nobelprize.org/nobel_prizes/medicine/laureates/1989/varmus-bio.html; 1989b.

Vogt PM. Spontaneous segregation of nontransforming viruses from cloned sarcoma viruses. Virology 1971;46:939–46.

Chapter 21.6

Cell-Mediated Immune Defense

1. THE PRIZE CITATION

The Nobel Prize in Physiology or Medicine 1996 was awarded jointly to Peter C. Doherty and Rolf M. Zinkernagel *for their discoveries concerning the specificity of the cell-mediated immune defense.*

2. THE ORIGINAL OBSERVATIONS

Peter Doherty and Rolf Zinkernagel were both postdoctoral fellows in the Department of Microbiology in the John Curtin School of Medical Research at the Australian National University in the early 1970s (Doherty, 1996b). At this time, the department was a leader in research on viruses and immunological mechanisms. Immune responses to viral infections presented an opportunity to study two questions which were at the forefront of immunobiological research at this time: First, how did the host animal distinguish between "self" (its own potentially antigenic molecules) and "nonself" (such as the tissues of another individual or the molecules of an invading parasite). Second, how did T lymphocytes attack target cells, which was clearly different from antibody-mediated immunity. Both questions bore on the enigmatic mechanisms of cellular immunity.

Doherty "inherited" from Cedric Mims the lymphocytic choriomeningitis virus (LCMV) as a model in mice. As a background to what follows, it is useful to know a bit about LCMV, which causes a unique spectrum of infections in mice, depending upon both virus and host variables. When injected intracerebrally into adult mice, some isolates of LCMV cause an acute CNS infection in which the mice develop fatal epileptic attacks. The seizures are due to an acute inflammation of the tissues that envelope the brain (choroid plexus and meninges). Doherty wished to use the LCMV model to pursue his investigations on the pathogenesis of viral infections of the CNS, an interest he had developed during his PhD research at the Moredun Institute in Scotland. To study inflammatory cells in the CNS, he had adopted a technique to obtain microliters of cerebrospinal fluid (CSF) from mice for examination.

Zinkernagel, coming from Lausanne, had learned the method developed by Brunner and Cerottini for the use of ^{51}Cr release as an in vitro assay for cell-mediated immunity. They teamed up to combine their methods—assaying CSF cells for their ability to lyse LCMV-infected targets—and soon were able to show that mice infected with LCMV had virus-specific cytolytic cells in their CSF, likely an important mediator of the fatal convulsive disorder (Zinkernagel and Doherty, 1973).

In retrospect, they were lucky in their original experiments because they happened to use "target" cells in their assay that expressed the same MHC haplotype (k) as the mice they had infected. (At this time they were using L-929 cells as their targets in the CTL assay, and these cells had been derived from mice that expressed the H-2^k haplotype.) Stimulated by the report by Oldstone and collaborators (1973) that different haplotypes of inbred mice differed in their susceptibility to LCMV, they decided to check whether there was a correlation between CTL activity in the CSF and susceptibility to LCMV disease. Using mice of haplotypes k and d, they observed that both inbred strains of mice were equally susceptible to LCMV disease, but only CSF cells from mice of the k haplotype (H-2^k) registered positive in the CTL test. (As noted above, this reflected the k haplotype of the L-929 target cells.)

This unpredicted discorrelation between susceptibility to disease and the CTL assay led them to consider several alternatives, one of which was "that our test was in some way inadequate." They quickly did experiments that showed that all mice that expressed the H-2^k haplotype generated effective CTLs. At this point they made an intuitive leap, and did the critical experiment (Figure 6), to test the hypothesis that haplotype compatibility was required for function of cytotoxic T lymphocytes (Zinkernagel and Doherty, 1974a). As presented in Figure 6, the crisscross experiment, with mice and target cells that expressed two different haplotypes (k and d), showed that CTL activity required haplotype compatibility. Also importantly, when heterozygotes that expressed both k and d were used, they also were functional, indicating that there was a requirement for compatibility—not a restriction by noncompatible haplotypes.

Zinkernagel and Doherty and their colleagues immediately realized that these results had important implications for the mechanisms of cellular immunity (Zinkernagel, 1996). Experiments by other investigators with different viral systems soon confirmed the generalizability of their

Table 1. Experiments demonstrating MHC-restriction specificity and positive selection in the thymus for antiviral cytotoxic T-cells.

	Stem cells (MHC, H-2)	Thymus (MHC, H-2)	Other host cells (MHC, H-2)	Virus-specific CD8$^+$ T-cells specific for (MHC, H-2)		Lethal CD8$^+$-T-cell-mediated chorio-meningitis after i.c. infection
I	*Original experiments with normal mice*			k	d	
1.	k	k	k	+	–	100 %
2.	b	b	b	–	–	100 %
3.	d	d	d	–	+	100 %
4.	k/d	k/d	k/d	+	+	100 %

FIGURE 6 Evidence that cytotoxic T lymphocytes are "restricted" by MHC (major histocompatibility complex or H-2) haplotype. Inbred mice that carried the k or d haplotype or heterozygote mice that expressed both k and d haplotype, were infected with LCMV (lymphocytic choriomeningitis virus) and their spleen cells were harvested for virus-specific CTLs (cytotoxic T lymphocytes). The MHC specificity of these effector T lymphocytes is shown in the columns labeled "Virus-specific CD8+ T cells." These T cells were tested for their ability to lyse target cells prepared from macrophages obtained from mice expressing the different haplotypes and infected in vitro with LCMV; the MHC specificity of the targets is shown in the left-hand three columns. As shown, T lymphocytes were effective against matched haplotype target cells (row 1 and 3). T cells from both k and d haplotype mice were able to lyse heterozygous k/d targets, indicating that the match was a requirement not a restriction (row 4). Targets prepared from mice expressing the H-2^b haplotype were not lysed by CTLs of either k or d haplotype, a necessary negative control (row 2). *Part of a table in Zinkernagel (1996), reproduced with permission.*

findings. What was unclear was the molecular basis that led to the requirement for MHC compatibility. Zinkernagel and Doherty (1974b) attempted to develop some ideas, expressed in diagrammatic form. With regard to "target" cells, they postulated (correctly) that viral antigens somehow were associated with the MHC complex to explain the dual requirement that target cells express both viral antigens and a specific MHC haplotype. In addition, the results required that the putative "receptor" on cytolytic T cells had to match the haplotype on the target cell; their diagram implied some type of reciprocal "lock and key" relationship between the receptor and the MHC molecule but could not be any more specific.

3. SIGNIFICANCE FOR VIRAL PATHOGENESIS

The finding that cytolytic T lymphocytes required an MHC match was a revelation that led to an explosion of work directed to dissecting the mechanisms of cellular immunity (Zinkernagel, 1996). These efforts culminated in the visualization of antigenic peptides in the "groove" of MHC molecules (Bjorkman et al., 1987; Stern and Wiley, 1994), which clearly demonstrated the molecular basis for the association of MHC and foreign antigen on target cells. The cognate T-cell receptors were also cloned, sequenced, and characterized as members of the immunoglobulin superfamily. Structural visualization (Garcia et al., 1996; Garboczi et al., 1996) indicated that the variable region of the T-cell receptor, like an immunoglobulin molecule, would bind to a large surface on the presenting MHC molecule that included both antigenic fragment and much of the adjacent MHC molecule itself (Baker et al., 2012). To generate a population of T cells bearing receptors cognate to the host's MHC repertoire, there is a selection—during ontogeny in the thymus—of T lymphocytes bearing the relevant TCR receptor molecules (Goldrath and Bevan, 1998).

The systematic elucidation of the molecular basis of cell-mediated immunity, initiated by the Zinkernagel–Doherty experiments, has provided a rational basis for understanding many aspects of viral pathogenesis as well as other immune-dependent pathological and clinical phenomena (Zinkernagel, 1996). One important example is the immunopathogenesis of LCMV infection itself. Variables that determine the outcome of both acute and persistent infection with LCMV include the nature and dynamics of the cellular immune response, variant isolates of the virus, and viral escape mutants. Furthermore, the LCMV model has proved a deft tool to explore the nature of T-cell ontogeny, memory, tolerance, and other important immunological phenomena. Insights into cellular immunity have also been central to understanding the pathogenesis of HIV/AIDS, and of the opportunistic infections that are its hallmark. Transplantation medicine, likewise, has depended upon manipulating cellular immune responses in both host and graft. The list is endless.

4. PETER DOHERTY (1940–PRESENT): BIOGRAPHICAL SKETCH

(Although both Doherty and Zinkernagel contributed equally to their Nobel-winning research, only one biographical sketch is presented.)

Peter Doherty grew up in the subtropical city of Brisbane, Australia, in a working-class family where there was a commitment to education as a route to a better life (Doherty, 1996a). After graduating from the local secondary schools, a chance or an opportunity led him to matriculate in the veterinary school of the University of Queensland. This was his first exposure to biological science and he soon developed an interest in research, but not in the practice of veterinary medicine. Following his professional training, he was required to serve in the Queensland Department of Agriculture. During his service, he spent time in several diagnostic laboratories, which honed his interest in research. Following a lead, he applied for a research fellowship at the Moredun Institute in Edinburgh, where he did a PhD project on the neuropathogenesis of louping ill, a tick-borne flavivirus infection of sheep in Scotland. This focused his interest in viral pathogenesis and immunity, and he returned to Australia to pursue further training in the Department of Microbiology in the John Curtin School of Medical Research at the Australian National University in Canberra.

In Canberra, Doherty decided to use LCMV infection of mice as a model to study viral infections of the CNS. When Rolf Zinkernagel arrived as a postdoctoral fellow to train with Bob Blanden, they teamed up to use their different laboratory skills to study the cellular immune response to LCMV. This quickly led to positive results, and their breakthrough findings, as described above.

After four very productive years in Canberra, in 1975 Doherty moved on, to accept an Associate Professorship at the Wistar Institute in Philadelphia, with the opportunity to focus on experimental research. During seven very productive years, he became established as an academic research scientist. Then he received a call to return to Canberra, as Head of the Department of Experimental Pathology, where he served from 1982 to 1988. For various "political" reasons, this did not work out well, and Doherty returned once more to the United States, to the St. Jude's Children's Research Hospital in Memphis, Tennessee. This institution offered a strong infrastructure, extensive research funding, and the opportunity to collaborate with other virologists and immunologists. Once again, Doherty thrived, focusing mainly on viral infections of the respiratory system. In recent years, Doherty has split his time between St. Jude's where he maintains a laboratory and the Department of Microbiology and Immunology in the University of Melbourne. He has also become a senior advisor to many scientific organizations in Australia.

As he comments in his Nobel biography, Peter Doherty has had an unconventional career, spanning three continents and many different institutions. But his love has always been experimental research, which he has practiced with outstanding success in these diverse environments.

REFERENCES

Baker BM, Scott DR, Blevins SJ, Hawse WF. Structural and dynamic control of T-cell receptor specificity, cross-reactivity, and binding mechanism. Immunological Reviews 2012;250:10–31.

Bjorkman PJ, Saper MA, Samraoui B, Bennett WS, Strominger JL, Wiley DC. The foreign antigen binding site and T-cell recognition regions of class I histocompatibility antigens. Nature 1987;329:512–8.

Garboczi DN, Ghosh P, Utz U Fan QR, Biddison WE, Wiley DC. Nature 1996;384:134–41.

Garcia KC, Degano M, Stanfield RL, et al. Science 1996;274:209–19.

Oldstone MB, Dixon FJ, Mitchell GF, McDevitt HO. Journal of Experimental Medicine 1973;137:1201–12.

Doherty PC. Nobel biographical. http://www.nobelprize.org/nobel_prizes/medicine/laureates/1996/doherty-bio.html; 1996a.

Doherty PC. Nobel prize lecture. http://www.nobelprize.org/nobel_prizes/medicine/laureates/1996/doherty-lecture.html; 1996b.

Goldrath AW, Bevan MJ. Selecting and maintaining a diverse T-cell repertoire. Nature 1998;402:255–62.

Stern LJ, Wiley DC. Antigenic peptide binding by Class I and class II histocompatibility proteins. Current biology. Structure 1994;2:245–51.

Zinkernagel RM. Nobel prize lecture. http://www.nobelprize.org/nobel_prizes/medicine/laureates/1996/zinkernagel-lecture.html; 1996.

Zinkernagel RM, Doherty PC. Journal of Experimental Medicine 1973;138:1266–9.

Zinkernagel RM, Doherty PC. Restriction of in vitro T cell-mediated cytotoxicity in lymphocytic choriomeningitis within a syngeneic or semiallogeneic system. Nature 1974a;248:701–2.

Zinkernagel RM, Doherty PC. Immunological surveillance against altered self components by sensitized T lymphocytes in lymphocytic choriomeningitis. Nature 1974b;251:547–8.

Chapter 21.7

Innate Immunity

1. THE PRIZE CITATION

The Nobel Prize in Physiology or Medicine 2011 was divided, one half jointly to Bruce A. Beutler and Jules A. Hoffmann *for their discoveries concerning the activation of innate immunity* and the other half to Ralph M. Steinman *for his discovery of the dendritic cell and its role in adaptive immunity.*

2. THE ORIGINAL OBSERVATIONS

Beutler's contributions were built on a background of research done by many investigators over almost a century (1880s–1980s), truly an example of the incremental nature of biomedical discovery (Beutler, 2011). In the 1880s, Richard Pfeiffer, working with Robert Koch (the discoverer of the causes of tuberculosis, anthrax, and cholera), found that an injection of heat-killed *Vibrio cholerae* killed guinea pigs. He suggested that the vibrio organism contained a "principle" of unknown nature, which he called "endotoxin." Subsequent research by many workers showed that endotoxin was a substance common to many Gram-negative bacteria, and identified it as a glycolipid that was a major structural component of the outer bacterial membrane. A long series of investigations resulted in the purification of the glycolipid, showed that it was a lipopolysaccharide (LPS), and—in the 1980s—characterized its chemical structure. In addition to the toxic effects, LPS was also found to act as a very potent adjuvant that enhanced the immunogenicity of protein antigens.

In parallel work, during the period 1960–1980, a few inbred mouse strains were shown to resist the toxic effects of LPS. The resistant mice were sensitive to other toxic microbial products, and were closely related to congenic inbred mice that were sensitive to LPS. From these observations it was inferred that there was a single receptor for LPS, which was "inactivated" by a putative mutation in the resistant mice. In a complex set of experiments with many strains of inbred and backcrossed mice, a locus on mouse chromosome 4 was identified to encode the postulated receptor (Watson et al., 1978). However, this locus spanned a very large segment of the chromosome and defied easy identification. Experiments with the LPS-resistant mice showed that they were highly susceptible to infections with Gram-negative bacteria, compared to other mice including congenic control animals. This observation led to the critical inference that responsiveness to LPS played an important role in resistance to infection, and that the LPS "pathway" might have major biological significance. Notably, this concept was somewhat counterintuitive, since it proposed that normal mice were "protected" from Gram-negative bacterial infections by their sensitivity to a Gram-negative bacterial toxin.

Enter Bruce Beutler, through a "side door." Following residency, Beutler did a research fellowship at the Rockefeller University. There he was given the project to isolate cachectin. At that time, cachectin was the name given to a "factor," which was secreted by macrophages upon exposure to LPS. Injected in mice, this factor induced cachexia, the wasting phenomenon that accompanied selected chronic diseases. For experimental purposes, cachectin was produced by cultured mouse macrophages that were treated with LPS. Following the implicit assumption that cachectin was a single molecule,

Beutler developed several protocols to purify the factor, and succeeded in identifying a single small protein whose N terminal sequence could be determined by Edman degradation. The sequence was found to resemble that of human tumor necrosis factor (TNF), which had just been isolated by other investigators. It appeared that cachectin was the mouse ortholog of human TNF. As evidence that the newly isolated molecule was a biological mediator of LPS toxicity, Beutler (1985) showed that antibodies against the purified cachectin would provide considerable protection against the lethal effect of LPS in mice (Figure 7).

TNF has an interesting history. In the early twentieth century William Coley, a surgeon, had noted that certain human cancers underwent remissions if the patients happened to undergo a Gram-negative infection. Coley therefore had—without consistent success—used killed Gram-negative bacteria in an attempt to treat patients with inoperable tumors. In the 1970s, Lloyd Old, working at the Sloan Kettering Cancer Research Institute, followed up on Coley's work. Old used mouse models to validate the principle discovered by Coley. He showed that LPS—a product of Gram-negative bacteria—would induce the secretion of a serum factor that could produce necrosis of certain transplantable mouse tumors (Carswell et al., 1975). He had purified this activity, which he dubbed TNF.

At that point, in the mid-1980s, these independent lines of research began to converge, revealing the elements of an important biological pathway. It appeared that the LPS component of the cell walls of Gram-negative bacteria stimulated the infected host to secrete TNF. In turn, TNF acted as a "master" cytokine, inducing many other cytokines and cellular activities, which together mounted a defense against the bacterial invader. Blockade of either LPS- or TNF-rendered mice highly susceptible to infection with Gram-negative bacteria as well as intracellular bacteria such as the tubercle bacillus. As stated by Beutler (2011) "TNF thus behaved as a clear executor of innate immunity." But this emerging view raised questions about the rest of the pathway, including the identity of the putative LPS receptor, where the innate immune response was initiated.

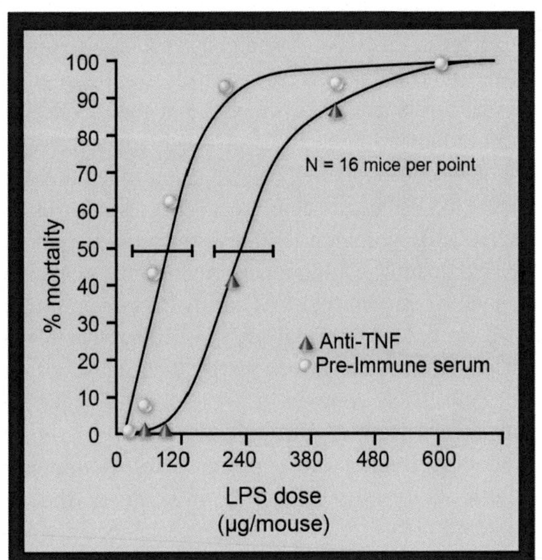

FIGURE 7 Evidence that tumor necrosis factor (TNF) is an important mediator of the lethal effects of LPS (lipopolysaccharide). The graph shows two dose–response curves, where mortality is plotted against graduated doses of LPS. Mice treated with an anti-TNF antiserum showed a marked reduction in mortality. *Copied from Beutler et al. (1985), with permission.*

Now began the hard work. Beutler determined to identify the gene encoding the LPS receptor. This presented a daunting challenge in the 1990s, since they were searching a large area of chromosome 4 in the absence of a sequence of the mouse genome. For 5 years (1993–1998), using then-current state-of-the-art gene cloning and sequencing methods, his team slogged on. In particular, they used the EST (expressed sequence tag) database to identify potential functional genes. (The EST database is a set of DNA sequences that were derived by reverse transcription of mRNA sequences. Therefore, they can be used to distinguish potentially functional genes from pseudogenes, introns, and "junk" DNAs.) Finally, they were rewarded (Poltorak et al., 1998) when they identified a gene that was homologous to Tlr4 (toll-like receptor 4).

Tlr4 was originally described by Jules Hoffman as a drosophila gene that played a role in host defense against fungal infection (Lemaitre et al., 1996; Hoffman, 2011). (Hoffman shared half of the Nobel Prize with Beutler.) Furthermore, Tlr4 was a component of the scheme for innate immunity that had been proposed by Charles Janeway (1989), in which the host constitutively expressed "pattern recognition receptors" (PRRs) on cell surfaces that would recognize pathogen-associated molecular patterns (PAMPs) or molecular components of invading parasites. Recognition in turn would trigger an "innate" defense response on the part of the host (Medzhitov, 2009). LPS (a PAMP) and TLR4 (a PRR) are now considered to be key molecules in one pathway in the innate immune system.

3. SIGNIFICANCE FOR VIRAL PATHOGENESIS

The identification of an LPS receptor had several immediate repercussions. First, it gave credence to Janeway's concept by identifying specific molecules involved in the innate recognition system. Also, this example explained important human disease observations, both with regard to bacterial infection and tumor suppression. Furthermore, by identifying a homolog of a drosophila gene, it supported the concept that innate immunity was an ancient system widely used by organisms across the biological spectrum. In addition, the activity of LPS as an adjuvant offered a conceptual bridge between innate responses and the induction of acquired immunity. (Synergistic to these insights was Ralph Steinman's discovery of the dendritic cell, which provided a cellular mechanism to explain the linkage between the two systems of defense.) Finally, the recognition of innate immunity as a second important system of host defenses opened the way for practical applications in both prevention and treatment of infectious and—possibly—other important human diseases. It is no surprise that Beutler's seminal paper (Poltorak et al., 1998) has had a vast number of citations (Beutler, 2011).

The discovery of the LPS/TLR4 pathway quickly led to the identification of a number of innate immunity pathways that utilize toll-like receptors 1–10. A number of these TLRs respond to viral ligands as PAMPs. As described in Chapter 4, these ligands include viral envelope proteins, single- and double-stranded viral RNAs, and CpG motifs. Beutler's research on bacterial LPS thus led to the recognition of a broad system of innate immunity that includes the first line of defense against viral invaders.

There is a counterintuitive aspect to the foregoing story. How can LPS simultaneously stimulate innate host defenses and act as a potentially lethal molecule? Dose matters. Modest amounts of LPS activate the innate system, but large amounts "over"stimulate TNF and other mediators to create a cytokine storm, characterized by increased capillary fragility leading to septic shock, destructive endothelial damage, and disseminated intravascular coagulation. Several lessons here: it's the host as well as the virus that determines the outcome of an infection; potent host defenses when excessive can cause disease; and the dynamics of the virus–host interaction are critical.

4. BIOSKETCH: BRUCE A. BEUTLER (1957–PRESENT)

In contrast to some future Nobel laureates, Bruce Beutler was clearly destined to be an experimental biologist from his earliest days. He was brought up in a successful professional family, and was close to his father who was a well-esteemed academic physician and a leader in clinical hematology. He was a somewhat precocious student, and entered college and medical school at an early age, graduating from the University of Chicago School of Medicine at age of 23 years, the youngest member of his class. He did 2 years of clinical training at the University of Texas Southwestern Medical Center in Dallas, which convinced him that he did not have a calling for clinical work and strengthened his resolve to train in the life sciences. Along the way, he had a number of experiences in research laboratories, and had already acquired a strong background in genetics, immunology, and protein biochemistry.

Beutler then moved to do a fellowship with Anthony Cerami at the Rockefeller University. There he had his first early success, isolating the protein cachectin from the supernates of macrophages that had been stimulated by LPS. He then accepted an appointment to return to Dallas as an HHMI (Howard Hughes Medical Institute) investigator, where he remained from 1985 to 2000. It was there that his group struggled and finally succeeded in identifying the gene that encoded the receptor for LPS, the mouse homolog of the drosophila gene, Tlr4. One striking aspect of Bruce Beutler's scientific career is his willingness to persevere against considerable obstacles in pursuit of a daunting challenge. This is a theme common to a number of Nobel laureates.

In 2000, Beutler moved to the Scripps Research Institute, but returned in 2011 to the University of Texas Southwestern Medical Center to set up a center to study the genetics of host defenses against infection. He continues to be an active investigator to this day.

REFERENCES

Beutler B. Nobel prize lecture. http://www.nobelprize.org/nobel_prizes/medicine/laureates/2011/beutler-lecture.pdf; 2011.

Beutler B, Milsark IW, Cerami A. Passive immunization against cachectin/Tumor Necrosis Factor (TNF) protects mice from the lethal effect of endotoxin. Science 1985;229:869–71.

Carswell EA, Old LJ, Kassel RL, Green S, Fiore N, Williamson B. An endotoxin-induced serum factor that causes necrosis of tumors. Proc Natl Acad Sci USA 1975;72:3666–70.

Hoffman JA. Nobel prize lecture. http://www.nobelprize.org/nobel_prizes/medicine/laureates/2011/hoffman-lecture.pdf#search='jules+hoffman'; 2011.

Janeway Jr CA. Approaching the asymptote? Evolution and revolution in immunology. Cold Spring Harb Symp Quant Biol 1989;54:1–13.

Lemaitre B, Nicolas E, Michaut L, Reichhart JM, Hoffmann JA. The dorsoventral regulatory gene cassette spätzle/Toll/cactus controls the potent antifungal response in Drosophila adults. Cell 1996;86:973–83.

Medzhitov R. Approaching the asymptote: 20 years later. Immunity 2009;30:766–75.

Poltorak A, He X, Smirnova I, et al. Defective LPS signaling in C3H/HeJ and C57BL/10ScCr mice: mutations in Tlr4 gene. Science 1998;282:2085–8.

Watson J, Kelly K, Largen M, Taylor BA. The genetic mapping of a defective LPS response gene in C3H/HeJ mice. J Immunol 1978;120:422–4.

Chapter 21.8

Discovery of the Dendritic Cell

1. THE PRIZE CITATION

The Nobel Prize in Physiology or Medicine 2011 was divided, one half jointly to Bruce A. Beutler and Jules A. Hoffman *for their discoveries concerning the activation of innate immunity* and the other half to Ralph M. Steinman *for his discovery of the dendritic cell and its role in adaptive immunity.*

2. THE ORIGINAL OBSERVATIONS

First, some background information (Nussenzweig, 2011). In 1967, Mishell and Dutton published a cell culture model to study immune responses, which permitted a dissection of the cells involved. In the Mishell–Dutton system, antigens were added to cultured lymphocytes and antigen-specific antibody production was assayed as the output. It was found that—in addition to lymphocytes—an additional "accessory cell" was required. This unknown cell was characterized by its adherence to glass and was presumed to present antigen to the lymphocytes.

In 1970 Ralph Steinman, on completion of medical school and residency, joined Zanvil Cohn's laboratory at the Rockefeller University as a postdoctoral fellow. Cohn had taken the study of macrophages as his central interest, and one question was their potential role as "accessory cells" in initiating the immune response. Most workers believed that the accessory antigen-presenting cells were macrophages, in view of their well-established ability to take up a variety of molecules and microbes. However, Cohn had his doubts, since he had shown that when macrophages captured proteins, they were degraded to their constituent amino acids. He put Steinman to work to test the macrophage hypothesis. Using horse radish peroxidase as a test antigen, they failed to detect it on the surface of macrophages where it should have been if macrophages were the sought-for antigen-presenting accessory cell.

They decided to look for the accessory cell by sorting for glass-adherent cells among a suspension of cells obtained from the spleens of mice. When they used the phase and electron microscope to look at the cells obtained in this manner, they were struck with differences from familiar macrophages. The adherent cells lacked vacuoles and ruffled membranes (characteristics of macrophages) and had tree-like processes that were actively mobile. At this point, Steinman and Cohn took a bold leap and asserted that they had discovered a new cell type which Steinman named "dendritic cells" after their elongated tree-like processes (Steinman and Cohn, 1973).

The identification of a putative accessory cell was only a hypothesis that required testing. As a first step, Steinman developed a method for purifying dendritic cells which could be tested for their functionality (Steinman and Cohn, 1974). This laborious and cumbersome method only yielded a small number of cells and was difficult to reproduce; as a result Steinman's experiments were not replicated by other workers for many years.

To test purified dendritic cells for their functionality, Steinman adapted the mixed leukocyte reaction (MLR), which had been developed to test for graft acceptance or rejection. It involved mixing white cells from two donors and determining whether or not they stimulated uptake of radioactive thymidine; uptake implied that each donor recognized the other as "nonself" and that a graft between them would be rejected. Steinman and colleagues set up an MLR using either unpurified white blood cells or purified dendritic cells from mice of one haplotype to stimulate T lymphocytes from mice of a different haplotype. They demonstrated that dendritic cells were about 100 times more effective than a mixture of white blood cells from the same donor (Figure 8). Since dendritic cells were only 1–2% of the total white blood cell population, they inferred that these were the major antigen-presenting cells in the MLR (Steinman and Witmer, 1978).

To quote Michel Nussenzweig's Nobel lecture on behalf of Ralph Steinman: "However, as William Paul pointed out in his commentary in Cell on the 2011 Nobel Prize in Physiology and Medicine: 'This report was initially received with some skepticism, based on the widely held view that the major antigen presenting cells were the far more numerous macrophages and on the uncertainty that many immunologists had about the assay that Steinman and Cohn used to establish the function of their dendritic cells'" (Paul, 2011).

Undeterred, Steinman persisted. He and his colleagues developed a more conventional in vitro immune response system, using haptenated cellular immunogens

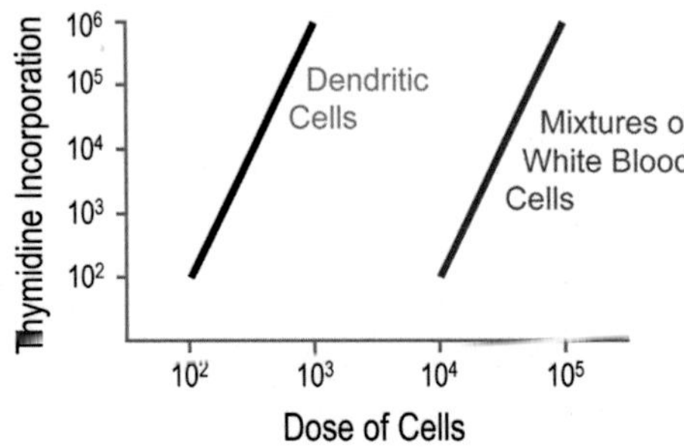

FIGURE 8 Experiment to show that dendritic cells are potent stimulators of the mixed leukocyte reaction. The graph shows a comparison of the stimulatory activity of dendritic cells and unfractionated spleen cells. *Reproduced from Steinman and Witmer (1978), with permission.*

(Nussenzweig et al., 1980; Shearer, 1974). In this syngeneic culture system, spleen cells were passed through Nylon wool to remove accessory cells, leaving T lymphocytes (potential responder cells). Stimulator cells were produced by incubation of a subset of spleen cells with trinitrobenzenesulfonate, which chemically modifies cellular surface proteins by addition of trinitrophenyl (the hapten), and were added to the culture. After 5 days of incubation, haptened ^{51}Cr-labeled spleen cells were added as targets, and released radioactivity was measured. This culture system could be assayed for the efficacy of different sources of added accessory cells, since hapten-specific cytolysis did not occur in their absence. In this system, purified dendritic cells were much more active than macrophages, leading to the statement "dendritic cells are the critical accessory cells, whereas macrophages regard-less of source or Ia (MHC II) are without significant activity" (Nussenzweig et al., 1980).

As a next critical step, the Steinman laboratory developed a monoclonal antibody that was specific for dendritic cells (Nussenzweig et al., 1982). This antibody could be used for depletion experiments, which showed that elimination of dendritic cells—even though these were only about 1% of the cell population—would markedly reduce T-cell responses. Also, it was now possible to visualize dendritic cells in mouse spleen and show that they were located in lymph nodes at the morphological nexus of antigen processing, as well as at mucosal surfaces where foreign microbes and other antigens might first be detected.

The next significant contribution from the Steinman laboratory was the concept that dendritic cells exist in two states, immature (steady state or inactive) and mature (activated). Immature dendritic cells cannot present antigen effectively. They "lie in wait" for signals from the innate immune system (that they receive through their toll-like receptors), which trigger their conversion to the activated state. Once activated, dendritic cells become effective as antigen presenters, associated with an increase of MHC Class II (major histocompatibility complex, the antigen-presenting molecular complex) molecules on their surface. Also, they become migratory and can carry antigens to lymph nodes where they interface with T cells. Mature dendritic cells also up-regulate costimulatory molecules on their surface, which activate the responding T cells carrying cognate antigen receptors.

Finally, Steinman's group found that dendritic cells play a role in immune "tolerance." Tolerance—the failure of "self" antigens (the host's own tissues) to induce a deleterious autoimmune response—has always been an enigmatic phenomenon. When immature quiescent dendritic cells capture self antigens in the absence of an activation signal, the cognate T cells fail to respond, either by cell death, silencing, or conversion into regulatory T cells (Hawiger et al., 2001).

3. SIGNIFICANCE FOR VIRAL PATHOGENESIS

First, it is important to point out that Ralph Steinman not only discovered and named the dendritic cell, but persevered for about 20 years (1973–1990s) to build experimental evidence for the reality and function of those cells in the face of skepticism from much of the community of research immunologists. A significant lesson.

The dendritic cell is now recognized to act as a bridge between the innate and acquired immunological systems. Dendritic cells require activation through their pattern recognition molecules (such as toll-like receptors), an essential component of the innate system, in order to initiate antigen presentation to T cells, a first step in the generation of acquired immunity. Furthermore, it appears that the dendritic cell plays a significant role in initiating tolerance. Both of these pathways are relevant to the immune response elicited by virus invaders. During primary infection with a virus, the acquired response is potentially capable of controlling the infection before it progresses to a lethal outcome, and is also required to clear the virus infection. The speed with which dendritic cells initiate the acquired response is critical in this race between virus replication and evolving host defense. Furthermore, for selected viruses, infection in utero or the neonatal period can lead to "tolerance" and persistent infection, while in adult animals the same virus elicits a brisk effective immune response. Dendritic cells play a key role in this "aberrant" response to foreign antigens.

4. BIOSKETCH: RALPH M. STEINMAN (1943–2011)

Ralph Steinman was born in Montreal to a comfortable middle-class Jewish family and developed an interest in science in his early age (Nussenzweig and Steinman, 2011). He was educated at McGill University and Harvard Medical School, during which he took a year out as a laboratory research fellow in cell biology. Following his clinical residency, he moved to the Rockefeller University in 1970 as a research fellow with Zanvil Cohn, where he spent the rest of his career. In 1973, during investigations on the role of macrophages on immune function, Steinman and Cohn discovered dendritic cells, which they first recognized through their unique morphological characteristics. For the remainder of his career, Steinman explored the functions of dendritic cells with the collaboration of many trainees and other colleagues. For about 20 years after their discovery, the importance of dendritic cells was underappreciated by the immunological community, but Steinman—unfazed—continued to pursue his studies. Eventually, his research and that of others led to the recognition of the key role of dendritic cells in antigen presentation and the initiation of the adaptive immune response.

In 2007, Steinman was diagnosed with pancreatic cancer but maintained an active professional life until his death in 2011. In an unprecedented happening, he died 3 days before his Nobel Prize was announced. However, the Nobel committee, which did not know of his death at the time, still conferred the prize, to the general approbation of the scientific community. His family donated the full amount of his prize to charitable causes.

5. CONCLUDING REMARKS

These preceding vignettes are interesting for the scientific stories they tell. In addition, they carry some significant lessons which are worth highlighting.

- Most major "discoveries" worthy of a Nobel Prize often create new paradigms and may change existing widespread scientific viewpoints. In that sense, these discoveries differ from the incremental impact of most biomedical research.
- A number of discoveries arise from unexpected experimental results, and require recognition by a "prepared mind" as famously stated by Pasteur. In some instances, they required an intuitive intellectual leap on the part of their discoverer.
- Many important discoveries were at first contested and in some instances rejected outright by the scientific community. Their champions often found themselves in a minority and had to maintain their views in the face of widespread skepticism or denial.
- Nobel-worthy discoveries often required a great deal of hard work over many years or even decades, with a focus and commitment by their champions.

REFERENCES

Hawiger D, Inaba K, Dorsett Y, Guo M, Mahnke K, Rivera M, et al. Dendritic cells induce peripheral T cell unresponsiveness under steady state conditions in vivo. J Exp Med 2001;194:769–79.

Mishell RI, Dutton RW. Immunization of dissociated spleen cell cultures from normal mice. J Exp Med 1967;126:423–42.

Nussenzweig MC. Nobel lecture. Ralph Steinman and the discovery of dendritic cells. http://www.nobelprize.org/nobel_prizes/medicine/laureates/2011/steinman-lecture.html; 2011.

Nussenzweig MC, Ralph M. Steinman—biographical. http://www.nobelprize.org/nobel_prizes/medicine/laureates/2011/steinman-bio.html; 2011.

Nussenzweig MC, Steinman RM, Gutchinov B, Cohn ZA. Dendritic cells are accessory cells for the development of anti-trinitrophenyl cytotoxic T lymphocytes. J Exp Med 1980;152:1070–84.

Nussenzweig MC, Steinman RM, Witmer MD, Gutchinov B. A monoclonal antibody specific for mouse dendritic cells. Proc Natl Acad Sci USA 1982;79:161–5.

Paul WE. Bridging innate and adaptive immunity. Cell 2011;147:1212–5.

Shearer GM. Cell mediated cytotoxicity to trinitrophenyl-modified syngeneic lymphocytes. Eur J Immunol 1974;4:527–33.

Steinman RM, Cohn ZA. Identification of a novel cell type in peripheral lymphoid organs of mice. I. Morphology, quantitation, tissue distribution. J Exp Med 1973;137:1142–62.

Steinman RM, Cohn ZA. Identification of a novel cell type in peripheral lymphoid organs of mice. II. Functional properties in vitro. J Exp Med 1974;139:380–97.

Steinman RM, Witmer MD. Lymphoid dendritic cells are potent stimulators of the primary mixed leukocyte reaction in mice. Proc Natl Acad Sci USA 1978;75:5132–6.

Chapter 22

What Lies Ahead?

Scientists Look into Their Crystal Balls

Chapter Outline

1. INTRODUCTION

Viral pathogenesis is a field in rapid evolution, reflecting the dynamic development of systems biology and the continuing introduction of new or improved methodologies. Therefore, this final chapter is dedicated to "futurism," a look at what lies ahead for this field. We have recruited a number of scientists to write short pieces where they are free to speculate on future developments in their respective areas of expertise.

Viral Pathogenesis. http://dx.doi.org/10.1016/B978-0-12-800964-2.00022-7

2. WHAT LIES AHEAD: INDIVIDUAL VISIONS

Chapter 22.1

Systems Analysis of Host–Virus Interactions

Ronald N. Germain

Ron Germain is Chief of the Laboratory of Systems Biology, NIAID, NIH, and Associate Director of the Trans-NIH Center for Human Immunology. Over the course of his career, he has studied the immune system at the molecular, cell, and more recently, organismal level. He is now using a combination of advanced imaging methods and systems approaches pioneered in his laboratory to better understand host antimicrobial defenses and their pathological manifestations.

Decades of detailed biological study have taught us that host defense against viral infection and spread relies on multiple components of the immune system, ranging from parenchymal cell production of interferons, to innate immune responses by both myeloid and lymphoid cells, to lymphocyte-dependent cellular and humoral adaptive immunity. In the evolutionary tug of war between host and pathogen, viruses have developed ways to circumvent or even to take advantage of the host response. It has also become clear that much of the damage and disruption of physiology caused by viral infections is not due to direct cytopathic effects of the virus itself, but to the side effects of misdirected or over-exuberant host immunity.

1. WHERE DO WE GO FROM HERE?

The multilayered nature of host defense to viruses, the complex interplay between the pathogen and host, and the immune system's impact on normal tissue function call for a more holistic approach to the study of viral pathogenesis going forward. Rather than drilling down to the function of specific viral proteins or looking for a singular putative mechanism of host protection as most studies have done for the past several decades, we need to think at a systems level, integrating information across biological scales (genes, proteins, cells, organs), across time, and across interaction directions (virus->host, host->virus, host->host).

How can this be accomplished? The exciting news is that we now have tools that permit us to collect quantitative data on each of these aspects of the host–pathogen interplay. In animal models, viruses that encode fluorescent proteins, advanced microscopes that allow deep, long-term, high-resolution imaging in many tissues and organs of living animals, and reporter systems that reveal the identity of immune cells and their molecular state, now permit investigators to observe viral spread and the concomitant host response in real time (Germain et al., 2012). Novel methods for tissue preparation allow even entire organs to be made accessible to high resolution confocal imaging in dozens of colors, enabling detailed phenotypic analysis of infiltrating immune cells, the location of virus, and the state of host cells (Gerner et al., 2012). Emerging tools for DNA and RNA in situ hybridization permits documentation of the transcriptional response to infection. The application of such tools can produce a detailed, four dimensional (volume and time) picture of how a virus spreads, and the nature of the host response. In animals given candidate vaccines or drugs, these tools can also be used to directly visualize the specific effects of the treatment on the infection and the host response, both immune and organ-related.

But as powerful as imaging has become, it is in combination with the emerging methods of systems biology that insight will be gained most rapidly. This is especially true for emerging or re-emerging infections that frequently have more severe effects on host homeostasis and for which new understanding of the infectious process and host response are most needed. Transcriptomic, proteomic, and metabolic analyses are becoming more and more accessible, powerful, and applicable even for the human host (Nakaya et al., 2012), allowing collection of data sets that can be compared with those from animal models to determine the extent to which the animal systems are suitable substrates for vaccine or drug development.

One sticking point going forward is the limited capacity of many investigators to integrate the enormous data sets these technologies can generate. It has become very challenging to interpret the data in the context of human genetic variation and microbiome differences that have critical impact on immune performance. Investigators just coming into the field have a remarkable opportunity to address questions of viral pathogenesis in ways few established scientists are able to employ. However, this will only be the case if these rising scientists embrace the need for training in quantitative analysis. Also, the scientific enterprise must provide suitable incentives for the large multidisciplinary teams needed to collect and mine the many types of information required for a systems-level understanding of the infectious process (Germain et al., 2011).

Reductionist biology is not nearing its extinction point, but more and more the future will be about analyzing biology in an integrated and holistic manner. Imaging will help us understand cell dynamics, migration, positioning, the cell–cell interactions critical to immune effector responses, and the range over which cytokine and chemoattractants work. These are all facets of immunity that are central to determining what occurs in the face of viral infection and how various cells and factors contribute to maintaining host homeostasis or to disrupting it. Informatic analysis and computational modeling will provide new ways to integrate disparate data sets so that the yin and yang of immune function during an infection can be better understood. More precise understanding of the tipping points where augmentation or blockade could play a key role has the potential to improve health outcomes. Rising virologists will need to become "physiologists" with an appreciation for how the entire organism behaves in the face of infection. Such insight will determine the levers that are needed to reduce morbidity and mortality from infection in the absence of fully protective vaccines, and indeed, to determine what components of the immune response are best suited to protection. The future is both exciting and demanding—we will need to do our science in new ways to move forward in the most efficient manner. Although challenging, the systems biology approach has the potential to vastly accelerate our capacity to understand the origins of virus-induced pathology and to develop novel ways to prevent it.

REFERENCES

Germain RN, Meier-Schellersheim M, Nita-Lazar A, Fraser ID. Systems biology in immunology: a computational modeling perspective. Annu Rev Immunol 2011;29:527–85.

Germain RN, Robey EA, Cahalan MD. A decade of imaging cellular motility and interaction dynamics in the immune system. Science 2012;336:1676–81.

Gerner MY, Kastenmuller W, Ifrim I, Kabat J, Germain RN. Histo-cytometry: a method for highly multiplex quantitative tissue imaging analysis applied to dendritic cell subset microanatomy in lymph nodes. Immunity 2012;37:364–76.

Nakaya HI, Li S, Pulendran B. Systems vaccinology: learning to compute the behavior of vaccine induced immunity. Wiley Interdiscip Rev Syst Biol Med 2012;4:193–205.

Chapter 22.2

New Cellular Assays

Holden T. Maecker

Holden Maecker is an Associate Professor of Microbiology and Immunology, and Director of the Human Immune Monitoring Center, at Stanford University. His research focuses on cellular immune responses to chronic infections and cancer, and immune correlates of protection.

Because cellular immunity is so critical to the control of viral diseases, our understanding of viral pathogenesis has always been somewhat limited by our ability to measure the cellular immune response. One of the first assays of T-cell function, the enzyme-linked immunospot (ELISpot) technique, was revolutionary for its ability to quantitate antigen-specific cytokine production on a single-cell level. This was followed by intracellular cytokine staining (ICS) by flow cytometry, which was similarly applied to the detection of antigen-specific T cells a few years later. Since then, the growth of multicolor flow cytometry capabilities has led to ICS assays that provide information on multiple cytokines and phenotypic markers in combination. MHC-peptide multimer staining, as well as degranulation via CD107 export, can also be assessed together with ICS.

But the number of T-cell cytokines and differentiation markers of interest has grown tremendously over these years of T-cell assay development. Of interest to the protective capacity of virus-specific T cells is their expression of exhaustion markers such as PD-1, or costimulatory markers like CD28. Cytokines of relevance have expanded from the traditional IFNγ, to include IL-17, IL-22, and others. Intracellular levels of granzymes and perforin can be important to T-cell function. And chemokine receptors such as CCR6, CXCR3, and CCR5 can give clues about T-cell cytokine profiles and functional patterns. There are also many markers to define regulatory T-cell subsets, which impact the function of effector T cells.

In addition to T-cell markers, recent evidence has shown that NK cells can have memory properties and contribute to control of infections. Dendritic cells are important for priming and sustaining T-cell responses, and may serve as vehicles of viral spread, in some virus infections such as dengue. In the B-cell lineage, the magnitude of plasmablasts has been correlated with dengue disease severity. Thus, there are reasons to quantitate and/or phenotype virtually every type of immune cell subset when looking for correlates of immunity and pathogenesis.

Fortunately, a technology has recently emerged that greatly increases the number of markers that can be simultaneously quantitated at the single-cell level compared to traditional flow cytometry. Mass cytometry, or CyTOF (for Cytometry by Time Of Flight), is based on the use of heavy metal ion labels, rather than fluorophores, for tagging antibodies and other probes (Tanner et al., 2008). The consequent readout by mass spectrometry yields two parallel benefits: many more channels can be simultaneously detected, with little to no spillover between them. The result is 40-plus parameter flow cytometry without the need to perform interchannel compensation.

Already, mass cytometry has been applied to the study of immune cell signaling capacities (Bendall et al., 2011; Bodenmiller et al., 2012), to phenotype the diversity of CD8 T cells (Newell et al., 2012), NK cells (Horowitz et al., 2013), and B cells (Bendall et al., 2014), to measure immune competence in cancer patients (Chang et al., 2014), and to find immune correlates of response to surgery (Gaudilliere et al.). A variation of the technique has even been adapted to reading cells in tissue sections, a kind of extension of immunohistochemistry with more than 40 parameters (Angelo et al., 2014).

With mass-tagged antibody conjugates now readily available for many markers, the ability to build high-dimensional panels has become easier, although expensive. Because of the relative lack of spillover, panel design is simpler, though not entirely fool-proof (Leipold et al., 2014). So what are the drawbacks of the technique? In addition to the cost and expertise required to set up and maintain the complex instrumentation, the two major drawbacks are acquisition speed and cell recovery. In other words, the CyTOF is slow in collecting cells, and the majority of cells are lost. A secondary drawback is sensitivity, since there is no channel in the mass cytometer that equals the sensitivity of phycoerythrin (PE) or similarly bright fluorochromes in fluorescence flow cytometry. Despite these drawbacks, the wealth of information collected by CyTOF is making it the technique of choice for both broad and deep profiling of immune cells. Many laboratories are using these methods to find cellular immune biomarkers of infection, pathogenesis, and immunity.

Of note, mass cytometry is not the only technique that can provide highly multiparametric data at the single-cell level. While it is technically possible to do whole-transcriptome RNAseq of single cells (Jaitin et al., 2014), a targeted version of this technology is also useful. Targeted RNAseq uses PCR to amplify a set of genes from cDNA of single cells. The amplified products from each cell are barcoded and pooled for deep sequencing. This provides an alternative to CyTOF in that it is not limited by the availability of good antibodies to target molecules. On the other hand, it generates transcript frequencies rather than protein abundance.

Parallel measurements on thousands of immune cells create complex data to decipher for potential biomarkers. The development of visualization and statistical algorithms to help interpret CyTOF data continues to grow. Currently published clustering algorithms that have been applied to mass cytometry include SPADE (Simonds et al., 2011), PCA (Newell et al., 2012), viSNE (Amir et al., 2013), Citrus (Bruggner et al., 2014), and ACCENSE (Shekhar et al., 2014). The next few years will likely see a refinement and selection of these algorithms for those that perform best for particular questions.

In summary, it may be predicted that future studies of viral interaction with the immune system will be dominated by big-data approaches such as mass cytometry and single-cell targeted RNAseq. Computational methods to decipher these large data sets will be key to finding the biomarkers hidden within them.

REFERENCES

Amir E-AD, Davis KL, Tadmor MD, Simonds EF, Levine JH, Bendall SC, et al. viSNE enables visualization of high dimensional single-cell data and reveals phenotypic heterogeneity of leukemia. Nat Biotechnol 2013;31:545–52.

Angelo M, Bendall SC, Finck R, Hale MB, Hitzman C, Borowsky AD, et al. Multiplexed ion beam imaging of human breast tumors. Nat Med 2014:1–9.

Bendall SC, Simonds EF, Qiu P, Amir EAD, Krutzik PO, Finck R, et al. Single-cell mass cytometry of differential immune and drug responses across a human hematopoietic continuum. Science 2011;332:687–96.

Bendall SC, Davis KL, Amir E-AD, Tadmor MD, Simonds EF, Chen TJ, et al. Single-cell trajectory detection uncovers progression and regulatory coordination in human B cell development. Cell 2014;157:714–25.

Bodenmiller B, Zunder ER, Finck R, Chen TJ, Savig ES, Bruggner RV, et al. Multiplexed mass cytometry profiling of cellular states perturbed by small-molecule regulators. Nat Biotechnol 2012;30:857–66.

Bruggner RV, Bodenmiller B, Dill DL, Tibshirani RJ, Nolan GP. Automated identification of stratifying signatures in cellular subpopulations. Proc Natl Acad Sci 2014;111:e2770–7.

Chang S, Kohrt H, Maecker HT. Monitoring the immune competence of cancer patients to predict outcome. Cancer Immunol Immunother 2014;63:713–9.

Gaudilliere B, Fragiadakis GK, Bruggner RV, Nicolau M, Finck R, Tingle M, Silva J, Ganio EA, Yeh CG, Maloney WJ, Huddleston JI, Goodman SB, Davis MM, Bendall SC, Fantl WJ, Angst MS, Nolan GP. Clinical recovery from surgery correlates with single-cell immune signatures. Sci Transl Med 6:1–12.

Horowitz A, Strauss-Albee DM, Leipold M, Kubo J, Nemat-Gorgani N, Dogan OC, et al. Genetic and environmental determinants of human NK cell diversity revealed by mass cytometry. Sci Transl Med 2013;5:208ra145.

Jaitin DA, Kenigsberg E, Keren-Shaul H, Elefant N, Paul F, Zaretsky I, et al. Massively parallel single-cell RNA-seq for marker-free decomposition of tissues into cell types. Science 2014;343:776–9.

Leipold M, Newell EW, Maecker HT. Multiparameter phenotyping of human PBMCs using mass cytometry. In: Shaw A, editor. Immunosenescence. 2014. p. 1–13.

Newell EW, Sigal N, Bendall SC, Nolan GP, Davis MM. Cytometry by time-of-flight shows combinatorial cytokine expression and virus-specific cell niches within a continuum of CD8+ T cell phenotypes. Immunity 2012;36:142–52.

Shekhar K, Brodin P, Davis MM, Chakraborty AK. Automatic classification of cellular expression by nonlinear stochastic embedding (ACCENSE). Proc Natl Acad Sci 2014;111:202–7.

Simonds EF, Bendall SC, Gibbs KD, Bruggner RV, Linderman MD, Sachs K, et al. Extracting a cellular hierarchy from high-dimensional cytometry data with SPADE. Nat Biotechnol 2011:1–8.

Tanner SD, Bandura DR, Ornatsky O, Baranov VI, Nitz M, Winnik MA. Flow cytometer with mass spectrometer detection for massively multiplexed single-cell biomarker assay. Pure Appl Chem 2008;80:2627–41.

Chapter 22.3

The Virome: Our Viral Commensals

Martin J. Blaser

Martin J Blaser is Professor of Medicine and Microbiology, and Director of the Human Microbiome Program at the NYU School of Medicine. His work over the past 30 years has focused on human pathogens, including Campylobacter species and Helicobacter pylori, *which also are model systems for understanding interactions of residential bacteria with their human hosts. Most recently, he wrote "Missing Microbes," a book targeted to general audiences.*

Animals have had resident microbes ever since there have been animals, at least for 500 million years. These have mostly been prokaryotic bacteria and archaea but wherever there are bacteria, there also are viruses that live with them. Whether in the ocean or in the human body (Reyes et al., 2012; Minot et al., 2011; Pride et al., 2012), bacteriophages are predators of their bacterial hosts.

These relationships are multifaceted and complex, marked by both competition and cooperation; the tension between these forces is ever-present and dynamic. Many bacterial species live within animal hosts, competing for niche, and viruses do the same within their bacterial hosts. In longstanding ecosystems, such as represented by the human body, viruses predictably affect the fitness of their immediate bacterial hosts (Reyes et al., 2013; Minot et al., 2013). Just as in the tundra, where the wolf keeps the caribou healthy, the inevitable force of viral predation affects the bacterial populations, and therefore the larger host in which they all reside (Reyes et al., 2010).

In addition to bacteriophages, which are the dominant viruses in humans, commensal viruses also live directly in our cells. New nucleotide-based approaches show that we are teeming with resident human viruses; between skin, mouth, nose, and vagina, and we each seem to be carrying 3–15 detectable DNA viruses at any time (Wylie et al., in press). Most of those with which we were familiar cause common, usually mild, but occasionally severe, infections. Cytomegalovirus and Epstein–Barr virus are two excellent examples. We focus on their ill effects, but most people are carrying these viruses silently for decades, essentially for life.

What are they doing? Are they just parasites, exploiting us for their own purposes, with relatively low biological cost to human fitness? Or are there benefits to our relationships, now mostly hidden? In the discipline of microbiology, most new organisms have been discovered as pathogens. "Pathogens" became our dominant mind-set for viruses, even for organisms that infect most of us, and persist for decades, 99% of the time without clinical consequence. Although some of the herpesviridae, JC virus, or certain papillomaviruses, for example, may cause illnesses and may be lethal, the net negative effects, integrated across the life span of the entire human population, are extremely low. When the cost side to human fitness is so low, it is possible that even a small benefit across most people may cumulatively be greater. Our drive toward "microbiological perfection," may be producing a paradoxical worsening of health, because we are removing their benefits as well.

We are endowed by immunity, now divided into adaptive and innate, but there is increasing evidence for a third class, which may be called "microbial." It is our commensals, our long-time partners, which among their functions, help protect the motherland, that is, the human host. Many observations indicate that our commensal bacteria are part of the protection against invading organisms (Bohnoff et al., 1954), and already there are reports that some "commensal viruses" may help control serious infections (Barton et al., 2007; Barr et al., 2013).

I predict that the evidence for such relationships will grow in the coming years. Medical science has focused on the horizontal, because consequences often are severe, but numerically, the vertical may be at least as important. It is in the evolutionary interest of commensals, whether bacterial or viruses, to protect us most of the time. The tension comes when their necessity to be transmitted to the next host exceeds their protective properties.

A final, seemingly unpleasant thought: we all must die. To some scientists, this is just the consequence of aging. An alternative view is that clock-like markers and phenomena of aging have been selected; those species that have orderly senescence fare better than those in which the herds are subject to culling regardless of age. Pathogens can sweep through a population and kill young and old alike, in some cases leading to the extinction of the entire group. Clock-imposed senescence appears safer, as mathematical modeling indicates.

Commensals can contribute to aging. In humans, they may contribute to the causation of cancers (think *Helicobacter pylori,* EB-virus, Hepatitis B) and degenerative diseases (JK virus and prions)—most of which are log-linearly age-related. Our commensals may contribute to killing us safely (in an age-related manner), without epidemic risk to our community (Blaser and Webb, 2014). What is bad for the individual may be good for the species.

If the above biological premises are correct, what should we do? One implication is that our commensals (viruses included) may be beneficial for us early in life through reproductive age, but costly later. Rather than try to prevent infection early, we might welcome at least a part of it, while controlling the untoward consequences, and focus on better late-in-life control. Less-virulent commensals may protect against higher virulence organisms, even of the same species. As one example, perhaps children should be exposed to varicella-zoster virus early, as they had been for eons, but now controlling for the serious ill effects with antiviral drugs. Then with aging, we might boost VZV immunity by vaccination to prevent serious consequences. Such an approach would need to be tested.

In conclusion, there clearly are viral pathogens—they mostly are exogenous—that have crossed from other animals. Historically, these are the most dangerous. As SARS, Ebola, MERS, and variant influenza have shown us, their introduction will continue. However, we are also endowed with many viral commensals, with new types being discovered with regularity (Reyes et al., 2010, 2013; Minot et al., 2013; Wylie et al., in press). Our biological relationships with them are complex, and the same commensal virus can be symbiotic or pathogenic, depending on host context and timing; the biosphere is full of contingency. Deep understanding of the underlying biology of viral commensalism will allow us to harness our friendly microbes to improve human health.

REFERENCES

Barr JJ, Auro R, Furlan M, Whiteson KL, Erb ML, Pogliano J, et al. Bacteriophage adhering to mucus provide a non-host-derived immunity. Proc Natl Acad Sci USA 2013;110:10771–6.

Barton ES, White DW, Cathelyn JS, Brett-McClellan KA, Engle M, Diamond MS, et al. Herpesvirus latency confers symbiotic protection from bacterial infection. Nature 2007;447:326–9. PMID: 17507983.

Blaser MJ, Webb GF. Host demise as a beneficial function of indigenous microbiota in human hosts. mBio 2014;5:1–9. http://dx.doi.org/10.1128/mBio.02262-14.

Bohnoff M, Drake B, Miller C. Effect of streptomycin on susceptibility of intestinal tract to experimental *Salmonella* infection. Proc Soc Exp Biol Med 1954;86:132–7.

Minot S, Sinha R, Chen J, Li H, Keilbaugh SA, Wu GD, et al. The human gut virome: inter-individual variation and dynamic response to diet. Genome Res 2011;21:1616–25. http://dx.doi.org/10.1101/gr.122705.111.

Minot S, Bryson A, Chehoud C, Wu GD, Lewis JD, Bushman FD. Rapid evolution of the human gut virome. Proc Natl Acad Sci USA 2013;110:12450–5. http://dx.doi.org/10.1073/pnas.1300833110.

Pride DT, Salzman J, Haynes M, Rohwer F, Davis-Long C, White 3rd RA, et al. Evidence of a robust resident bacteriophage population revealed through analysis of the human salivary virome. ISME J 2012;6:915–26. http://dx.doi.org/10.1038/ismej.2011.169.

Reyes A, Haynes M, Hanson N, Angly FE, Heath AC, Rohwer F, et al. Viruses in the faecal microbiota of monozygotic twins and their mothers. Nature 2010;466:334–8. http://dx.doi.org/10.1038/nature09199. PMID: 20631792.

Reyes A, Semenkovich NP, Whiteson K, Rohwer F, Gordon JI. Going viral: next-generation sequencing applied to phage populations in the human gut. Nat Rev Microbiol 2012;10:607–17.

Reyes A, Wu M, McNulty NP, Rohwer FL, Gordon JI. Gnotobiotic mouse model of phage bacterial host dynamics in the human gut. Proc Natl Acad Sci USA 2013;110:20236–41. http://dx.doi.org/10.1073/pnas.1319470110. Epub 2013 November 20.

Wylie KM, Mihindukulasuriya KA, Zhou Y, Sodergren E, Storch GA, Weinstock GM, Metagenomic analysis of double-stranded DNA viruses in healthy adults. BMC Biology 2014;12:71–81.

Chapter 22.4

Forward Genetics and Viral Pathogenesis

Bruce Beutler

Bruce Beutler directs the Center for the Genetics of Host Defense at UT Southwestern Medical Center in Dallas. He analyzes immune function in mammals by random germline mutagenesis. In 2011, he shared the Nobel Prize in Physiology or Medicine for discoveries concerning the activation of innate immunity.

How are we to understand precisely how viruses work, and how are we to defeat them? Genetics continues to lead the way to understanding the precise molecular interactions between virus and host. A number of remarkable tools have become available to probe precisely how viruses work.

Today it is possible to ask, "what proteins of the host are essential for virus X to complete its life cycle?" In mice, at the whole organism level, genetics can deliver answers to questions of this kind with greater speed than ever before. As soon as a new phenotype is observed (for example, instances of mice in which a normally virulent virus fails to proliferate), one can confidently state which mutation in the host genome is responsible for this phenotype. This new reality has stemmed from the development of massively parallel sequencing platforms, methods for high-speed genotyping, and new computational tools for this express purpose (Moresco and Beutler, 2013; Bull et al., 2013; Wang et al., 2015). The latter have been developed in our laboratory, and while a detailed description cannot be offered in this short essay, real-time forward genetic analysis in the mouse works approximately as follows.

Male mice of known sequence, homozygous at all loci (usually C57BL/6J background) are exposed to the mutagen ENU (N-ethyl-N-nitrosourea) which induces single base pair changes in the genomic DNA of spermatogonia. Several thousand mutations are transmitted to each sperm, causing coding changes in about 70 genes per haploid genome. These mutations are transmitted to G1 males by breeding the mutagenized sire to a C57BL/6J female. About 40 G1 males are produced each week, and each G1 male is subjected to deep, whole exome sequencing to detect all mutations that cause (or may cause) coding change. Sperms are preserved from this male, which is then used to produce G2 daughters. Ten G2 daughters are then backcrossed to the G1 sire to yield between 30 and 50 G3 progeny. Before the G3 progeny are released for screening, they are genotyped at all mutant sites to determine whether they are homozygous WT (REF), heterozygous (HET), or homozygous variant (VAR). Foreknowledge of genotype assures that when phenotype is measured, an immediate computational determination of linkage can be made, using recessive, semi-dominant, or dominant inheritance models. In general, linkage will be observed only if a phenotype has a genetic basis. And if linkage is observed, it often indicates unambiguous cause and effect (i.e., it shows that a particular mutation is responsible for the phenotype).

Because 40 G1 mice are processed each week, and each G1 mouse bears about 70 mutations, about 2800 mutations are investigated each week for their phenotypic consequences (and many phenotypes can actually be studied in parallel). Annually, more than 100,000 mutations can be examined. Over time, every gene is struck repeatedly by mutation, and many alleles can be tested for phenotypic effects. It does not take long until all genes have been examined in considerable depth, so that they may be considered "implicated" or "exonerated" in the phenomenon of interest. Where putative null alleles are concerned (those that cause premature truncation of a protein or aberrant splicing), three observations in the homozygous state are considered an ample test of the importance of a gene in a phenomenon of interest.

Not only simple (Mendelian) traits can be linked to mutations, but complex traits involving mutation at multiple loci can be solved as well, particularly if pedigrees of a large size are constructed for the purpose. As contrasted with quantitative trait locus mapping, ENU mutagenesis tends to produce monogenic phenotypes, but when complex phenotypes are observed, they are more easily solved owing to the limited number of mutations under investigation. The very fact that complex traits do occur, even with so few as 70 mutations causing coding change in a given pedigree, suggests that interactions between naturally occurring mutations are abundant, and may represent a major source of a phenotype as it is observed in wild populations, or in humans. This, of course, would presumably apply to all phenotypes, including viral susceptibility phenotypes.

Will ENU mutagenesis offer an explanation of all the workings of a complex biological system and tell us precisely how we fight viruses, or how viruses take advantage of the host? It can only give us a good start. The future may call for even more sophisticated methods than forward genetics as we presently practice it. Already widespread is the use of CRISPR/Cas9 technology to modify the genome. It is quite possible to create many targeted mutations within a given model organism genome to probe interactions between pathways that might be thought redundant in function. Yet this innovation too will run its course, and in the future, we may rely upon synthetic genomics. Already it is possible to create designer microbes, with genomes that are modified as the investigator chooses. A complete chromosome of a eukaryotic organism (yeast) has been synthesized as well. Will the day come when we may synthesize the genome of a mouse, modified just as we choose? Very likely yes, and very likely such mice will answer questions that cannot be addressed using the current generation of genome-modifying technologies. Undoubtedly, a cascade of future technical advances will continue to elucidate the interactions between host and pathogen.

REFERENCES

Bull KR, Rimmer AJ, Siggs OM, Miosge LA, Roots CM, Enders A, et al. Unlocking the bottleneck in forward genetics using whole-genome sequencing and identity by descent to isolate causative mutations. PLoS Genetics 2013;9(1):e1003219.

Moresco EM, Beutler B. Going forward with genetics: recent technical advances and forward genetics in mice. Am J Pathol 2013;182:1462–73.

Wang T, Zhan X, Bu CH, Lyon S, Pratt D, Hildebrand S, Choi JH, Zhang Z, Zeng M, Wang KW, Turer E, Chen Z, Zhang D, Yue T, Wang Y, Shi H, Wang J, Sun L, SoRelle J, McAlpine W, Hutchins N, Zhan X, Fina M, Gobert R, Quan J, Kreutzer M, Arnett S, Hawkins K, Leach A, Tate C, Daniel C, Reyna C, Prince L, Davis S, Purrington J, Bearden R, Weatherly J, White D, Russell J, Sun Q, Tang M, Li X, Scott L, Moresco EM, McInerney GM, Karlsson Hedestam GB, Xie Y, Beutler B. Real-time resolution of point mutations that cause phenovariance in mice. Proc Natl Acad Sci USA February 3, 2015;112(5):E440–9. http://dx.doi.org/10.1073/pnas.1423216112. Epub 2015 January 20.

Chapter 22.5

The Future of Synthetic Virology

Eckard Wimmer, Yutong Song, Oleks Gorbatsevych

Eckard Wimmer is Distinguished Professor of Molecular Genetics and Microbiology, Stony Brook University. For decades, he has studied the replication and pathogenesis of human RNA viruses, particularly poliovirus, and is known for the de novo chemical synthesis of poliovirus, the first organism generated on the basis of sequence information in the public domain.

Yutong Song is a Research Assistant Professor of Molecular Genetics and Microbiology, Stony Brook University School of Medicine, Stony Brook. His research focuses on translational and replication control of viral RNA, and pathogenesis of human RNA viruses such as poliovirus and Hepatitis C virus.

Oleks Gorbatsevych is a scientist at Stony Brook University. He developed software for designing synthetic genes, and his research is focused on the evolution of synthetic viruses.

The first test tube synthesis of a virus (Cello et al., 2002) caused a global uproar. One axiom in biology holds that proliferation of cellular organisms or viruses depends on the presence of a functional genome. The cell-free synthesis of poliovirus has violated this law: no natural template was required to recreate this organism. In 2002, few people were prepared to accept the new reality that viruses exist

TABLE 1 List of Virus Genomes in Chronological Order that Were Synthesized either Partially, or in toto in the Absence of Natural Templates

I.	2002	Poliovirus[a]
II.	2003	Phage PhiX174[b]
III.	2005	Recreation of the 1918 influenza Virus[c]
IV.	2005	Refactoring bacteriophage T7[d]
V.	2006	Codon deoptimized polioviruses[e,f]
VI.	2007	Reconstitution of an infect. Human endogenous retrovirus[g]
VII.	2007	Generation of infectious molecular clones of HIVcpz[h]
VIII.	2008	Codon pair deoptimized polioviruses[i]
IX.	2008	Bat SARS-like Coronavirus[j]
X.	2010	Codon pair deoptimized influenza virus[k]
XI.	2010	West Nile virus[l]
XII.	2012	Poliovirus: Discovery of novel regulatory elements by recoding[m]
XIII.	2013	Influenza virus HA and NA codon pair deoptimization[n]
XIV.	2014	Tobacco Mosaic virus[o]
XV.	2014	Respiratory syncytial virus: attenuation by recoding[p]

[a]*Cello J, et al. Chemical synthesis of poliovirus cDNA: generation of infectious virus in the absence of natural template. Science 297:1016–1018.*
[b]*Smith HO, et al. Generating a synthetic genome by whole genome assembly: phiX174 bacteriophage from synthetic oligonucleotides. PNAS 100:15,440–15,445.*
[c]*Tumpey TM, et al. Characterization of the reconstructed 1918 Spanish influenza pandemic virus. Science 310:77–80.*
[d]*Chan LY, et al. Refactoring bacteriophage T7. Mol Syst Biol 1:2005 0018.*
[e]*Burns CC, et al. Modulation of poliovirus replicative fitness in HeLa cells by deoptimization of synonymous codon usage in the capsid region. J Virol 80:3259–3272.*
[f]*Mueller S, et al. Reduction of the rate of poliovirus protein synthesis through large-scale codon deoptimization causes attenuation of viral virulence by lowering specific infectivity. J Virol 80:9687–9696.*
[g]*Lee YN, et al. Reconstitution of an infectious human endogenous retrovirus. PLoS Pathog 3:e10.*
[h]*Takehisa J, et al. Generation of infectious molecular clones of simian immunodeficiency virus from fecal consensus sequences of wild chimpanzees. J Virol 81:7463–7475.*
[i]*Coleman JR, et al. Virus attenuation by genome-scale changes in codon pair bias. Science 320:1784–1787.*
[j]*Becker MM, et al. Synthetic recombinant bat SARS-like coronavirus is infectious in cultured cells and in mice. PNAS 105:19,944–19,949.*
[k]*Mueller S, et al. Live attenuated influenza virus vaccines by computer-aided rational design. Nat Biotechnol 28:723–726.*
[l]*Orlinger KK, et al. An inactivated West Nile Virus vaccine derived from a chemically synthesized cDNA system. Vaccine 28:3318–3324.*
[m]*Song Y, et al. Identification of two functionally redundant RNA elements in the coding sequence of poliovirus using computer-generated design. PNAS 109:14,301–14,307.*
[n]*Yang C, et al. Deliberate reduction of hemagglutinin and neuraminidase expression of influenza virus leads to an ultraprotective live vaccine in mice. PNAS 110:9481–9486.*
[o]*Cooper B. Proof by synthesis of Tobacco mosaic virus. Genome Biol 15:R67.*
[p]*Le Nouen C, et al. Attenuation of human respiratory syncytial virus by genome-scale codon-pair deoptimization. PNAS 111:13,169–13,174.*

as infectious particles in nature as well as entries in a database[1]. Not surprisingly the response of laymen and experts alike included praise, ethical concerns, ridicule, and fierce condemnation (Wimmer, 2006; Wimmer and Paul, 2011).

Total synthesis of viruses, meanwhile, has come of age. Yet the number of synthetic viruses (Table 1) is still modest because the technology needed to produce large segments of DNA that can be stitched together is far from efficient. Moreover, the price of error-free synthetic oligonucleotides (>100 nucleotides) is still significantly greater than predicted in 2004 (20,000 bp for $1). A revolution, however, is brewing (Notka et al., 2011; Kosuri and Church, 2014) even though the new technologies have not yet reached the commercial sector. It can be predicted that new strategies will drastically change research

1. http://www.ncbi.nlm.nih.gov/genomes/GenomesHome.cgi?taxid = 10239.

in molecular genetics and biological engineering: the tedious steps—construction of vectors, of site-specific mutants, etc.—will soon be replaced by fast and affordable DNA synthesis.

Template-free synthesis of RNA viruses, of course, relies on the chemical synthesis of genome-complementary (double stranded) DNA, referred to as "cDNA." cDNA, in turn, can be enzymatically transcribed into viral RNA and ultimately converted to infectious virus. The chemical synthesis of cDNA discussed here is reminiscent of the enzymatic synthesis of cDNA by retrovirus reverse transcriptase that was introduced by Charles Weissmann and his colleagues in 1978 (Taniguchi et al., 1978). Weissmann's strategy, termed "reverse genetics," has led to the revolutionary method allowing DNA-based genetic manipulations of RNA viruses. However, whereas the Weissmann strategy requires natural viral isolates to generate cDNA (Figure 1(A)), chemical synthesis only needs the genome sequence information readily available on the Internet (Figure 1(B)). Assuming the sequence of an RNA is known, we believe that the enzymatic synthesis of cDNAs will soon be replaced by chemical synthesis (Wimmer and Paul, 2011).

Owing to their properties as "quasi species" (very high spontaneous mutation frequency during each cycle of RNA replication), RNA virus genomes are generally much smaller (around 10,000 nucleotides) than DNA virus genomes (between 3000 and 1,500,000 bp). Moreover, among the terrestrial viruses, RNA viruses outnumber DNA viruses by a ratio of 3:1. Hence, it is not surprising that the chemical synthesis of viruses has so far targeted predominantly RNA viruses (Table 1). The expected revolution in DNA synthesis, and exciting novel strategies for manipulating and assembling large synthetic DNA molecules (Gibson et al., 2010), will likely result in the chemical syntheses of many more DNA viruses.

Table 1 summarizes examples of synthetic viruses in roughly chronological order. The question lingers: what was/is the purpose of synthesizing viruses independently of their natural isolates? Most importantly, chemical synthesis of viral genomes allows investigation of the structure and function of the organism's biology to an extent hitherto impossible. However, different virus synthesis projects have had different objectives that we will only briefly mention. Wimmer and Paul, 2011 have discussed most of these projects in some detail.

Syntheses I and II (Table 1) served as proof of principle. Regrettably, in synthesis I all discussion of societal implications of the work and possible applications were cut by the *Science* editors (Cello et al., 2002; Wimmer, 2006). The generation of phage PhiX174 described in Synthesis II "improved upon the methodology and dramatically shortened the time required for accurate assembly of 5–6 kb segments of DNA from synthetic oligonucleotides"—the entire synthesis of the phage consumed only 2 weeks (II, Table 1).

Syntheses III, VI, VII, and IX served to identify beyond doubt the history, identity and/or pathogenesis of important human pathogens (Wimmer and Paul, 2011). These include most notably the 1918 Spanish Flu Influenza virus, an organism that disappeared in the years after the devastating pandemic of 1918/1919 (III); the infectious Simian Immunodeficiency Virus SIVcpz (VII); and the infectious Bat SARS-Like CoronaVirus (IX). The latter studies "demonstrated the usefulness of genetics and whole-genome synthesis in the investigation of

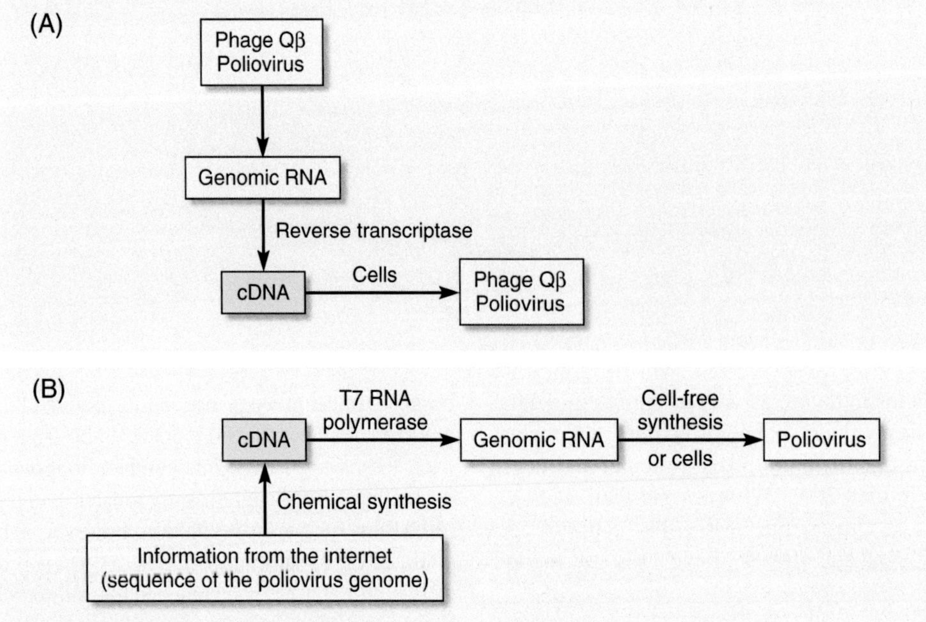

FIGURE 1 Two different strategies for generating RNA viruses via complementary DNA (cDNA) intermediates. (A) Synthesis of cDNA catalyzed with reverse transcriptase followed by transfection of the cDNA into host cells (Taniguchi et al., 1978). (B) Chemical synthesis of cDNA followed by transcription and incubation of the infectious viral RNA in a cell-free extract. *Modified from Ref. Cello et al. (2002).*

the trans species movement of zoonoses." The reconstitution of the Infectious Human Endogenous Retrovirus HERV-K (VI) from remnants in our human genome may yield valuable clues to the impact of endogenous retroviruses on human evolution.

Syntheses IV, V, VIII, X, XIII, XV, and XVI describe "designer viruses" in which large segments (20–40%) of the genome have been recoded with the purpose of either studying function (IV) or modifying expression of viral genetic elements. The latter has focused on expression of proteins by recoding open reading frames either through changing codon bias (V) or codon pair bias (VIII, X, XIII, XV, XVI). The authors of this commentary view codon pair deoptimization as a promising new strategy to design new vaccine candidates (VIII, Table 1). In synthesis XII, the entire ORF of the poliovirus genome was recoded, thereby mutating 1304 of 6249 nucleotides encoding the polyprotein. This has led to the discovery of two redundant RNA regulatory elements involved in RNA replication. In synthesis XI, seed virus for the development of vaccines was synthesized to avoid licensing problems by regulatory authorities. This proved to be an important bypass for creating a commercial product targeting human disease.

Finally, Cooper reported the synthesis of tobacco mosaic virus, TMV (synthesis XIV), which was the first plant virus originally discovered by Beijerinck in 1898. TMV is one of the most researched viruses of all time yet Cooper chose to title his publication: "Proof by synthesis of Tobacco Mosaic Virus." Curiously, he refers to the classical code of chemists who considered synthesis the ultimate proof for any deciphered chemical structure. Cello and his colleagues also made reference to this code (Cello et al., 2002), which will be significant in "proof reading" by synthesis the nucleotide sequences deposited in sequence databases (see also ref. 7).

We conclude by reminding the reader that synthetic biology represents a dual use dilemma in which the same technologies can be used legitimately for the benefit of humankind and misused for terrorism, a quandary referred to as "dual use research" (Wimmer and Paul, 2011; National Research Council, 2014). Nothing better matches this definition than the template-free synthesis of viruses. On the one hand, it advances our understanding of these organisms and leads to new methods to protect us from viral disease; yet, on the other hand it could be exploited with malicious intent. With optimism and enthusiasm, we predict that future applications of synthetic virology will be vastly more constructive than destructive.

REFERENCES

Cello J, Paul AV, Wimmer E. Chemical synthesis of poliovirus cDNA: generation of infectious virus in the absence of natural template. Science 2002;297:1016–8.

Gibson DG, Glass JI, Lartigue C, Noskov VN, Chuang RY, Algire MA, et al. Creation of a bacterial cell controlled by a chemically synthesized genome. Science 2010;329:52–6.

Kosuri S, Church GM. Large-scale de novo DNA synthesis: technologies and applications. Nat Methods 2014;11:499–507.

National Research Council. Biotechnology research in an age of terrorism. Washington (DC): National Academies Press; 2014. 164 pp.

Notka F, Liss M, Wagner R. Industrial scale gene synthesis. Methods Enzymol 2011;498:247–75.

Taniguchi T, Palmieri M, Weissmann C. QB DNA-containing hybrid plasmids giving rise to QB phage formation in the bacterial host. Nature 1978;274:223–8.

Wimmer E, Paul AV. Synthetic poliovirus and other designer viruses: what have we learned from them?. Annu Rev Microbiol 2011;65: 583–609. 65.

Wimmer E. The test-tube synthesis of a chemical called poliovirus. The simple synthesis of a virus has far-reaching societal implications. EMBO Rep 2006;7(Spec No):S3–9.

Chapter 22.6

Precision Medicine: Applications of Genetics to Pathogenesis and Treatment of Viral Diseases

D. Gary Gilliland

Gary Gilliland is President and Director of the Fred Hutchinson Cancer Research Center. He has served as Professor of Medicine at Harvard Medical School, Head of Global Oncology at Merck Research Laboratories, and Vice President at the University of Pennsylvania Perelman School of Medicine, where he was responsible for the Precision Medicine Initiative. His research focuses on the genetic basis of blood cancers, and he is a leader in the development of programs in precision medicine.

Genome-wide association studies (GWAS) and high-throughput DNA sequencing, often referred to as next generation sequencing (NGS), offer exciting new opportunities for the diagnosis and treatment of viral disease. These evolving technologies are leading to a new era of precision (also known as personalized) medicine. This approach is based on the fact that common variants in host genomes may have profound influence on the course of viral infections and the response to treatment. Two examples illustrate the potential of precision medicine: (1) host and viral genome variation in determining response to current standard of care for hepatitis C; and (2) use of NGS as a diagnostic test to identify microbial pathogens.

1. HOST GERMLINE VARIATION PREDICTS RESPONSE TO THERAPY IN HEPATITIS C

Until recently, the mainstay for standard of care for Hepatitis C had been antiviral drugs that required 48 weeks of therapy with pegylated interferon (PegIntron) and ribavirin, which were associated with significant side effects. Some patients achieved sustained viral response (SVR) with PegIntron/ribavirin, but many did not. Furthermore, patients of African-American and Hispanic descent did not respond as well, and it had not been possible to identify those individuals among all populations who were most likely to respond.

GWAS comparing responder and nonresponder populations identified a single nucleotide polymorphism near the interferon responsive gene IL28B that segregates with SVR and explains a significant proportion of the variation in response (Ge, D et al. Nature (2009) 461: 399). Patients homozygous for the C/C nucleotide showed a 78% successful outcome while those homozygous for the T/T nucleotide had a 26% successful outcome. The T/T genotype was more frequent in African-Americans, explaining their lower response rate.

These data are not new and therapy for hepatitis C is rapidly evolving toward potential cures. However, even with newer RNA polymerase and protease inhibitors, there will be variation in response based on host and viral genotype. Thus, use of GWAS is likely to have value in responder stratification for treatment of hepatitis C, and also for other virus diseases.

2. GENETIC VARIATION IN PATHOGENS TO PREDICT RESPONSE TO THERAPY: HEPATITIS C GENOTYPE AS AN EXAMPLE

There are six Hepatitis C sequence-specific genotypes, and these variants differ in their response to newly available antiviral therapies. These include viral RNA polymerase inhibitors such as sofosbuvir, and protease inhibitors such as simeprevir that impair viral entry into mammalian cells. As shown in Table 1 below, Hepatitis C genotype determines the best therapeutic strategy. Hepatitis C thus provides a valuable case study of how both host and pathogen genetic variation influence treatment choices and outcomes.

TABLE 1 Hepatitis C Viral Genotype Determines Therapeutic Approach to Treatment

Population	Recommended Regimens	FDA Approved?	Alternative Regimens	FDA Approved?
Genotype 1 interferon eligible	Sofosbuvir 400 mg daily + PEG/RBV × 12 weeks	Yes	Simeprevir 150 mg daily + PEG/RBV × 12 weeks then PEG/RBV × 12 weeks (genotype 1b or 1a without Q80K)	Yes
Genotype 1 interferon ineligible	Simeprevir 150 mg daily + sofosbuvir 400 mg daily + (with or without RBV) × 12 weeks	No	Sofosbuvir 400 mg daily + RBV × 24 weeks	Yes
Genotype 2	Sofosbuvir 400 mg daily + RBV × 12 weeks	Yes	None	
Genotype 3	Sofosbuvir 400 mg daily + RBV × 24 weeks	Yes	If interferon eligible: Sofosbuvir 400 mg daily + PEG/RBV × 12 weeks	No
Genotype 4 interferon eligible	Sofosbuvir 400 mg daily + PEG/RBV × 12 weeks	Yes	Simeprevir 150 mg daily + PEG/RBV × 12 weeks Then PEG/RBV × 12–36 weeks	No
Genotype 4 interferon ineligible	Sofosbuvir 400 mg daily + RBV × 24 weeks	No		
Genotype 5 or 6	Sofosbuvir 400 mg daily + PEG/RBV × 12 weeks	No	PEG/RBV × 48 weeks	Yes

3. USE OF NGS FOR IDENTIFICATION OF MICROBIAL PATHOGENS: IS IT POSSIBLE TO MOVE TO "ONE TEST" MICROBIAL DIAGNOSTICS WITH NGS?

Infectious disease has been a flagship for personalized diagnostics in medicine, with diverse strategies for identification of the microbial pathogen and sensitivities to antibiotics. In many cases, these diagnostic tests take days—for example at least 48 h of growth of blood cultures, followed by additional days required to culture the organism and determine sensitivities.

For GWAS, bodily tissues or fluids are obtained, and a DNA library is prepared representing all DNA in the sample—both host and microbial pathogen. Unbiased massively parallel sequencing is performed. Human DNA sequences are removed from consideration, and microbial sequences are aligned with a comprehensive database containing DNA sequences for all known human microbial pathogens. This enables rapid identification of the organism, and often predicts sensitivity to antibiotics.

The following example focuses on the use of NGS to identify a bacterial pathogen in an immune compromised host. However, viral sequences of nonpathogenic strains were also identified, and demonstrate that this approach should be equally applicable for rapid diagnosis of viral pathogens.

A 14-year-old boy with severe combined immunodeficiency disease developed fever and progressively worsening headache. A comprehensive diagnostic work-up including blood cultures, serology, and cerebrospinal fluid (CSF) was negative. Laboratory data were consistent with either a viral meningitis or potential autoimmune encephalitis, and he was discharged without antibiotic therapy.

The patient subsequently progressed, and was readmitted with progressive severe neurological disease. At this point, with informed consent, cerebrospinal fluid was obtained for unbiased massively parallel DNA sequencing. A total of 475 sequences among 3,063,784 total reads corresponded to the spirochete Leptospira, known to be sensitive to penicillin. NGS also identified several other microbes, including nonpathogenic viral species such as Anelloviridae, demonstrating the ability to detect viral sequences. After a total of 7 days of high-dose penicillin, the patient showed evidence of clinical improvement, and was ultimately discharged home with nearly full recovery to his premorbid state.

This single case report suggests the possibility of a future where a single test—unbiased massively parallel sequencing of relevant tissue or bodily fluid—could be used for rapid identification of microbial pathogens and sensitivities. It seems unlikely that this approach would supplant all conventional approaches to microbial pathogen detection. However, it may be a useful adjunct in cases such as this one. And to be provocative, it is conceivable that NGS could ultimately be the "single test" performed for microbial identification and sensitivity testing.

Chapter 22.7

Oncolytic Viruses

Stephanie L. Swift, David F. Stojdl

Stephanie Swift is a scientist and science writer at the Children's Hospital of Eastern Ontario Research Institute, Canada. She has a research background exploring viruses as tools to prevent or treat disease. She writes about exciting new microbe research on her blog, mmmbitesizescience.com.

David Stojdl is a senior scientist at the Children's Hospital of Eastern Ontario Research Institute and an Associate Professor at the University of Ottawa, Canada. He has a long-standing interest in oncolytic viruses, and has translated them into clinical immunotherapies as a co-founder of both Jennerex Biotherapeutics and Turnstone Biologics.

Viruses that selectively replicate in and lyse cancer cells, but leave normal cells intact, are known as "oncolytic." In fact, oncolytic viruses (OVs) are multimodal therapeutics capable of not only lysing cancer cells, but also modulating the tumor microenvironment and collapsing tumor vasculature. This ability to engage multiple therapeutic pathways has clear potential benefits, since cancer treatments designed to apply therapeutic pressure against a single tumor target can trigger an antigenic shift and, ultimately, the development of tumor resistance.

Only relatively recently have we begun to understand that the immune responses triggered by OVs also have a critical role in their antitumor efficacy. Consequently, strategies designed to enhance OV immune engagement represent a key research focus.

Wild-type OVs are ideal agents to shift the tumor immune microenvironment from tolerogenic to antigenic, lysing tumor cells to create inflammatory conditions loaded with damage- and pathogen-associated molecular signals whose level increases as replication proceeds. Indeed, replication is required for efficacy, since nonreplicating viruses and virus-like particles typically fail to achieve tumor control. Ultimately, replicating OVs represent a self-limiting infection, since susceptible tumor cells are a finite population whose availability declines as oncolysis begins.

Systemically delivered OVs can target both primary and disseminated metastatic disease (Breitbach et al., 2011; Russell et al., 2014), and can penetrate both peripheral and lymphoid compartments, unlocking the potential to activate different subtypes of memory CD8+ T-cell responses. While the duration of therapy from systemic OV infusion can be limited by pre-existing antiviral immunity, the de novo development of functional neutralizing antibodies against some OVs is surprisingly delayed, allowing virus to survive in the bloodstream for over a month during repeat intravenous administrations. Expanding the clinical development of such self-protective viruses has clear therapeutic value.

Immune responses are generated not only against the OV itself (ultimately leading to virus control, an important safety feature), but also against the virus-infected tumor cells. Indeed, OV-activated antitumor immunity can mediate objective clinical responses: in phase III trials with Amgen's T-VEC, a modified oncolytic herpes simplex virus encoding GM-CSF, tumor regression was observed both in directly injected tumors and distant noninjected tumors that harbored no detectable virus (Kaufman et al., 2010). Such immune activation can culminate in the establishment of memory populations that protect against tumor rechallenge. Finding new ways to enhance CD8+ T-cell memory formation will be critical for establishing long-term durable responses in patients.

One exciting approach that magnifies the differential immune response in favor of tumor over viral targets encodes a tumor-associated antigen (TAA) into the OV genome to create a bona fide OV vaccine (Bridle et al., 2010). Oncolytic vaccines can be designed to expand naturally established memory CD8+ T cells that recognize tumor-specific targets. For example, human cytomegalovirus (HCMV) antigens have been detected in several tumour types, including glioma and neuroblastoma, and approximately 70% of cancer patients have HCMV-specific memory CD8+ T cells established during a past infection that can be activated by an oncolytic vaccine encoding HCMV TAAs. Alternatively, tumor-reactive T cells can be artificially primed with another vaccine vector expressing a shared TAA, followed by OV infusion in a "prime-boost" regimen (Bridle et al., 2010).

Boosting immune system engagement will continue to be a central theme in the field of oncolytic virotherapy. While OVs can naturally condition the tumor microenvironment to attract T cells and maintain their local activity (Nishio et al., 2014), pairing OVs with other immunomodulatory therapies to promote antitumor immune responses is an exciting approach. Partner therapies may range from the more traditional chemo- and radiotherapies to cutting edge techniques, such as antibody-mediated immune checkpoint inhibition and adoptive cell therapy (ACT). Pairing an OV with anti-CTLA4 has already been shown to improve the control of metastatic melanoma beyond that observed with either treatment alone (Zamarin et al., 2014). Similarly, ACT has shown clinical benefit as a stand-alone treatment for multiple malignancies, and is now showing preclinical promise in combination with OVs (Rommelfanger et al., 2012). While either therapy alone may fail, coadministration of both can compensate for solo deficiencies and render heterogeneous tumors susceptible to therapy.

Oncolytic virotherapy represents an exciting approach for the treatment of malignant disease. Further efforts are needed to consolidate the ability of OVs to (1) activate immune cells against appropriate antigenic tumor targets, (2) improve recruitment and infiltration of immune cells into the tumor microenvironment, and (3) maintain their in situ activity. New therapeutic targets and strategies continue to be uncovered by host-virus screening programs (Mahoney et al., 2011), or informed by mathematical models (Bailey et al., 2013; Le Boeuf et al., 2013). Greater efforts to map the complex interactions between OVs and their host cells, in the context of an expanding immune response, will uncover new therapeutic opportunities. These initiatives will undoubtedly generate anticancer responses that not only substantially enhance tumor regression, but also extend tumor-free survival by protecting against regrowth and relapse.

REFERENCES

Bailey K, Kirk A, Naik S, Nace R, Steele MB, Suksanpaisan L, et al. Mathematical model for radial expansion and conflation of intratumoral infectious centers predicts curative oncolytic virotherapy parameters. PLoS One 2013;8(9):e73759.

Breitbach CJ, Burke J, Jonker D, Stephenson J, Haas AR, Chow LQ, et al. Intravenous delivery of a multi-mechanistic cancer-targeted oncolytic poxvirus in humans. Nature 2011;477(7362):99–102.

Bridle BW, Stephenson KB, Boudreau JE, Koshy S, Kazdhan N, Pullenayegum E, et al. Potentiating cancer immunotherapy using an oncolytic virus. Mol Ther 2010;18(8):1430–9.

Kaufman HL, Bines SD. OPTIM trial: a Phase III trial of an oncolytic herpes virus encoding GM-CSF for unresectable stage III or IV melanoma. Future Oncol 2010;6(6):941–9.

Le Boeuf F, Batenchuk C, Vaha-Koskela M, Breton S, Roy D, Lemay C, et al. Model-based rational design of an oncolytic virus with improved therapeutic potential. Nat Commun 2013;4:1974.

Mahoney DJ, Lefebvre C, Allan K, Brun J, Sanaei CA, Baird S, et al. Virus-tumor interactome screen reveals ER stress response can reprogram resistant cancers for oncolytic virus-triggered caspase-2 cell death. Cancer Cell 2011;20(4):443–56.

Nishio N, Diaconu I, Liu H, Cerullo V, Caruana I, Hoyos V, et al. Armed oncolytic virus enhances immune functions of chimeric antigen receptor-modified T cells in solid tumors. Cancer Res 2014;74:2195–205.

Rommelfanger DM, Wongthida P, Diaz RM, Kaluza KM, Thompson JM, Kottke TJ, et al. Systemic combination virotherapy for melanoma with tumor antigen-expressing vesicular stomatitis virus and adoptive T-cell transfer. Cancer Res 2012;72(18):4753–64.

Russell SJ, Federspiel MJ, Peng KW, Tong C, Dingli D, Morice WG, et al. Remission of disseminated cancer after systemic oncolytic virotherapy. Mayo Clin Proc 2014;89(7):926–33.

Zamarin D, Holmgaard RB, Subudhi SK, Park JS, Mansour M, Palese P, et al. Localized oncolytic virotherapy overcomes systemic tumor resistance to immune checkpoint blockade immunotherapy. Sci Transl Med 2014;6(226):226ra232.

Chapter 22.8

Prions and Chronic Diseases

Stanley B. Prusiner

Stanley Prusiner is Director of the Institute for Neurodegenerative Diseases and professor of neurology at the University of California, San Francisco. Prusiner discovered prions—proteins that cause neurodegenerative diseases in animals and humans, for which he received the 1997 Nobel Prize in Physiology or Medicine.

Prions are proteins that adopt alternative conformations, which become self-propagating (Prusiner, 2013). Generally, one conformation is rich in β-sheet, a conformation that is prone to polymerization into amyloid fibrils.

Looking into the future of prion biology and diseases, a rich universe of previously unknown biology is beginning to emerge (Prusiner, 2014). The number of different physiological prions, particularly in mammals, is steadily increasing. Physiological prions play a role in various normal functions, ranging from long-term memory to innate immunity to metabolic adaption to fungal incompatibility (Xu et al., 2014). Similarly, the list of neurodegenerative diseases caused by prions is expanding (Jucker and Walker, 2013; Prusiner, 2013). Over the last 5 years, evidence has continued to accumulate, arguing that Alzheimer's and Parkinson's diseases as well as multiple system atrophy, the tauopathies, and Huntington's disease are caused by prions (Stöhr et al., 2014; Watts et al., 2013, 2014). It seems likely that in addition to Huntington's disease, some of the other polyQ (polyglutamine) disorders including the spinocerebellar ataxias will be prion diseases.

Several important new concepts have emerged from the study of prions. First, prions create a novel mechanism whereby physiological functions can be regulated almost instantaneously, by shifting the conformation of a protein from one structural state to another (Prusiner, 2014). Second, prions often feature in diseases where a particular protein accumulates inside cells, as found in neurofibrillary tangles, Lewy bodies, glial cytoplasmic inclusions, and nuclear inclusions. In other disorders, prions accumulate outside of cells, like the plaques in Alzheimer's and Creutzfeldt–Jakob diseases. Third, strains of prions with different phenotypes represent distinct conformations of these alternatively folded proteins. Fourth, the late onset of heritable neurodegenerative diseases seems likely to be explained by the conversion of the mutant causative protein into a prion as the precipitating event. Fifth, prion diseases are age dependent; this is likely due to the protein quality control machinery, which slowly becomes less efficient as organisms age.

While many explanations for late onset, heritable neurodegenerative diseases have been offered to explain their manifestations in the fifth, sixth, or even seventh decade of life, it seems more likely that the initial event is the formation of a sufficient number of prions to stimulate sustainable self-propagation (Prusiner, 2013). Although mutations in patients with familial neurodegenerative diseases have been demonstrated to cause these disorders by genetic linkage studies, explaining the late onset of these illnesses has remained problematic. One explanation is that a stochastic event results in a sufficient number of prions accumulating to initiate a sustainable infection. With aging, an increase in the frequency of random events that produce prions, in tandem with a decline in the protein quality control machinery, conspire to produce sustainable prion infections. This mechanism is applicable to both the inherited and sporadic prion diseases.

These new concepts may offer some novel approaches to developing both early diagnostics and effective therapeutics. Currently, there are no drugs that halt or even slow any neurodegenerative disease (Prusiner, 2014). Developing PET (positron emission tomography) reporters that can be used to establish the diagnosis early in the course of disease will be critical. Accurate and early diagnoses are likely to be critical in choosing appropriate therapeutics.

Although attempts to develop effective therapeutics for Alzheimer's and Parkinson's diseases have been both costly and unsuccessful, the discovery that these diseases are caused by prions offers new strategies for drug discovery. Notably, some point mutations have been found in the PrP protein that causes Creutzfeldt–Jakob disease, which have been shown to be dominant negatives (Prusiner, 2013). Deciphering the structural changes that such mutations initiate may give important insights that could inform novel therapeutic approaches. Drugs have been developed that extend the lives of wild-type and transgenic mice inoculated with mouse-passaged scrapie prions and chronic wasting disease prions, respectively (Berry et al., 2013), which gives promise of future therapeutics.

How many systemic diseases will be found to be caused by prions is unknown. Certainly, there is much interest in the possibility that adult-onset type II diabetes may be caused by prions. In such patients, the β-islet cells are often filled with amyloid fibrils composed of the protein amylin.

The area of prion biology and diseases is certainly "ripe" for increased investigation. Our knowledge of most physiological prions is in its infancy. Learning how such prions propagate is likely to offer novel approaches to therapeutics for such diseases like Alzheimer's and Parkinson's that are already prevalent and predicted to increase as human life expectancy continues to rise.

REFERENCES

Berry DB, Lu D, Geva M, Watts JC, Bhardwaj S, Oehler A, et al. Drug resistance confounding prion therapeutics. Proc Natl Acad Sci USA 2013;110:E4160–9.

Jucker M, Walker LC. Self-propagation of pathogenic protein aggregates in neurodegenerative diseases. Nature 2013;501:45–51.

Prusiner SB. Biology and genetics of prions causing neurodegeneration. Annu Rev Genet 2013;47:601–23.

Prusiner SB. Madness and memory. New Haven: Yale University Press; 2014.

Stöhr J, Condello C, Watts JC, Bloch L, Oehler A, Nick M, et al. Distinct synthetic Aβ prion strains producing different amyloid deposits in bigenic mice. Proc Natl Acad Sci USA 2014;111:10329–34.

Watts JC, Giles K, Oehler A, Middleton L, Dexter DT, Gentleman SM, et al. Transmission of multiple system atrophy prions to transgenic mice. Proc Natl Acad Sci USA 2013;110:19555–60.

Watts JC, Condello C, Stöhr J, Oehler A, Lee J, DeArmond SJ, et al. Serial propagation of distinct strains of Aβ prions from Alzheimer's disease patients. Proc Natl Acad Sci USA 2014;111:10323–8.

Xu H, He X, Zheng H, Huang LJ, Hou F, Yu Z, et al. Structural basis for the prion-like MAVS filaments in antiviral innate immunity. Elife 2014;3:e01489.

Chapter 22.9

Pathogenesis Research and the HIV/AIDS Pandemic

Anthony S. Fauci, Hilary D. Marston

Anthony S. Fauci is Director of the National Institute of Allergy and Infectious Diseases at the U.S. National Institutes of Health, and chief of the NIAID Laboratory of Immunoregulation. He has made many contributions to the understanding of the pathogenesis of HIV/AIDS and to antiretroviral therapy, and he was a principal architect of the President's Emergency Plan for AIDS Relief (PEPFAR).

Hilary D. Marston is a Medical Officer and Policy Advisor for Global Health at the National Institute of Allergy and Infectious Diseases, U.S. National Institutes of Health. She is an internal medicine physician who has worked with Partners in Health and the Bill and Melinda Gates Foundation.

The global HIV/AIDS pandemic continues to exact an enormous toll, claiming 1.2 million lives in 2014 alone and 34 million since AIDS was recognized more than three decades ago. Worldwide, 37 million individuals live with HIV/AIDS. Despite these daunting statistics, the global deployment of proven treatment and prevention strategies has slowed the onslaught of HIV/AIDS, with both incident infections and deaths falling by more than one-third over the past decade. These successes did not come easily. They were the result of decades of innovation beginning with fundamental basic research that led to successful interventions. Indeed, basic research on HIV/AIDS is inexorably linked to the development of effective interventions for the disease. In this regard, the development of new tools for the treatment and prevention of HIV has been facilitated by a detailed understanding of HIV pathogenesis.

Studies on viral pathogenesis underpin all HIV research, but perhaps most tangibly the field of therapeutics. Detailed knowledge of the replication cycle of HIV provided the first targets for antiretroviral drugs (ARVs). The first FDA-approved ARV, the nucleoside analog zidovudine, targeted the reverse transcriptase enzyme, a critical component of the HIV replication cycle that converts viral RNA to proviral DNA, thus allowing integration into the host cell genome. Zidovudine was first synthesized as an antineoplastic agent years prior to the discovery of HIV and was found to have activity against HIV during a screening process. As knowledge of the HIV replication cycle improved, new classes of therapeutic agents were developed. Notably, inhibitors of the HIV protease and integrase enzymes were synthesized and optimized based on detailed crystal structures of the target proteins. Similar investigations have yielded more than 30 licensed ARVs and ARV combinations, all dependent on an intimate knowledge of the viral replication cycle. By March of 2015, 15 million people were receiving ARVs, averting an estimated 7.8 million deaths between 2000 and 2014 (UN Joint Programme on HIV/AIDS (UNAIDS), 2015).

ARVs have similarly revolutionized HIV prevention, notably through prevention of mother-to-child transmission (PMTCT) programs, which averted 1.4 million infections since 2000 (UN Joint Programme on HIV/AIDS (UNAIDS), 2015). ARVs have also proved effective when used for pre-exposure prophylaxis (PrEP) for uninfected individuals, and as "treatment as prevention" or TasP when taken by HIV-infected people. Oral PrEP demonstrated more than 90% efficacy in preventing viral acquisition when taken as prescribed (Haberer et al., 2013). The landmark study HPTN 052 demonstrated the value of TasP, whereby antiretroviral therapy (ART) given to the infected partner in a serodiscordant couple lowers his or her viral load, thus reducing the risk of transmitting the virus to the uninfected sexual partner by 96% (Cohen et al., 2011). These ARV-based prevention modalities, combined with other interventions such as condoms and voluntary medical male circumcision, provide the building blocks of comprehensive prevention programs.

Understanding the pathogenic mechanisms of HIV infection and the details of the immune response to the virus are also critical to the development of a safe and effective vaccine to prevent HIV infection. In this regard, the development of an effective HIV vaccine is a formidable challenge due to the fact that the natural immune response to HIV is inadequate in controlling and certainly in eliminating the virus. Decades of disappointing clinical trials, first with vaccines aimed at the induction humoral immunity and then cell-based, "T-cell" vaccines, failed to produce an effective immune response. The RV-144 trial in Thailand offered the first signs of clinical efficacy, with a 31% reduction in viral acquisition (Rerks-Ngarm et al., 2009). The RV-144 regimen (canarypox vector prime, recombinant gp120 boost) appears to have elicited non- or weakly neutralizing antibodies against the V1V2 region of the envelope trimer. RV-144 represented the first moderately successful study of an HIV vaccine candidate in humans and reinvigorated the field of HIV vaccinology.

Simultaneously, researchers have deepened their understanding of the immune response to HIV, discovering broadly neutralizing antibodies (bNAbs) induced over the course of chronic infection. It is curious that these bNAbs are produced by only a minority of infected individuals (~20%), and usually after 2 or more years of infection (Liao et al., 2013). Serial blood samples taken from an acutely infected donor revealed the coevolution of viral mutations and a broadening humoral immune response to the virus. At the end of more than 2 years, a hypermutated, broadly reactive anti-HIV antibody evolved together with a highly mutated virus that was continually trying to escape the immune response. Therefore, in trying to evade the evolving antibodies, the virus ultimately stimulates bNAb production, a paradox of HIV immunology. Current research is aimed at determining if these bNAbs can prevent or treat HIV infection. If successful, the major challenge in HIV vaccinology will be to induce these antibodies via appropriate immunogens. Recently, the potential for bNAbs, whether infused directly or produced in vivo by gene inserts administered via viral vectors, to prevent or treat infection has been demonstrated in animal models (Balazs et al., 2014; Shingai et al., 2013; Barouch et al., 2013). In this regard, vaccinologists are attempting to recapitulate in a more rapid and expeditious manner the natural immune response seen in the minority of infected individuals through prime-boost vaccination regimens. Specifically, they are identifying viral envelope epitopes that induce bNAbs in natural infection, and expressing those as immunogens to induce B-cell maturation toward bNAb-producing cells. Thus, the understanding of the complexities of HIV pathogenesis has offered new hope for a moderately effective HIV vaccine.

Research toward a cure for HIV infection is also closely linked to an understanding of HIV pathogenesis. In pursuing a cure, it is important to first define the goal of the work. Simply put, a "cure" is an indefinite remission of disease following cessation of ART. In HIV, this could come in one of two forms—viral eradication or sustained virologic remission (SVR); for the latter the virus would remain at low levels in the absence of daily ART. The now famous case of the "Berlin Patient" offers some evidence that eradication is possible. Still, reliance on hematopoietic stem-cell transplant is unlikely to be feasible and certainly will be risky for the majority of infected individuals. Novel scientific approaches have employed gene-editing techniques to engineer ex vivo mutations into the CCR5 coreceptor for HIV on autologous T cells followed by re-infusion into the donor host, thus rendering them "resistant" to HIV infection (Tebas et al., 2014). These efforts are at an early stage of discovery; however, they merit attention.

The case of the "Mississippi Child" offers evidence that an SVR is possible. The infant was infected with HIV in utero and started on ART within 30 h of birth. ART was continued through 18 months at which point the child was lost to follow up and stopped treatment. When the child represented to care 5 months later, there were no traces of replication competent virus, a status that persisted for 27 months before the virus ultimately rebounded (Persaud et al., 2013; National Institute of Allergy and Infectious Diseases (NIAID)). The case indicates that treatment soon after infection can minimize, but not eliminate HIV reservoirs resulting in long-term SVR. Translating this experience into adults could be possible, as evidenced by 14 patients in France treated during acute HIV infection who maintained an SVR after treatment interruption (Sáez-Cirión et al., 2013). However, these results need to be confirmed in additional studies. Establishing an SVR in the absence of continual ART for a broader population will likely require adjuvant therapies such as therapeutic vaccines or passive infusion of bNAbs following the initial suppression of viremia.

In summary, the development of effective interventions for the prevention and treatment of HIV infection, critical to any hope of controlling and ultimately ending the HIV/AIDS pandemic, is heavily dependent on an in-depth understanding of the viral and immune pathogenesis of HIV disease.

REFERENCES

Balazs AB, Ouyang Y, Hong CM, et al. Vectored immunoprophylaxis protects humanized mice from mucosal HIV transmission. Nat Med 2014;20:296–302.

Barouch DH, Whitney JB, Moldt B, et al. Therapeutic efficacy of potent neutralizing HIV-1-specific monoclonal antibodies in SHIV-infected rhesus monkeys. Nature 2013;503:224–8.

Cohen MS, Chen YQ, McCauley M, et al. Prevention of HIV-1 infection with early antiretroviral therapy. N Engl J Med 2011;365:493–505.

Haberer JE, Baeten JM, Campbell J, et al. Adherence to antiretroviral prophylaxis for HIV prevention: a substudy cohort within a clinical trial of serodiscordant couples in East Africa. PLoS Med 2013;10:e1001511.

Liao HX, Lynch R, Zhou T, et al. Co-evolution of a broadly neutralizing HIV-1 antibody and founder virus. Nature 2013;496:469–76.

National Institute of Allergy and Infectious Diseases (NIAID). "Mississippi baby" now has detectable HIV, Researchers Find. Available at: http://www.niaid.nih.gov/news/newsreleases/2014/Pages/Mississippi-BabyHIV.aspx [accessed 10.07.14].

Persaud D, Gay H, Ziemniak C, et al. Absence of detectable HIV-1 viremia after treatment cessation in an infant. N Engl J Med 2013;369:1828–35.

Rerks-Ngarm S, Pitisuttithum P, Nitayaphan S, et al. Vaccination with ALVAC and AIDSVAX to prevent HIV-1 infection in Thailand. N Engl J Med 2009;361:2209–20.

Sáez-Cirión A, Bacchus C, Hocqueloux L, et al. Post-treatment HIV-1 controllers with a long-term virological remission after the interruption of early initiated antiretroviral therapy ANRS VISCONTI Study. PLoS Pathog 2013;9:e1003211.

Shingai M, Nishimura Y, Klein F, et al. Antibody-mediated immunotherapy of macaques chronically infected with SHIV suppresses viraemia. Nature 2013;503:277–80.

Tebas P, Stein D, Tang WW, et al. Gene editing of CCR5 in autologous CD4 T cells of persons infected with HIV. N Engl J Med 2014;370:901–10.

UN Joint Programme on HIV/AIDS (UNAIDS). How AIDS changed everything — MDG6: 15 years, 15 lessons of hope from the AIDS response. 2015. Available at: http://www.unaids.org/sites/default/files/media_asset/MDG6Report_en.pdf [accessed 8.24.15].

Chapter 22.10

The Future of Viral Vaccines

Gary J. Nabel

Gary Nabel is Chief Scientific Officer, Sanofi, a global pharmaceutical company. Dr Nabel served as Director of the Vaccine Research Center of the National Institute of Allergy and Infectious Diseases, 1999–2012, where he guided research on the development of novel vaccine strategies against HIV and other emerging infectious diseases.

Progress in understanding the pathogenesis of viral infections has stimulated innovative approaches to the development of vaccines. This work builds upon insights from the basic sciences, including virology, microbiology, immunology, molecular biology, and genetics. In addition, advances in biotechnology are generating alternative platforms to elicit specific immune responses that facilitate the development of next-generation vaccines. Finally, the tools of molecular medicine provide new insights into immune responses induced by vaccines, as well as associated adverse reactions. This understanding will accelerate clinical vaccine development and the identification of biomarkers that predict a successful protective response. Together, the underlying science lays the groundwork for the development of promising vaccines that can be both safe and efficacious in protecting against a variety of pathogens.

1. BUILDING ON BASIC SCIENCE

In the basic sciences, our understanding of T- and B-cell differentiation, as well as the details of antigen recognition and signaling, is leading to insights into the pathways of immune maturation. Recent research has also helped to identify structural features of antigens that are required to elicit a specific immune response. For example, it is now possible to identify the determinants of antigens that will engage germ-line B cells and promote their differentiation into memory B cells that produce broadly neutralizing antibodies to influenza (Lingwood et al., 2012). Similarly an understanding of T-cell biology, both maturation and signal transduction following interactions with antigen-presenting dendritic cells, have pointed to rational approaches to immunogen design. This has led to improved adjuvants, and ways to modulate the balance of Th1 or Th2 immunity. Progress in understanding innate immunity, including the TLR signaling, interferon activation and Rig-I stimulation provide tools to selectively activate or suppress these pathways as needed. In the future, the ability to develop small-molecule agonists or antagonists against these or immune-suppressive targets, such as checkpoint or TGF-beta inhibitors will further empower these efforts.

The ability to utilize biologics therapy, such as monoclonal antibodies, nanoparticles, or peptides, provides diverse and more effective ways to elicit durable T-cell responses that mediate protective cellular immunity. These approaches will increase the likelihood of success against many challenging infectious disease agents, such as cytomegalovirus, Ebola and Marburg viruses, or HIV-1. Similarly they can be harnessed to elicit protective antibody responses. Such tools could enable the development of a universal influenza vaccine that will better protect public health against strain drift and the emergence of new influenza virus pathogens from animal reservoirs.

Advances in DNA sequencing have already enabled more rapid and rational responses to evolving outbreaks and to the identification of the mutations that render vaccines ineffective. Such genetic analysis is now routinely performed for HIV-1, influenza, and Ebola viruses. As more is understood about the structural implications of these mutations, rational design of new vaccines to counter viral resistance will be enhanced. In addition, as the molecular epidemiology is further understood, it will become increasingly possible to develop preemptive vaccination strategies, as proposed in the past for avian influenza (Yang et al., 2007). This will be based on an understanding of viral evolution, human immunity, and predictive patterns of viral mutation. In addition to genetic sequencing of pathogens, much will be learned from the genetic polymorphisms of humans. When human genome sequencing becomes affordable and routine, it will undoubtedly provide insights into optimizing immunogenicity. It will also help to probe the causes of adverse responses that may limit the use of some vaccines or immunotherapies.

2. IMPROVEMENTS IN VACCINE DELIVERY

Viral pathogenesis, particularly the study of viral assembly, has contributed to a better understanding of synthetic biology, which will facilitate new approaches to the production of safe and more effective vaccines. The increased success of virus-like particles as vaccines is encouraging, because the immunogen closely resembles native virus and elicits effective immunity while the particle is unable to replicate and cause adverse responses. Next-generation improvements, involving the use of synthetic biology and nanotechnology (Kanekiyo et al., 2013), are expected to create scaffolds and antigen presentation surfaces that will better expose specific epitopes on viral protein. This will make it possible to target responses to subdominant epitopes not normally recognized in the immune response. These advances may permit targeting of highly conserved and vulnerable structures—normally protected by the virus—that can induce broad protective responses immune responses (reviewed in refs (Nabel and Fauci, 2010; Nabel, 2013)). Finally, as the methodology for expressing engineered molecules improves, it is increasingly possible to modulate immune function through the expression of novel molecules, or through genetic delivery of antibodies (Johnson et al., 2009; Balazs et al., 2011). Bi-specific antibodies allow for dual targeting against different epitopes on the surface of pathogens (Byrne et al., 2013). They also provide a mechanism to redirect immune cells that normally do not respond to a specific antigen. The tool of gene delivery can be used to produce selected antibodies in subjects who cannot be induced to make them. For example, this will enhance the induction of neutralizing influenza antibodies in the elderly (Limberis et al., 2013; Balazs et al., 2013), or by modulating the specificity of T cells to recognize antigenic determinants using chimeric antigen receptors for cancer vaccines (Jensen and Riddell, 2014). While the T-cell chimeric antigen receptor approach has currently been directed to cancer immunotherapy, it may also prove efficacious against infectious disease targets.

3. ADVANCED ANALYTICS AND THE HUMAN IMMUNE RESPONSE

The ability to interrogate human immune responses has expanded greatly in recent years. Powerful technologies—such as flow cytometry and nanofluidics—now enable detailed qualitative and quantitative interrogation of human immune responses. Similarly, the ability to perform deep sequencing of selected tissues or immune cells provides insight into the ontogeny of the adaptive immune response. At the same time, modern imaging techniques have been applied in vivo and are facilitating an understanding of the trafficking of these cells in response to specific stimuli. This information will lead both to a better understanding of the mechanisms of immune protection as well as to the definition of correlates of immunity against specific pathogens. It will also facilitate the development of new vaccines, or improvements of existing vaccines, which will confer broader reactivity and fewer adverse effects in the large numbers of people who can benefit from them.

Finally, it is important to recognize that global surveillance of viral infections, as well as rapid international vaccine distribution, will be greatly assisted by advances in pathogen detection, molecular definition of microbial resistance, and infectious disease surveillance. The ability of electronic and Web-based monitoring will facilitate efforts to distribute vaccines to populations at risk, particularly when unexpected outbreaks occur. The recent epidemic spread of Ebola virus in western Africa highlights the rapidity with which emerging pathogens can spread across international borders. Vaccines represent an essential tool to counter the pandemic spread of infectious diseases and preserve the public health. The welfare of people throughout the world has become increasingly dependent on our ability to provide effective countermeasures against emerging infectious disease threats. In the future, the more effective and efficient development of vaccines, as well as more timely distribution, will increasingly protect the public against these pathogens.

REFERENCES

Balazs AB, et al. Antibody-based protection against HIV infection by vectored immunoprophylaxis. Nature 2011;481:81–4.

Balazs AB, Bloom JD, Hong CM, Rao DS, Baltimore D. Broad protection against influenza infection by vectored immunoprophylaxis in mice. Nat Biotechnol 2013;31(7):647–52.

Byrne H, Conroy PJ, Whisstock JC, O'Kennedy RJ. A tale of two specificities: bispecific antibodies for therapeutic and diagnostic applications. Trends Biotechnol 2013;31(11):621–32.

Jensen MC, Riddell SR. Design and implementation of adoptive therapy with chimeric antigen receptor-modified T cells. Immunol Rev 2014;257(1):127–44.

Johnson PR, et al. Vector-mediated gene transfer engenders long-lived neutralizing activity and protection against SIV infection in monkeys. Nature Med 2009;15:901–6.

Kanekiyo M, Wei CJ, Yassine HM, McTamney PM, Boyington JC, Whittle JR, et al. Self-assembling influenza nanoparticle vaccines elicit broadly neutralizing H1N1 antibodies. Nature 2013;499(7456):102–6.

Limberis MP, Adam VS, Wong G, Gren J, Kobasa D, Ross TM, et al. Intranasal antibody gene transfer in mice and ferrets elicits broad protection against pandemic influenza. Sci Transl Med 2013;5(187):187ra72. 29.

Lingwood D, McTamney PM, Yassine HM, Whittle JR, Guo X, Boyington JC, et al. Structural and genetic basis for development of broadly neutralizing influenza antibodies. Nature 2012;489(7417):566–70.

Nabel GJ, Fauci AS. Induction of unnatural immunity: prospects for a broadly protective universal influenza vaccine. Nat Med 2010;16(12):1389–91.

Nabel GJ. Designing tomorrow's vaccines. N Engl J Med 2013;368(6):551–60.

Yang Z-Y, Wei C-J, Kong W-P, Wu L, Xu L, Smith DF, et al. Immunization by avian H5 influenza hemagglutinin mutants with altered receptor binding specificity. Science 2007;317(5839):825–8.

Chapter 22.11

Emerging Viruses

W. Ian Lipkin

W. Ian Lipkin is the John Snow Professor of Epidemiology, Professor of Neurology and Pathology, and Director of the Center for Infection and Immunity at Columbia University. He pioneered the use of molecular methods in identifying viruses and other infectious agents in acute and chronic diseases, and in responding to infectious disease outbreaks including West Nile encephalitis, SARS, MERS, and Ebola.

In the late 1990s, I attended a retirement symposium for an eminent virologist at which a speaker of the same vintage bemoaned the end of the golden age of virology. In a more recent symposium honoring the life of the late Hilary Koprowski, a nonvirologist speaker asked whether viruses are alive. As you are reading the epilogue of this book, you would have firm evidence to dispute the waning status of virology and to respond "no" to the second. You would likely add that although viruses are not living things, they are critical to life. Viral sequences comprise 8% of our genomes and are far from inert. The endogenous retroviral element syncytin, for example, is essential for placental development and embryo survival (Mi et al., 2000). The field of medical virology is alive and vibrant. New viruses continue to emerge, posing threats to public health, food security, and commerce. Viral databases are rapidly expanding as investigators survey animals and environments using ever more efficient and inexpensive sequencing platforms. With the introduction of new antiviral drugs, therapeutic antibodies and vaccines, viral diagnosis has become more than an arcane academic exercise. Evidence is mounting that viral infections contribute to chronic diseases including neurodevelopmental disorders and some forms of cancer. Viruses have been harnessed in oncology (Miest and Cattaneo, 2014) and gene replacement therapy. On a global scale, there is growing appreciation that viruses contribute to elemental cycling and oceanic carbon sequestration through effects on phytoplankton (Suttle, 2007). Given the allotted space and the focus of this book on pathogenesis, I can only touch on a few predictions with medical applications but encourage the reader to think more broadly about the implication of viruses and virology.

Globalization of travel and trade, loss of wildlife habitats, growth of megacities, mass migrations due to economic privation and political instability and changes in the distribution of mosquitoes due to climate change will continue to enable the emergence of viruses that might otherwise remain sequestered (Lipkin, 2013). We recently estimated that mammals alone harbor more than 300,000 new viruses (Anthony et al., 2013). At present, we have no way of ascertaining from viral sequence data alone which of them poses substantive threats to humans, wildlife, or domestic animals. However, investments in sequencing, bioinformatics, and systems biology may translate into algorithms that allow us to assess potential for host switching and pathogenicity. At minimum, knowledge of which viruses are circulating and where will enable targeted surveillance in populations at risk for exposure. Surveillance will also become increasingly efficient as diagnostic capacity improves in the developing world. It will be critical to address potential concerns regarding sovereignty and intellectual property if we are to ensure that investigators will be amenable to sharing data and isolates. Public enthusiasm for social media may lead to a global viral equivalent of the American Gut Project wherein citizen scientists share fecal samples and data to develop a human bacteriome database linked to information concerning diet, geography, season, health, and disease (http://humanfoodproject.com/americangut/). A key question that remains to be addressed is whether there are commensal or symbiotic viral flora. Do viruses, like bacteria, have a role in priming or regulating the immune system? Do viruses (primarily bacteriophage) regulate the composition or abundance of bacterial flora? If so, is there an optimum viral microflora and how is it established? Can it be modified?

Improvements in diagnostics and insights into the role of viruses in health and disease will provide incentives for the development of antiviral drugs and vaccines. Our armamentarium for herpesviruses, HIV and HCV will expand to include drugs that address not only chronic infections like HPV and HBV that have life threatening sequelae but also acute, self-limited infections (e.g., rhinoviruses) that interfere with activities of daily living and productivity. At present, antiviral discovery typically begins with screening of massive compound libraries. Compounds with activity are then modified and optimized through medicinal chemistry. This brute force approach will ultimately give way to more elegant strategies for rational drug design based in genomics, proteomics, structural biology, and cellular biology. These drugs will target not only the viruses themselves, but also host responses that contribute to viral replication, morbidity, and mortality. Insights into viral biology and evolution and host response will profoundly impact vaccine research. Vaccines will be optimized to expedite the development of protective immune responses, enhance immunity in the very young and the very old, and to increase the duration of protection. Vaccines will target conserved conformational domains to enable immunity to representatives of higher order viral taxa rather than only specific strains. New platforms will be established that facilitate inexpensive, rapid production and atraumatic immunization.

Finally, one wonders what Peter Medawar, reported to have described viruses as bad news wrapped up in protein, would make of the deliberate use of viruses in medicine. Viruses are ideal vectors for intracellular delivery of genetic information. As extracellular and intracellular determinants of tropism are defined, I anticipate that viruses will become increasingly important tools for targeted destruction of neoplastic cells and expression of RNA and proteins that enhance cell function.

In summary, virology is alive and well. As Timbuk3 sang in 1986, "The future's so bright, I gotta wear shades."

REFERENCES

Anthony SJ, Epstein JH, Murray KA, Navarrete-Macias I, Zambrana-Torrelio CM, Solovyov A, et al. A strategy to estimate unknown viral diversity in mammals. mBio 2013;4:e00598–00513.

Lipkin WI. The changing face of pathogen discovery and surveillance. Nature Rev Microbiol 2013;11:133–41.

Mi S, Lee X, Li X, Veldman GM, Finnerty H, Racie L, et al. Syncytin is a captive retroviral envelope protein involved in human placental morphogenesis. Nature 2000;403:785–9.

Miest TS, Cattaneo R. New viruses for cancer therapy: meeting clinical needs. Nature Rev Microbiol 2014;12:23–34.

Suttle CA. Marine viruses–major players in the global ecosystem. Nature Rev Microbiol 2007;5:801–12.

Chapter 22.12

Pandemics: What Everyone Needs to Know

Peter Doherty

Peter Doherty works at the University of Melbourne and St Jude Children's Research Hospital. He has broad interests in virus pathogenesis, latterly influenza, and shared the 1996 Nobel Prize in Physiology or Medicine for discoveries about T-cellmediated immunity. In his spare time, he writes science books for lay audiences, including "Pandemics: What Everyone Needs to Know" and "Their Fate is Our Fate: How Birds Foretell Threats to Our Health and Our World."

Back in 2012 I was invited to write for the *What everyone needs to know* series published by Oxford University Press (Doherty, 2013). The subjects range from: *China in the twenty-first Century*, to *The Catholic Church*, to *Food Politics*, and so on. My charge was: *Pandemics*. I thought about this for a while as, being a laboratory scientist rather than a public health medico or an epidemiologist, I wondered whether I was the right person for the job. But then, perhaps in a triumph of optimism over rationality, I decided that, after working for 50 years on viruses and immunity, I might just have something useful to say. Also, since back in 1996, the Swedes made me more famous than I deserve to be, I have been covering a much broader remit in public lectures, talking to legislators and so forth than was the case when I was just known in my academic field.

It also occurred to me that any informed individual is likely to take a different angle. Moreover, I do know something about the influenza A viruses that are, after all, the most likely cause of future pandemics. Another motivation was that this gave me a chance to talk about how great the science of infectious disease has become over the past decades. I am less than enthusiastic about the fear mongering, "shock horror be afraid" scenarios that drive many books and movies on emerging pathogens. On the other hand, I also wanted to make the case that, in these days of cost-cutting governments, it is essential that we keep our public health services strong and that we continue to fund research on dangerous pathogens at a good level. Perhaps it is our evolutionary history, but we seem more enthusiastic about military spending to keep "bad guys" in their place than we are about preparing for an attack by "bad bugs." It is also important that, with all the media hype about "gain-of-function experiments" we do not impose excessive regulation that discourages talented people from working with exotic viruses.

Writing a "lay" book on such a subject has its challenges. It helped that my medical infectious disease colleagues were happy to talk and review whatever I wrote. Then this particular series is done in a Question and Answer format: it is odd to sit in front of a computer alone, invent questions, and then provide answers. This is probably how paranoids and conspiracy theorists operate! And you are subject to editors trained in a literature rather than science. That led, for example, to a challenging chapter summarizing infection and immunity for a general reader.

One surprise was to find that there is no universal definition of what constitutes a pandemic versus an epidemic. A pandemic alert is sounded for a novel influenza A virus when it spreads between two WHO regions. Look at a WHO map and you will realize this is a pretty arbitrary definition. When an escape mutant of a currently circulating flu virus goes global that is described as a "seasonal" epidemic though, for the same situation with the noroviruses, it is called a pandemic.

The disease I got half wrong was Ebola. Going on past history, my view was that any Ebola outbreak would be jumped on fast and quickly contained. We now know that this is not necessarily so. The lesson: we cannot expect already overstretched missionary and volunteer organizations, like Doctors Without Borders, to handle something like this. Was the dilatory Western response a direct consequence of financial cutbacks? I do not know, but we just did not get enough well qualified "boots on the ground" soon enough. I have also realized that the world needs a new economic model to bring antiviral drugs and vaccines to the post development/human trial phase for potential pandemic risk pathogens, so we can go into immediate, large-scale production in the face of a dangerous outbreak. I am in no doubt that could have been "ready to go" for Ebola, but there just was not the money to do it. This is a global responsibility, and we cannot expect that "big pharma" will, without financial compensation, take up the challenge.

Pandemics: what everyone needs to know was published in October 2013. Even with all the publicity around the Ebola outbreak, sales "grumble slowly." *China in the twenty-first century* is doing a lot better! It is hard to get people to engage with science and, as anyone who frequents bookstores knows, the whole science section is usually smaller than that for "alternative medicine." I think *Pandemics* is readable, honest, and informative, but it will never hit the sales heights of the "be terrified" genre!

REFERENCE

Doherty PC. Pandemics: what everyone needs to know. New York: Oxford University Press; 2013.

Chapter 22.13

One Health

Thomas P. Monath

Tom Monath is Chief Science Officer of BioProtection Systems/NewLink Genetics Corporation, where he is developing a vaccine against Ebola virus. His career spans 20 years in vaccine development in the biotechnology industry, preceded by 25 years at CDC and USAMRIID leading research on arbovirus epidemiology and pathogenesis.

One Health is a conceptual framework that seeks to establish "collaborative efforts of multiple disciplines..." including physicians, veterinarians, environmental and climate scientists and others "...working locally, nationally and globally to attain optimal health for people, animals and our environment." While not a new concept, One Health has gained considerable traction in the past 5 years, and has been embraced by academic institutions, professional societies, and governments (AVMA, 2008).

Underlying this momentum is the fact that zoonoses (diseases transmissible from animals to humans) represent ~60% of all infectious pathogens of human beings and 70% of all emerging infectious diseases. At this writing, a horrific epidemic of Ebola virus disease, a virus carried by fruit bats but capable of interhuman contact spread, has riveted the world's attention. Among emerging infections, viral pathogens are over-represented, since these agents are associated with host species having high population turnover and density surges; evolve genetic changes rapidly, permitting adaptation to new hosts and vectors (species jumping); may be infectious at very low doses due to their high replicative capacity; and are often shed in secretions. The importance of zoonotic viral infections to human and animal health, the complexity of virus life cycles, and the multifactorial causes of disease emergence, underlie the need to integrate a variety of scientific disciplines in their study.

In addition to the viruses that are known to infect both animals and humans, there are many viruses that cause disease only in animals or that circulate silently in animals. These include some that are related genetically to human pathogens, such as the hepatitis C-related flavivirus in dogs. Understanding these agents and their natural history can prepare us for new disease emergences and requires collaborative efforts across scientific disciplines.

Viral pathogenesis is a key field of investigation in a One Health approach to zoonotic and emerging diseases. This field encompasses the evolution of viruses, including genetic changes through immune pressure, mutation, recombination, and reassortment that may change transmission, receptor usage and host range, vector competence, and virulence. Examples are too numerous to cite in this brief account, but influenza viruses, SARS coronavirus, New World arenaviruses, and chikungunya virus provide examples. Minor changes in viral genes encoding ligands for cell receptors may result in a shift in cell tropism and host range from an animal reservoir to humans, or a shift in vector competence, causing increased virus transmission to humans. Elucidating the factors underlying such changes requires collaborative efforts of molecular virologists and cell biologists, as well as experts on the responses to infection of individual organisms and species of vertebrates, insects, and ticks. The same conclusions concerning a multidisciplinary approach apply to understanding disease expression through systems biology. Signal transduction pathways cause pro-inflammatory changes following viral infections, as well as the innate and adaptive immune responses to viral proteins, topics covered extensively in this book.

Animal models are widely used in the study of viral pathogenesis. This is the realm of Comparative Medicine, a distinct discipline of experimental medicine designed to translate information from animal models to human disease, and arguably the clearest example of One Health principles. There are also numerous examples of important diseases only affecting animals that provide model systems for understanding disease emergence and pathogenesis. For example, bluetongue and related orbiviruses (Epizootic Hemorrhagic Disease of deer, and African horse sickness) provide interesting model systems for study of virus movement, chronic infection, hemorrhagic fever pathogenesis, and virus evolution and antigenic variation. Porcine respiratory and reproductive syndrome, an important pathogen of swine, is a model for studies of the molecular basis for viral virulence, which can explain the emergence of epizootics.

A few important animal diseases are mentioned in this book, especially as they relate to analogous diseases in humans (e.g., prion diseases), but the omission of many important examples perhaps reflects the need for a broader, One Health approach to the study of pathogenesis. Similarly, arthropod vectors of viral infections remain a relatively understudied area of viral pathogenesis. Vectors are critical to an understanding of virus transmission, persistence in nature, and evolution. Once again, this illustrates the need to integrate disciplines of entomology and insect taxonomy, physiology, pathology, and ecology.

Prevention and control through vaccines, antiviral therapy, vector control, and other strategies also rely on effective One Health interactions. A recent review emphasizes how vaccine development and utilization can benefit from such interactions (Monath, 2013). Hopefully, in the future, the One Health vision will increasingly inform both basic and applied research in viral pathogenesis.

REFERENCES

AVMA. One health: a new professional Imperative; AVMA one health initiative task force report. Available from: https://www.avma.org/KB/Resources/Reports/Documents/onehealth_final.pdf. 2008. See www.onehealthinitiative.com.

Monath TP. Vaccines against diseases transmitted from animals to humans: a one health paradigm. Vaccine 2013;31:5321–38.

Chapter 22.14

Controversial Policy Issues

Michele S. Garfinkel[1]

Michele Garfinkel manages the European Molecular Biology Organization Science Policy Program. Her research focuses on societal concerns associated with discoveries in synthetic genomics, and crafting options to mitigate associated risks.

Public policy attempts to match the concerns of the public with the work that scientists do. The goal of policy analysis is to inform and—in some cases—influence decision makers, including politicians, research administrators, and companies that develop biomedical products. How does policy intersect with viral pathogenesis, particularly research in this field? In this brief commentary I will focus on some examples where controversy has developed at the nexus of research and public health. Salient issues discussed are: dual use research; biosecurity; gain-of-function experiments; and public policies including those regarding "Select Agents."

As defined by the U.S. National Institutes of Health, "Dual Use Research of Concern (DURC) is life sciences research that, based on current understanding, can be reasonably anticipated to provide knowledge, information, products, or technologies that could be directly misapplied to pose a significant threat with broad potential consequences to public health" (NIH OSP, ND). Dual use research poses a difficult problem for scientists, especially virologists, because it is poorly defined. One approach has

1. The opinions expressed in this commentary are those of the author and not necessarily those of EMBO.

been provided by the Fink Committee report of the US National Academy of Sciences, which delimited dual use research to a fairly specific set of experiments of concern, including experiments that involve microbes (Committee on Research Standards and Practices, 2004).

Others have approached the problem of biosecurity by focusing on the trade-offs between potential benefits and potential risks, including ways to mitigate some of those risks (Garfinkel et al., 2007; AAAS, 2013; Baltimore et al, 2015; Dupres et al, 2015). The idea that benefits should be shared has been accepted, though it remains difficult to implement. But the idea that risks should be equally shared is more difficult to analyze; some work has been done on this problem by ethicists, but policy options for sharing risk remain scarce. Stated broadly, there are clearly risks of doing nothing against the risks of doing something. For example, there is almost no question that we must be doing more in the area of influenza virus research. But what are the trade-offs in researching smallpox, poliovirus, or Ebola?

Going further than naturally occurring microbes, gain-of-function experiments provide a current example of these issues. Recent experiments with influenza virus have sought to identify the genetic determinants of influenza virus that are required to produce a global pandemic. But this work led to an explosion of controversy, whether they should have been permitted in the first place, and in the second place, whether they should be published (Fouchier et al., 2013; Lipsitch and Galvani, 2014). At what level should such experiments be subject to oversight and approval? Is the local institutional biosafety committee sufficient? Are there societal concerns that would be mitigated by having such approvals at a national level?

The National Science Advisory Board for Biosecurity (NSABB) was established by the US government in 2004 to deal with general biosecurity issues. Recently, there has been a shake-up in the composition of the board, and an apparent move away from oversight of experiments, particularly gain of function (Begley, 2014). At the same time, it is clear that the overall safety of laboratories with respect to microbiological work is excellent. How this may change as experiments become more complex and more new researchers enter the field remains to be seen.

One approach to risk mitigation is the control of potential dangerous microbes. In the United States, the Select Agent Program lists viral, bacterial, and toxic agents that require special efforts and oversight (National Select Agent Registry, ND). The lists and accompanying rules were drafted during a time of understandable concern about new terrorist attacks following the events of September 11, 2001. Naturally, this list and accompanying rules is highly controversial. For instance, which agents should be on those lists? Certainly most researchers would agree some kind of additional oversight is important. But should the list of agents be expanded? Using what kind of analysis?

These broad policy issues do not respect borders, and international treaties present interesting and complex policy issues. For example, under the Convention on Biological Diversity (CBD, ND), treaties relating to biosafety broadly defined (Cartagena Protocol, ND) and to access and benefit sharing (Nagoya Protocol, ND), plus technical assessment processes to inform the main Convention (SBSTTA, ND), may have direct impacts on some scientific research. Although, the United States is not a party to the Convention, it is important for scientists to understand the basics of such Conventions if they are doing any work outside of national borders.

REFERENCES

AAAS (American Association for the Advancement of Science). Bridging science and security. 2013. Available at: http://www.aaas.org/page/bridging-science-and-security.

Begley S. US rolls back oversight of potentially dangerous experiments. Reuters 2014. Available at: http://www.reuters.com/article/2014/08/13/us-health-anthrax-oversight-insight-idUSKBN0GD08L20140813.

Baltimore D, et al. A prudent path forward for genomic engineering and germline gene modification. Science 2015;348:36–8.

Cartagena Protocol on Biosafety (Cartagena Protocol), http://bch.cbd.int/protocol/; No date.

Committee on Research Standards and Practices to Prevent the Destructive Application of Biotechnology, National Research Council of the National Academies. Biotechnology research in an age of bioterrorism. Washington (DC): The National Academies Press; 2004.

Convention on Biological Diversity (CBD), http://www.cbd.int/; No date.

Dupres WP, Fouchier RAM, Imperialie MJ, et al. Gain-of-function experiments: time for a real debate. Nat Rev Microbiol 2015;13:58–61.

Fouchier RAM, Kawaoka Y, 20 co-authors. 2013. Correspondence: Avian flu: gain-of-function experiments on H7N9. Nature 500:150–151.

Garfinkel MS, Endy D, Epstein GL, Friedman RM. Synthetic genomics: options for governance. 2007. Available at: http://www.jcvi.org/cms/fileadmin/site/research/projects/synthetic-genomics-report/synthetic-genomics-report.pdf.

Lipsitch M, Galvani AP. Ethical alternatives to experiments with novel potential pandemic pathogens. PLoS Med 2014;11:e1001646.1–16.

NP Nagoya Protocol on access and benefit-sharing (Nagoya Protocol), http://www.cbd.int/abs/; No date.

National Institutes of Health, Office of science policy. Dual Use Research of Concern. Available at: http://osp.od.nih.gov/office-biotechnology-activities/biosecurity/dual-use-research-concern; No date.

National Select Agent Registry, http://www.selectagents.gov/index.html; No date.

Subsidiary Body on Scientific. Technical and Technological Advice (SBSTTA), http://www.cbd.int/sbstta/; No date.

Glossary and Abbreviations

17D A strain of yellow fever virus that is attenuated and used as a vaccine
ACE2 Angiotensin converting enzyme 2
ADAR IFN-inducible dsRNA-specific adenosine deaminase
ADCC Antibody-dependent cell-mediated cytolysis
ADV Aleutian disease virus
AE Adverse events
AIDS Acquired immunodeficiency syndrome, caused by HIV
Aleutian disease A disease caused by a parvovirus that is particularly pronounced in the Aleutian strain of mink
Alpha herpesviruses Herpesviruses are classified as alpha, beta, or gamma based on their cellular tropism
ALS Amyotrophic lateral sclerosis
ALT Alanine transaminase, a liver enzyme whose level in the serum reflects liver function
Anti-HBs Antibody against HBs antigen
APC Antigen-presenting cell
APOBEC3 Apolipoprotein B mRNA editing enzyme, catalytic polypeptide-like, a protein that catalyzes the degradation of HIV DNA
APOBEC3G Apolipoprotein B editing catalytic subunit 3G
Apoptosis Programmed cell death
Arbovirus Arthropod-borne virus
ASC Antibody-secreting cell
ASC Apoptosis-associated specklike protein containing a caspase-recruitment domain
Asibi The name of the patient from whom a virulent strain of yellow fever virus was obtained
Av01 An attenuated mutant rabies virus selected by a neutralizing monoclonal antibody
AZT Azidothymidine
B cell Lymphocyte that has matured in bone marrow
B7.1, B7.2 Peripheral membrane proteins found on activated antigen-presenting cells that, when paired with either a CD28 or CD152 (CTLA-4) surface protein on a T cell, can produce a costimulatory or inhibitory signal, respectively. B7.1 is also known as CD80 while B7.2 is also known as CD86
B8R A poxvirus protein homologous to the IFNγR
BALB/c Black albino, an inbred mouse strain
Basic reproductive number (R_0) Average number of new infections initiated by a single infectious individual in a completely susceptible population over the course of that individual's infectious period
Bcl-2 Cellular protein involved in the regulation of apoptosis. Specifically, bcl2 has antiapoptotic activities, preventing apoptosis by binding and sequestering proapoptotic proteins
BHK-21 A continuous line of baby hamster kidney cells
BK polyomavirus A human polyomavirus, named after a patient from whom it was isolated
BrdU Bromodeoxyuridine, a nucleotide used to label cells
C1qB Complement gene C1qB
C3, C1q, C4b Complement proteins
C4b A complement protein, an intermediary in the complement cascade
C4b-BP Plasma protein that binds C4b
CA Capsid protein
CC Cysteine–cysteine chemokines
cccDNA Covalently closed circular DNA
CCR5 Chemokine receptor 5, also a coreceptor for HIV-1
CD11a Adhesion molecule
CD127 IL-7 receptor α chain
CD25 Component of IL-2 receptor
CD27 TNF receptor superfamily
CD28 Costimulatory receptor
CD4 lymphocytes A subset of T lymphocytes identified by the CD4 antigenic marker on their surface
CD4, CD8 Subsets of T lymphocytes identified by their surface markers that act as helper or effector cells, respectively
CD4+, CD8+ Cluster of differentiation; surface proteins that mark subsets of T lymphocytes
CD43, CD44 Adhesion molecules
CD62L Lymph node homing receptor
CD69 Early activation marker
CD8 A marker on T lymphocytes that subserve effector function
CDC Centers for Disease Control and Prevention
cDNA Complementary DNA; synthesized from an mRNA template in a reaction catalyzed by reverse transcriptase
CDV Canine distemper virus, a morbillivirus of dogs
CHIKV Chikungunya virus
CJD Creutzfeldt–Jakob disease
Class I, Class II These are two primary classes of major histocompatibility complex (MHC) molecules
CMV Cytomegalovirus, a herpesvirus of humans and animals
CNS Central nervous system
CoV Coronavirus

CpG motif Unmethylated CpG motifs in plasmid DNA provide a T_H1-biased adjuvant effect through toll-like receptors
CPV Canine parvovirus
CRISPR/Cas9 Clustered regularly interspaced short-palindromic repeats/CRISPR associated protein-9 nuclease
CRPV Cottontail rabbit papillomavirus
CSF Cerebrospinal fluid
c-src (src kinase) Proto-oncogene that in its role in cell-signaling pathways phosphorylates specific tyrosine residues in other proteins
CTL Cytolytic T lymphocytes, effector T cells that usually carry the CD8 marker
CVS Challenge virus standard, a brain passaged highly neurovirulent strain of rabies virus
CXC Cysteine amino acid x cysteine chemokines
CXCR4 A chemokine receptor 4, also a coreceptor for HIV-1
Cyclin D2 Cellular protein involved through interaction with cyclin-dependent kinases in the Rb-pathway's regulation of cell cycle progression from G1 to S phase
DAA Direct-acting antiviral
DAI DNA-dependent activator of IRFs
DAP-12 DNAX-activating protein of 12 kDa
DC Dendritic cell
DCI Enoyl-CoA Delta Isomerase one; a mitochondrial enzyme involved in the beta-oxidation of unsaturated fatty acids
DDX41 A DNA helicase family member
delta 32 mutation Mutation in CCR5 gene that inactivates it
DHF/DSS Dengue hemorrhagic fever/dengue shock syndrome
Digital transcriptome subtraction Computational method to detect the presence of transcripts belonging to novel pathogens. Total cDNA from a sample is sequenced in an unbiased high-throughput manner before host transcripts are subtracted from the file. The remaining sequence is then analyzed for pathogenic transcripts
DPPT4 Cellular receptor dipeptidylpeptidase 4
DTH Delayed-type hypersensitivity
DTS Digital transcriptome subtraction
E1A, E3 Early proteins of adenovirus
E6, E7 Oncogenes of human papillomavirus
E6AP Human papillomavirus E6-associated protein
EBNA1 Epstein–Barr virus nuclear antigen 1
EBNA3C Epstein–Barr virus nuclear antigen 3C
EBV Epstein–Barr virus, a human herpesvirus
Effective population size The number of replicating individuals in an idealized population that would show the same amount of dispersion of allele frequencies under random genetic drift
EIAV Equine infectious anemia virus, an equine lentivirus that causes a characteristic acute anemia
EIF2A Eukaryotic translation initiation factor 2
ELISA Enzyme-linked immunosorbent assay
Elispot An assay for functional CD8 cells that measures the production and secretion of cytokines such as IFNα in response to a specific antigen
env Envelope gene
ErbB-1 Cellular growth factor
EST Expressed sequence tag
FACS Fluorescent-activated cell sorter
FcR Receptor for the Fc domain of immunoglobulin molecules
FcγIIIR Fcγ receptor
FIV Feline immunodeficiency virus, a lentivirus of cats
FOXP3 A member of the FOX protein family, FOXP3 appears to function as a master regulator (transcription factor) in the development and function of regulatory T cells
FPV Feline panleukopenia virus
G1–G4 Rotavirus glycoproteins
GAF IFN-γ activation factor
gag gene the gene that encodes the major internal structural proteins of retroviruses
gag Group antigen, a polyprotein that is cleaved into four smaller peptides
GALT Gut-associated lymphoid tissue
GAS IFN-γ activated sequences
gE, gI, gC Glycoproteins of herpes simplex virus
Generation time Average period between infection in one individual and transmission to another
Geometric replication A mode of virus replication whereby a copy of the template genome is used as a template to produce new genomes, which are then used as a template for copying and so on
GLEAM Global epidemic and mobility model
GM-CSF Granulocyte-Macrophage colony-stimulating factor
GOARN Global Outbreak Alert and Response Network
gp120 The envelope glycoprotein of HIV (also known as the SU or surface protein) that binds to the primary receptor, CD4
gp160 HIV envelope glycoproteins prior to cleavage into gp41 and gp120
gp41 The second envelope protein of HIV also known as the TM or transmembrane protein
GPHIN Global Public Health Intelligence Network
GPS Global positioning system
GWAS Genome-wide association study
H Hemagglutinin of influenza virus
H1N1 Influenza viruses are recognized by their hemagglutinin and neuraminidase subtypes (as well as strain and location of isolation). There are many hemagglutinin and neuraminidase subtypes, most of which have not been associated with human infection
H-2 MHC for the mouse
HAART Highly active antiretroviral therapy

HADHB Hydroxyacyl-CoA dehydrogenase/3-Ketoacyl-CoA thiolase/Enoyl-CoA hydratase; a mitochondrial enzyme that catalyzes the last three steps in the beta-oxidation of long-chain fatty acids

HBcAg The core antigen of HBV

HBIG Hepatitis B immune globulin

HBsAg Hepatitis B surface antigen, the viral envelope protein

HBV Hepatitis B virus

HBX Hepatitis B virus X-protein

HCV Hepatitis C virus

Herd immunity Protection conferred on unimmunized members of a partially vaccinated population because virus transmission is reduced in the immunized members of the group

HESN Highly exposed seronegative

HHV8 Human herpesvirus 8

HIV Human immunodeficiency virus, a lentivirus

HIV/AIDS Human immunodeficiency virus/acquired immunodeficiency syndrome

HIV-1, HIV-2 The two major types of human immunodeficiency virus

HLA Human leukocyte antigen

HPV Human papillomavirus, a cause of cervical carcinoma

HR1, HR2 Heptad repeats in HIV envelope protein

HR-HPV High-risk human papillomavirus

HSV Herpes simplex virus

HTLV I Human T-cell leukemia virus type I, a cause of T-cell leukemia and spastic tropical paraparesis

ICE Interleukin one converting enzyme

ICOS inducible co-stimulatory ligand

ICP"n" Infected cell protein, a term used to designate individual proteins of HSV

ICS Intracellular cytokine staining

ID50 50% infectious dose

IE Immediate early, E (early), and L (late) genes of HSV; a listing of viral genes according to the time of their expression during replication

IFI16 IFN-γ-inducible protein 16

IFN Interferon

IFNγ interferon gamma, or immune interferon

IFNγR cellular receptor for IFNγ

Ig Immunoglobulin

IgA Immunoglobulin A, an immunoglobulin that is secreted by B cells found in mucosal tissues

IgG Immunoglobulin, once called gamma globulin

IIV Inactivated influenza virus (vaccine)

IL Interleukin

IL-1 Interleukin 1

IL-2 Interleukin 2 (also called T-cell growth factor)

IL-4 Interleukin 4

ILC Innate lymphoid cells

Immortalization The ability of a cell to undergo unlimited population doublings

In situ PCR A method for the histochemical identification of specific nucleic acids using sequence amplification followed by in situ hybridization

IN Integrase, a viral enzyme essential for the integration of viral DNA into host DNA

Incidence Number of new cases per period of observation (e.g., 1000 person-years)

IND Investigational new drug

Infectious period Average duration of infectiousness

iNOS Inducible nitric oxide synthase

IPV Inactivated poliovirus vaccine

IRES Internal ribosomal entry site

IRF Interferon regulatory factor

ISG15 Interferon-stimulated gene 15

ISGF3 IFN-stimulated factor 3

ITAM Intracellular immunoreceptor tyrosine-based activating motif

ITIM Immunoreceptor tyrosine-based inhibitory motif

JAK Janus tyrosine kinase

JC polyomavirus A human polyomavirus, named after a patient from whom it was isolated

Kaposi's sarcoma Cancer of connective tissue often associated with AIDS

KIR Killer inhibiting receptor, receptors in NK cells that inhibit perforin-mediated killing

Knockout mice Mice in which a specific gene has been inactivated using a method that involves homologous DNA recombination in embryonic stem cells

KS Kaposi's sarcoma

KSHV Kaposi's sarcoma-associated herpesvirus

Kuru A fatal wasting disease epidemic among the Fore peoples of the highlands of eastern New Guinea

L1 A surface capsid protein of HPV, used in HPV vaccines

L929 cells A murine cell line

LAIV Live attenuated influenza virus (vaccine)

LANA Latency-associated nuclear antigen

Lansing A strain of poliovirus

LAT Latency-associated transcript, a term used for RNA transcripts of the HSV genome that are produced during latency

Latent period Average time from infection to the onset of infectiousness

LCMV Lymphocytic choriomeningitis virus, an arenavirus of mice

LD50 50% lethal dose

LDA Limiting dilution assay, used to quantify CTL precursor or memory cells

LDV Lactic dehydrogenase virus

loxP-Cre Cre-Lox recombination is a site-specific recombinase technology, widely used to carry out deletions and insertions

LPS Lipopolysaccharide

LRRF1P1 Leucine-rich repeat (in Flightless 1) interacting protein 1

LTR Long terminal repeat, a noncoding sequence at the ends of the retroviral genome
M cells Microfold cells found in epithelium overlying MALT
M, N, O Main, new, outlier, subgroups of HIV-1
M2 Matrix protein 2
MALT Mucosa-associated lymphoid tissue
MAR Monoclonal antibody resistant, virus selected by growth in the presence of a neutralizing monoclonal antibody
MAVS mitochondria antiviral signaling protein
MBP Myelin basic protein
MDA5 Melanoma differentiation-associated gene 5
MERS Middle East respiratory syndrome
MERS-CoV Middle East respiratory syndrome coronavirus
MHC Class I protein Major histocompatibility proteins that are divided into two groups, class I and class II
MHC Major histocompatibility complex
MHV Mouse hepatitis virus, a nidovirus
MOI Multiplicity of infection is defined as the ratio between the number of infecting particles and the number of cells
MRKAd5HIV Shorthand name for a candidate Merck HIV vaccine
MT-2 A continuous cell line of human T lymphocytes in which T-cell-adapted HIV-1 strains can be grown
Mu-LT Murine polyomavirus large T antigen
MuLV Murine leukemia virus, an oncogenic retrovirus of mice
Mu-MT Murine polyomavirus middle T antigen
MuPyV Murine polyomavirus
Mu-sT Murine polyomavirus small T antigen
Mutation rate Defined as the number of errors per site per replication. In viruses, a replication cycle is commonly defined as the infection cycle, but may also refer to a strand copying event
MVA modified virus Ankara, an attenuated vaccinia virus
MyD88 myeloid differentiation factor 88
N Neuraminidase of influenza virus
NA A continuous cell line derived from a mouse neuroblastoma (nervous system tumor)
NC Nucleocapsid protein
Nef Negative factor, an accessory protein with pleiotropic effects
NFκB Nuclear factor κB, a transcriptional activator
NGS Next-generation sequencing
NHP nonhuman primate
NIH National Institutes of Health (USA)
NK cell Natural killer cell
NKR NK cell receptor
NLR NOD-like receptor
NNRTI Nonnucleoside reverse transcriptase inhibitor
Nod-like receptor Nucleotide-binding oligomerization domain receptors are pattern-recognition receptors
NRTI Nucleos(t)ide reverse transcriptase inhibitor
NS1 Nonstructural protein 1 of influenza virus
NYVAC An attenuated vaccinia virus
Oncogene A gene that when overexpressed can promote oncogenesis
Oncogenesis The generation of cancer
OPV Oral poliovirus vaccine (aka Sabin vaccine)
Oropharyngeal carcinoma Carcinoma of the head and neck, specifically the oropharynx, the middle part of the pharynx or throat behind the mouth including the back third of the tongue, the soft palate, the side and back walls of the throat and the tonsils
p15e 15kD envelope protein of some retroviruses
p53 Tumor suppressor involved in signaling apoptosis, repair, cell cycle arrest, or senescence in response to multiple-stress stimuli
PAGE Polyacrylamide gel electrophoresis
PAMP Pathogen-associated molecular pattern
PBMC Primary blood mononuclear cells
PCR Polymerase chain reaction, a method to amplify specific DNA or RNA sequences
pCTL Precursor cytolytic T lymphocytes, assayed for lytic activity after culturing dilutions of harvested cells in the presence of antigen
pDC Plasmacytoid dendritic cell
PEG-IFN Pegylated interferon
PFU Plaque forming unit
Phylodynamics A combined phylogenetic and epidemiological approach to study transmission dynamics of viruses
Pol III RNA polymerase III
Pol Polymerase gene
PR Protease, a viral enzyme
Pr60^{gag} The gag protein encoded by the MAIDS virus, a variant of the normal gag protein
pRB Retinoblastoma protein is a tumor suppressor protein that regulates cell cycle progression by binding and sequestering transcription factors necessary for transition into the S-phase of the cell cycle
Prevalence Number of all cases in a defined population and time period
Principal component analysis A statistical procedure used to reduce the dimensionality of high-throughput data while retaining the variation of the data set. This method makes it possible to visually assess similarities and differences between samples without plotting thousands of data points for each sample
Prion Proteinaceous infectious protein
Proteinase K A proteolytic enzyme
PrP, c, and Sc Forms of the prion protein; c, cellular conformer; Sc, scrapie conformer
PRR Pattern recognition receptor
Rev Regulator of expression of viral proteins, regulates splicing of viral messages and their transport to cytoplasm
RFC/B.5 An attenuated bunyavirus mutant selected by passage in cell culture

RIG-I Retinoic acid-inducible gene I is a pattern recognition receptor
RISC RNA-induced silencing complex
RLR RIG-I-like receptor
RNA Ribonucleic acid
RNAi RNA interference
RSV Respiratory syncytial virus
RSV Rous sarcoma virus
RT Reverse transcriptase
RTI Reverse transcriptase inhibitor
RV194-2 An attenuated mutant rabies virus selected by a neutralizing monoclonal antibody
SA Sialic acid
SARS Severe acute respiratory syndrome
SARS-CoV SARS coronavirus
Scrapie A subacute spongiform encephalopathy of sheep
Senescence of cells Permanent exit from the cell cycle
SHIV Simian human immunodeficiency virus, a chimeric virus that usually has the env gene of HIV inserted into a SIV backbone
SI Stimulation index
Singular value decomposition-multidimensional scaling (SVD-MDS) A combination of mathematical algorithms used to reduce the dimensionality of high-throughput data and to visualize complex data in a graphical environment
siRNA Small interfering RNA
SIV Simian immunodeficiency virus, a lentivirus of nonhuman primates
SNV Sin nombre virus
SSPE Subacute sclerosing panencephalitis, a chronic progressive fatal infection of humans caused by measles and—rarely—rubella virus
Stamping machine replication A mode of virus replication whereby one template is used to create multiple copy genomes
STAT Signal transducer and activator of transcription, a component of the intracellular signaling pathway
STING Stimulator of IFN genes
SU Surface protein, one of the two viral glycoproteins (gp120)
SV40 Simian virus 40
T20 Compound that binds to heptad repeats on HIV
T cell Lymphocyte that has matured in the thymus
$t_{1/2}$ half life
tat Transactivator of transcription, an HIV gene that increases transcription of viral DNA
TBK Tank-binding kinase one
TCD Tissue culture dose, adequate to infect a cell culture
TCL T-cell line, a cell line derived from transformed T lymphocytes
TCM Central memory T cells
TCR T-cell receptor
Td mutant Transformation defective mutant
Telomerase Ribonucleoprotein responsible for adding repetitive sequence to the ends of chromosomes
TEM Effector memory T cells
Tfh CD4+ cells secreting IL-21 and providing help for B cells
T_H1 CD4+ cells secreting IL-2 and IFNγ and inducing cellular immune responses
T_H1, T_H2 Helper T lymphocytes that direct the immune response toward cellular or humoral immune induction, respectively
Th17 CD4+ cells secreting IL-17 and IL-22
T_H2 CD4+ cells secreting IL-4, IL-5 and IL-13 and associated with allergy
Th9 CD4+ cells secreting IL-9 and IL-10
TLR toll-like receptor
Tlr4 gene for toll-like receptor of insects
TM Transmembrane, a protein that crosses the viral envelope (gp41)
TNF Tumor necrosis factor
TNFRS17 A B-cell growth factor
TNFα Tumor necrosis factor α, a cytokine with cytopathic properties
TRAIL TNF-related apoptosis inducing ligand
Transformation Process by which cells acquire the properties of cancer, including immortalization, anchorage-independent growth, loss of contact inhibition, and being oncogenic upon transplantation into another organism
TRIF TIR-domain-containing adapter-inducing interferon-β
Trigeminal ganglion A sensory ganglion of cranial nerve V, containing cell bodies of second order sensory neurons
Trigeminal nerve Cranial nerve V, the cranial nerve that provides sensory innervation to the face, and that may act as a site for latent HSV genomes
TRIM Tripartite motif family of proteins
TRM Resident memory T cells
Ts mutant Temperature-sensitive mutant
Ts Temperature sensitive, a viral variant that can replicate at “standard” temperatures such as 37C, but is restricted at elevated temperatures such as 40C (where wild-type viruses can usually replicate well)
Tumor suppressor A protein that protects cells from transformation
Tumorigenesis The process of producing or forming tumors
Twitter Free social networking service that enables users to exchange text-based messages of up to 140 characters known as tweets
UCSF University of California in San Francisco
UNC93B1 Unc-homolog B1
V1, V2 Variable loops on the spike of HIV, targets of some neutralizing antibodies
VAP Virus attachment protein
v-bcl2 Viral homolog of the cellular bcl2 protein expressed by Kaposi’s sarcoma-associated herpesvirus
VCP Vaccinia complement control protein, a protein that binds to C4b

v-cyclin Viral homolog of the cellular cyclin D2 protein expressed by Kaposi's sarcoma-associated herpesvirus
Vif Viral infectivity factor, an accessory protein of HIV
Virogene Gene of Rous sarcoma virus that encodes the viral enzymes and proteins
VMV Visna-maedi virus, an ovine lentivirus that causes interstitial pneumonitis and demyelination
v-onc Viral-encoded homolog of a cellular oncogene
VP4, VP7 Surface proteins of rotavirus, targets of rotavirus vaccine-induced neutralizing antibodies
Vpr Virion protein R, an accessory protein that is required for nuclear import of the preintegration complex in non-dividing cells
Vpu Virion protein U, an accessory protein that enhances virion release from infected cells
v-src Viral homolog of c-src expressed by Rous sarcoma virus. This homolog has been mutated into a permanently active form of the cellular protein
VZV Varicella zoster virus, a human herpesvirus
WHO World Health Organization
Wikipedia Free, online encyclopedia written collaboratively by users
WNV West Nile virus

Index

Note: Page numbers followed by "f", "t" and "b" indicate figures, tables and boxes respectively.

C

D

E

F

G

H

I

J

K

L

M

N

O

P

Q

R

S

W

X

Y

Z